INVITATIONS TO SCIENCE INQUIRY

Second Edition

Tik L. Liem

SCIENCE EDUCATION CONSULTANT

SCIENCE INQUIRY ENTERPRISES

14358 Village View Lane
Chino Hills, CA 91709

First Edition copyright © 1981 by Tik L. Liem.
All rights reserved.

Second printing 1982.
Third printing 1984.
Fourth printing 1985.
Fifth printing 1985.
Sixth printing 1986.

Second Edition copyright © 1987 by Tik L. Liem.
All rights reserved.

Second printing, Second Edition 1989
Third printing, 1990

Permission in writing must be obtained from the publisher before any part of this work may be reproduced or transmitted in any form or by any means, electronic or mechanical, including photocopying and recording, or by any information storage or retrieval system.

Science Inquiry Enterprises
14358 Village View Lane
Chino Hills, California 91709

ISBN: 1-878106-00-7

Printed in the United States of America

TO THOSE TEACHERS
WHO WANT TO PUT SOME EXCITEMENT
IN THEIR TEACHING

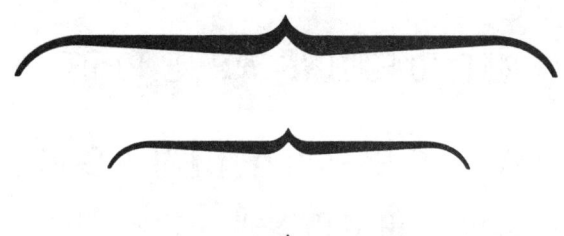

CAUTION

USE EXTREME CAUTION FOR THE FOLLOWING EVENTS:

EVENT	TITLE	PAGE	PRECAUTION
5.7	Grind a Cracker	141	DO NOT USE MORE THAN 0.5 GRAM OF EACH OF THE CHEMICALS !
5.21	The Chemical Flag	155	DO NOT SPILL THE PHOSPHOROUS SOLUTION ON HANDS OR OTHER MATERIAL ! DISPOSE ANY LEFT OVER PHOSPHOROUS SOLUTION OUTDOORS OR BURN OFF !
5.22	The Barking Dog	156	
5.23	Color the Flames	157	DO NOT USE MORE THAN THE PRESCRIBED AMOUNTS OF CHEMICALS ! CARRY OUT UNDER FUME HOOD OR OUTDOORS !
6.4	The Hot Acid	182	
7.20	The Cool Flame	218	DO NOT USE AN EXCESSIVE AMOUNT OF THE FLAMMABLE LIQUID IN THE MIXTURE !

A GENERAL SAFETY RULE TO FOLLOW: ADHERE TO THE INSTRUCTIONS !

IF UNSURE ABOUT THE AMOUNTS OF CHEMICALS TO BE USED, ALWAYS START WITH VERY SMALL QUANTITIES !!!

THE AUTHOR OF "INVITATIONS TO SCIENCE INQUIRY" IS NOT RESPONSIBLE FOR ANY ACCIDENTS RESULTING FROM IRRESPONSIBLE OR INCOMPETENT USE OF THE DEMONSTRATIONS AND ACTIVITIES SUGGESTED IN THIS BOOK !

TIK L. LIEM

PREFACE

This book is written and designed for teachers of science, particularly for those at the upper elementary and junior high or intermediate level, for college students preparing to teach science, and for all those individuals who are interested in science.

In the teaching of a science concept, it is most essential for the teacher to arouse the students' curiosity. Unless the student wants to know what the teacher has to say, it is most likely that time and effort in trying to teach the student would have been completely wasted. The first task of a teacher is to attract the attention of the student. This goes for all levels of learning, whether it is primary grade or college. A student without interest is the desperation of any teacher. In fact, the main objective of formal education is the arousing of the students' interest, if nothing else. Once curiosity is aroused the student will learn much more on his/her own than the teacher can ever teach him/her. Learning only takes place if the student wants to learn.

The use of discrepant events in the teaching of science is one of the best methods to do just that: to arouse interest and curiosity. The discrepant or counterintuitive event in science, sometimes also called the **mind-capture** or the **intuition-offending** event, is an occurrence or happening which goes counter to what one thinks likely. This event is set up in such a way that it poses a question to the student and asks him/her to come up with the explanation. There is evidence that students will retain science concepts longer when they were engaged in experiencing counterintuitive events. This book is a collection of thoroughly tested discrepant events. They can be used to initiate or sustain a lesson in virtually any topic of science at the upper elementary or intermediate level. They can also be used as reinforcement activities for the students or as challenging problems for further inquiry. Each of the events is a focus for inquiry teaching, a most appropriate way of reflecting the true nature of science in the classroom. The students are required to put into practice most of the science processes when engaged in inquiry.

The description of each of the discrepant events is organized in such a way as to provide the teacher with the maximum guidance for conducting a science inquiry lesson. Emphasis has been placed on the use of simple material so that most of the events may be carried out with things that are found in everyday life or may be bought in local stores.

Although each of the events have been tested and the procedure, questioning, and explanation constructed by the author, many of the ideas have come, probably without realizing it, from others. The author is very conscious of his indebtedness to numerous persons -- mostly teachers, colleagues, and students -- and wishes it were possible to thank each one individually. Perhaps his gratitude is best expressed by his desire to share with a larger audience any contribution he has made as a result of their inspiration.

DR. TIK L. LIEM
14358 VILLAGE VIEW LN.
CHINO HILLS, CA 91709
TEL: 714-590-4618

CONTENTS

Preface

Introduction   xix

SECTION I : ENVIRONMENT   1

CHAPTER 1   WHAT ARE THE PROPERTIES OF AIR?   3

1.1.	The Air Catcher	5
1.2.	The Bottle and the Bag	6
1.3.	Pour Air Under Water	7
1.4.	Keep Paper Dry Under Water	8
1.5.	The Empty Box Candle Snuffer	9
1.6.	The Refusing Funnel	10
1.7.	The Inverted Glass of Water	11
1.8.	The Straw Drinking Race	12
1.9.	The Upwards Falling Test Tube	13
1.10.	The Mysteriously Rising Water (I)	14
1.11.	The Mysteriously Rising Water (II)	15
1.12.	The Collapsing Can	16
1.13.	The Balloon in the Flask	17
1.14.	Fountain in a Flask	18
1.15.	The Sticking Cup of Water	19
1.16.	The Bionic Finger (I)	20
1.17.	The Bionic Fingers (II)	21
1.18.	Stop the Leak	22
1.19.	The Perpetual Fountain?	23
1.20.	Transfer Water with a Straw	24
1.21.	The Watertight Cloth	25
1.22.	The Cups and the Balloon	26
1.23.	Inflate a Balloon by Sucking	27
1.24.	The Heavy Newspaper	28
1.25.	The Magic Beaker	29
1.26.	The Egg and the Milk Bottle	30
1.27.	The Two Bottles in Love	31
1.28.	The Inverted Paper Bag Balance	32
1.29.	Squeeze the Glass Bottle?	33
1.30.	The Paper Merry-Go-Round	34
1.31.	The Dancing Penny	35
1.32.	The Live Balloon	36
1.33.	What Causes the Water to Rise? (I)	37
1.34.	What Causes the Water to Rise? (II)	38
1.35.	The Candle Under the Jar	39
1.36.	The Water-Attracting Steel Wool	40
1.37.	The Balancing Balloons	41
1.38.	The Ball that Gains Weight	42

1.39.	The Plastic Bag Air Lift	43
1.40.	The Obedient Diver	44
1.41.	Turn a Little Water Into a Lot of Lemonade	45
1.42.	Make an Air Bullet Shooter	46
1.43.	The Smoke Ring Race	47

CHAPTER 2   WHAT CAN FLOWING AIR DO?   49

2.1.	The Lifting Paper	51
2.2.	The Stubborn Paper Card	52
2.3.	The Funnel and the Ball	53
2.4.	The Attracting Spheres	54
2.5.	The Clanging Soda Pop Cans	55
2.6.	The Floating Card	56
2.7.	The Floating Ball	57
2.8.	Lift Water by Blowing	58
2.9.	The Mysteriously Moving Flame	59
2.10.	Which Way Will the Flame Flicker?	60
2.11.	The Paper Wing	61
2.12.	The Paper Airplane Contest	62
2.13.	The Cork Race	63
2.14.	The Paper Fan	64
2.15.	The Leaping Egg	65
2.16.	Lift a Moth Ball with Streaming Water	66
2.17.	The Self Priming Siphon	67
2.18.	The Self-Directing Cards	68

CHAPTER 3   WHAT FACTORS INFLUENCE THE WEATHER?   69

3.1.	A Cloud in a Bottle (I)	71
3.2.	A Cloud in a Bottle (II)	72
3.3.	The Jar Barometer	73
3.4.	The Bottle Barometer	74
3.5.	Which is the Best Windvane?	75
3.6.	How Fast Does the Wind Blow?	76
3.7.	Find Wind Speed and Direction Together	77
3.8.	Create Wind Currents	78
3.9.	Make Your Own Thermometer	79
3.10.	At What Temperature is Dew Formed?	80
3.11.	Recycle the Water	81
3.12.	The Frosty Can	82
3.13.	The Wet and Dry Thermometer	83
3.14.	The Jar Hygrometer	84
3.15.	The Hair Hygrometer	85
3.16.	Make a Fire Cyclone	86
3.17.	Why Longer or Shorter Days?	87
3.18.	The Cool Winter Sun	88

CHAPTER 4	WHAT ARE THE CHARACTERISTICS OF MATTER?	89
4.1.	Can the Container Hold More?	91
4.2.	The Shrinking Balloon	92
4.3.	The Shrinking Mixture of Liquids	93
4.4.	Is the Container Leaking?	94
4.5.	The Cooling Rubber Band	95
4.6.	The Vanishing Ice Cubes	96
4.7.	The Invisible Flame Extinguisher	97
4.8.	The Disappearing Liquid	98
4.9.	Make a Purple Gas	99
4.10.	Changing Gas into Crystals	100
4.11.	Make Milk from Water and Oil	101
4.12.	The Invisible Steam	102
4.13.	Lift an Ice Cube with Salt	103
4.14.	Cut Through Ice with a Wire	104
4.15.	Heat Water Above Its Boiling Point	105
4.16.	Boil Water with Cold Water	106
4.17.	Drink Muddy Water	107
4.18.	The Clinging Water Streams	108
4.19.	The Smaller, The Stronger	109
4.20.	Pour Water Along a String	110
4.21.	Where Does the Cork Float?	111
4.22.	How Many Pennies Can Go In?	112
4.23.	Float Metal on Water	113
4.24.	Sink the Pepper	114
4.25.	The Detergent Propelled Boat	115
4.26.	The Water-Filled Cups	116
4.27.	The Detergent-Fearing Powder	117
4.28.	Which Bubble is Stronger?	118
4.29.	The Thread Circle in the Soap Film	119
4.30.	The Strong Soap Film	120
4.31.	The Floating Oil Sphere	121
4.32.	The Funny Water	122
4.33.	Which Will Float Where?	123
4.34.	The Bobbing Moth Balls	124
4.35.	Which Grape is Heavier?	125
4.36.	The Different Clay Sticks	126
4.37.	Is the Crown Made of Pure Clay?	127
4.38.	Float the Egg with Salt	128
4.39.	Can You Make Clay Float?	129
4.40.	Make Your Own Letter Scale!	130
4.41.	The Water Candle?	131
4.42.	The Floating Soap Bubble	132

CHAPTER 5		HOW DO CHEMICALS BEHAVE IN OUR DAILY LIFE?	133
	5.1.	The Charcoal Sausage	135
	5.2.	Walk Through a Hole in Ordinary Notebook Paper	136
	5.3.	The Blue and Red Cabbage	137
	5.4.	The Funny Colors	138
	5.5.	Draw with Fire	139
	5.6.	Turn a Red Rose into a White One	140
	5.7.	Grind a Cracker	141
	5.8.	Which Gas is Which?	142
	5.9.	Turn Water into Wine, Milk, and Beer!	143
	5.10.	Make a Smoke Screen	144
	5.11.	The Returning Color	145
	5.12.	Dissolve More Salt By Cooling?	146
	5.13.	The Three-Layered Liquid	147
	5.14.	The Disappearing Precipitate	148
	5.15.	The Ammonia Fountain	149
	5.16.	Make a Silver Tree	150
	5.17.	The Confused Blue Solution	151
	5.18.	The Fiery Water	152
	5.19.	The Glowing Penny	153
	5.20.	The Rising Suds	154
	5.21.	The Chemical Flag	155
	5.22.	The Barking Dog	156
	5.23.	Color the Flame	157
	5.24.	The Magic Wand	158
	5.25.	The Rocket Exhaust	159
	5.26.	The Angry Bucket	160
	5.27.	Burn Paper with Ice?	161
	5.28.	The Potassium Chlorate Bombs	162
	5.29.	Burn a Piece of Metal in Water	163
	5.30.	The Air-Burning Flask	164
	5.31.	The Human Flame Thrower	165
	5.32.	The Magic Candle	166
	5.33.	The Glowing Aluminum	167
	5.34.	Touch a Cracker!	168
	5.35.	The Sticky Board	169
	5.36.	Shaking the "Blues"	170
	5.37.	The Color-Absorbing Bacon	171
	5.38.	Make a Nylon Thread Out of Two Liquids	172
	5.39.	The Sticky Matches	173
SECTION II :		ENERGY	175
CHAPTER 6		WHAT FORMS OF ENERGY ARE THERE?	177
	6.1.	The Hot Bolt	179
	6.2.	Melting Ice Below Freezing?	180
	6.3.	Warming Breezes	181
	6.4.	The Hot Acid	182

6.5.	The Lighter-Fuel Cannon	183
6.6.	Warm a Bottle by Shaking	184
6.7.	The Wire Heater	185
6.8.	Use Heat to Turn a Wheel?	186
6.9.	Magnetic Perpetual Motion?	187
6.10.	Fission and Fusion	188
6.11.	The Test Tube Greenhouse	189
6.12.	Start a Fire with a Magnifying Glass	190
6.13.	Does a Book Have Energy?	191
6.14.	Does Water in a Lake Have Energy?	192
6.15.	The Spinning Rings	193
6.16.	Will the Heavy Brick Hit Your Nose?	194
6.17.	The Twin Pendulum	195

CHAPTER 7 HOW DOES HEAT AFFECT THINGS? 197

7.1.	The Coil Candle-Snuffer	199
7.2.	The Charless Cotton	200
7.3.	The Scorching Paper	201
7.4.	The Heat Race (I)	202
7.5.	The Heat Race (II)	203
7.6.	Can Ice-Water Boil?	204
7.7.	The Convection Tester	205
7.8.	The Confused Bottles	206
7.9.	Interesting Currents	207
7.10.	The Mysteriously Rising Napkin	208
7.11.	Which is the Warmer Color?	209
7.12.	Which Coin Will Stay on Longer?	210
7.13.	The Rising Juices	211
7.14.	Withdrawing Juices	212
7.15.	The Curving Tape	213
7.16.	The Moving Rod	214
7.17.	The Heavy Heat	215
7.18.	The Weakening Wire	216
7.19.	Boil Water in a Paper Cup	217
7.20.	The Cool Flame	218

CHAPTER 8 HOW DOES MAGNETISM WORK? 219

8.1.	The Floating Paper Clip	221
8.2.	Which Pole Is Attracted?	222
8.3.	The Magic Dancer	223
8.4.	The Floating Discs	224
8.5.	Seeing Magnetic Lines	225
8.6.	Which One Is the Magnet?	226
8.7.	The Mysteriously Moving Needle	227
8.8.	The Test Tube Magnet	228
8.9.	Magnetic Confusion	229
8.10.	Make a Needle Compass	230
8.11.	The Temporary Magnet	231

CHAPTER 9		WHAT IS STATIC ELECTRICITY?	233
	9.1.	The Imaginary Shelf	235
	9.2.	The Magnetic Ruler	236
	9.3.	The Paper Jumping Jacks	237
	9.4.	Separate the Pepper from the Salt	238
	9.5.	The Pepper Reproduction	239
	9.6.	Bend the Water Stream	240
	9.7.	The Electric Meter Stick	241
	9.8.	The Aluminum Foil Electroscope	242
	9.9.	The Confused Pithball	243
	9.10.	The Balloon Electroscope	244
	9.11.	The Induced Charge	245
	9.12.	The Spark-Producing Finger	246
	9.13.	The Electrostatic Storage	247
	9.14.	The Water Cup Spark Collector	248
CHAPTER 10		HOW IS CURRENT ELECTRICITY CREATED?	249
	10.1.	Make a Conductivity Tester	251
	10.2.	The Secret Burglar Alarm	252
	10.3.	Light the Bulb	253
	10.4.	Regulate the Current with a Pencil	254
	10.5.	The Dimmer and Brighter Bulbs	255
	10.6.	The Aluminum Foil Fuse	256
	10.7.	The Balloon Fuse	257
	10.8.	The Inference Boards	258
	10.9.	Change Electricity into Heat	259
	10.10.	The Liquid Battery	260
	10.11.	Electricity from a Lemon	261
	10.12.	The Coin Battery	262
	10.13.	Electricity from a Coil?	263
	10.14.	The Nail Magnet	264
	10.15.	The Deflecting Compass	265
	10.16.	The Simple Galvanometer	266
	10.17.	The Simple Telegraph	267
	10.18.	Turn Electricity into Sound	268
CHAPTER 11		HOW DOES LIGHT BEHAVE?	269
	11.1.	Make a Pinhole Camera	271
	11.2.	The Cool Candle Flame	272
	11.3.	Funny Reflections	273
	11.4.	How High to Place the Mirror?	274
	11.5.	Look at Yourself as Others See You	275
	11.6.	The Swollen Finger	276

11.7.	The Reappearing Coin	277
11.8.	The Invisible Penny	278
11.9.	The Waterdrop Magnifier	279
11.10.	The Glass Rod Magnifier	280
11.11.	The Broken Pencil	281
11.12.	Produce an Image of the Window	282
11.13.	Why Do We See Two Coins?	283
11.14.	Use Water as a Mirror?	284
11.15.	The Reflecting Brick Wall	285
11.16.	Make Rainbow Colors	286
11.17.	Make Your Own Rainbow	287
11.18.	The Simulated Sunset	288
11.19.	The Confused Flashlight	289
11.20.	The Peek-a-Boo Paper	290

CHAPTER 12   WHAT ARE THE PROPERTIES OF SOUND?   291

12.1.	Which Amplifier Works Best?	293
12.2.	The Reversing Pitch	294
12.3.	The Straw Oboe	295
12.4.	Pluck a Rubber Band	296
12.5.	The Pipe and Straw Trombones	297
12.6.	The Singing Glass	298
12.7.	The Tuned-In Washers	299
12.8.	The Swinging Book	300
12.9.	The Singing Bottle	301
12.10.	The Soundless Bell	302
12.11.	Listen to the Popping Bubbles	303
12.12.	The Soda Can Telephone	304
12.13.	The Coat Hanger Church Bell	305
12.14.	The Resonating Bar	306
12.15.	The Multiple Tuning Pipe	307
12.16.	The Twirling Bugle	308
12.17.	The Two-Note Tube	309
12.18.	Measure the Speed of Sound	310
12.19.	The Echo Speed Calculator	311
12.20.	The Elliptical Wonder	312

SECTION III :   FORCES AND MOTION ON EARTH AND IN SPACE   313

CHAPTER 13   HOW DO FORCES AFFECT THINGS?   315

13.1.	Rolling Uphill?	317
13.2.	Will Paper Fall Like a Stone?	318
13.3.	The Falling Pennies	319
13.4.	Hit the Falling Can	320
13.5.	The Mysteriously Moving Steel Ball	321
13.6.	The Balancing Pins	322

13.7.	The Balancing Act	323
13.8.	The Plate Carousel	324
13.9.	Hang a Hammer on a Loose Ruler	325
13.10.	Are Women Stronger Than Men?	326
13.11.	The Unreachable Cup	327
13.12.	Stuck to the Wall?	328
13.13.	The Center-Seeking Paper	329
13.14.	The Standing Matchbox	330
13.15.	Stand a Dollar Bill on Your Finger	331
13.16.	The Wobbling Circles	332
13.17.	Kick a Straight Line	333
13.18.	Stand a Raw Egg on Its Head	334
13.19.	The Floating Belt Hanger	335
13.20.	The Fork and Spoon Act	336
13.21.	The Weighted Pipe	337
13.22.	The Golden Middle	338
13.23.	The Magic Strip of Newspaper	339
13.24.	The Invisible Glue	340
13.25.	The Cardboard Bottom	341
13.26.	The Squirting Water Holes	342
13.27.	The Outpour Race	343
13.28.	Hit the Bottle on the Back Swing?	344
13.29.	How Many Swings Can You Get?	345
13.30.	The Magic Come-Back Can	346
13.31.	The Incredible Stick	347
13.32.	Is the Ball Repelled?	348
13.33.	Pierce a Potato with a Straw?	349
13.34.	The Loose Knife Supports	350
13.35.	The Dollar Bill Bridge	351
13.36.	How Long Can You Hold the Burning Paper?	352
13.37.	The Confused Twirling Paper	353
13.38.	The Yip-Yip Stick	354
13.39.	Tilt a Heavy Load with One Finger	355
13.40.	Is the Hammer a Lever?	356
13.41.	The Third Class Bicep	357
13.42.	Which One Is Heavier?	358
13.43.	Which One Will Be Moved?	359

CHAPTER 14   WHAT DO WE LEARN WHEN DEALING WITH SPACE SCIENCE?   361

14.1.	Put the Coin in the Cup	363
14.2.	Pull the Tablecloth	364
14.3.	The Immovable Penny	365
14.4.	Get the Egg in the Glass	366
14.5.	Get the Chalk in the Bottle	367
14.6.	Break the String Wherever You Want	368
14.7.	What Breaks the Thread?	369
14.8.	The Apple and the Knife	370
14.9.	The Falling Washers	371
14.10.	The Balloon Race	372

14.11.	The Match Missile	373
14.12.	The Milk Carton Sprinkler	374
14.13.	The Straw Rocket	375
14.14.	Blow Your Own Sail?	376
14.15.	The Recoiling Skate	377
14.16.	The Sticky Penny	378
14.17.	The Flying Wine Glasses	379
14.18.	A Measure of Centripetal Force	380
14.19.	The Gravity Machine (I)	381
14.20.	The Gravity Machine (II)	382
14.21.	The Funny Marbles	383
14.22.	The Crashing Skates	384
14.23.	The Colliding Steel Balls	385
14.24.	How High Will the Ball Bounce?	386
14.25.	The Test Tube Cannon	387
14.26.	The Spinning Planets	388
14.27.	The Paper Card Boomerang	389
14.28.	The Spinning Football	390
14.29.	The Human Gyroscope	391
14.30.	The Tin Can Race	392
14.31.	The Cup of Coffee Drop	393
14.32.	The Falling Elevator	394

CHAPTER 15 WHAT ARE SOME PHENOMENA ABOUT THE EARTH AND THE MOON? 395

15.1.	Make Stalactites and Stalagmites	397
15.2.	Can Stones Dissolve?	398
15.3.	Can Plants Break Rocks?	399
15.4.	How Can Water Break Rocks?	400
15.5.	Why Do Eroded Mountains Keep Rising?	401
15.6.	How Does a Geyser Work?	402
15.7.	How Can We Determine the Rock's Volume?	403
15.8.	Simulate a Volcano Eruption	404
15.9.	What Causes the Phases of the Moon?	405
15.10.	What Causes an Eclipse?	406

SECTION IV : LIVING THINGS 407

CHAPTER 16 WHAT VARIABLES ARE AFFECTING THE GROWTH OF PLANTS? 409

16.1.	How Do Seeds Germinate?	411
16.2.	The Bending Plant	412
16.3.	The Crooked Root	413
16.4.	The Water-Sucking Roots	414
16.5.	Make an Egg Osmometer	415
16.6.	The Swollen Egg	416
16.7.	How Does Gravity Affect Growth?	417
16.8.	Make Soot Prints of Leaves	418

16.9.	Do Leaves Give Off Water?	419
16.10.	What Do Green Leaves Breathe Out? (I)	420
16.11.	What Do Green Leaves Breathe Out? (II)	421
16.12.	How Is the Green in the Leaves Produced?	422
16.13.	Can Air Enter Through a Leaf?	423
16.14.	Grow Sweet Potato and Carrot Leaves	424
16.15.	Make a Red-Blue Carnation	425
16.16.	The Toothpick Star	426
16.17.	How Do Molds Reproduce?	427
16.18.	Pick up a Carafe with a Straw	428

CHAPTER 17 WHAT CAN WE LEARN ABOUT THE HUMAN BODY? 429

17.1.	The Reversed Image	431
17.2.	Are We Partially Blind?	432
17.3.	Why Do We Need Two Eyes?	433
17.4.	Do We Really Use Both Eyes?	434
17.5.	Are You Left- or Right-Sighted? (I)	435
17.6.	Are You Left- or Right-Sighted? (II)	436
17.7.	How Do We Perceive Color?	437
17.8.	What Gives Us the Illusion?	438
17.9.	The Swaying Cardboard	439
17.10.	See a Hole in Your Hand	440
17.11.	The Floating Piece of Finger	441
17.12.	The Hand is Quicker than the Eye	442
17.13.	Put the Bird in the Cage	443
17.14.	The Elliptical Pendulum Swing	444
17.15.	How Many Points Are Touching?	445
17.16.	Is the Water Warm or Cold?	446
17.17.	Catch the Dollar Bill	447
17.18.	How Fast Can You React?	448
17.19.	The Stimulus-Response Action	449
17.20.	How Fast Does Your Heart Beat?	450
17.21.	Which Contains More Carbon Dioxide?	451
17.22.	Measure the Capacity of Your Lungs	452
17.23.	How Do We Breathe?	453
17.24.	The Uncontrollable Foot	454
17.25.	The Kicking Frog Legs	455
17.26.	Are Women More Agile than Men?	456
17.27.	Turn a Cracker into Sugar	457
17.28.	Try Drinking While Standing on Your Head	458

APPENDIX 459

FURTHER READINGS 460

INDEX 461

ORDER FORM 469

Be a Happy and

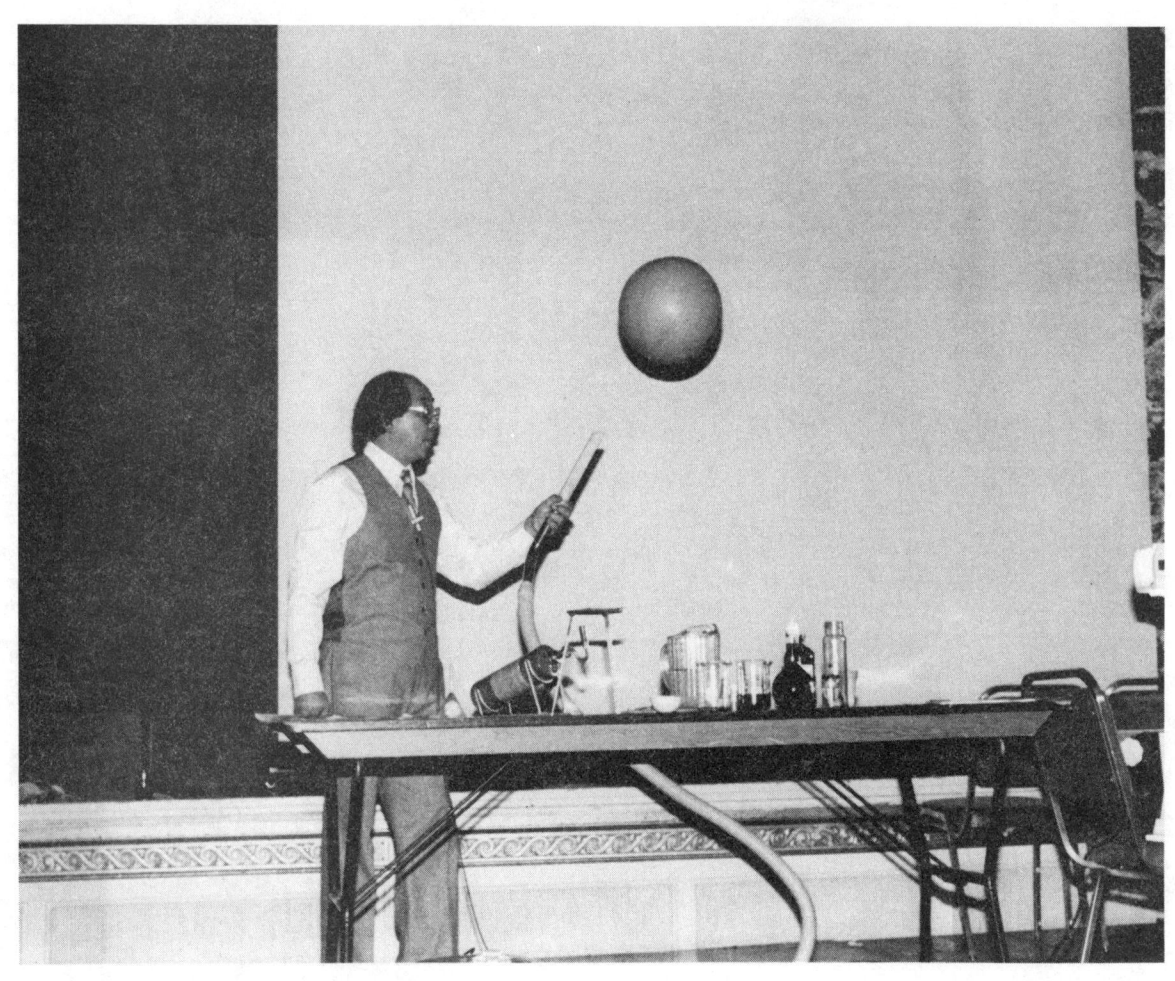

Enthusiastic Teacher!

INTRODUCTION

A point of concern that dedicated science teachers will sooner or later face in their career, is how to arouse the students´ interest in the subject matter that he or she is teaching. This is the age old problem that will confront every single teacher, whether at the primary, the secondary, the university, or at the post graduate level. It is a problem of motivation. The question is: how do we teachers arouse the students´ interest? In modern terminology: how do we turn the students on to whatever subject matter we teach, in our case science?

How do we turn someone on to learn to play the piano? - or a toddler to eat his/her meal? - or anybody to learn to swim? Harry Overstreet, in his illuminating book: **Influencing Human Behavior,** said: "Action springs out of what we fundamentally desire.... and the best piece of advice which can be given to would-be persuaders, which you and I are, whether it is in the home, in business, in school or in politics, is: First arouse in the other person an eager want. He who can do this will have the whole world with him. He who cannot, will walk the lonely way."

We certainly do not want to be placed in the latter category. How can we then create this **eager want,** in our case an eager want in the student to learn? The answers to the above questions are thus quite obvious. By showing and letting the person listen to beautiful piano playing we arouse in the other person a want to play the piano. By showing the toddler that you yourself enjoy eating whatever food you want him/her to eat, an **eager want** to eat is created in the young person. By showing that swimming is a lot of fun, **the want to learn** to swim is aroused in the other person.

Similarly, an **eager want** to learn science can be created in the student by showing the student that doing science can be very enjoyable. Unless there is a need, or a want in the person, or a reward for the other person, we cannot turn anybody on to do anything. Romey 1) stated that "the science teacher´s first task is to **establish a need to know**". How can you make someone try to taste a foreign dish? Would you just describe the dish or force the person to take a bite? Neither! Unless you yourself show the other person that you enjoy eating the particular dish and really show it (smacking your lips accompanied with exclamations of ecstacy!), that person most likely won´t touch the strange dish.

So it goes with learning science or any other subject matter. In order to get students **fired up** about something, first you yourself have to be **enthused** about it. Like in a wood fire, in order for you to give others a spark, first you yourself have to be aglow and burning. Thus, showing **enthusiasm,** to be dealt with under Science Teacher´s Characteristics, is one of the conditions for good teaching.

In the process of teaching and learning of science in the classroom, and the interaction between teacher and student, in order for the teaching to be effective and for the learning to be optimal, several key questions need to be answered:

1) What are the characteristics of the really successful and outstanding science teachers?
2) What is the nature of science itself as subject matter to be learned?
3) What are the characteristics of the learner and the conditions that have to be met in order for learning to take place?
4) Is there a specific device, method or technique that would create the **need to know** or the **eager want** to learn in the student?

This introductory section attempts to bring out the answers to the above questions in a logical progression, with the assumption that you, the reader, are already motivated yourself to read further.

1) Science Teacher's Characteristics

The features that will be discussed here are the special attitudes of a teacher that were seldom encountered in the average teacher, but that were always present in successful and effective science teachers.

a) Enthusiasm

Do you show any enthusiasm in your own science teaching? Have you ever asked yourself whether the degree of your enthusiasm in science that you show in class is high enough for it to spill over to your students? High enough to turn your students on to science?

Several works have been published on the subject of **enthusiasm**. One is by the author of "The Power of Positive Thinking", Norman Vincent Peale's 2): "Enthusiasm Makes the Difference", in which he states: "It is a proven law of human nature that as you imagine yourself to be and as you act on the assumption that you are what you see yourself as being, you will in time strongly tend to become, provided you persevere in the process." In other words, if you want to be enthusiastic, **act enthusiastic!** And you'll eventually be enthusiastic!

If you think back of your own education and ask yourself how you became interested in science or how you became a science teacher, you almost certainly would be able to trace it back to one or two teachers, whether it was at the elementary, secondary or college level, who were enthusiastic enough in science so that he/she made you want to continue a career in that field.

Enthusiasm is a funny thing. Although you cannot teach it to the students, they can actually catch it! "Enthusiasm, like measles, mumps and the common cold, is highly contagious", says Emory Ward. One of the most enthusiastic teachers of science I know is Hubert Alyea 3), originator of T.O.P.S. (Tested Overhead Projection Series) and Armchair Chemistry. His boundless zest, self-reliance and overflowing enthusiasm for chemistry teaching is very contagious. As a graduate student I once visited him at Princeton University to interview him about TOPS. Although he was very busy at that time, he made a special effort and insisted that I should see one demonstration that he would like to do for me. This was the 'clock reaction' where a colourless liquid would turn to orange and blue on command! He did it with such enthusiasm that I caught some of it and could actually weave some zest into my own presentations to pre- and in-service science teachers.

Another enthusiastic science educator is Vernon Rockcastle, Professor of Science Education at Cornell University. His vitality and zest for science teaching has made him not only an author of science texts, but also one of the most outstanding science education presenters at conferences. His science workshops offered at professional meetings are always some of the most enjoyable ones to attend.

Other science educators who stand out in my mind are: Paul F. Brandwein, researcher and author of numerous Science Methods textbooks; Harry K. Wong, Junior High School Science Teacher, author, and most sought after speaker in the science education world; Irwin Talesnick, Chemistry Professor at Queen's University and most dynamic demonstrator of chemistry; George VanderKuur, scientist and physics educator at the Ontario Science Centre; Alan McCormack, Professor of Science Education and Zoology at the University of Wyoming; and many more that I could mention.

What is a common characteristic that all these zestful and dynamic speakers possess? It is **enthusiasm!** Moreover, they are not afraid or inhibited to show it. They actually move faster, talk faster, smile more, and have a good time themselves! What are other features that they have in common?

b) **A High Self-esteem**

Every single science educator or teacher mentioned above, possesses a **high self-esteem.** They feel good about themselves. They value themselves highly. They know that they have something of value to contribute and this shows in their stance, poise, manner, and the voice volume and intonation that they use. In other words, their **attitude** is right!

How can you **develop the right attitude** as a science teacher? Attitudes not only show through but also "sound" through to the students. When you come in the classroom, it is not only your clothing that students will look at, but they will also listen to the voice and tone that you are using, which all determine your attitude. Expressions, voice volume, and inflections that you are using will give away your attitude towards science and science teaching in general. It will tell the students whether you like science, whether you are truly enjoying your occupation as a science teacher.

David J. Schwartz 4), in his book The Magic of Thinking Big suggests to make your attitudes your allies in everything you do. In the chapter that has the same title as the suggestion, he talks about growing the following three attitudes to win for you in every situation:

i. **Develop the Attitude of "I´m Activated"**

Let me illustrate this with the chemistry class that I was enrolled in at the Grade 12 level. At that point I actually already developed an interest in chemistry because of the previous years´ chemistry courses, which were offered by rather good teachers. Then came this grade 12 teacher (Mr. A), who was a middle-aged gentleman, apparently knowing the chemistry content very well, but nevertheless, was pathetically dull. He wrote constantly on the board all the reactions that he talked about, but never showed us any demonstration. It was amazing how he could change an exciting subject like chemistry into one that seemed so terribly dull and boring.

You may well imagine what the effect of such a boredom was on the students. Almost no one paid attention, some were sleeping and some were reading comic books, and actually almost none of the students learned anything.

Well, one day this teacher did not show up. As a matter of fact he completely disappeared from school - I really did not know why - but I think it was a lucky break for us students, because he could have turned everybody off in chemistry, actually he was already in the process of doing so. If he had stayed on, I know that I would certainly not be in the field of science education.

With the chemistry teacher missing, the grade 12 class had to go to the medical college, where a chemical engineer (Mr. B) was offering our grade 12 students special chemistry classes. This teacher then was completely the opposite of Mr. A. He was enthusiastic and he accompanied and illustrated every single reaction that he put on the board with a demonstration. How exciting! Chemistry became alive to me! Everybody was spellbound and our eyes were glued to the teacher and what he did. Questions were invited at any time and discussions were quite frequent. Everybody was learning!

The most remarkable thing is that I still remember some of the details of that course and Mr. B himself. I can still remember (and this is 34 years ago) his face, his smile, and his name: Ir.(Engineer) Van Gyn; and the most outstanding demonstrations of spontaneous combustion reactions, most of which are described in this book under "Chemistry".

Can you see the difference between Mr. A and Mr. B? In the case of Mr. A the students had no interest whatsoever in what the teacher was saying, because the teacher himself was not interested - at least he did not show it - in the subject matter. He must have been bored with chemistry and it certainly showed through.

The second teacher, Mr. B, just was himself and showed interest in his subject matter. He showed to his students that he liked what he was doing, that he enjoyed himself, and voila - his students became interested! His enthusiasm spilled over and was caught by his students! Thus the lesson from these examples is: **TO ACTIVATE OTHERS, YOU MUST FIRST ACTIVATE YOURSELF!** In order to get people enthusiastic, you must first be enthusiastic yourself. Before you can give off a spark, you must first be aglow yourself.

ii. Develop the Attitude of "You are Important"

Every human being, no matter whether rich or poor, young or old, backwards or sharp, living in New York City or Timbuktu, has a desire to feel important. Man's most compelling, non-biological craving is this desire to be important.

John Dewey, one of America's most profound philosophers, said that the deepest urge in human nature is **"the desire to be important"**.

Consider the following two classes and imagine yourself as a student coming into these classes the first day of school:

Class A
The teacher starts the class by saying: "Boys and girls. I am your science teacher. I am the boss and you have nothing to say. You are here only to learn from me. Science is important for our future, that is why we are here to learn science."

Class B
The teacher starts the class by saying: "Ladies and gentlemen. I am your science teacher. You are our future generation. You will be the leaders of our country 10-20 years from now. Since science plays such an important role in our future, you are here to learn what science is all about."

In which of the two classes would you feel more at home? In which of the two classes would you feel more important? In which would you likely learn more? It is quite obvious, of course in class B. Why? Because the teacher made you feel that you will be the leader of the country and leaders must be knowledgeable about science.

Why do we want to make others feel important? Because not only will people do more for you, but when you help others feel important, you will help yourself feel important too.

In the classroom we teachers can do this by praising the students whenever they are doing the right thing, or whenever they respond with the correct answer. Give praise at every opportunity you have. Respond to correct answers with: "Excellent!" or "Bravo!". Build students up and make them feel that they are contributing valuable information to the discussion. By praising the students you make them feel good, and good feeling students will never give you trouble in class. It is those students that have either bad feelings towards you or those that hate to be in your class, that will give you discipline problems.

So **MAKE YOUR STUDENTS FEEL IMPORTANT** by praising them and looking at the positive things that they did. In doing so, the students will feel good and love to be in your class.

iii. **Develop the Attitude of "I Care for You"**

The third attitude to develop to get the right teacher's attitude is showing others, the students in this case, that you sincerely care for them. This attitude should be quite apparent to the students. They should be able to sense and feel that you actually love your students. Students should know that you, the teacher, are there to help the students to learn in any way you possibly can, because you love them.

This aspect of "love" is the most neglected one in the attitude of the teacher in our educational system, whether it is at the elementary, secondary, or at the college level. The kind of love we are talking about here is comparable to parental love, or brotherly/sisterly love (when the age difference between teacher and student is still small). The love of a parent that is totally and unequivocally dependable. The love that a child feels coming from a parent, that gets him/her through thick and thin in life.

Harry K. Wong 5) expresses his love to the students by actually touching them. The kind of touch he is talking about and uses are the following: a) shake hands, b) a pat on the back, and c) a punch against the shoulder. This kind of touch he uses to enforce his appreciation or approval after the student has achieved something or has done something correctly, accompanied with a comment like: "Hey Tom, I knew I could count on you!" or "That-a-girl Susan, that's the way to do it!", builds the students' self-esteem.

Love in the classroom as expressed by the teacher for his/her students, is the unconditional acceptance of the students, whether they are yellow, white, red, or black, whether they have a 75 or 140 IQ. It is always "looking for the good" in the student and noticing the positive side of things.

Denis Waitley 6) in his book <u>Seeds of Greatness</u> states: "With love, there can be no fear. Love is natural and unconditional. Love asks no questions - neither preaching nor demanding; neither comparing nor measuring. Love is - pure and simple - the greatest value of all."

So bring love into your classroom, but before you can do that you must be able to totally accept yourself. Waitley calls this one of his best kept secrets of total success: **"WE MUST FEEL LOVE INSIDE OURSELVES BEFORE WE CAN GIVE IT TO OTHERS"**

In developing a healthy self-esteem we must understand that the word **esteem** means to appreciate the value of. We must realize that each of us are unique individuals and that **we are a masterpiece of creation.** Waitley says: <u>"There will never be a person who is more important than any other person, no matter how they look and no matter what kind of work they do. Each of us is as valuable and worthwhile as any other person."</u>

c) **Creativity**

Another common characteristic that the successful science teacher possesses is **creativity.** Among all living organisms on earth, only the human being is endowed with creativity or creative imagination. Napoleon once stated: "Imagination rules the world." Einstein said: "Imagination is more important than knowledge, for knowledge is limited to all we now know and understand, while imagination embraces the entire world, and all there ever will be to know and understand."

A study that was recently conducted by a research team of Stanford University revealed that: "what we watch is having an effect on our imaginations, our learning patterns, and our behaviors". This means that we have to be selective in what we feed our minds with. Our mind is like a computer: feed it with good stuff and good stuff will come out of our mouths. In the computer world there is a term: **GIGO** meaning - garbage in, garbage out. Feed a computer with garbage and you can't get more than garbage out of it.

Denis Waitley 7) has written a whole chapter on creativity related to success in life. He mentioned a few characteristics of creative individuals. How many of them fit your personality?

* Optimistic about the future
* Constructive discontent with status quo
* Highly curious and observant
* Open to alternatives
* Daydreamer, projecting into future
* Adventurous, with multiple interests
* Ability to recognize and break bad habits
* Independent thinker
* Whole-brain thinker (innovative ideas into practical solutions)

Have you met any individuals with most of the above mentioned features? If so, make friends with them, make them your role models, associate with them, and learn as much as you can from them.

d) Responsibility

Most teachers do possess this characteristic, which is one of the most valuable features for a teacher to have in the teaching world and actually for anyone to have in life in general. A responsible person is dependable, can be trusted, is a person in whom other people have faith. In other words, his word is to be trusted. When he says that he´ll be present at a certain place for an appointment at such and such a time, he should be there at that particular time, unless he has some legitimate reason for not showing up. This goes especially for business people in the business world. Can you imagine a business man making a very important appointment and the next day he comes with the excuse: "Oh, I forgot!" Will he have credibility with his colleagues? Will his colleagues put trust in his words?

What does this mean for the teaching world? In teaching the teacher´s word should be believed by the students. The teacher should be consistent and should actually do what he says he would do, whether it´s in giving back papers or whether it is in carrying out penalties for not doing homework, or any other thing. It is just a matter of remembering what you yourself said (or promised) and carrying out/performing what you said. This means that one should actually do what one preaches.

Many teachers (especially at the college level) say one thing and do another. For instance, one might be lecturing about the excellent advantages/effectiveness of the activity approach in teaching science, but one keeps teaching the students science by lecturing! This way, the whole thrust of the course might be in jeopardy. Students immediately lose trust in the teacher. Students will not put any value to the ideas that the teacher may have to offer. By not putting into practice what you, as a teacher, are saying - in theory -, your preaching about the theory is mostly in vain!

Another aspect of this "doing what you preach" is the role model part. You, the teacher, are usually the students´ role model. You give the example to the students in your whole attitude towards science and learning, and in living your life in general. So, if you want your students to develop responsibility, you yourself have to demonstrate all the desired characteristics of a responsible, dependable person of high integrity. So, your motto should be: **"Never preach what you don´t practice!"**

In order to develop responsibility in your students, you need to give them opportunities to do so. Let them show you that they can be responsible individuals. Give them tasks that demand some responsibility, like organizing books in class, cleaning and organizing glassware and materials in the laboratory, making posters, taking care of animals and plants, etc. The more we do for the students things that we like for them to be able to do, the less they can do for themselves.

As a father of four children and a teacher of teachers, I know from experience that the most lasting gifts that parents can give to their children, or that teachers can give to their students, are **roots and wings: Roots of Responsibility** and **Wings of Independence**. When you, as a teacher, can leave your students on their own, and learning is still continuing, then you have actually taught your students responsibility and independence.

e) **A Sense of Humour**

A teacher should have a good sense of humour. This does not mean that he must be a comedian or constantly have to come up with funny stories. A good sense of humour means that the teacher must be able to look at his own mistakes and laugh at himself.

Before you can laugh at yourself and the mistakes you made, you have to be able to accept and admit that you have made a mistake. We are all human beings and being human means that we all are bound to make mistakes. By admitting your mistakes in front of the students you actually show them that you, as a teacher, are human too. Students would rather see you admit your mistakes than see you try to cover them up.

Not only should you be able to admit your mistakes, but also the answers to questions that you don´t know! Paul Brandwein once stated that teachers should not say: "I don´t know", they should say: "It´s not in my field". He was of course just joking when he made that statement. You may comfortably say: "I don´t know" and the students do expect you to say it to some of the questions that they pose, but you might be able to suggest where they can look for the answer. You see, a teacher is just another human being that knows a little bit more than the students, but he certainly is not a walking encyclopedia.

So be optimistic, put a smile on your face and get excited about the fact that you are so privileged to be able to determine the thinking in our younger generation and thus the future of our country!

f) **Communication**

Successful science teachers and educators are always also successful communicators. They seem to be able to capture the attention of their audiences, keep them awake for an extended period of time, and get across whatever concepts or messages they wanted their audiences to know.

In order to get things across, we may want to communicate in different languages if we are dealing with people of different nations. I was once in Europe and saw a sign in front of a restaurant. It said: "All Languages Spoken Here", so I stepped in and asked the waitress: "Who speaks all languages?" She said: "The customers!" But even in the same English language, it can sometimes be difficult to communicate well. Like this woman that wanted a divorce and saw her pastor. He asked her: "Do you have any grounds?" "Yes", she said, "We own about 15 acres north of town". "No, that is not what I meant", said the pastor. "I mean, do you have any grudge?" "No", she answered, "But we do have a carport in our driveway". "No, no", said the pastor, "How can I put it? Does your husband beat you up?" To this question she reacted: "Beat me up? I almost always wake up much earlier than he does every morning". Quite exasperated he said to the woman: "Lady, would you please listen carefully! Does you husband give you any trouble?" "Yes", she said, "He just cannot communicate!"

Nido R. Qubein 8) in his book "Communicate Like A Pro" stated: ....**sincere, honest, enthusiastic, and positive people make the best communicators.** <u>The better you get to know them, the more you like and trust them; and the more you are willing to listen to what they have to say.</u> This certainly is true for the effective teacher and his success in communication with his students.

In science, however, above and beyond the general features above, successful teachers have a particular way of communicating a concept to the students. They do not only try to describe the concept with words, but they usually show the students an object or event which applies the particular concept. A Law of Teaching Science which Paul Brandwein 9) coined is: **A CONCEPT IN SCIENCE IS SYNONYMOUS WITH A CORRESPONDING SET OF OPERATIONS.**

This means that whenever we want to convey or communicate or teach a science concept to the students, we should show or demonstrate a corresponding set of operations in which that concept is applied. In other words, we should do a science demonstration which applies the concept or principle.

In science, therefore, (and in my opinion even in other disciplines) if you want to teach a concept, do also a demonstration which applies the concept. This is why the demonstration is such an important tool for the science teacher. A shorter version of Brandwein's Law is: **DEMONSTRATION IS TEACHING.**

SUMMARY: SCIENCE TEACHER'S CHARACTERISTICS

    a) DEVELOP ENTHUSIASM

    b) DEVELOP A HIGH SELF-ESTEEM

        i. DEVELOP THE ATTITUDE OF "I'M ACTIVATED"

        ii. DEVELOP THE ATTITUDE OF "YOU ARE IMPORTANT"

        iii. DEVELOP THE ATTITUDE OF "I CARE FOR YOU"

    c) BE CREATIVE

    d) SHOW RESPONSIBILITY

    e) DEVELOP A GOOD SENSE OF HUMOUR

    f) DEVELOP YOUR COMMUNICATION SKILLS

2) The Nature of Science and the Scientist

In order to teach science properly, we have to know what particular characteristics science has, and also the particular features that scientists display.

a) The Nature of Science

If we look in Webster's Dictionary, there are several definitions for science, but the closest one to what teachers are concerned with in school is this: "Science is a branch of study that is concerned with observation and classification of facts and especially with the establishment or strictly with the quantative formulation of verifiable general laws chiefly by induction and hypotheses".

The science philosopher Benjamin 10) defined science as: "That **mode of inquiry** which attempts to arrive at information about our world (universe) by the method of observation and by the method of confirmed hypotheses based on observation."

In both definitions we can see that science is a process as well as a product. In the latter one we can see more of an activity rather than just the study of facts. If we hear the word "inquiry", what do we immediately think of? A process of questioning and answering to search for the truth of the matter. It is that part of the inquiry process which tries to obtain truthful information about our universe by using our senses, by gathering data, by formulating hypotheses, and by confirming these hypotheses on the basis of what we obtained from observation.

What are the implications in teaching this particular discipline in the classroom? The best way to teach a particular subject is to reflect the true nature of that subject. In our case, reflecting the nature of science is by doing science activities in the classroom. By having the students involved in observing, measuring, calculating, formulating hypotheses, gathering data, etc. in other words, involving them in all the science processes.

In dealing with the processes, it is virtually impossible to avoid dealing with science concepts. We have to observe some event, hypothesize why something is happening. Put in other words: there has to be a focus or science concept around which students can practice the processes. This book provides the teacher with the needed focuses - **discrepant events** which are interest arousing and challenging.

By presenting these **discrepant events** to the students in the proper way (see section under the discrepant event), it is very natural and logical for the teacher to involve the student in the process of inquiry. Since the event is so unexpected or against our intuition, it is easy for the teacher to ask questions about it. The teacher can ask questions requiring either simple answers or more complicated answers, depending on how deep he wants to go into the subject. The teacher can therefore adjust each of the **discrepant events** to his own teaching level.

b) **The Nature of the Scientist**

We teachers are especially interested in what makes a scientist work on one topic with such dedication. How do scientists become curious about a particular topic and start a series of investigations about it that could sometimes last for a whole lifetime? What is it that triggers their attention to such a degree that they can become so engrossed in it? In other words: what is it that makes them tick?

Let us consider a couple of great discoveries and let us take a closer look at their discoverers. Just before the turn of the century Henri Becquerel discovered radioactivity and a few decades after the turn of the century Sir Alexander Fleming discovered penicillin.

Becquerel was fascinated by Rontgen´s discovery of X-rays and wondered if the natural luminescence, or glow, of certain minerals might also be accompanied by similar x-ray emission. He took a thin crust of a uranium crystal and placed it on a photographic plate wrapped in lightproof paper. The whole wrapping was exposed to sunlight for several hours. When developed, an outline of the crystals showed up on the photographic plate. Subsequently, in a similar experiment he found that where some uranium crystals had been left in a dark drawer, unexposed to light, together with a photographic plate, they left even a stronger outline on the developed plate! At that moment he must have been very surprised and could have exclaimed: "Hey! There is a **discrepancy** here!" He became so interested in this phenomenon that during the following years he became totally engrossed in his work on this strange radiation.

Fleming was working in the laboratory with cultures of Staphylococci, bacteria that causes boils and blood poisoning. He had also cultures of Penicillium mould around and one day some of his bacteria cultures were contaminated with the Penicillium. While he was in the process of discarding the contaminated bacteria cultures his attention and interest was aroused by **something unusual**: there were certain spots in the bacteria cultures that were blank. In other words, the poisoning causing bacteria did not grow where it was contaminated with penicillium. From that moment on the investigation and further extraction of penicillin from the penicillium mould continued.

From the above examples we can see that in both cases the major thing that triggered the scientists´ attention and interest was a **discrepancy**. Something that they did not expect to happen at all, something that is against their intuition. For most people in general something that is out of the ordinary may on the whole attract one´s attention. Also something that is unusual generally can be remembered for a longer period of time.

What are the implications in the teaching of science and making it interesting for the students? Confronting the students with **discrepant events** and meeting all the conditions for learning would be the ideal things for teachers to remember. This brings us to the question which has pertinent influence on the learning process:
    a) How can new information be retained longer by students?
    b) What are the conditions for information retention?

Scheme 1. PROCESS OF INFORMATION RETENTION AND RECALL

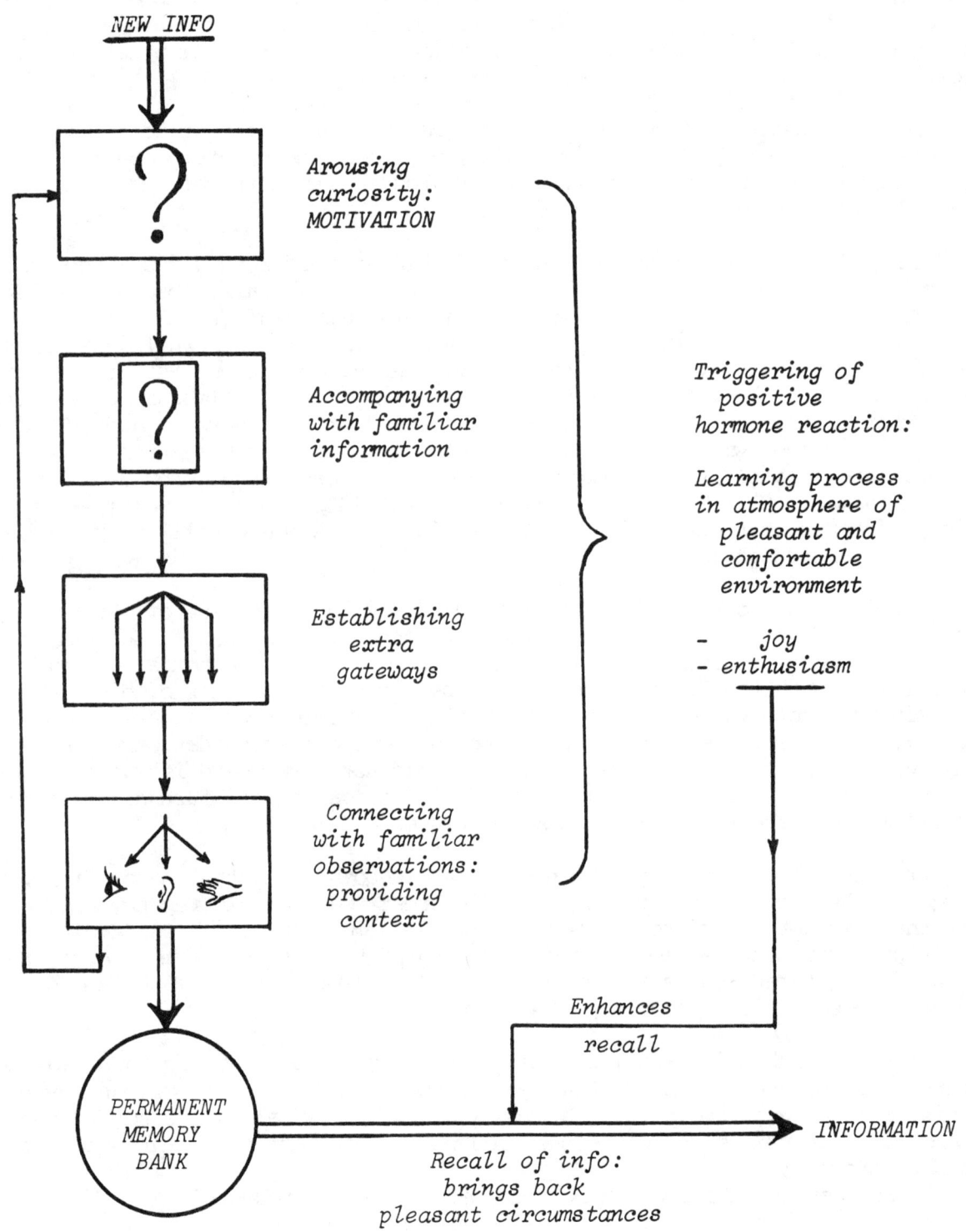

3) The Learning Process

In considering the Learning Process or the Process of Information Retention as described by Vester 11), it can be seen that several conditions have to be met in order for new information to penetrate our Permanent Memory Bank (see flowsheet on page 12).

Vester mentioned the following conditions to be most essential in order for new information to penetrate the Permanent Memory Bank:

a) Arousing curiosity: MOTIVATION
b) Accompanying with familiar information.
c) Establishing extra gateways.
d) Connecting with familiar observations: providing context.
e) Providing an atmosphere of JOY and ENTHUSIASM.

Let us take a closer look at each of these conditions:

a) **Arouse Curiosity**

Of the above mentioned conditions, the most important one is the arousing of curiosity in the learner. Baez 12) mentioned that curiosity is one of the four important traits in people that should enrich the quality of life. He called them the four C´s, which stands for: "Curiosity, Creativity, Competence, and Compassion". About curiosity he further states: "Curiosity is the motor that drives the scientist.....curiosity is the source of discoveries in science and technology.....The spark of curiosity ought to be fanned into a flame by teachers and parents. It can make learning a pleasurable experience, but it is sometimes stifled by uninspired teachers who find it easier to demand rote learning."

How can we fan the spark of curiosity into a flame? What can the science teacher do to arouse the student´s cusiosity? Romey stated that: "One of the best ways to stimulate interest is to **offend the student´s intuition** in some way or to confront him with a situation that is not readily acceptable". Sund and Trowbridge 13) are of the opinion that..."another technique for developing motivation and interest in a discussion is to use **pictorial riddles**".

What better way to offend the student´s intuition in science than by doing a demonstration? A demonstration of which the outcome is not expected or of which the performance is thought impossible: a demonstration of a **discrepant event**. Gagne 14) is of the opinion that in order for learning to occur, the student´s **brain must be awake**. The learner must have a state of alertness that corresponds to the common sense word ATTENTION.

b) **Use Simple Materials**

The use of simple materials or accompanying the question with familiar information is also one of the essential conditions for the learning process. In order for students to learn something new, the teacher has to start with something that the student already knows and is familiar with.

Especially in presenting a demonstration to the student with the purpose of showing a **discrepant event,** it will not be successful unless the demonstration is carried out with simple materials which the student is familiar with. Brandwein 15) says: "Unless an object or event is recognized, a problem is not recognized". In other words, a discrepancy or discrepant event is not recognized unless an object or event is recognized."

c) **Use all Gateways**

With these we mean gateways into the human brain. There are of course, five gateways or five entrances into our brain, each opened by the use of our five senses: sight, hearing, touch, smell, and taste.

When teachers strictly lecture, they make use of only one gateway, which is the students´ hearing. As soon as teachers show or demonstrate something, sight is utilized as an entrance into the brain. But it is only when teachers give the students an opportunity to do things themselves, it is then that they also utilize the other three gateways into the brain: touch, smell, and taste. This means that when teachers illustrate their lectures with a **demonstration,** and then involve the students with a **follow-up activity** on the same concept, they make maximum use of gaining entrance into the students´ brain.

d) **Provide Context**

Any science concept or principle when taught isolated from our daily life, becomes completely meaningless and thus much harder to learn.

Therefore, **give examples** of where that particular concept presents itself in our daily life, or where that particular **principle applies.** Instead of talking only about the First Law of Newton, ask: "Why do you hold on to something solid when stepping into a bus that is just about to leave?" (First part of First Law) "Why do you do the same when the bus is about to stop?" (Second part of First Law).

It is when science concepts or principles are placed in context and connected with familiar experiences of the student, that they will become much more meaningful to the learner, and thus much easier to be learned.

e) **Do it with JOY and ENTHUSIASM**

Can you imagine yourself learning science from a teacher who thinks him/herself that science is boring? No, most likely not, as you probably would not learn any science, because it was portrayed as a boring discipline.

On the other hand, imagine yourself as a student coming into a classroom and seeing the teacher whistling to himself while preparing for a demonstration, jumping around to get the materials from the shelf here and from the shelf there, looking forward to conducting the lesson and just enjoying himself during the few minutes before the bell rings while waiting for the class to fill up. Would you expect to learn science in this classroom?

The answer to the above question is: "Of course!" The teacher's characteristic of <u>having enthusiasm</u> is therefore a very important one to possess. How do we acquire enthusiasm?

Not only is it a desirable teacher's characteristic, enthusiasm will also improve your personality. If you really want to change your personality, you can and it is not that hard to do! Here are a few excerpts from Norman Vincent Peale's **Enthusiasm Makes the Difference**:

>.......It is the change from apathy to enthusiasm, from indifference to exciting participation; it is an astonishing personality change which sensitizes the spirit, erases dullness and infuses the individual with a powerful motivation that activates enthusiasm and never allows it to run down.......
>
>.......You can develop enthusiasm, and of a type that is continuous and joyous in nature.......
>
>.......It has been established by repeated demonstration that a person can make of himself just about what he wants to, provided he wants to badly enough and correctly goes about doing it. A method for deliberately transforming yourself into whatever type of person you wish to be is first to decide specifically what particular characteristic you desire to possess and then to hold that image firmly in your consciousness. Second, proceed to develop it by acting as if you actually possessed the desired characteristic. And third, believe and repeatedly affirm that you are in the process of self-creating the quality you have undertaken to develop........
>
>.......It is a proven law of human nature that as you imagine yourself to be and as you act on the assumption that you are what you see yourself as being, you will in time strongly tend to become, provided you persevere in the process.

From the last two paragraphs we can conclude the following:
If you want to be enthusiastic, act enthusiastic!

SUMMARY: CONDITIONS FOR LEARNING

Remember the acronym: **INSIGHT**

IN = INTEREST; DEVELOP INTEREST, AROUSE CURIOSITY

SI = SIMPLE; USE SIMPLE MATERIALS

G = GATEWAYS; USE AS MANY GATEWAYS AS POSSIBLE

H = HINGE; HINGE IT WITH EXAMPLES AND PUT IT IN CONTEXT

T = TIE; TIE IT ALL TOGETHER WITH JOY AND ENTHUSIASM!

4) The Discrepant Event

When properly used and presented to students, the **discrepant event** will encompass all five conditions for learning.

a) Psychological Background

The **discrepant event** in science finds its base in the Theory of Cognitive Dissonance by Festinger [16], in which he proposes two basic hypotheses:

1. The existence of dissonance being psychologically uncomfortable, will motivate the person to try to reduce the dissonance and achieve consonance;

2. When dissonance is present, in addition to trying to reduce it, the person will actively avoid situations and information which would likely increase the dissonance.

It is especially the first statement that applies to using discrepant events to initiate learning. Festinger is further of the opinion that the **discrepant event** functions by causing dissonance between what is physically observed to occur and what one thinks should occur. Since it is impossible to change what is physically observed to occur already, the only alternative is to begin to seek information which will logically explain the occurrence.

Waetjen [17] stated that the **discrepant event** is a dissonant situation which results in arousal of conflict with a consequent need for the learner to assimilate or articulate the unknown, incongruous, unfamiliar material into his/her cognitive structure. To do this, he/she engages in exploratory behaviour.

The **discrepant event** has also been called Disconfirmation of Expectancy. It is a situation contrary to what the learner expects, and actually a state necessary for cognitive development, for it moves the learner to try to resolve the discrepancy between what was expected and what actually happened.

It has also been called Conceptual Conflict. When the learner faces a situation which is in conflict with what he expects, the doubt, perplexity, contradiction, and incongruity play an important role in stimulating the learner's curiosity.

Piaget [18] says that: "This state (of perplexity and doubt) is a necessary first step in learning!", and further states that "puzzles are excellent sources for learning simply because they unsettle the learner, upset his intellectual equilibrium, and incite him to change or adapt his existing intellectual scheme...... The learner who meets such a challenge develops and assimilates new skills that make him or her a cognitively richer individual."

The **discrepant event** then, whether it is a demonstration by the teacher or an activity performed by the student, is mainly used for its motivational effect on the learner. It creates the **eager want** in the student to know more about the event, and thus do other activities that are aimed at dissonance reduction and subsequent knowledge-seeking behaviour is easily aroused and strongly reinforced.

b) **What is a Discrepant Event?**

Everybody has seen objects fall downwards. The fact is hardly surprising or unusual. However, if one were to see an object fall upwards, it would be an entirely different matter. It would be an event which defies gravity, and in this case the moving of the test tube against the force of gravity is a **discrepancy** (See The Upwards Falling Test Tube).

Most people know that men are stronger than women. However, if one were to see men not being able to pick up a chair from a bent position and women doing it with ease, what would one think? Some feeling of surprise and curiosity would be aroused. This would especially be true when one gets involved in trying to perform the event oneself (See Are Women Stronger Than Men?).

All children know that a newspaper is so light in weight that it can be easily blown away by the wind. However, if one sheet of a newspaper placed on a long flat wooden stick (protruding over the edge of the table) would hold the stick down, even if one would hit the protruding end of the stick, what thoughts are being aroused in one's mind? (See The Heavy Newspaper).

One would expect that objects would fall down and not upwards, men are usually stronger than women and the chair is certainly not so heavy that men would not be able to lift it. One sheet of newspaper would surely not be heavy enough to hold down a stick, even to breaking! These examples of **discrepant events** are often described as **surprising**, **counter-intuitive**, **unexpected**, **paradoxical**, and **intuition-offending**.

The **discrepant event** has the tendency to arouse strong feelings within the observer or the participant. Generally, one will have a sudden urge of wanting to know more about the event. One gets curious and wishes to solve the discrepancy in one's mind. Children as well as adults will demonstrate a strong desire to resolve the unexpected. The enthusiasm of the first will certainly be much greater. Children will simply not stop asking questions until they find out why certain events occurred the way they did!

When students' interest is thus hightened, he/she will more likely become motivated to learn. **Discrepant Events** are therefore excellent means to create in the student an **eager want** to learn more about science. **Discrepant Events** capitalize on the students' own curiosity, already present within the person, helping him/her to gain a better understanding and retention of science concepts.

c) **How to Use Discrepant Events?**

In order the achieve an atmosphere of inquiry in the classroom, it is important that the teacher presents the discrepant event as a science problem to be investigated or as a puzzle to be solved. The discrepant event or intuition-offending demonstration must be presented in such a way that the science principle or concept underlying the event is not immediately revealed. The teacher using the discrepant event should in general lines follow the procedure below:

Teaching Procedure in Using Discrepant Events

1. Presentation
    Present to the learner or involve the learner with the discrepant event by describing or commenting on the names of objects and operations only and not mentioning the reasons for the occurrence. In other words, the teacher may tell the student what he/she is doing and what materials he/she is handling, but not why something is happening.

2. Interaction
    Ask the learner questions that eventually will lead him/her to the main reason for the occurrence. In doing this the students will be engaged in science inquiry and actually practicing the science processes of observing, measuring, inferring, predicting, interpreting data, identifying and controlling variables, hypothesizing, and experimenting.

3. Involvement
    Participate the learner in similar (and simpler) discrepant events or counterintuitive activities illustrating and based on the same science concept. This will reinforce the learning and retention of that particular concept. Students may work individually or in pairs, or in groups depending on the availability of materials.

It is essential that the presentation of the intuition-offending demonstration or the involvement of the students in the discrepant event takes place under the following conditions (see also The Learning Process on page 13):

1. Arouse Interest
    The event should confront the observer with a perplexing problem. It should be presented almost like the way a magician would do a magic act.

2. Use Simple Materials
    The event should be performed with materials that are familiar to the learner, in other words use simple daily life materials.

3. Use All Gateways
    The students should have the opportunity to observe as well as carry out the events themselves. They should be allowed to totally experience the discrepant event.

4. Hinge It With Examples
    When dealing with the concept underlying the event, examples and applications of the concept in our daily life should be mentioned to make it more meaningful to the learner.

5. Tie It With Joy And Enthusiasm
    The teacher should show genuine enthusiasm in presenting the puzzling event and reveal his/her enjoyment in the subject matter in general.

When all of the above elements are present during the learning process, the student will be able to store or retain new information with much more ease. The learning experience becomes more meaningful and enjoyable, resulting in a much better retention of the science concept. The recalling process of the stored information will bring back all the pleasant circumstances during which it was learned, which makes it a nearly effortless mental activity.

The above conditions for learning a science concept may be attained when the teacher is genuinely showing concern for the students, uses discrepant events to initiate science inquiry, and most importantly is enthused about what he/she is doing. Once this optimum condition is reached, teacher and students will enjoy the teaching-learning process to a much greater extent.

d) **The Empirical Evidence**

Suchman 19) indicated in his study that the teaching technique with discrepant events positively influence content achievement. The study showed that this content achievement is affected by the arousal of student motivation, which is directed at the comprehension of the causes of the observed events.

Marlins' 20) findings were confirmed by Liem 21), that "upper elementary school students taught using the demonstration-discussion method with counterintuitive events, had significantly higher retention scores compared to those taught without counterintuitive events."

I followed up my studies on the effect of discrepant events on science concept retention at the junior high level 22). I subjected one group of students (Control) to only discussion and reading of a text containing descriptions of discrepant events positively influences content achievement. The study showed that this content achievement is affected by the arousal of student motivation, which is directed at the comprehension of the causes of the observed events. measures were administered to both groups:
The Pretest: administered before the lessons.
Post-test A: administered immediately after the lessons.
Post-test B: administered one month after the lessons.
Post-test C: administered three months after the lessons.

The results and means of the tests were then compared and subjected to the t-test described by Best 23). The following t-values were obtained: Pretest: 0.27; Test A: 1.96; Test B: 1.78; Test C: 2.72. This shows that there was a significant difference between the means of Tests A and B at the .05 level, and of Test c at the .01 level, and also that there was no significant difference between the means of the Pretests.

From these findings one would be led to the conclusion that the larger the time lapse between the lessons and the administering of the test, the more significant the difference becomes between the mean score of the Experimental compared to that of the Control Group. In other words, the group taught using the discussion method whereby discrepant events were demonstrated by the teacher and experienced by the students, retained the science concepts longer compared to the group taught using the discussion method whereby discrepant events were only read.

The findings of the studies conducted thus far would seem to justify the use of **discrepant events** for teacher demonstration and student activities to increase retention and understanding of science concepts at the elementary and intermediate levels. One is also led to infer that the **discrepant events** lose their **discrepancy** or **motivating effect** on the student, when they are merely described in a text and read by the student.

THE FORMAT

This book is organized in such a way that it offers to teachers the optimum guidance for teaching science concepts and engaging the students in science inquiry. The chapters of the book are organized under an introductory section and four sections: Environment, Energy, Space and Forces, and Living Things. The introductory section provides the teacher with the psychological background for effective science teaching and the use of **discrepant events.** Each of the latter four sections is preceded by a short introduction providing the teacher with some information on the science concepts to be dealt with in the chapters.

The discrepant events, making up the main body of each chapter, are preceded by a list of objectives. These are stated in behavioral terms in order that student and teacher may work towards achieving these behaviors.

The discrepant events are invitations to science inquiry, which are all fitted on one page each, for simplicity in format and easy location. Each event is consistently constructed in the following format:

1. Upper left corner: Chapter Science Concept.
2. Upper right corner: Science Sub-concept or Science Properties.
3. Intriguing, curiosity arousing title of event.
4. Materials needed for the demonstration or activity.
5. Sketch of materials set-up.
6. Procedure, in step by step fashion.
7. Questions for teacher's use during demonstration.
8. Explanation, providing background information for the teacher.

The book ends with a list of 'Further Readings' in which similar or more detailed information may be obtained. A general index follows the reference list, where users of the book may find concepts and sub-concepts and the relevant events that are appropriate to use in teaching those concepts to students.

References

1. Romey, W.D., INQUIRY TECHNIQUES FOR TEACHING SCIENCE, Prentice-Hall, Inc., 1968, p 16.

2. Peale, Norman Vincent, ENTHUSIASM MAKES THE DIFFERENCE, Fawcett Crest, New York, 1982.

3. Alyea, Hubert, TESTED OVERHEAD PROJECTION SERIES, & Dutton, Frederic B., TESTED DEMONSTRATIONS IN CHEMISTRY, J. of Chem. Ed., Easton, Penn. 1965.

4. Schwartz, David J., THE MAGIC OF THINKING BIG, Audio Tape, Sound Ideas, Simon & Schuster Audio Publishing Division, 1230 Ave of the Americas, New York, N.Y. 10020

5. Wong, Harry K., HOW YOU CAN BE A SUPER SUCCESSFUL TEACHER, Cassette Audio Tape Series, 1536 Queenstown Court, Sunnyvale, CA 94087.

6. Waitley, Denis, SEEDS OF GREATNESS, Fleming H. Revell Co., 1983.

7. Ibid., pp 43-61.

8. Qubein, Nido R., COMMUNICATE LIKE A PRO, Berkley Books, NY 1983.

9. Brandwein, Paul F., THE METHOD OF INTELLIGENCE, Presentation to NSTA Convention, Toronto, 1968.

10. Benjamin, Abram Cornelius, SCIENCE TECHNOLOGY AND HUMAN VALUES, Columbia, Univ. of Missouri Press. 1965.

11. Vester, Frederic, HOE WIJ DENKEN, LEREN EN VERGETEN (Original title: DENKEN, LERNEN, VERGESSEN. Deutsche Verlags Anstalt GmbH, Stuttgart, 1975), Bosch en Keaning, Baarn, 1976.

12. Baez, Albert V., "Curiosity, Creativity, Competence and Compassion - Guidelines for Science Education in the Year 2000", WORLD TRENDS IN SCIENCE EDUCATION, Atlantic Institute of Education, Halifax, NS, 1980.

13. Sund, R.B. and Trowbridge, L.W., TEACHING SCIENCE BY INQUIRY IN THE SECONDARY SCHOOL, 2nd ed., Charles E. Merrill, 1973.

14. Gagne, R.M., THE CONDITIONS OF LEARNING, Holt, Rinehart and Winston, Inc., New York, 1965.

15. Brandwein, Paul F., THE METHOD OF INTELLIGENCE, Presentation to NSTA Convention, Toronto, 1968.

16. Festinger, Leon, A THEORY OF COGNITIVE DISSONANCE, Row, Peterson and Co., 1957, p.3.

17. Waetjen, Walter B., "Learning and Motivation: Implication for the Teaching of Science", READING IN SCIENCE EDUCATION FOR THE SECONDARY SCHOOLS, New York: The MacMillan Co., 1969, p.91.

18. Piaget, Jean, "The Child and Reality: Problems of Genetic Psychology", translated by Arnold Rosin, London: Frederic Muller, 1974.

19. Suchman, J.R., "Inquiry Training in the Elementary School", THE SCIENCE TEACHER, Vol. 27, Nov. 1960.

20. Marlins, James G., "A Study of the Effects of Using the Counterintuitive Event in Science Teaching on Subject-matter Achievement and Subject-matter Retention of Upper-elementary School Students", Doctoral Thesis, The American University, 1973.

21. Liem, Tik L., "A Study of the Effects of Using Discrepant Events in Science Teaching on Concept Retention of Upper Elementary School Students", WORLD TRENDS IN SCIENCE EDUCATION, McFadden, C.P. (Ed), Atlantic Institute of Education, Halifax, 1980, pp. 287-293.

22. Liem, Tik L., "Effects of Using Discrepant Events on Science Concept Retention of Junior High School Students", Paper presented to the National Co-Educators Conference, Winnipeg, October 1980.

23. Best, J.W., RESEARCH IN EDUCATION, Prentice-Hall, New Jersey, 1977.

SECTION I

ENVIRONMENT

This section consists of five chapters dealing specifically with the properties of matter in our immediate environment. This includes the following:

1. The air around us in its stationary form. It occupies space, it exerts pressure, it expands when heated, it has weight, and it contains oxygen.

2. The air around us when it moves or flows. Flowing air exhibits properties that are unexpected, namely a lower pressure compared to stationary air, and the faster the flow, the lower the pressure exerted.

3. The weather and all the variables influencing it, like: high and low pressures, the wind direction and velocity, the temperature and humidity, altitude and climate, and location on earth.

4. The characteristics of matter. This chapter deals with the molecular spacing of matter, cohesion and adhesion, surface tension, the three states of matter, boiling and melting point.

5. The chemicals in our environment and how they behave. Acids and bases, salt formation, catalytic reactions, spontaneous combustion, substitution reactions and electrochemical reactions are dealt with.

Science ......

for Young and Old

CHAPTER 1

WHAT ARE THE PROPERTIES OF AIR?

OBJECTIVES

After dealing with and studying the concepts in this chapter, the students should be able to:

a. recognize the correct explanation of an observed event based on each of the sub-concepts;
b. explain in their own words which of the sub-concepts is determining the course of an event;
c. distinguish true from false statements concerning each one of the sub-concepts;
d. identify the correct explanation of an event in daily life applying one of the sub-concepts;

all in relation to the following sub-concepts:

-- Air exists all around us.
-- Air occupies space.
-- Air exerts pressure.
-- Air expands when heated, and rises as a result.
-- Air contains oxygen.
-- Air has weight.

AIR   OCCUPYING SPACE

1.1. THE AIR CATCHER

Materials: 1. A medium plastic garbage bag.
            2. One plastic sandwich bag per student.

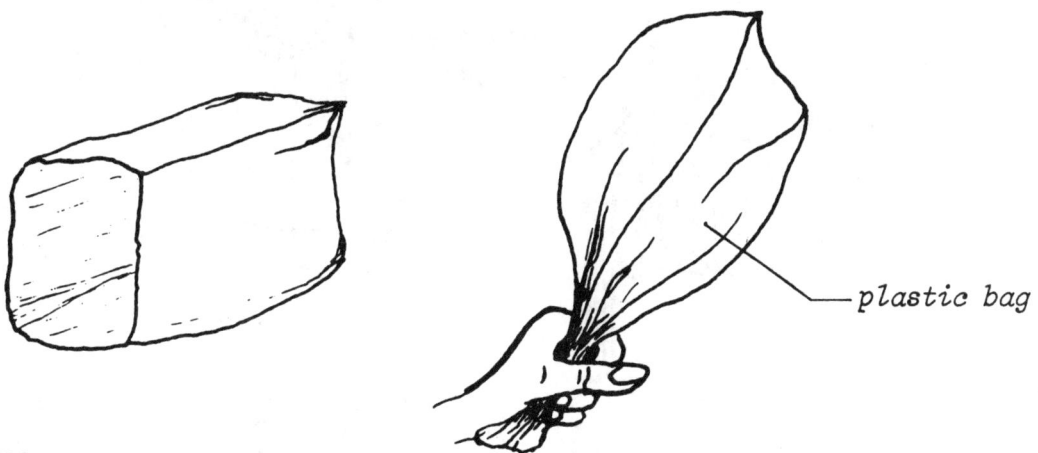

Procedure:
1. Take the medium size garbage bag, open its mouth and ask the students: "What is in the bag?" (Anticipated answer: 'nothing').
2. Move the bag now with two hands back and forth (like wanting to catch a bug in the bag), then quickly close the mouth of the bag with a twisting motion.
3. Ask the students: "What do I have in the bag now?"
4. Distribute the sandwich bags to the students and let them try to catch air in their own seats, without blowing into the bag.

Questions:
1. What was filling the bags?
2. Can we catch air under the bench or behind the door?
3. Is the air the same everywhere?
4. How else can we fill the bag?
5. Would the material in the bag be the same if we blew in it?
6. How can we keep the bag inflated?
7. What would happen if we hit the inflated small plastic bag with the palm of the other hand?

Explanation:
Air is found everywhere. The plastic bags may be filled with air above the table, under the table, behind the door or anywhere else with the same air. The bags can also be inflated by blowing in them, but then the bags would contain exhaled air, which has a higher percentage of carbon dioxide (CO_2) and more water vapor.
When the filled bag is slammed between the two palms of the hands it will burst with a loud pop. This explosion is caused by the sudden expansion of the air rushing out of the torn plastic bag.
An example of this is a popping balloon.

AIR — OCCUPYING SPACE

1.2. THE BOTTLE AND THE BAG

Materials:
1. One or two plastic sandwich bags.
2. One or two large wide-mouthed glass jars (pickle jars).
3. Masking or transparent adhesive tape.

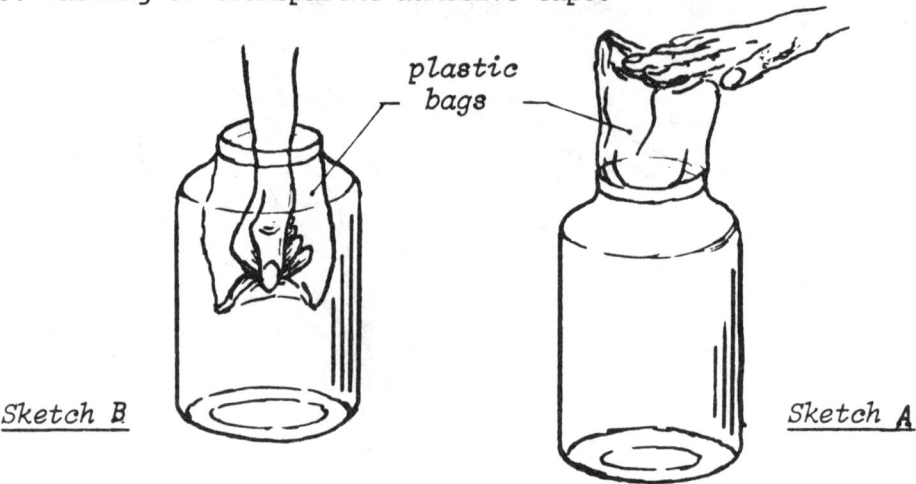

Procedure:
1. Invert the bag over the mouth of the jar, blow a little air in the bag such that it stays inflated over the jar (see Sketch A).
2. Tape the bag air-tight against the jar.
3. Now ask one of the students to push the bag into the jar (without tearing it): It won't work!
4. Place another plastic bag inside another wide-mouthed jar (or use the same bag and jar) and let the edge of the bag hang over the jar rim (see Sketch B).
5. Tape it air-tight against the jar and let a student try and take the bag out of the jar (without tearing it): It won't work!

Questions:
1. Before putting the plastic bag on the jar, ask: "What is inside the jar? Inside the plastic bag?"
2. What is holding the bag out of the jar? (when trying to push it in).
3. What is holding the bag inside the jar? (when trying to take it out).
4. How could we get the bag inside the jar without making a hole in it?

Explanation:
   It is the air occupying the space in the jar which kept the bag from going inside after it had been taped air-tight against the jar. In trying to push the bag in, the pressure increased (because the volume decreased) and this held the bag out.
   When trying to take the bag out of the jar, the air pressure inside the jar decreased, because the volume increased, and this kept the bag inside. The outside air pressure kept the bag inside the jar. We encounter the first situation often when we try to fold up a plastic air mattress or inflatable plastic toy.

AIR   OCCUPYING SPACE

1.3. POUR AIR UNDER WATER

Materials: 1. Two glass or transparent plastic cups.
             2. A large transparent container (a small aquarium).

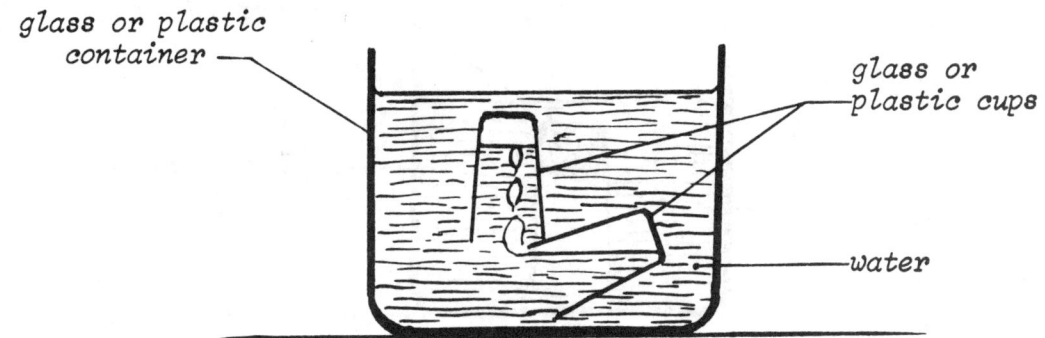

Procedure:
1. Fill the plastic or glass container about 3/4 full with water.
2. Hold one cup in each hand upside down and push them under water.
3. Fill one of the cups with water by holding it slanted and thus releasing the air bubbles (do not leave any bubbles).
4. Now we have one cup filled with water in the one hand, and one cup with air in the other (still under water).
5. Now push the cup with air a little lower than the other, and **pour** the air in the other cup by slowly slanting it (catch the air bubbles: see Sketch).
6. We can repeat this **pouring** from one cup into the other.

Questions:
1. Before immersing the cups, ask: "What is in the beakers or cups?" (anticipated answer: 'nothing').
2. At the time of immersing the inverted cups, ask: "Why doesn't the water enter the cups?"
3. Why do the bubbles rise and not sink?
4. Can the cup with water be held partly above the water level without letting the water run out of the cup?

Explanation:
   Air **occupies space** and also the space in the cups. At the time the cups were immersed under water, they were filled with air, and that is why the water could not enter the cups. By holding one cup slanted the air bubbles were free to escape and thus the water could take its place. Air is much lighter in weight than water and that is why air bubbles rise and not sink in water. The water-filled cup can be held above the water level without letting the water run out, because the atmosphere pressure is pushing on the water surface.

AIR OCCUPYING SPACE

1.4. KEEP PAPER DRY UNDER WATER

Materials:
1. A dry glass or transparent plastic cup.
2. A large beaker or transparent plastic container (large enough to fit a person's hand).

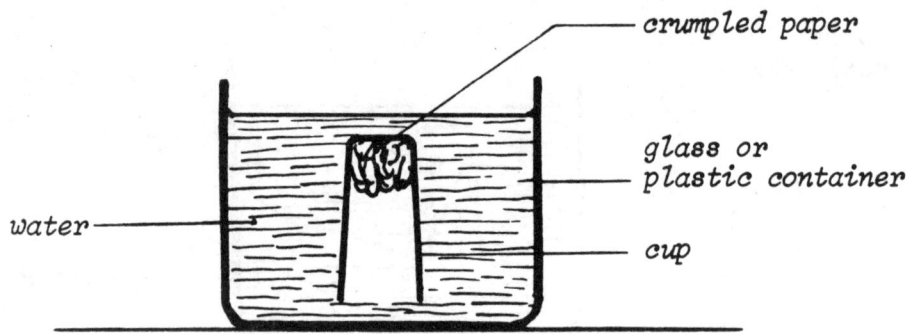

Procedure:
1. Fill the large container about 2/3 full with water.
2. Crumple a piece of dry paper and squeeze it to the bottom of the glass or plastic cup.
3. Invert the glass (making sure that the crumpled paper stays up in the cup) and immerse it completely under water, holding it as vertically as possible.
4. Take the cup back out of the water and let the water drip off (do not shake off!).
5. Take the crumpled paper out of the cup with a dry hand and let the students feel and check whether it is dry or not.

Questions:
1. Before inserting the crumpled paper, ask: "What is in the cup?" (anticipated answer: 'nothing').
2. Before immersing the glass under water, ask: "What else besides the paper is in the cup?"
3. While immersing the cup: "Why doesn't the water enter the cup?"
4. Why does the paper have to be crumpled?

Explanation:
Air is **space occupying.** The glass is therefore filled with air, no matter whether it is right side up or upside down. Besides the crumpled paper there was air in the cup. This is why the water could not enter the cup during the immersion process. The paper stayed therefore completely dry.

Applications of this characteristic of air can be found when people have to work under water. Air is then pumped in and around the area where the people are working, enclosed by a water-tight wall.

AIR OCCUPYING SPACE

1.5. THE EMPTY BOX CANDLE SNUFFER

Materials: 1. One empty shoe box.
            2. A birthday candle & matches.
            3. Masking tape.

small hole

candle

Procedure:
1. Show the open shoe box to the students and ask: "What is in the box?" (anticipated answer: 'nothing').
2. Make a small hole in the side of the long end of the box (about 1/2 cm in diameter) at the same height from the bottom as the candle length, and tape the top to the box.
3. Light the candle and place it about 5 cm away from the box in front of the hole (see Sketch).
4. Hit the box top with a sudden tap with the open hand (make sure that the hole is in line with the candle flame).
5. The lighting and snuffing of the candle may be repeated by doing points 3 and 4.

Questions:
1. What was blowing the candle flame out?
2. What did the tap do to the volume of the box?
3. How far can you hold the candle from the box and yet be blown out?
4. What would happen to the flame if you pushed the box top in gently instead of giving it a sharp tap?

Explanation:
   The shoe box is **occupied with air** and by tapping the top of it, the air was forced through the little hole, blowing the candle flame out, just like when we pucker up our lips to blow out a candle. By tapping the box, the volume of it suddenly becomes smaller for a short moment and this action forces the air out. Pushing the box in gently is like blowing very lightly against the flame. This demonstration shows that air is **occupying all the space** around us, also in an 'empty' box.

AIR — OCCUPYING SPACE

1.6. THE REFUSING FUNNEL

Materials:
1. Two identical glass or plastic funnels with narrow stem.
2. One two-hole stopper & one one-hole stopper.
3. Two identical empty jars or bottles.

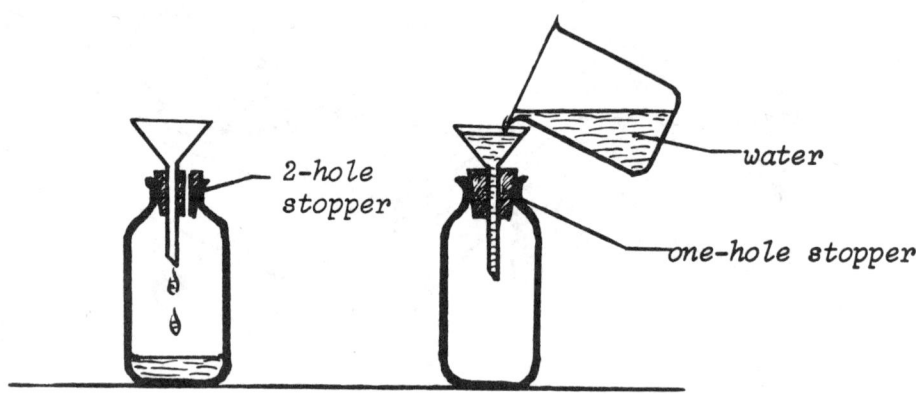

Procedure:
1. Set up the funnels with the stoppers on the bottles as shown in the sketch above (do not tell that one of the stoppers has two holes).
2. Fill the funnels with water: in one the water will run through (2-hole stopper) and in the other the water will stay in the funnel.
3. Take the bottle with the 2-hole stopper by the neck and put casually the forefinger on the open hole, and pour water in the funnel.
4. Take the bottle with the 1-hole stopper, squeeze the stopper open and the water will run through.

Questions:
1. Before pouring the water: "What is in the bottle?"
2. Why is the water only running through in one bottle?
3. After step 3 of procedure: Why is this funnel now also refusing?
4. What is holding back the flow of the water?
5. Before doing step 4 of procedure, ask: "How can we let the water run through this funnel?"

Explanation:
The bottles were filled with air in the first place. In the bottle with the two-hole stopper the air can escape through the second hole and the water will run through. But in the bottle with the 1-hole stopper, there is no way for the air to escape and this will hold back the water.

An example of this phenomenon we find when trying to fill a perfume bottle with water or other liquid, or any narrow mouthed bottle with water by holding it under a large stream of water from the tap. There is only one passage way for the water to come in and for the air to go out, in other words, the air is blocking the way for the liquid to come in.

AIR — EXERTING PRESSURE

1.7. THE INVERTED GLASS OF WATER

Materials: 1. A transparent glass or plastic cup.
2. A paper card (slightly larger than the mouth of the cup).

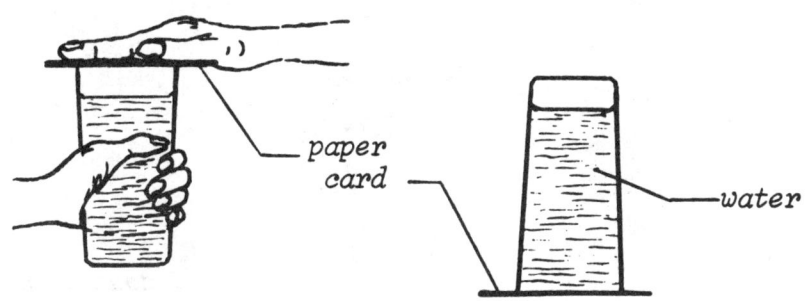

Procedure:
1. Fill the cup half way or full with water.
2. Place the paper card on the cup.
3. Put one hand on top of the card and invert the cup holding the card in place (do this over a sink or large container to catch any water drippings; also make sure that the hand holding the card is dry).
4. Take the hand that was holding the card slowly away.

Questions:
1. Why does the paper have to be rather stiff?
2. Why do we have to make sure that the hand holding the card on the cup is dry? What will a wet hand do?
3. What is keeping the water in the inverted cup?
4. Can we hold the cup slanted without letting the water pour out?
5. Will we be able to do the same thing with other liquids? (f.i. alcohol, oil, carbonated drink, etc.?)

Explanation:
When the cup is completely filled with water, there is no air left in the cup and thus no air pressure. The inverted cup can therefore hold the water up, because the **atmospheric pressure** is working against the under-side of the cup.

In case of a partially filled cup of water, we can explain it as follows: During the process of inverting, some of the water is dripping out, this increases the volume of the air pocket without increasing the amount of air, thus decreasing the pressure of the air pocket above the water. Again, the **atmospheric pressure** is therefore larger and thus holding the water inside the cup.

Alcohol and oil will also be held up inside the inverted cup, but the carbonated drink will not, because the carbon dioxide exerts pressure inside the glass above the liquid and prevents a partial vacuum from forming.

AIR EXERTING PRESSURE

1.8. THE STRAW DRINKING RACE

Materials:
1. Two identical drinking straws (one punctured with a few needle holes over the whole length).
2. Two small cups and some soda drink.

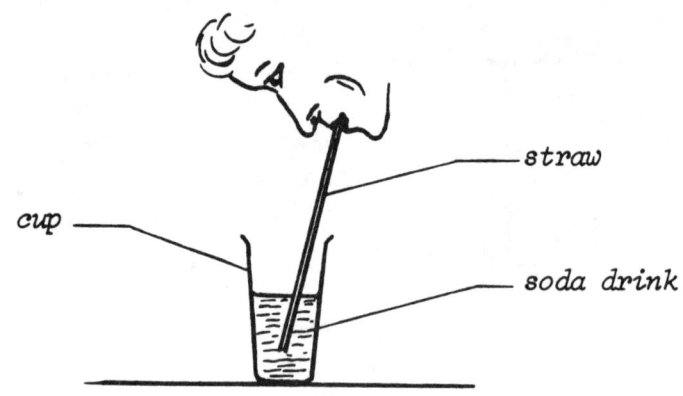

Procedure:
1. Fill the two small cups halfway with soda pop.
2. Ask two students to come up and comment: "Who would be the better 'sucker'?", and give each of them a straw (one of them with holes in it, but **do not** mention this).
3. Let them start drinking (sucking) on the count of three.
4. Give another pair of students a turn to race (the one who has the straw with the holes always loses).

Questions:
1. What makes the liquid go up the straw when we drink?
2. Why is it so hard to drink through a leaky straw?
3. What do we actually create when we suck through a straw?
4. Would we be able to drink through a straw if there was no air pressure around us?
5. Would an astronaut in space or on the moon be able to drink liquid through a straw?

Explanation:
By sucking we are creating a partial vacuum or a lower pressure in the straw above the liquid that we drink. The higher pressure in the atmospheric air pushes the liquid up the straw in our mouth.

The student with the leaky straw sucks in air and thus cannot create a vacuum above the liquid, so the liquid is not pushed up.

If there were no air pressure in the atmosphere, we would not be able to suck liquids through a straw. If the mouth of an astronaut was connected to a straw through his space suit to an open cup with liquid on the moon surface, he would not be able to drink the liquid by sucking through the straw, because there is no pressure on the liquid surface that will push it up the straw.

AIR — EXERTING PRESSURE

1.9. THE UPWARDS FALLING TEST TUBE

Materials: 1. Two test tubes, one just fitting into the other.
2. A little colored water.

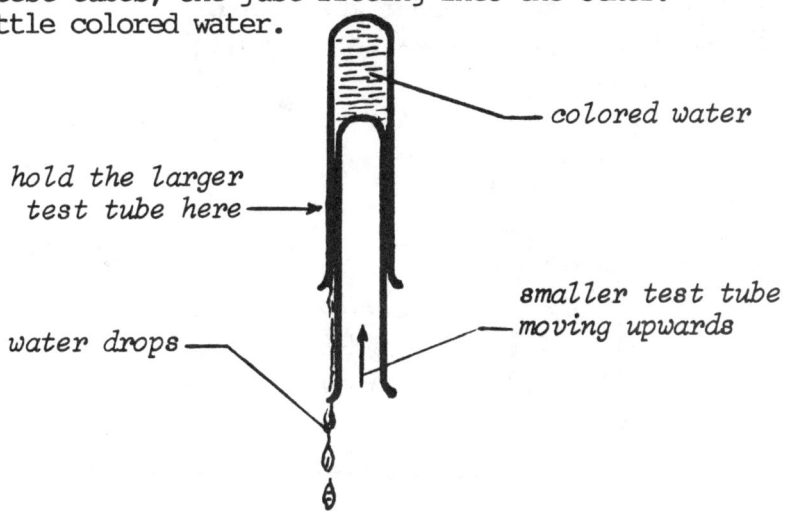

Procedure:
1. Fill the larger test tube half way with colored water.
2. Let the smaller test tube float on the water; push it a little further in so the water will overflow.
3. Now invert both test tubes over a small container to catch the dripping water and hold only the larger test tube (you may need to push the smaller tube up to start the motion).

Questions:
1. What happened to the volume of the water after inverting?
2. What made the smaller test tube move upwards?
3. What would be the total force pushing the test tube upwards?
4. Would this 'falling upwards' also succeed with a much smaller test tube floating on the water of a much larger test tube and then inverting it? Why? Why not?

Explanation:
   The smaller test tube has to be pushed a little further into the larger one till the water overflows, so that no air will be left between the two test tubes. After inverting the set of test tubes, the water is dripping out **decreasing the volume of the water.** Since there is no air above the smaller tube, the outside air pressure is pushing it up. The total force upwards is 1 kg per cm^2. For a test tube with about 1.2 cm diameter this upward force is about 1 kg. This event of 'falling up' will not succeed with two test tubes that have a larger difference in diameter, because the air will be able to replace the dripping water and thus equalize the pressure above and under the smaller test tube.
   Other thought questions would be:
-- Would this event work on the surface of the moon?
-- Would this event work in a pressurized satellite around the earth?

AIR — EXERTING PRESSURE

1.10. THE MYSTERIOUSLY RISING WATER (I)

Materials:
1. One small beaker (100 ml) & one large beaker (400 ml).
2. One hot plate or burner and stand.
3. Boiling chips and food coloring.

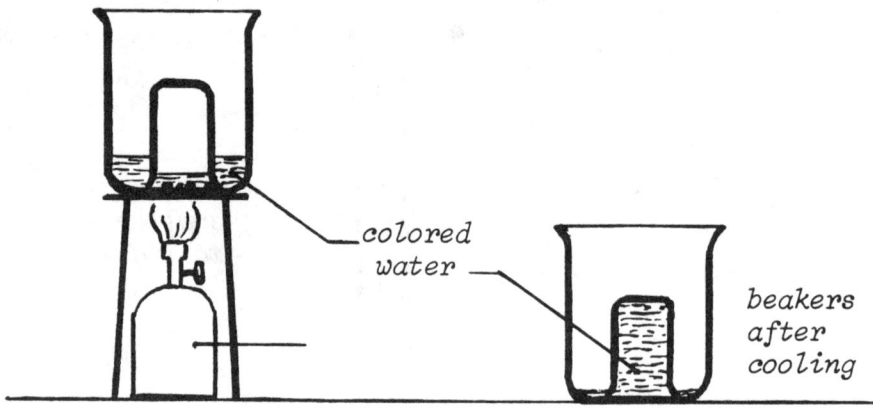

Procedure:
1. Put about 110 ml of water in the large beaker, a few drops of food coloring and a few boiling chips.
2. Heat this beaker and the water to a boil, while the small beaker is placed upside down inside the larger beaker.
3. Let the water boil for at least one full minute, taking care that the small beaker does not tip over.
4. Take the two beakers off the fire and let them cool to room temperature. Observe the water level in the small beaker.

Questions:
1. What is inside the inverted small beaker before the heating?
2. What happens to the water when we boil it?
3. Why does the small beaker keep bobbing up and down?
4. What do the bubbles consist of?
5. Why does the water level rise in the small beaker?
6. What would happen if after the heating, we put a few drops of cold water on the small inverted beaker?

Explanation:
   By boiling the water, it is changing from the liquid state into the gaseous or vapor state. This water vapor is also formed under the inverted beaker and it replaces the air under this beaker. The longer we let the water boil, the more air will be replaced by water vapor. The cooling process makes this water vapor condense and turn back into water, thus reducing the pressure inside the small beaker. The water is therefore pushed up inside the inverted beaker by the higher atmospheric pressure. By letting a few drops of cold water drip on the inverted beaker, the cooling process is speeded up and also the rising of the water.

1.11. THE MYSTERIOUSLY RISING WATER (II)

Materials:
1. One test tube & test tube holder.
2. One large beaker (400 ml) or other water container.
3. One Bunsen or alcohol burner & food coloring.

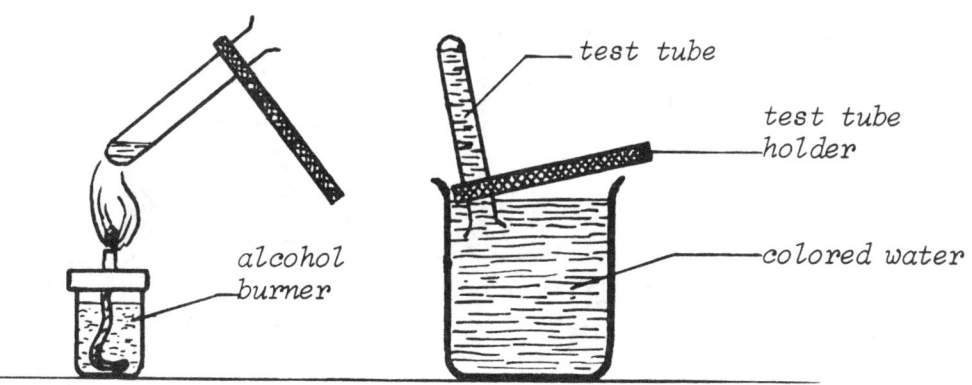

Procedure:
1. Fill the large beaker with cold water and color.
2. Put a little water in the test tube (about 3 ml) and boil it (A).
3. After the water has been boiling vigorously for about 10 seconds, invert this test tube immediately in the colored water (making sure that the test tube mouth stays under water: see Sketch B).

Questions:
1. What is in the test tube besides the water before heating?
2. What is in the test tube during the heating?
3. What happens during inverting of the test tube in the colored water?
4. What does water do when it is boiled?
5. What does water vapor do when it is cooled?
6. How did the temperature of the test tube change when it was inverted?
7. Why did the colored water rise in the inverted test tube?

Explanation:
Before the heating, the test tube was filled with a little water and the rest was air. Water vapor was produced by boiling the water and this water vapor pushed out the air in the test tube (vigorous boiling of the water). After inverting the test tube in the colored cool water, the hot water from the test tube poured out into the large container, the test tube slowly cooled off, causing the water vapor to condense. This decreases the pressure inside the test tube and thus the water is pushed up the tube by the existing **atmospheric air pressure.**

Other very similar events are: **The Mysteriously Rising Water (I) and The Fountain in a Flask.** We encounter this kind of an event in daily life sometimes when, after boiling water or soup in a pan with a well fitting top, cooling the pan afterwards and trying to open the cover, only to find the cover to be sticking to the pan.

AIR   EXERTING PRESSURE

1.12. THE COLLAPSING CAN

Materials: 1. One empty gallon (or 4-5 litre) can, or any other tin can that can be closed off air-tight.
2. A hot plate or burner and tripod.

can after cooling

Procedure:
1. Put about 20 ml of water in the can (just enough to cover the bottom) and heat it over the hot plate or burner.
2. Let the water boil vigorously for about 2 minutes (vapors should come out of the can).
3. Take the can with the boiling water off the heat (don't burn your fingers!) and immediately close off with the cap very tightly.
4. Let it stand upright on the table and cool off to room temperature, or for faster results: cool off with wet towel.

Questions:
1. What was in the can besides the water?
2. What happens when water is boiled?
3. What will the air in the can do when much water vapor is formed?
4. What would happen if we did not close the can very tightly?
5. What is the total force that is working on the outside of the can?

Explanation:
   Before heating, the can was filled with water and air. By boiling the water, it changed states, from liquid to gaseous state (water vapor). The water vapor or steam pushed the air that was inside, out of the can. In closing off the can with an air-tight cap, we are actually trapping the air out of the can: we are preventing it from going back into the can. The cooling **condenses** the water vapor back to water. All the vapor which took up the interior space of the can before, is now turned into a few drops of water, which take up much less space. This causes the pressure to drop and the atmospheric pressure is therefore pushing on the can and crushing it.
   The total force working on the outside of the can is the total of the can's surface area in cm^2 multiplied by 1 kg.

AIR   EXERTING PRESSURE

1.13. THE BALLOON IN THE FLASK

Materials: 1. An Erlenmeyer flask (150-200 ml).
             2. A round balloon with a large mouth (uninflated).
             3. A hot plate or burner and stand.

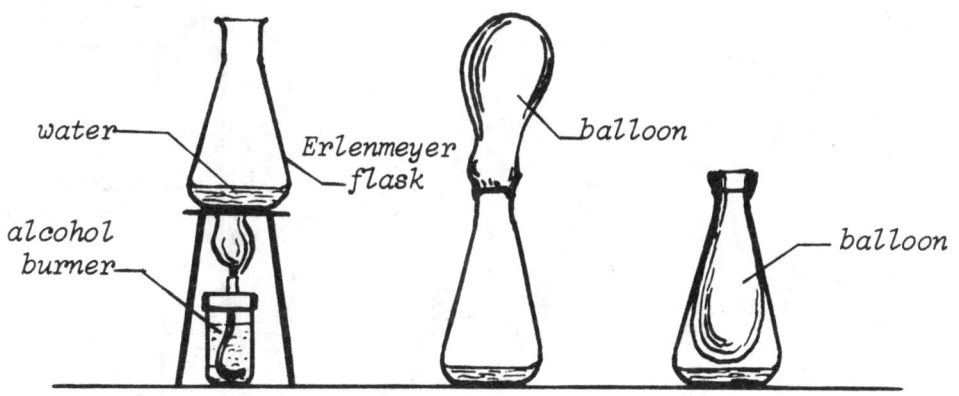

Procedure:
1. Put a little water (about 20 ml) in the flask and heat it to a boil (use a few boiling chips)
2. Let the water boil vigorously for at least one full minute.
3. Take the flask off the fire and immediately place the balloon with the mouth over the flask's mouth.
4. Let cool slowly at room temperature (the balloon will be sucked inside out into the flask).

Questions:
1. What is in the flask besides the water?
2. What is water doing if we boil it?
3. What is the steam doing to the air in the flask?
4. Why did the balloon go inside the flask?
5. Why did the balloon continue to expand inside the flask?

Explanation:
   By boiling the water in the flask, it was changed from the liquid state into the vapor state. Water vapor or steam is formed and this pushes the air that was originally present in the flask, out of the flask. The longer we let the water boil, the less air will be left in the flask. After the flask is closed off with the balloon, the air cannot get back into the flask. The cooling of the flask will slowly **condense** the water vapor and thus create a **partial vacuum** in the flask. This will cause the sucking in of the balloon and the **atmospheric air pressure** will further blow up the balloon inside the flask.
   During the cooling of the flask and the sucking in of the balloon, care has to be taken that the balloon's mouth is not pinched closed (this will prevent the whole balloon from getting into the flask; only part will go in and probably burst).

AIR   EXERTING PRESSURE

1.14. FOUNTAIN IN A FLASK

Materials: 1. One round- or flat-bottomed flask (400 ml) & holder.
             2. A glass tube drawn at one end in a 1-hole stopper.
             3. A beaker (500 ml), rubber tubing, boiling chips.
             4. A hot plate or burner and stand.

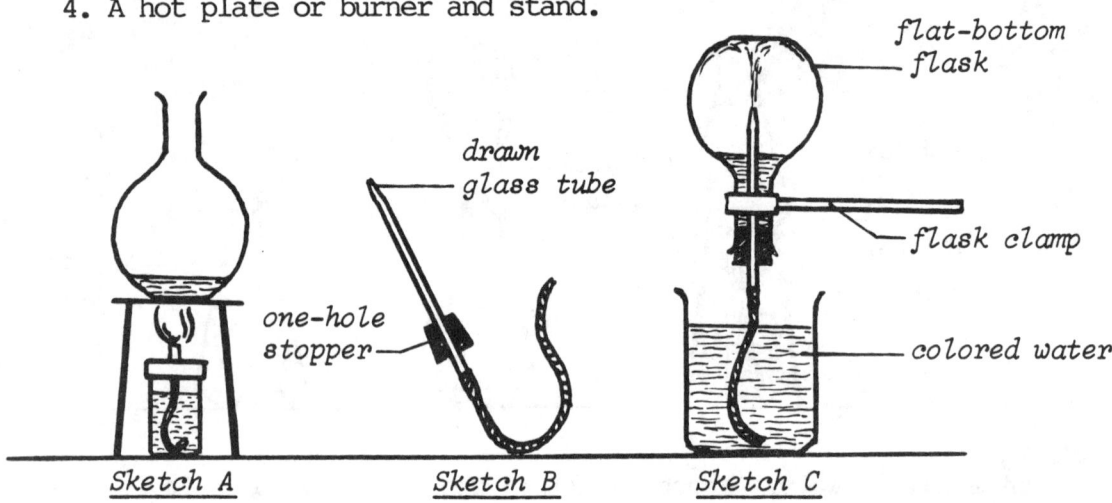

Procedure:
   1. Put about 30 ml of water in the flask and heat it over the fire.
   2. Let it boil for at least one full minute (see Sketch A).
   3. Get ready with the glass tube, stopper and rubber tubing to close off the flask (see Sketch B).
   4. Hold the flask by the flask holder, take it off the fire, and insert the stopper with the drawn tube on the flask.
   5. Immediately invert the stoppered flask and dip the end of the rubber tubing into the colored water in the beaker (see Sketch C).

Questions:
   1. Why is water needed in the flask?
   2. Would the water also be drawn up if the flask was heated without any water in it?
   3. What caused the water to go up the tube?
   4. Why did the water go up the tube so slowly in the beginning? And why did it suddenly speed up after the water hit the flask?

Explanation:
   Boiling the water produced water vapor in the flask and this caused the air to be pushed out. By closing the flask off with the stopper, the air was trapped out. The cooling of the flask after inverting caused the water vapor to condense slowly and to **contract**, with a decrease of pressure as a result. The water is therefore sucked up the tube slowly. As soon as the cool water hits the bottom of the flask, this latter cools off much faster and the water vapor condenses much faster, creating a sudden partial vacuum, thus sucking up the water vigorously. (Bromthymol Blue in the beaker and acid in the flask gives color change).

AIR   EXERTING PRESSURE

1.15. THE STICKING CUP OF WATER

Materials: 1. A petri dish (or other shallow dish with smooth rim).
            2. A smooth surface, ceiling, or underside of a table.

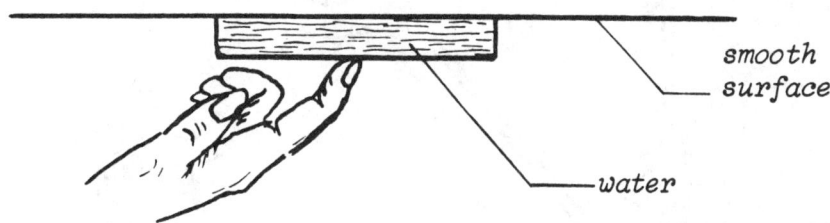

Procedure:
1. Fill the petri dish brim-full with water.
2. Bring the dish to the smooth surface: push it against the surface making sure that there are no air bubbles left in it.
3. At this point make sure that the dish will stick to the surface, then ask a student to hold it up or do as if you need to get something and have to leave the dish.
4. Give the student permission to let go of the dish or say: "Maybe I can let go of the dish" (if holding it yourself).

Questions:
1. Why do we need a smooth surface?
2. Why does the dish stick to the surface?
3. How much force is holding the dish up?
4. How heavy a dish can we stick to the surface?
5. How long will the dish keep sticking to the surface?

Explanation:
   By filling the dish completely full with water, there is no air left and thus there is no air pressure working down on the dish. The only force working down on the dish is **gravity** and thus the weight of the dish plus the water. The force holding the dish against the surface is equal to the air pressure of 1 kg per cm^2 of dish surface area. A dish with a 3 cm radius will have a force of about 27 kg holding it up minus the weight of the water and the dish itself. Once the dish is sticking to the surface, it will stay up for quite a long time until some water evaporates and air seeps into the dish.
   The water, in this case, acted as a seal preventing the air from coming into the dish. An application of this principle is when we wet suction cups with water to make them stick better to smooth surfaces.

AIR OCCUPYING SPACE
EXERTING PRESSURE

1.16. THE BIONIC FINGER (I)

Materials: 1. One heavy rubber plunger (to unplug a sink).
            2. A stool or chair with a smooth seat.

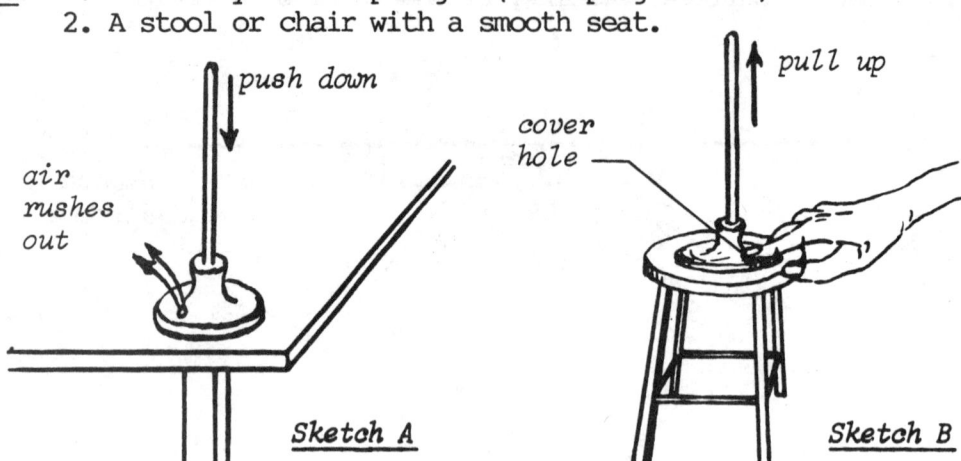

Sketch A   Sketch B

Procedure:
1. Make a small hole in the plunger with a scissor's point.
2. Show the students the plunger and ask: "What is under the plunger when I place it on the table?" (anticipated answer: 'nothing').
3. Ask one of the students to come up and put his/her cheek close to the hole in the plunger.
4. Push the plunger in: air rushes out and blows against the cheek! **AIR OCCUPIES SPACE!** (see Sketch A).
5. Show the students the plunger on top of a stool.
6. Tell them that you possess a bionic finger and that you can hold down the plunger against the stool with one finger.
7. Push down on the plunger and hold it down with one finger covering the hole (a wet finger will work better), and ask a student to come up and pull the plunger up (see Sketch B).
The whole stool will stick to the plunger and be lifted!

Questions:
1. What was under the plunger?
2. What was helping my finger to keep the plunger down?
3. How much force is pressing down on the plunger?
4. How heavy can the stool be and still be lifted up?

Explanation:
   There was air under the plunger and it rushed out when it was pushed in. When holding the plunger down with one finger, the hole was covered and this prevented the air from coming back in under the plunger, causing a lower pressure under it. A moist finger works better to plug the hole, because the water acts as a seal.
   The force holding down the plunger is equal to the surface area of the plunger multiplied by 1 kg (about 75 kg for a plunger with a 10 cm diameter).

AIR EXERTING PRESSURE

1.17. THE BIONIC FINGERS (II)

Materials: 1. Two heavy rubber plungers (to unplug a sink).
(one of them with a small hole in it).

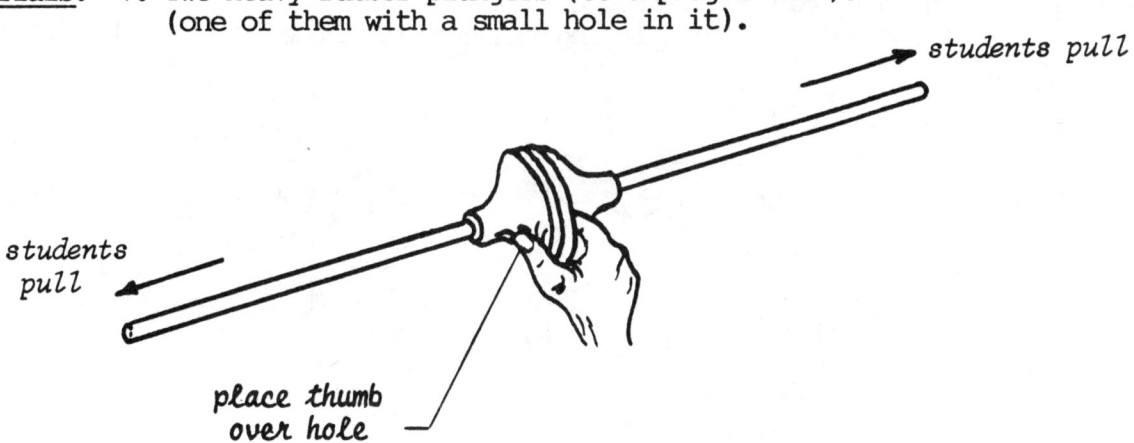

Procedure:
1. Ask two students to come up and push the two plungers against each other.
2. Ask them now to pull them apart (can be easily done).
3. Tell the students now that you have **bionic fingers** to keep the two plungers together.
4. Let the students push the two plungers together again and this time hold the plungers with thumb and forefinger, making sure that you cover the hole with your thumb (a wet finger works better).
5. Now let the students try to pull the plungers apart (most likely the wooden handle will come off first!).

Questions:
1. Why was it so easy to pull the plungers apart before I held them?
2. What did actually hold the plungers together?
3. Did I have to hold the plungers in a special way?
4. How much force was keeping the two plungers together?
5. Would larger or smaller plungers stick together better?

Explanation:
The plungers were quite easy to separate before holding them together because there was a hole in one of them, allowing the air to come inside between the two plungers. By covering the hole, the air was prevented from coming in (water on the finger acts as a seal) and when the students tried to pull them apart, the volume increased between the plungers, thus decreasing the pressure. It was again the atmospheric air pressure that was holding the two plungers together. The total force that was holding them together can be calculated from the total surface area of the two circles multiplied by 1 kg (about 150 kg for plungers with a 10 cm diameter). A similar experience was done with large steel half-spheres and horses pulling on each side: Maagdenburg spheres.

AIR — EXERTING PRESSURE

1.18. STOP THE LEAK

Materials:
1. An empty can or plastic bottle.
2. A one-hole stopper fitting the neck of the can or bottle.

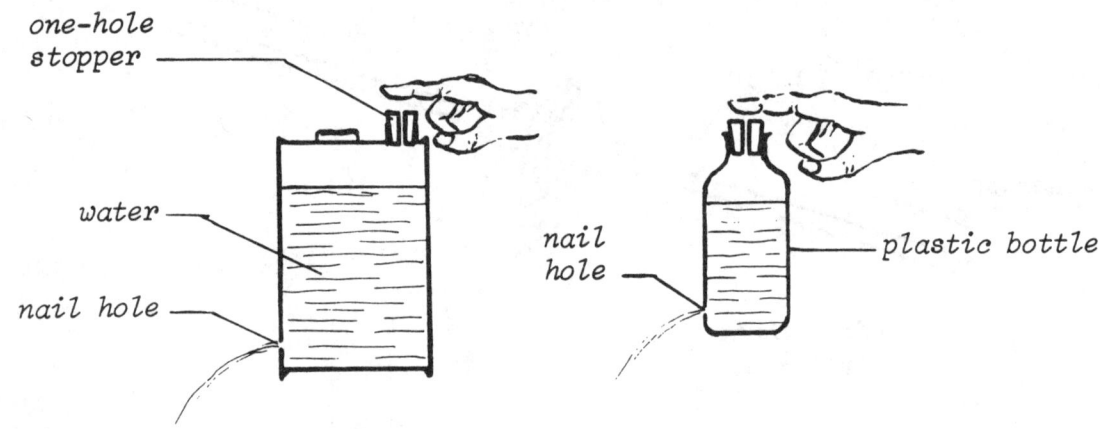

Procedure:
1. Punch a hole in the side near the bottom of the can or bottle with a small nail.
2. Fill the can or bottle with water and show the leaking container.
3. Ask: "How can I stop the leak without wetting my finger?"
4. Put the one-hole stopper on the can or bottle and cover the hole with one finger: the leak is stopped!
5. Release your finger from the stopper: leak will start again.

Questions:
1. Why did the water stop flowing out of the can?
2. How could we stop the leak without the one-hole stopper?
3. Does the water stop flowing immediately after the hole is covered?
4. How does the air pressure inside the can compare with the atmospheric air pressure after the water stops flowing?
5. What is it that we prevent from entering the can by covering the hole?

Explanation:
   The water does not stop flowing immediately after covering the stopper, but it still keeps dripping out of the can for a while. This increases the volume of the air pocket above the water. The amount of air stays the same because air is prevented from coming into the hole in the stopper. The increase in volume causes a decrease in pressure (Boyle's Law). The outside atmospheric air pressure pushes against the water and prevents it from flowing out.
   This is why we always punch two holes in a can of evaporated milk in order to pour the milk out. Also on a gallon frying-oil can, it usually is recommended to punch a hole in a corner opposite the pouring spout. This hole will allow air to enter the can while oil pours out. It will promote a smooth flowing of oil out of the can.

AIR — EXERTING PRESSURE

1.19. THE PERPETUAL FOUNTAIN?

Materials:
1. One medium size jar and two large bottles.
2. A 2-hole stopper fitting the jar, 2 glass tubes (the longer one drawn to a point at one end).
3. Two lengths of rubber tubing (about 30 cm).

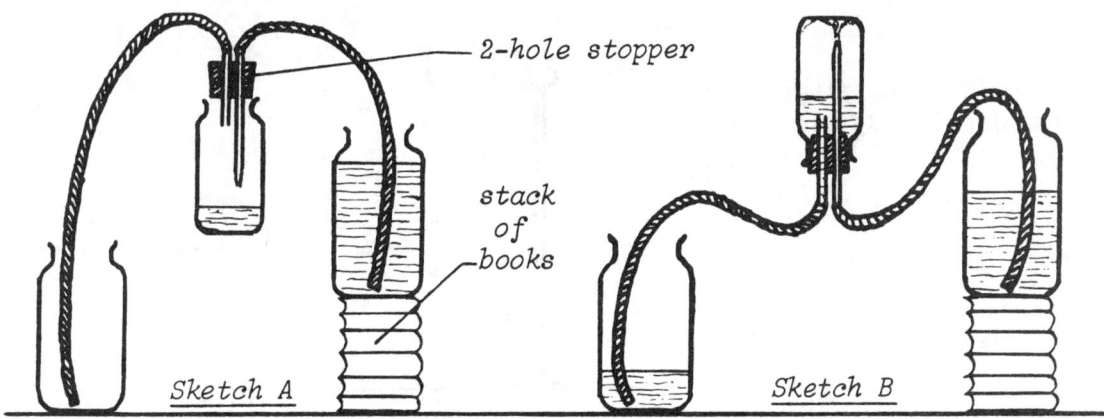

Procedure:
1. Push the two pieces of glass tubing through the 2-hole stopper such that the one with the point extends farther than the other (see Sketch).
2. Connect the two lengths of rubber tubing to the glass tubes.
3. Fill one of the two large bottles full with cold water and set up the flask and the two bottles as shown in Sketch A (Hold the jar).
4. Pour about 100 ml of water in the jar and insert the stopper tightly.
5. Invert this jar, making sure that the two rubber tubes stay in the two bottles (the end of the tube in the bottle with water should stay under water at all times.

Questions:
1. What was the first thing that happened after inverting the jar?
2. What happened to the volume of the air pocket above the water in the jar as the water poured in the empty bottle?
3. Why was the water drawn up into the jar?
4. Why does the water-filled bottle have to stand higher than the other?
5. How can we make the fountain flow harder to slower?

Explanation:
The water in the jar was needed to 'prime' the siphoning action. immediately after inverting the jar, the water ran down and out into the empty bottle. This caused an increase in volume of the air pocket above the water in the jar, thus decreasing the pressure. This lower pressure caused the sucking up of the water from the water-filled bottle. In other words, the atmospheric air pressure is pushing the water up into the jar. The larger the difference in height of the water levels in two bottles the stronger the flow of water. As soon as the water levels are the same height in both bottles, the water flow will stop. This is the same principle of the siphon.

AIR — EXERTING PRESSURE

1.20. TRANSFER WATER WITH A STRAW

Materials:
1. One drinking straw for each group of three students.
2. Two small cups for each group of three students.

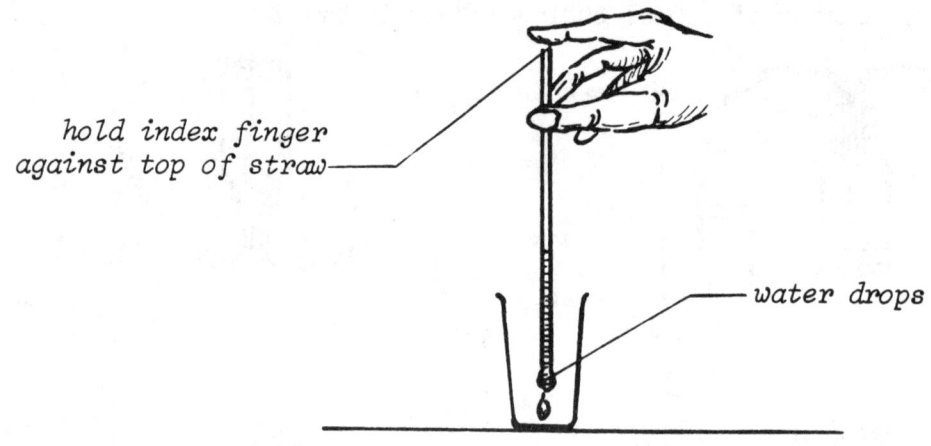
hold index finger against top of straw
water drops

Procedure:
1. Divide the class in groups of three students.
2. Provide each group with one straw and two cups.
3. Let one student hold a water-filled cup, another an empty cup and the third transfer the water from one into the other cup.
4. Ask the students how they can do this without tilting the cups.
5. Demonstrate how the water can be held in the straw:
   - dip the straw vertically in the water-filled cup;
   - hold forefinger against top end of straw and lift;
   - move straw with the water in it over the empty cup and release finger: water drips out!

Questions:
1. How can we transfer the water without tilting the cups?
2. How can we hold water in the straw without sucking through it?
3. What holds the water in the straw when your forefinger is closing off the top end of the straw?
4. What happens if we try to hold the water in the straw and not close the top end of the straw tightly?
5. What is prevented from coming in the straw at the end where our finger is pressing?
6. What is holding the water back if we close the top end of the straw with our finger before immersing it in the water?

Explanation:
By closing off the top end of the straw with our forefinger, we prevent the air from coming in. Some water will drip out of the lower end of the straw, enlarging the volume of the air pocket above the water and thus creating a lower pressure inside the straw. The outside atmospheric pressure is holding the water up in the straw. When the finger is released, air is allowed to enter and the pressure is equalized.

AIR

**EXERTING PRESSURE
COHESIVE FORCES**

1.21. THE WATERTIGHT CLOTH

Materials: 1. A drinking glass.
2. A piece of cheese cloth (or any other thin cloth) large enough to cover the mouth of the glass.

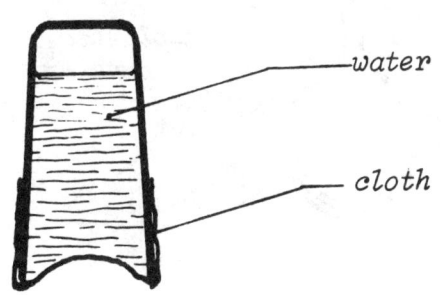

Procedure:
1. Fill the glass half way or full with water.
2. Wet the cloth under the tap and show students that water will flow easily through the cloth.
3. Place the cloth now over the mouth of the glass and push the fringes against the side of the glass.
4. Hold the fringes with one hand against the glass; with the other hand hold the bottom part of the glass without holding the cloth and invert.
5. Let go of the first hand: cloth and water will stay up in the glass!

Questions:
1. Why would this not work with a dry cloth?
2. Why does the wet cloth cling to the sides of the glass?
3. Why does some water flow out of the glass in the beginning?
4. Why does the water stop flowing out?
5. What is keeping the water and the cloth up?
6. Can we hold the glass slanted or sideways without letting the water flow out?
7. What shape does the cloth take when we hold the glass vertically upside down?

Explanation:
The cloth needed to be wet with water so that it could adhere to the sides of the glass (adhesive forces between water and cloth, and water and glass). During the process of inverting, some water poured out because the cloth is porous, but this caused an increase of the air pocket volume above the water, which in turn reduced the pressure inside the glass. This demonstration can also be used to show the cohesive forces between the water molecules, as a film of water molecules is formed in the little pores of the cloth.

AIR — EXERTING PRESSURE

1.22. THE CUPS AND THE BALLOON

Materials: 1. An uninflated round balloon.
2. Two small plastic or glass cups (with a smooth rim).

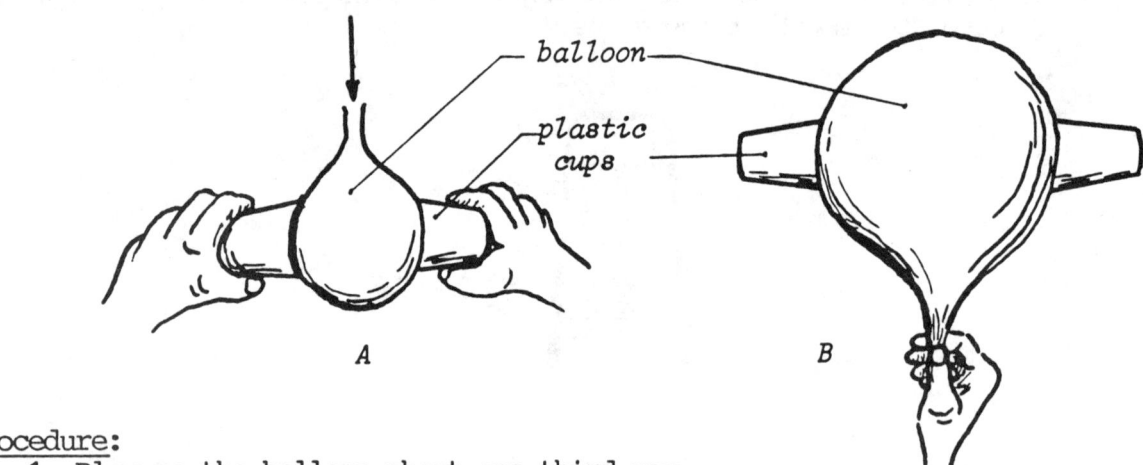

Procedure:
1. Blow up the balloon about one-third way.
2. Tell the students that you will hold the two cups against the sides of the balloon and blow it further up.
3. Hold one cup in each hand, hold the cups against opposite sides of the balloon (while the balloon is in the mouth) and blow further (until about twice the size).
4. Let go of the two cups (they will stick to the balloon; if not, moisten the cup rims) and hold the balloon in one hand by the mouth.
5. Show the students that the cups are not glued to the balloon by releasing the air slowly out of the balloon (the cups will fall off as the balloon gets smaller).

Questions:
1. Why did the cups stick to the balloon?
2. Did the air pressure in the balloon have anything to do with it?
3. How does the volume of the cups change from stage A to stage B?
4. What does the pressure inside the cups have to be to stick to the balloon, compared to the atmospheric pressure (in stage B)?
5. How much larger does the balloon have to be in stage B compared to stage A for the cups to stick to it?

Explanation:
   Because of the increase in curvature or actually a flattening of it from A to B, the volume in the cups increased, thus causing the pressure to decrease. The air pressure inside the balloon had nothing to do with the sticking of the cups to the balloon.

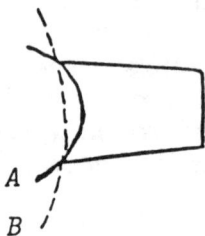

AIR   EXERTING PRESSURE

1.23. INFLATE A BALLOON BY SUCKING

Materials: 1. An empty glass jar & a fitting 2-hole stopper.
             2. Two lengths of glass tubing (one with a 90° bend).
             3. A small balloon & tape or rubber band.

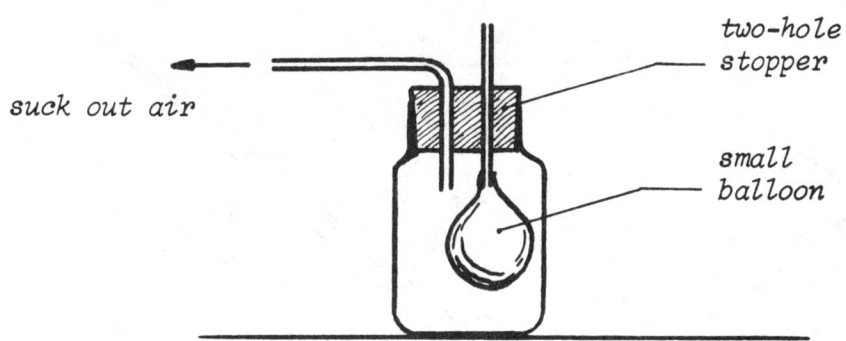

Procedure:
1. Insert the glass tubing into the stopper and tie or tape the small balloon to the end of the straight tube.
2. Place the stopper with the tubing and the balloon over the jar, insert tightly, and set up the equipment as shown in the sketch above.
3. Suck through the bent glass tube until the balloon is inflated and close off the end of the bent tube with your finger (prevent the air from getting back into the jar).
4. Ask the students all the questions.

Questions:
1. What was I doing to inflate the balloon?
2. How does the pressure inside the jar compare to the pressure outside the jar during inflation of the balloon?
3. Why doesn't the balloon deflate? (finger over end of bent tube).
4. How else can the balloon be inflated?
5. What would happen if I blew through the bent tube?
6. How does the pressure inside the jar compare to that of outside during inflation of the balloon by blowing through the straight tube?

Explanation:
   By sucking through the bent tube, the pressure inside the jar is decreased and the atmospheric air pressure is inflating the balloon. By placing the finger over the end of the bent tube (after inflating the balloon), the air is prevented from reentering the jar. A lower pressure inside the jar (outside the balloon) is now maintained, which causes the balloon to stay inflated even with an open mouth. Another way to inflate the balloon is to blow through the straight tube, and by putting a finger over the bent tube it can be kept inflated. This way, during inflation the pressure inside the jar is higher than the atmospheric pressure.

1.24. THE HEAVY NEWSPAPER

Materials:
1. One or two full sheets of an ordinary newspaper.
2. A stick of pine wood (.3x3x75 cm or 1/4"x1"x2')
   (a piece of wood paneling or old ruler is excellent).

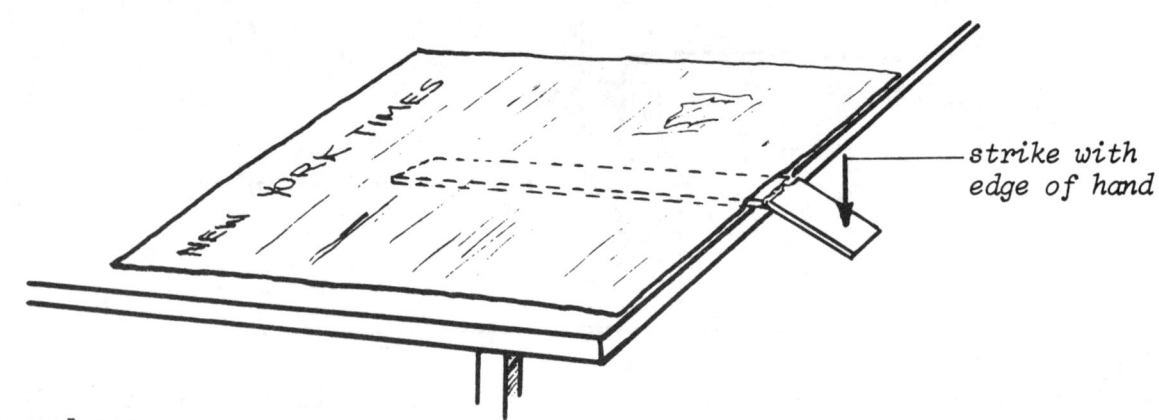
strike with edge of hand

Procedure:
1. Place the stick on a table with a smooth surface and let it protrude over the edge about 8 cm.
2. Ask: "What will happen if I hit this protruding end of the stick?" (anticipated answer: 'stick will fly up').
3. Strike it and let the students catch the flying stick.
4. Place the stick back on the table like in point 1, and cover it with the newspaper flush with the edge of the table.
5. Ask: "What do you think will happen now if I hit it again?" (anticipated answer: 'the paper will fly up' or 'paper will tear').
6. Smooth down the paper with your left hand and strike the protruding end of the stick with your right hand (a sudden sharp blow with the edge of the palm): stick breaks!
7. By pulling the stick out another 8 cm after breaking, the cycle of smoothing the paper and breaking (point 6) may be repeated.

Questions:
1. What did I do with my left hand?
2. Why was it necessary to smooth the paper before hitting?
3. What would happen if the protruding end of the stick was slowly pushed down?
4. How much weight was actually holding the stick down?

Explanation:
By smoothing the paper down, there was almost no air under it, but a whole column of air exists above the paper, pushing down on the paper with the atmospheric pressure. This is about 1 kg/cm^2 (14.6 or rounded 15 lbs per sq. in.). The total weight or force pushing down on a 60 x 80 cm or 20" x 30" paper is roughly: 60 x 80 x 1 kg = 4800 kg, or 20 x 30 x 15 lbs = 9000 lbs, which is close to the weight of two large stationwagons! It is therefore impossible to lift it with the thin stick!

AIR

**EXPANSION BY HEAT
EXERTING PRESSURE**

1.25. THE MAGIC BEAKER

Materials: 1. A small jar (50 ml) or small wine glass.
2. A piece of paper and matches.

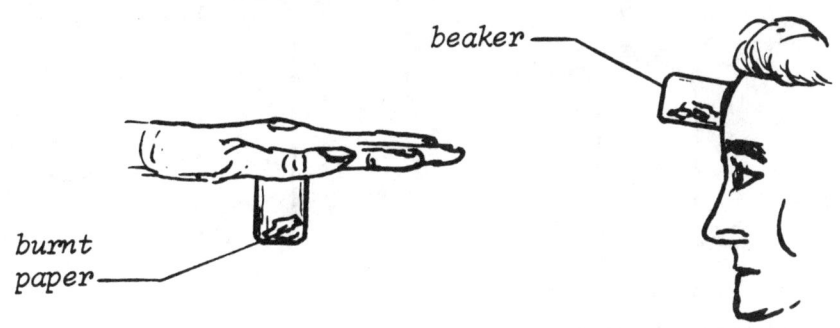

Procedure:
1. Show that the jar or wine glass is empty by inverting it.
2. Crumple a small piece of paper.
3. Strike a match and start burning the paper.
4. Place the burning paper in the jar or wine glass and immediately press the jar against your forehead or against the palm of your hand (keep pressing until the fire is out or until you feel a suction).
5. When you feel a suction inside the jar, slowly let go of it (it will stay stuck against your forehead or palm).

Questions:
1. What was inside the jar or glass before burning the paper?
2. Why did the flame inside the jar go out?
3. What does the heat of the flame do to the air inside the jar?
4. What was left inside the jar after the flame went out?
5. What made the jar or glass stick to the forehead or palm?

Explanation:
    Although you showed the pupils that there was 'nothing' in the jar by holding it upside down, there was naturally air in it. By placing the burning piece of paper in the jar, the air inside the jar was **heated and expanded.** The expansion made the air rush out of the jar and when pressed against the forehead or palm of the hand, the jar was closed off from the outside air. This caused the flame to go out, because there was no more oxygen to feed the burning process. The extinguishing of the flame caused the air inside the jar to cool off and thus to contract. The pressure inside the jar was therefore getting smaller than the atmospheric pressure outside the beaker, and this higher air pressure kept the jar against the forehead or palm. In other words, a **partial vacuum** was created inside the jar, which kept the jar stuck against the skin surface.
Warning: The partial vacuum will leave a red spot on the skin, which will disappear in a few hours or longer, depending on how long the beaker has been left against the skin!

AIR

EXPANSION BY HEAT EXERTING PRESSURE

1.26. THE EGG AND THE MILK BOTTLE

Materials: 1. One hard-boiled egg.
2. One empty glass milk bottle.

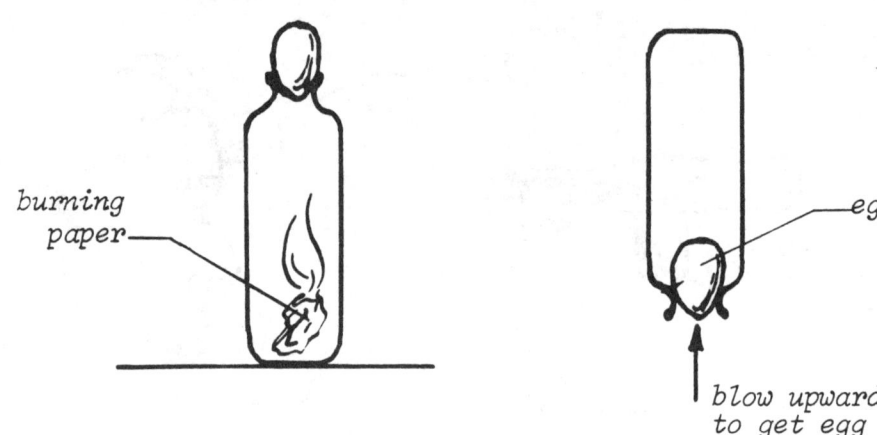

Procedure:
1. Peel the shell off from the hard-boiled egg and place the egg on the mouth of the milk bottle.
2. Ask: "What will happen if I put a burning piece of paper in the bottle and place the egg back on the bottle?" (anticipated answer: 'egg might jump off', 'fire goes out', 'the bottom of the egg will get black', 'the bottle might crack', etc.)
3. Burn a small piece of paper, lift the egg up, put the burning piece of paper in the bottle and place the egg immediately back on the bottle's mouth (egg will be sucked into the bottle).
4. Ask question number 4.
5. To get the egg out of the bottle whole: invert the bottle and let the egg fall in the bottle neck, blow a short spurt of air up into the inverted bottle and catch the falling egg.

Questions:
1. Why did the egg get pushed inside the bottle?
2. What did the burning paper do to the air inside the bottle?
3. What did the egg do before it went into the bottle? (Observe!)
4. How can we get the egg out of the bottle without cutting it up?
5. Is there another way to get the egg out whole?

Explanation:
   The burning paper is heating the air in the bottle and expanding it, and this is why the egg was vibrating before it was sucked into the bottle. Some of the air slipped under the egg out of the bottle and thus there was less air pressure in the bottle. An added cause for decreasing the pressure inside the bottle is, that the burning of the paper took away the oxygen of the remaining air, turning it into CO_2 and water vapor. The latter condensed against the cold bottle. Another way to get the egg out is, to warm the bottle while the egg sits in the bottle neck.

AIR

EXPANSION BY HEAT
EXERTING PRESSURE

1.27. THE TWO BOTTLES IN LOVE

Materials: 1. Two identical bottles with thick smooth rims (medicine bottle).
2. A piece of filter paper or blotting paper.

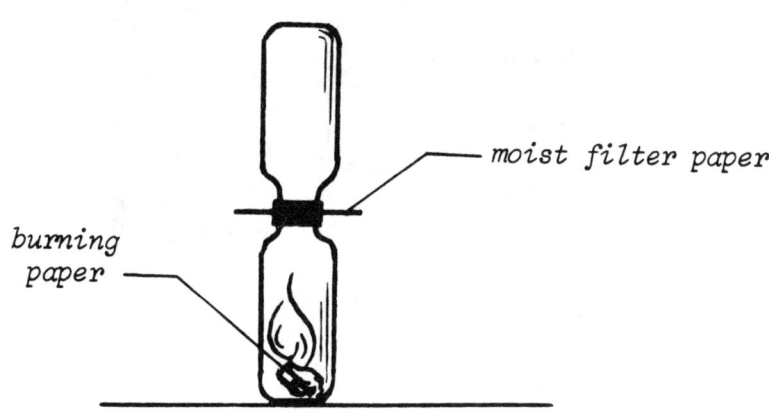

Procedure:
1. Soak the filter paper in water and place it over one of the bottles.
2. Twist a small piece of paper and make sure that it will fit through the mouth of the other bottle.
3. Burn this twisted piece of paper, put it in the open bottle and immediately cover this bottle with the filter-paper-covered bottle.
4. Press the top bottle for a few seconds longer on the lower one (until the flame of the burning paper is completely extinguished).
5. Now lift the two bottles up by holding on only to the top bottle.
6. Hold both bottles, invert the set, and let go of the lower bottle.

Questions:
1. What did the burning paper do to the air in the bottle?
2. Why did the filter paper have to be soaked in water?
3. Why did the two bottles stick to each other?
4. Would the bottles stick to each other without the filter paper? Why?
5. How long will the two bottles stick to each other?

Explanation:
The burning paper heated up the air in the open bottle, which caused the air to expand. Thus some of the air in this bottle escaped. By covering this bottle immediately with the other bottle, the escaped air was trapped outside. The air pressure outside the bottles is keeping the set pressed together. The moist filter paper functions as a seal between the two bottles making them almost airtight. Without this filter paper, the air would seep between the two bottles and the trick will not work. The two bottles will stick together as long as the pressure inside is lower than the atmospheric pressure. Slowly air will seep in through the filter paper and as soon as the air pressures inside and outside the bottles are equalized, the bottles come apart.

AIR EXPANSION BY HEAT

1.28. THE INVERTED PAPER BAG BALANCE

Materials:
1. Two small identical paper bags.
2. Two drinking straws and a pin.
3. Two short lengths of thread, tape and matches.

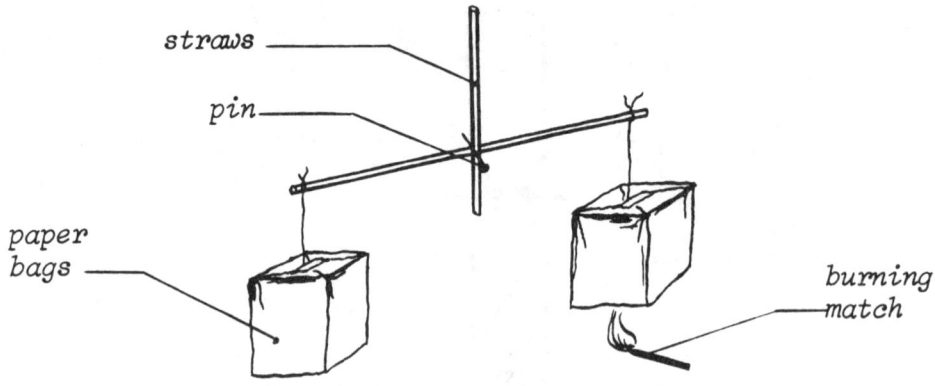

Procedure:
1. Open the two small paper bags, invert, and attach the thread to the center of the bottom with a piece of tape.
2. Hang the two bags upside down on the ends of one of the straws, and balance this on a pin attached to the other straw as shown in the sketch (make sure that the straws move freely around the pin).
3. Let one of the students hold the vertical straw, strike a match and hold the flame under one of the bags (be careful not to set the bag on fire).
4. This bag moves up. Take the flame away and the equilibrium is restored (both bags will be in balance again).
5. Hold a flame under the other bag and this side will now move up.

Questions:
1. What is in the bags?
2. What is air doing if it is heated?
3. Why did the heated bag move up?
4. What happened after the flame was taken away from under the bag?
5. In which direction does heated air move?

Explanation:
   The flame under the bag heats up the air in the bag and expands it. This leaves less and lighter air in the bag. This lighter air pushes under against the bag and the bag moves up. When we take the flame away, the warm air cools down and slowly the balance of the two bags is restored.
   This principle of making air lighter by heating is applied in **hot air balloons.** Air inside the balloon is heated, it expands and becomes lighter and the balloon is pushed up by the lighter air (compared to the surrounding atmospheric air). The larger the balloon, the larger the force upward and thus the more weight it can lift.

AIR   EXPANSION BY HEAT

1.29. SQUEEZE THE GLASS BOTTLE?

Materials: 1. An empty glass bottle.
2. Glass tubing (about 30 cm) in a fitting 1-hole stopper.
3. Water and food coloring in a small beaker.

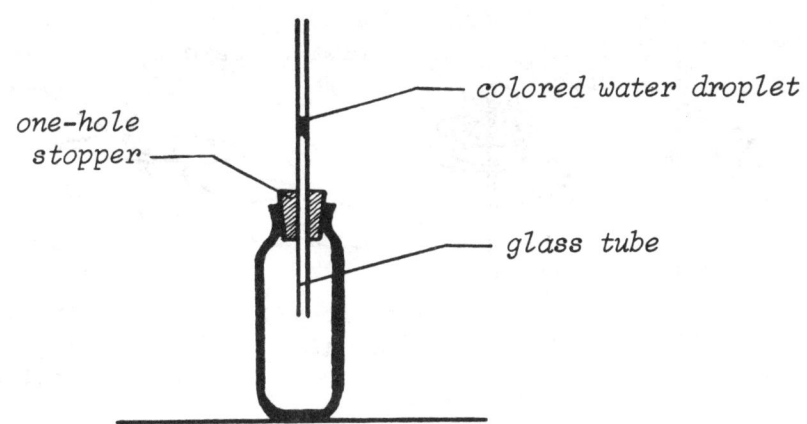

Procedure:
1. Color a few millilitres of water with food coloring in the beaker.
2. Dip the end of the glass tubing in the colored water and close the other end off with a finger. Take the tube out of the water, hold the tube horizontal and let the waterdrop ride in the tube until close to the stopper. (close off the end again with the finger).
3. Place the stopper in the bottle neck and insert tightly (waterdrop will move up some).
4. Mark off the position of the waterdrop with a grease pencil, masking tape, or rubber band.
5. Hold bottle in both hands: what happens to the waterdrop?
6. Let the bottle stand on the table: what happens to the waterdrop?

Questions:
1. What is in the bottle?
2. Why does the waterdrop not slide down into the bottle?
3. Why does the waterdrop move up when inserting the stopper?
4. What made the waterdrop move up when the bottle was held?
5. How can we get the waterdrop to move down the tube?

Explanation:
The 'empty' bottle is of course filled with air. In closing off the bottle with the stopper and tube, the waterdrop was pushed upwards by the air. It is because of the presence of air inside the bottle that the waterdrop in the tube cannot slide down by itself. It is actually held up in the tube by the air.
By holding the bottle in our hands, our body heat is warming up the bottle, which in turn warms the air. This air expands and pushes the waterdrop up the tube. By cooling the bottle with cool water and blowing, the waterdrop can be made to move down the tube.

AIR — EXPANSION BY HEAT

1.30. THE PAPER MERRY-GO-ROUND

Materials:
1. Two (5" x 8") paper cards.
2. A needle or pin, masking tape, scissors.
3. A pencil or straw, a lamp or candle and matches.

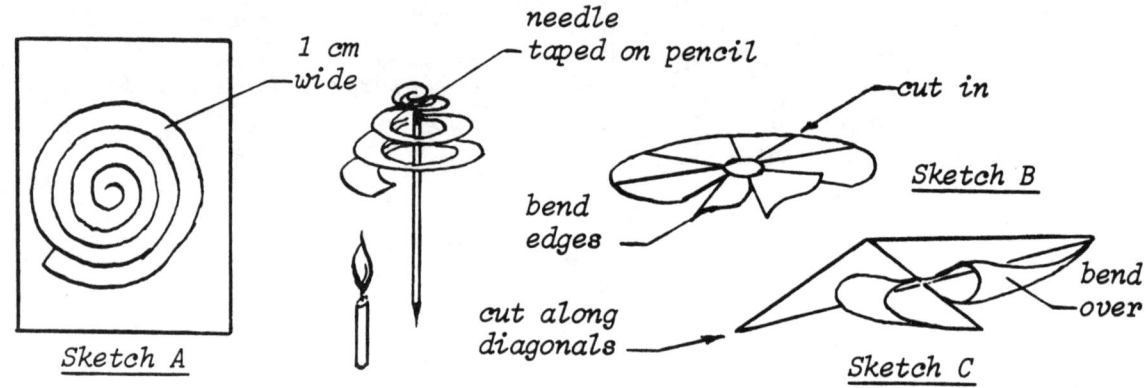

Procedure:
1. Draw a spiral on the paper card of about 1 cm width and cut out the spiral with scissors following the line (Sketch A).
2. Tape the needle or pin to the pencil or straw.
3. Balance the paper spiral on the needle point and hold it about 10 cm above a lit candle or lamp: the spiral turns!
4. Other types of merry-go-rounds can be made from paper cards or light paper (construction paper) (see Sketch B & C).

Questions:
1. What does the burning candle do to the air?
2. What will heated air do?
3. What does a windmill need to turn?
4. Why did the spiral turn?
5. Can we make other paper cut-outs that will spin?

Explanation:
  The lamp or burning candle was the source of heat which heated the air. Hot air takes up a larger volume than cold air and is therefore lighter in weight. Thus hot air rises and by holding the paper spiral in this stream of air, it flows along the spiral and makes it turn.
  Examples that are based on this same principle are the windmills, which only turn when there is a wind blowing. The air stream here, however, is flowing in a horizontal direction. The air stream which was moving the paper spiral moved in a vertical direction. Other paper cut-outs are drawn in figure B and C, which will also spin in a stream of air.
  The needle point is the center point of rotation and it serves as a near frictionless set up for the rotation of the object. A very slow moving airflow would therefore be detectable.

1.31. THE DANCING PENNY

Materials:
1. One empty bottle with a narrow neck.
2. One large beaker or container for warm water.
3. A penny or dime, or other small coin.

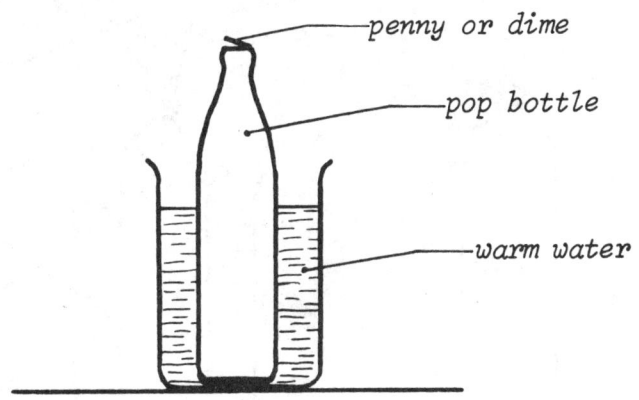

Procedure:
1. Moisten the opening of the bottle and place the penny or dime flat on the bottle mouth.
2. Fill the large beaker with hot water (not steaming, so that it would look like cold water; if possible fill this container before doing the demonstration).
3. Immerse the bottle, which is covered with the penny, in the water and observe the dancing coin.

Questions:
1. What was in the bottle before covering it with the coin?
2. What kind of water was in the beaker?
3. What was the moisture on the bottle opening necessary for?
4. Why did the penny or dime go up and down (vibrate)?
5. Would it also vibrate without moisture on the opening of the bottle?

Explanation:
The bottle was filled with air before covering it with the coin. The moisture on the opening of the bottle functions as a seal between the inside and outside of the bottle. When the bottle is placed in the hot water, the air inside the bottle is heated and this causes the air inside the bottle to expand. The only way it can escape from the bottle is through the opening, and thus it has to lift the coin. The coin falls back, more air expands and lifts up the coin again. When this sequence of events happens quickly, a vibration of the dime is caused.

Without the moisture on the opening of the bottle, the coin does not seal off the air, so that the escaping air from inside the bottle could just seep under the coin out into the open without lifting the coin. The coin would thus not vibrate.

EXPANSION BY HEAT

1.32. THE LIVE BALLOON

Materials:
1. A small balloon (one that has been inflated before).
2. A large bottle with a narrow neck.
3. A large beaker or container for hot water.

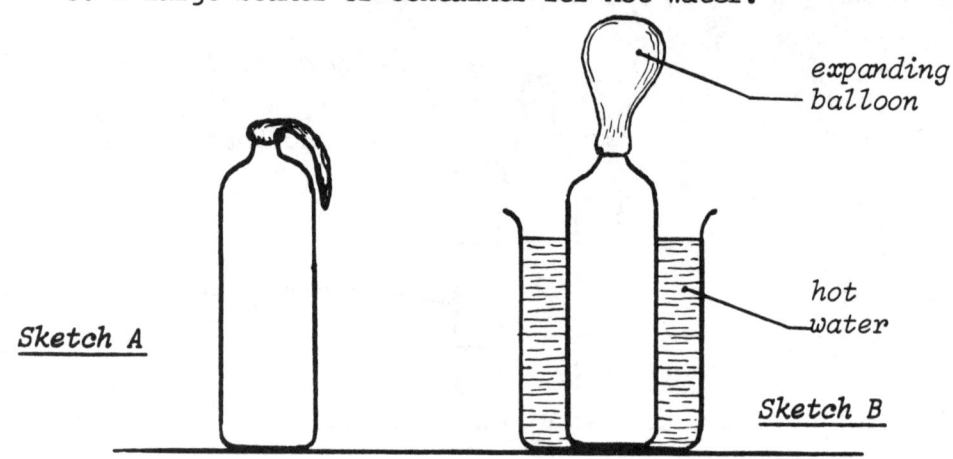

Procedure:
1. If another identical bottle is available put a few small things in this bottle; show the bottles to the students and ask them what is in the bottles (anticipated answer: 'nothing' in one of the bottles).
2. Place the small balloon over the mouth of the 'empty' bottle and let it hang limp against the side of the bottle (see Sketch A).
3. Fill the container with hot water (not steaming hot, if possible do not show students that it is hot water).
4. Immerse the bottle with the balloon on it in the hot water (the balloon will slowly fill with the expanding air) (see Sketch B).

Questions:
1. Before demonstrating: What is in the bottle?
2. Why did the balloon inflate itself?
3. What kind of water do you think was in the large container?
4. How can we deflate the balloon without taking it off the bottle?
5. What does air do when it is heated?
6. Is there more or less air in the warm bottle compared to the cold one?

Explanation:
The bottle was filled with air at the moment that the balloon was put over it. The immersion of the bottle in the hot water heated the air inside the bottle, which made it expand and fill the balloon. The larger the bottle, the more air will expand, and the quicker the balloon will inflate. Taking the bottle out of the warm water and leaving it to stand on the table will cool it off and contract the air. This will deflate the balloon again. The amount of air is not changing, whether it is hot or cool, as no outside air was allowed in or out of the bottle. It is just the volume of air that is changing.

AIR   EXPANSION BY HEAT

1.33. WHAT CAUSES THE WATER TO RISE? (I)

Materials: 1. Three identical saucers or one large flat tray.
            2. Three identical glass cups.
            3. Six small birthday candles and matches.

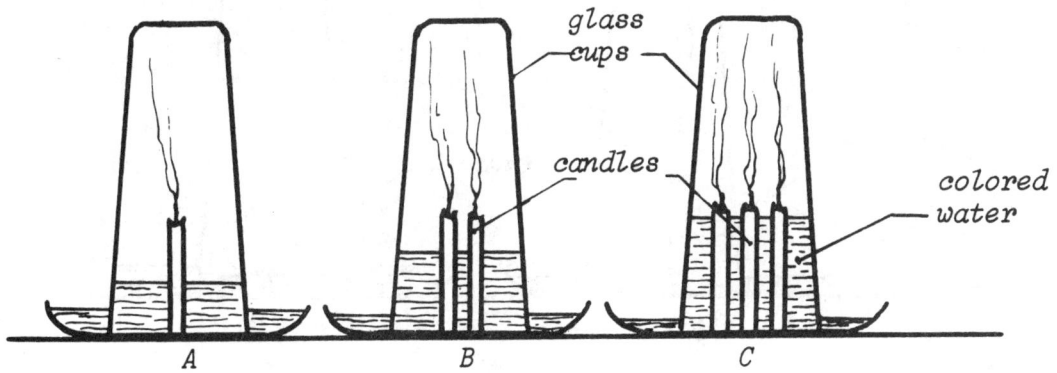

Procedure:
1. Attach one, two, and three candles to the center of each saucer or at different spots of the tray (in case of using a flat tray).
2. Fill each saucer with at least 3/4 cup of water.
3. Light all six candles and wait till they all burn evenly.
4. Place the three cups at the same time over the candles on the saucer like illustrated (let a student help you).

Questions:
1. Under which cup does the water level rise highest?
2. Why did we take identical saucers, cups and candles?
3. Do we need to pour the same amount of water in the saucers?
4. Which variable is manipulated or changed in comparing A, B, and C?
5. Above which saucer did heat develop most?
6. Was the amount of air trapped under the cups the same for all three?
7. Why did the water level rise highest in case C?

Explanation:
   In all three cases the saucers, cups, and candles were identical. This meant that these variables did not influence the rising of the water level. The only variable that was changed or manipulated was the number of candles, and thus the amount of heat. This was the main factor causing the air under the cup to expand just before it hit the water, thus under cup C the trapped air was the least and thus exerting the lowest pressure. This is the reason why the water is pushed up the highest in cup C. Most students will think that the three candles will burn more oxygen, which is not the case. The use of oxygen for the burning process also helped in decreasing the pressure under the cups, but the amount of oxygen was the same for all three cases.

AIR EXPANSION BY HEAT

1.34. WHAT CAUSES THE WATER TO RISE? (II)

Materials:
1. A small saucer and a glass cup.
2. A birthday candle or wooden matches and a coin.

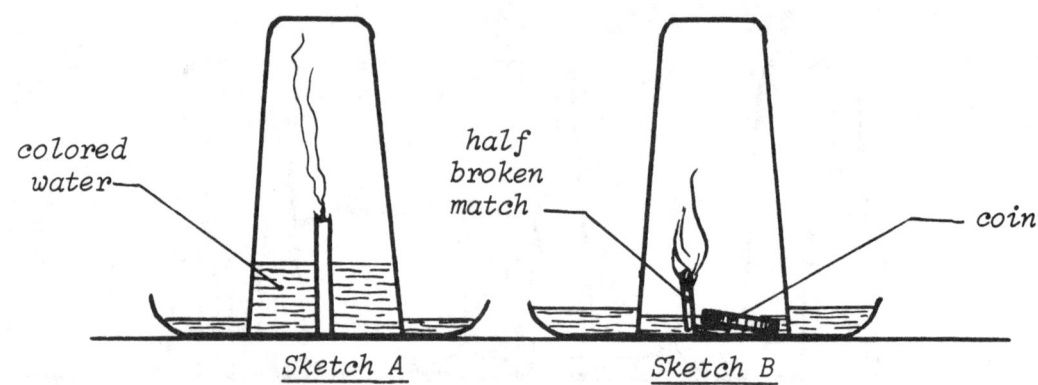

Sketch A   Sketch B

Procedure:
1. Attach the candle in the center of the saucer with a drop of melted wax (see Sketch A), or if no candle is available: break a short end of a couple of matches halfway and place them vertically on the saucer (using a coin as a weight to support them: see Sketch B).
2. Fill the saucer with about half a cup of water.
3. Light the candle or the two matches and cover it immediately with the inverted cup. Observe the water level!

Questions:
1. Why did the water level under the cup rise?
2. What does the candle need in order for it to burn?
3. Did the water level under the cup rise immediately after covering?
4. What did the heat of the flame do to the air under the cup?
5. Would the burning of two or three candles bring up the water level under the cup to the same height?
6. How would the size of the cup influence the rising of the water level? How can we set up experiments to test this?

Explanation:
The burning of the candle needs oxygen and is therefore taking away all the oxygen under the cup. The flame is extinguished as soon as all the oxygen is used up. As the space under the cup does not contain any oxygen any more, it is exerting less pressure compared to the atmospheric air. The water is therefore pushed into the space under the cup.

Another major factor contributing to the decrease of pressure inside the cup, is the fact that the heat of the flame expanded the air under the cup, just before it hit the water. At that moment air escaped from under the cup. After the flame extinguished, the remaining air cooled off and contracted and in so doing it sucked up the water.

1.35. THE CANDLE UNDER THE JAR

Materials:
1. Three or more jars of different sizes.
2. Three or more birthday candles (depending on number of jars).
3. A large clock or watches for timing the burning candles.

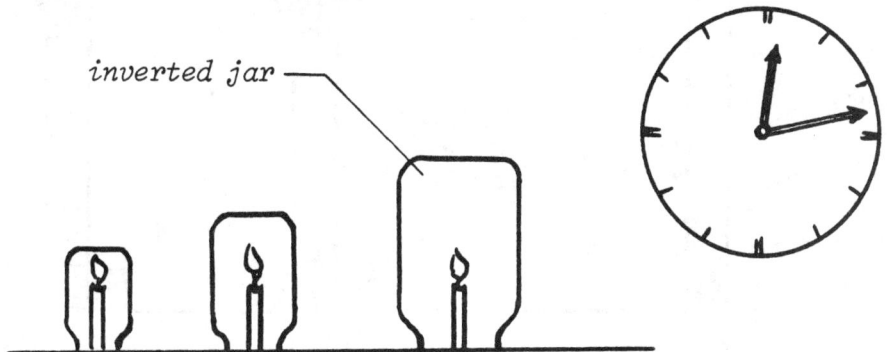

Procedure:
1. Divide the class in small groups of three to six students.
2. Give each group one jar, one candle and matches.
3. Make sure that the students can see the large clock or that each group of students have a watch for timing.
4. Ask the groups to measure the burning time of the candle under the jar: Start timing as soon as the jar is put over the candle and stop timing when smoke evolves from the wick (repeat several times, make sure to replenish the air in the jar).
5. Compare size (volume) of jar with the burning times (plot on graph).

Questions:
1. Why do we measure the burning time more than once?
2. What should we do to the gas in the jar before repeating the measuring of the burning time of the candle?
3. What do you think the burning time of the candle would be, if we did not replenish the air in the jar after the first reading?
4. Why did the candle stop burning under the jar?
5. How can we measure the volume of the jar?
6. What is the relationship between size or volume of the jar and the burning time of the candle?

Explanation:
Air or actually oxygen in the air (about 20%) is needed to sustain the burning of the candle. After the first burning, the jar has no more oxygen. All the oxygen is replaced by carbon dioxide and water vapor, which can be seen condensing on the cold inside surface of the jar. Before doing a next timing of the burning candle, the air or oxygen in the jar has to be replenished. This can be done by pushing crumpled paper in and out of the jar several times. A straight line relationship exists between the volume of the jar and the burning time of the candle when the data are graphed.

AIR   CONTAINING OXYGEN

1.36. THE WATER-ATTRACTING STEEL WOOL

Materials: 1. Two graduated cylinders or two tall glasses.
            2. Two shallow dishes or trays.
            3. A small batton of steel wool.

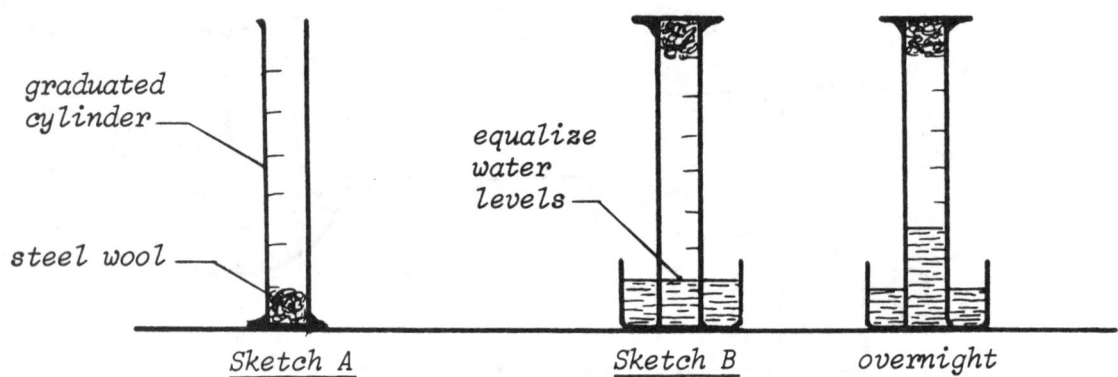

Procedure:
1. Fill both shallow dishes with water.
2. Place the steel wool batton close to the bottom of one of the cylinders or tall glasses. (see Sketch A)
3. Wet this steel wool thoroughly by pouring some water in the cylinder.
4. Fill the cylinder one quarter full with water, cover it with a small paper card, place it upside down on the dish, and take the card out from under the inverted cylinder.
5. Adjust the water level in the cylinder evenly with the water level in the dish by letting an air bubble in the cylinder (holding it somewhat slanted) if the water level inside is too high (see Sketch B).
6. Repeat steps 4 and 5 with the other cylinder without the steel wool.
7. Place them side by side and leave them overnight. Observe water level!

Questions:
1. Why did the water level under the cylinder with the steel wool rise?
2. Did the steel wool change color? What happened to it?
3. What part of the volume in the cylinder did the water level rise?
4. What other chemical can we use in place of the steel wool?

Explanation:
   The steel wool in the cylinder reacted with the oxygen in the air forming iron rust (Fe_2O_3). When all the oxygen under the cylinder has reacted with the iron, the water level should have risen about one-fifth of the volume, since 20% of the air consists of oxygen. Having taken away the oxygen under the cylinder, the pressure of the remaining air is decreased, forming a partial vacuum and thus sucking the water level higher up.
   Magnesium ribbon or a chunk of white phosphorus may be used in place of the steel wool to bind the oxygen out of the air.

AIR   HAS WEIGHT

1.37. THE BALANCING BALLOONS

Materials: 1. Two drinking straws.
             2. Three pins or needles & two pieces of thread.
             3. Two identical uninflated balloons.

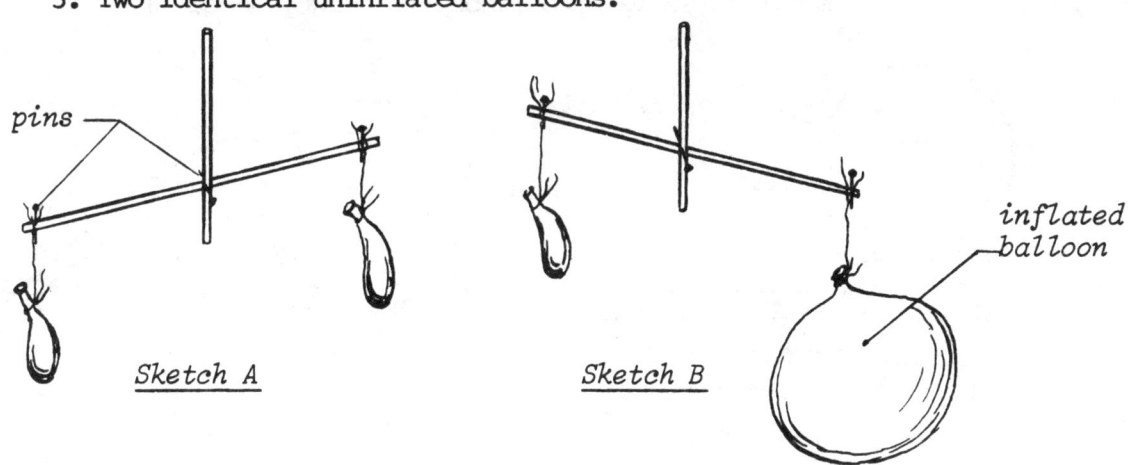

Sketch A   Sketch B   inflated balloon

Procedure:
1. Tie a piece of thread to each of the two balloons and tie the threads to the two ends of one of the straws.
2. Balance this straw on your finger, push a pin through the straw where it is balancing and attach it to the other straw (see Sketch A).
3. Make sure that the straws are moving freely around the needle; balance the horizontal straw, then push a pin through at the spot where the threads are attached (to prevent them from sliding).
4. Make sure that the two uninflated balloons are in perfect balance; then blow air in one of them and tie a knot in the mouth. The balance will tip down at the end of the inflated balloon! (Sketch B).

Questions:
1. What is inside the uninflated balloons?
2. What kind of air was blown in one of the balloons?
3. What could happen if no pins were placed on the ends of the horizontal straw where the threads were attached?
4. What does the balance indicate after inflating one balloon?
5. What would you expect the balance would do if the other balloon was also inflated?
6. How else could we show that air has weight?

Explanation:
   The straw balance may be adjusted by moving the threads further or closer to the end of the straw. In order to keep these attached threads from sliding, we need the pins. The air that was blown in the balloon was exhaled air, which is containing some water vapor but, for our purposes, may be neglected. By inflating the other balloon, the balance should be in equilibrium again.

AIR HAS WEIGHT

1.38. THE BALL THAT GAINS WEIGHT

Materials:
1. A basketball or volleyball (with a valve).
2. A handpump (to pump up the ball).
3. A technical scale or balance (to weigh the ball).

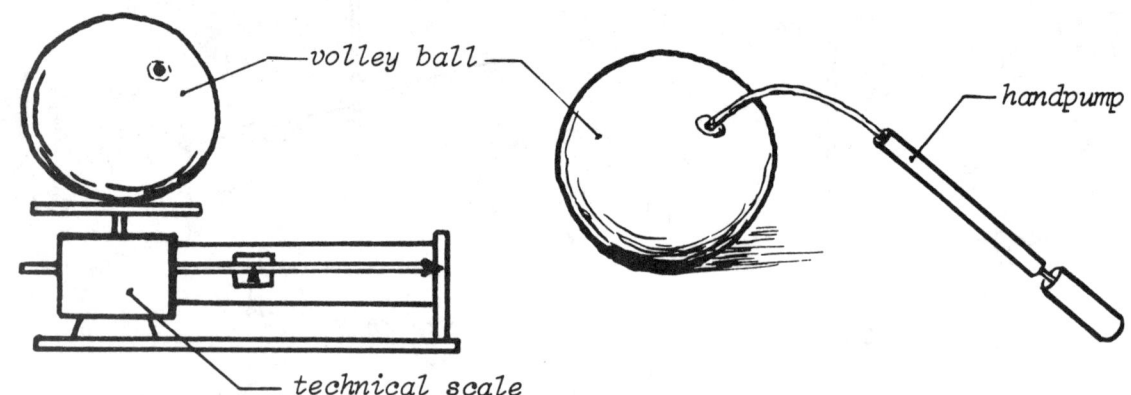

Procedure:
1. Place a rather soft basketball or volleyball on the pan of the technical scale and determine the weight.
2. Connect the handpump to the ball and pump ten strokes of air into the ball.
3. Disconnect the pump and read off the weight of the ball. How much did the ball gain in weight?
4. Repeat steps 2 and 3, and have students predict what the gain in weight would be after 5, 10, 15, 20, 25 strokes of the pump.

Questions:
1. What made the ball gain in weight?
2. What can we stay about the relationship between the number of pump strokes and the gain in weight of the ball?
3. How can we make the ball lose weight?
4. How much would a beachball gain in weight when pumped with 5, 10, or 15 strokes of the same handpump?
5. Would an airtight bottle gain weight if air were pumped into it?

Explanation:
   This demonstration shows that **air has weight.** By adding air to the ball, it increases in weight. The same number of pump strokes should result in the same gain in weight. Half the number of pump strokes gives half the gain in weight. The number of pump strokes is therefore directly proportional to the increase in weight. If the data were plotted on a graph, a straight line relationship would be obtained between the number of pump strokes and the weight of the ball. Whether a ball or an air mattress or an airtight bottle is pumped, the increase in weight should be the same, provided that the same pump is being used and the same number of strokes is applied.

AIR TECHNOLOGICAL APPLICATION

1.39. THE PLASTIC BAG AIR LIFT

Materials: 1. Twelve to twenty medium size garbage bags (plastic).
2. Two identical flat top tables.

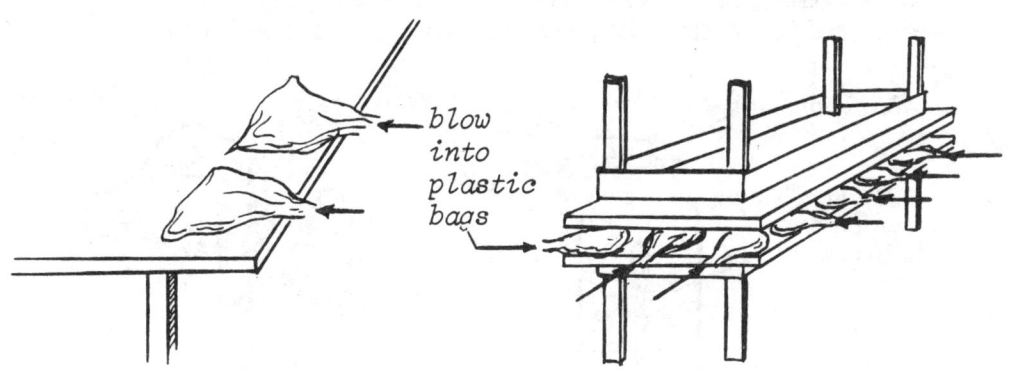

Procedure:
1. Ask as many students as can possibly stand around one of the tables to stand around the table and give them each a plastic bag.
2. Let them spread the bags out on the table and hold the bag's mouth in their hands to get set to blow air in them (let the students stay in a squatting position around the table).
3. Make sure that all students are ready to blow air into the bags with their hands and fingers away from the table top!
4. Ask two or four other students to lift the other identical table, turn it upside down and put it slowly on the first table (this has to be done carefully as it has to move over the heads of the students!).
5. Ask one or two students to climb up and sit on top of the set of tables.
6. Let the squatting students now blow air in the plastic bags all together on the count of three.

Questions:
1. Did you expect a heavy weight like that to be lifted by air?
2. What made the top table rise?
3. How did the pressure of the air inside the plastic bags compare to the outside atmospheric air pressure?
4. Where do we find applications of this principle?

Explanation:
By blowing in the plastic bags, air is being compressed. This compressed air is exerting pressure underneath the inverted table causing the table to rise. This principle is being applied when pumping tires of a bicycle or automobile, or compressing air in air lifts (at gas stations or garages). Tire pressures are twice or four times as high as the atmospheric pressure, and in air lifts these pressures go as high as 20 to 50 atmospheres.

AIR

COMPRESSIBILITY
DENSITY OF MATERIAL

1.40. THE OBEDIENT DIVER

Materials:
1. A milk bottle or other jar, and a large beaker.
2. A medium size test tube.
3. A medicine dropper, wooden match, a large balloon.

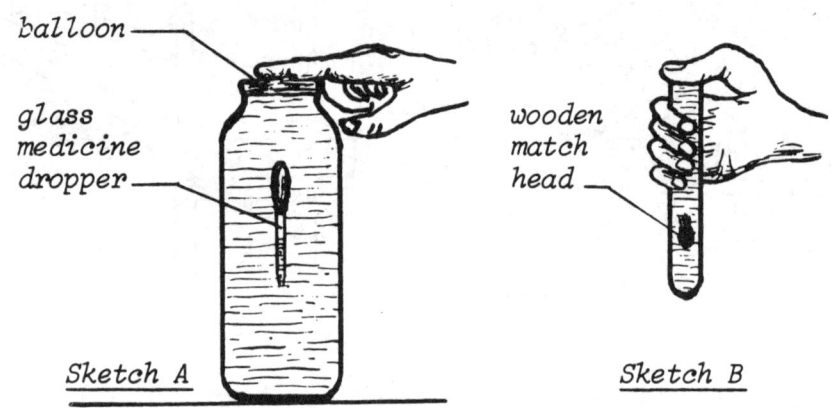

Procedure:
1. Fill the beaker with water; fill the dropper half way with water and test it in the beaker: it should just float (more or less water in the dropper will make it sink or float).
2. Fill the bottle brimfull with water and tie an uninflated balloon flat on the bottle opening (do this after you transferred the dropper from the beaker into the bottle).
3. Put your finger on the stretched balloon and control the sinking or floating of the dropper by putting pressure on it.
4. Ask the students to say "up" or "down" or "stop" and the dropper will respond to their command!
5. A broken off wooden match head in a water-filled test tube can be similarly controlled by pressure of the thumb (see Sketch B).

Questions:
1. What made the dropper or the match head sink? Or float?
2. What did the water level inside the dropper do when it dove down?
3. Did the air inside the dropper increase or decrease in volume during the diving? During the floating?
4. Is water or air more compressible?

Explanation:
By putting pressure on the stretched balloon with the finger, or with the thumb on the test tube, the air inside the dropper (or inside the wood fibers of the wood match) gets compressed, the water level rises and thus the whole dropper gets heavier and sinks. By releasing the pressure, the water is pushed out again, the dropper gets lighter and floats. This principle is applied in submarines, where water is pumped in to submerge or pumped out to surface. This demonstration is also suitable to initiate inquiry on **sinking and floating or on density**.

AIR

TECHNOLOGICAL APPLICATION
TRANSFERABILITY OF PRESSURE

1.41. TURN A LITTLE WATER INTO A LOT OF LEMONADE

Materials: 1. Two identical gallon (4 litre) tin cans.
2. Two 2-hole stoppers (to fit the opening of the cans).
3. Glass tubing, rubber tubing, and a glass funnel.

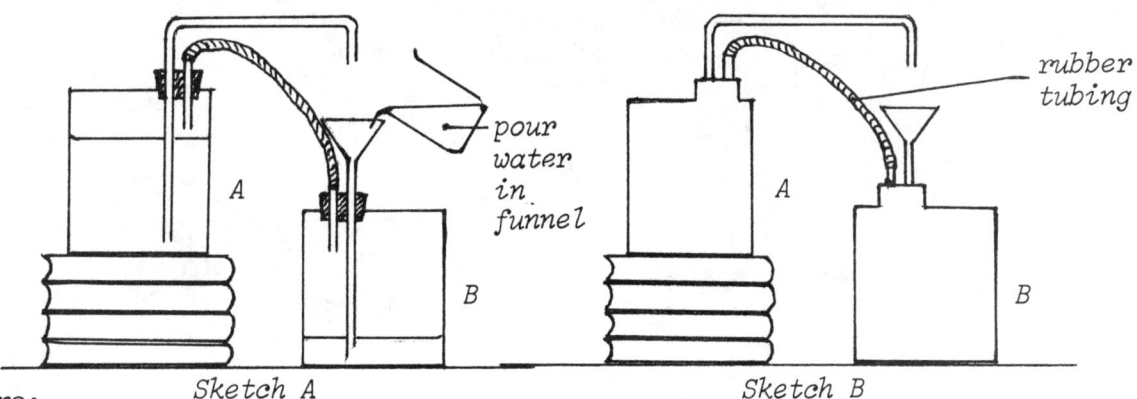

Sketch A   Sketch B

Procedure:
1. Fit the glass and rubber tubing, and the glass funnel in the two hole stoppers as shown in the sketch above (See Sketch A).
2. Fill Can A about 3/4 full of water (add red coloring for lemonade effect), and pour about 100 ml of water in Can B.
3. Make sure that the stoppers are pressed tightly in the can openings. Now you are ready for the demonstration (do not reveal the construction of the tubes inside the cans - students only see things as in Sketch B).
4. Pour some water in the funnel and say: "I am turning a little water into a lot of wine/lemonade", "Can you find out what the structure of the glass tubes inside the cans have to be to make it work like this?"
5. Have students work in groups, and have each group come up with their hypothesis and explanation.

Questions: (Before seeing the inside structure)
1. Did you observe any water flowing from Can B to Can A?
2. What would cause the water to flow from Can A?
3. What would adding water to the funnel do to the water level in Can B?
4. Do either of the two tubes in Can B have to extend into the water level or not, or both?
5. The same question (4) for Can A?
(After seeing the correct inside structure):
6. What would happen if: the funnel did not extend into the water?
7.   "   "   "   : the glass tube in Can B also extended into the water?
8.   "   "   "   : the long tube in Can A did not extend into the water?
9.   "   "   "   : both tubes in Can A extended into the water?

Explanation:
With the correct structure in Sketch A above, after pouring some water into the funnel, the water level in Can B rises, increasing the pressure of the air above the water, thus also increasing the air pressure in Can A. This will push down the water level and thus push the water up in the long tube. The water drips into the funnel and the cycle is repeated again. The flow of water will stop as soon as either the water level in Can A gets lower than the opening of the long tube, or the water level in Can B reaches the opening of the short tube.

AIR AERODYNAMICS

1.42. MAKE AN AIR BULLET SHOOTER

Materials:
1. A medium sized cardboard box.
2. A heavy polyethylene sheet (to cover one end of the box).
3. Four long heavy rubber bands, a rubber stopper.
4. Two small petrie dishes, conc. ammonia, conc. hydrochloric acid.

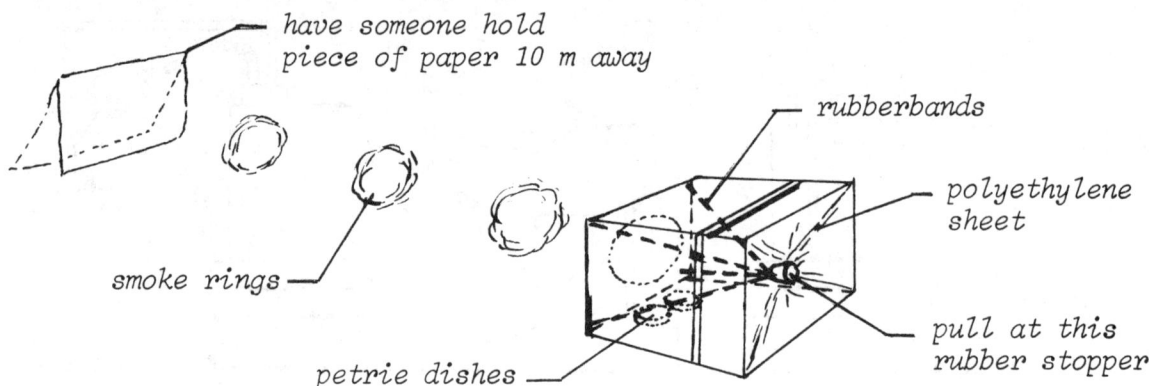

Procedure:
1. Cut a circular hole of about 15 cm diameter in the bottom of the box, and flip the flaps of the open side inwards.
2. Staple the long rubber bands to the four corners of the side with the hole, and staple the other end of the bands to the rubber stopper.
3. Cover the open end of the box rather loosely with the polyethylene sheet and tape the edge of the sheet tightly against the box.
4. Push the rubber stopper against the center of the sheet, wrap the sheet around the stopper, and tie it with a small thick rubber band.
5. Now the air shooter is ready. Have someone hold a paper or cloth sheet about 10 meters away and shoot air bullets at the sheet by holding the box, aiming the circular hole at the sheet, pulling the rubber stopper and release it suddenly (see Sketch above).
6. The air shooter can also be used to produce smoke rings. Tape the two small petri dishes next to each other inside the box, place a few drops of conc. ammonia in one and a few drops of conc. HCl in the other dish, and shoot smoke rings!

Questions:
1. What makes it possible to blow at the paper sheet from 10 m away?
2. Would a square or triangular opening have the same effect?
3. What shape would the "air bullet" have?
4. What did the smoke consist of?

Explanation:
The circular opening in the box made it possible for the air to form ring-shaped "bullets" that can move through the stationary air much faster and farther. Other shapes of the opening in the box would hamper this natural travel of fluids. The smoke was formed by the gases or vapours of ammonia and hydrochloric acid: $NH_3 + HCl \longrightarrow NH_4Cl$, forming ammonium chloride, which is a solid (smoke consists of finely devided solid particles in a gas).

Fill the air shooter box with air freshner mist or perfume, and shoot perfume "bullets" to the people in the audience!

AIR   AERODYNAMICS

1.43. THE SMOKE RING RACE

Materials: 1. The air bullet shooter (see Event 1.42)
2. Cigarettes (if you are able to make/blow smoke rings).

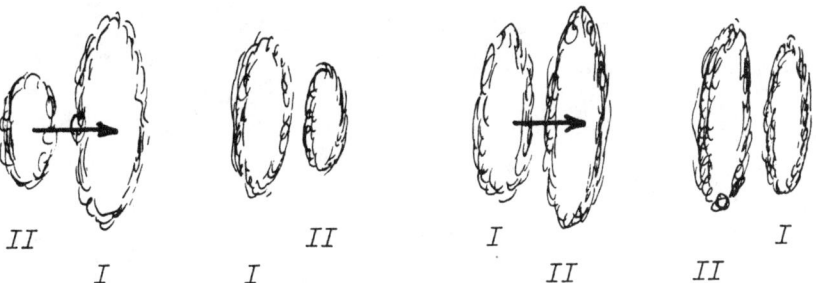

Procedure:
1. Place a few drops of conc. ammonia in one petri dish, and a few drops of conc. hydrochloric acid in the other dish in the air shooter box.
2. Make a slow moving smoke ring by pulling the stopper halfway or not as far out of the box and releasing it.
3. Immediately after making this slow moving ring, produce a fast moving smoke ring by pulling the stopper out much farther and releasing it. When making this second ring, do not change the direction of the box opening, in other words, shoot directly at the first smoke ring.
4. Observe what the rings are doing!
5. If you are able to blow cigarette smoke rings: make a large ring, then blow a smaller ring through this first one by holding your mouth opening a little narrower.
(If the rings made with the air shooter are to even sized, hold a cardboard with a smaller circular opening in front of the air shooter to make the second smaller smoke ring).

Questions:
1. Do you notice that the rings are chasing each other?
2. What did the first ring do when the second ring went through it?
3. Which of the rings actually won the race?
4. Did the rings stay the same size during the race?
5. Which size rings would travel faster? The smaller or the larger ones?
6. Why would a ring be formed in the first place?

Explanation:
The circular opening in the box (or the mouth) makes the air in the center of the circle move fastest. The outside of the circle of air is held back somewhat, thus a circular motion is created, causing a ring to be formed. The smaller second ring will flow with the air in a circular path and thus become larger as soon as it has gone through the first ring, while the first ring gets smaller and passes through the second, etc.

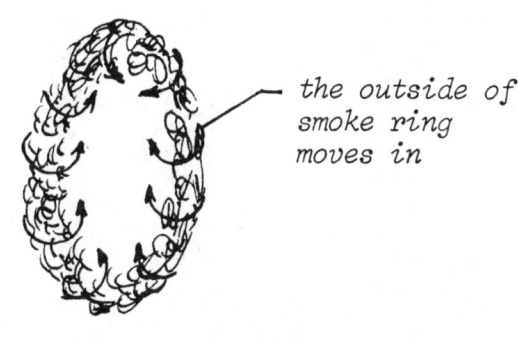

the outside of smoke ring moves in

CHAPTER 2

WHAT CAN FLOWING AIR DO?

OBJECTIVES

After dealing with and studying the concepts in this chapter, the students should be able to:

a. recognize the correct explanation of an observed event based on the Principle of Bernoulli;

b. distinguish true from false statements concerning events applying the Principle of Bernoulli;

c. distinguish between air flow along a surface and air flow against a surface (wind energy);

d. identify the correct explanation of an event in daily life applying the Principle of Bernoulli.

FLOWING AIR — BERNOULLI'S PRINCIPLE

2.1. THE LIFTING PAPER

Materials: 1. Strips of paper of about 15 x 3 cm (for each of the students).

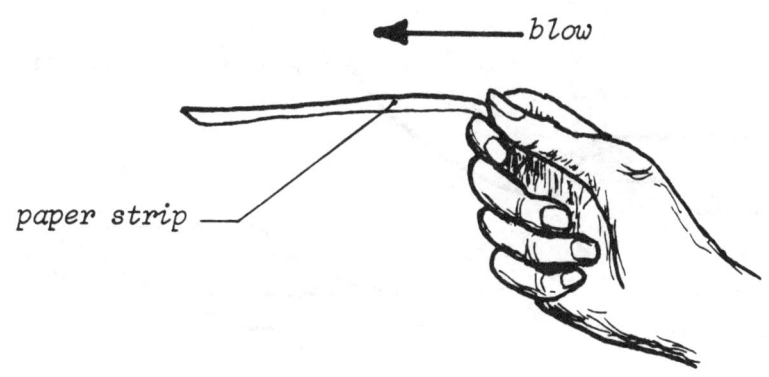

Procedure:
1. Hand out a strip of paper to each of the students.
2. Have them make a fold at one end of the paper strip.
3. Holding the paper strip by the folded end, let them blow against the underside of the paper: what do you observe?
4. Now hold the strip near the chin and blow over it (let the students try and predict what will happen to the strip).

Questions:
1. What did you observe in the first instance? (blowing under it).
2. What did you observe in the second case? (blowing over it).
3. What do you know about the properties of stationary air?
4. What is different about the air on top of the paper compared to that underneath the paper, when we blow over it?
5. What effect does the moving air have on the gravity that is working on the strip of paper?

Explanation:
Stationary air exerts pressure in all directions. When we blow over the top of the paper strip, we are creating a faster moving flow of air above the strip. The faster the flow of a fluid (in this case the air), the lower the pressure it exerts (Bernoulli's Principle):

$$P + \frac{K.E.}{V} = \text{Constant}$$, where P = Pressure, K.E. = Kinetic Energy and V = Volume.

This last ratio can be equated with the velocity of the moving fluid. The effect of the moving air above the strip of paper is, that it reduces the pressure above the strip. Thus the higher pressure under the strip pushes the paper up into the lower pressure region.

FLOWING AIR — BERNOULLI'S PRINCIPLE

2.2. THE STUBBORN PAPER CARD

Materials: 1. Paper cards (3" x 5") for each of the students.

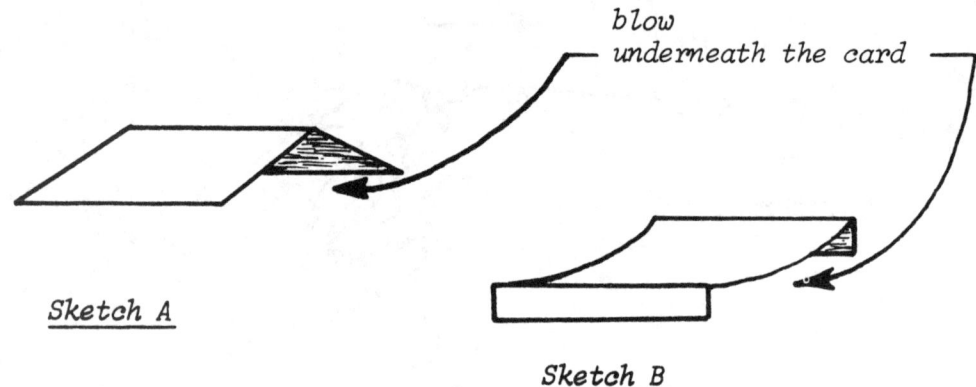

Procedure:
1. Distribute the 3 x 5 cards to each of the students.
2. Fold the cards either in the center (Sketch A) or fold them at each end about 1 cm from the edge (see Sketch B).
3. Place the folded card on the table or desk and try to blow the card off the table by blowing underneath it. What do you observe?

Questions:
1. What did you observe when blowing underneath the paper?
2. What properties does stationary air have?
3. What is different about the air underneath the paper compared to that above the paper, when blowing underneath it?
4. Why do you think the paper did not fly off the table?
5. What is different about flowing air compared to stationary air?

Explanation:
   This activity can be a follow-up to 'The Lifting Paper' in dealing with Bernoulli's Principle. Among other properties, stationary air exerts pressure. Flowing air exerts less pressure as compared to stationary air. The faster the flow, the lower the pressure it exerts. By blowing underneath the card, we actually created less pressure underneath the paper, so that the pressure above the paper became larger than below the paper, and this is why the card got pressed down against the table.
   In terms of moving molecules, we can see that by blowing underneath the card, the air molecules are pushed away from under the card. It is as if a partial vacuum is created at the spot where one is blowing. The air molecules that surround the card, those on the side as well as those above the air stream, are sucked towards this spot. In rushing to replace the molecules that were blown away, these air molecules push on the upper side of the card, thus bending it down towards the table.

FLOWING AIR   BERNOULLI'S PRINCIPLE

2.3. THE FUNNEL AND THE BALL

Materials: 1. One long-stem funnel (glass or plastic).
            2. One ping-pong ball.

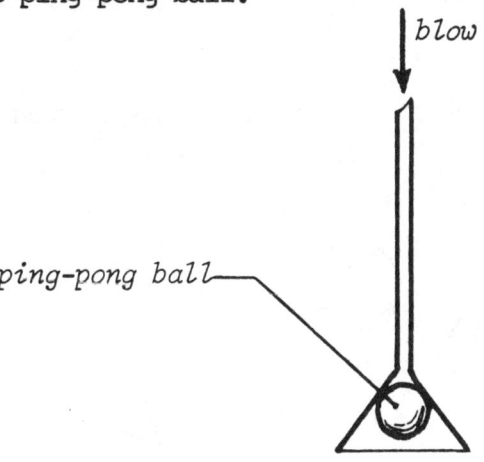

Procedure:
1. Place the funnel and the ball next to each other on the table.
2. Ask: "How can I pick up the ball with the funnel without sucking through it? I may not touch the ball."
3. Pick up the funnel by the stem, place it over the ball and blow through the stem, lift the funnel while blowing.
4. Hold one hand under the funnel and stop blowing (ball drops).
5. Place the ball in the funnel and have a student try to blow it out of the funnel (he/she will not succeed).

Questions:
1. How did we pick up the ball with the funnel without sucking through it?
2. What happened when we stopped blowing?
3. Is it possible to blow the ball out of the funnel?
4. Where is the air moving fastest when we blow through the funnel?
5. What is flowing air creating that stationary air doesn't?
6. What is the difference about the inside compared to the outside of the funnel when blowing through it?

Explanation:
   The ball can be picked up from the table with the funnel by blowing through it. By blowing through the funnel, we create a lower pressure inside the funnel, especially at the spot where the stem is attached to the conical shape of the funnel. Here the fastest flow of air occurs, because the air molecules have suddenly more space to move about. **The faster the flow of air, the lower the pressure.** This is why the ball is sucked into the funnel by blowing, and for the same reason it is not possible to blow the ball out of the funnel. The harder we blow through the funnel, the lower the pressure gets in the mouth of the funnel.

FLOWING AIR   BERNOULLI'S PRINCIPLE

2.4. THE ATTRACTING SPHERES

Materials: 1. Two plastic spheres (tennis balls or apples will work).
             2. Two lengths of thread (about 40 cm) & adhesive tape.

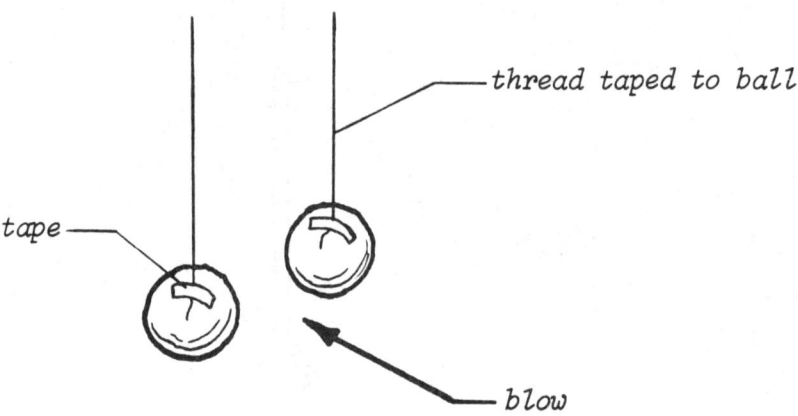

Procedure:
1. Tape the two pieces of thread to the two spheres.
2. Hang the two spheres about 3 cm apart from a horizontal support, or let someone hold them.
3. Ask: "What would you expect the balls to do when I blow in between them?" (anticipated answer: fly apart).
4. Blow in between the two hanging spheres and observe the movement of the two spheres (hard plastic spheres will click together).

Questions:
1. What do you observe when blowing straight against one ball?
2. What did you expect the two spheres would do when blowing between them?
3. What did you observe the two spheres were doing when blowing in between?
4. What did we create between the balls by making the air flow?
5. What is different about the air between the two spheres as compared to the surrounding air?

Explanation:
   When blowing head on against one sphere, we observe that the sphere moves away from the source of the air flow. This is because the air molecules bump directly against the surface of the sphere, the kinetic energy is transferred to the sphere and it is thus pushed away. When blowing in between the two spheres, we are actually pushing away the air molecules and thus creating a lower pressure. The stationary air surrounding the two spheres on the outside exerts a higher pressure and pushes them against each other.
   Applications of Bernoulli's Principle with spheres are encountered in playing tennis or baseball. A top spin in tennis creates a stronger and faster air flow under the ball and curves the path down (in court). A spin throw of a baseball curves the ball path in the direction of the spin.

FLOWING AIR   BERNOULLI'S PRINCIPLE

2.5. THE CLANGING SODA POP CANS

Materials: 1. Two empty soda pop cans.
             2. About two dozen straight drinking straws.

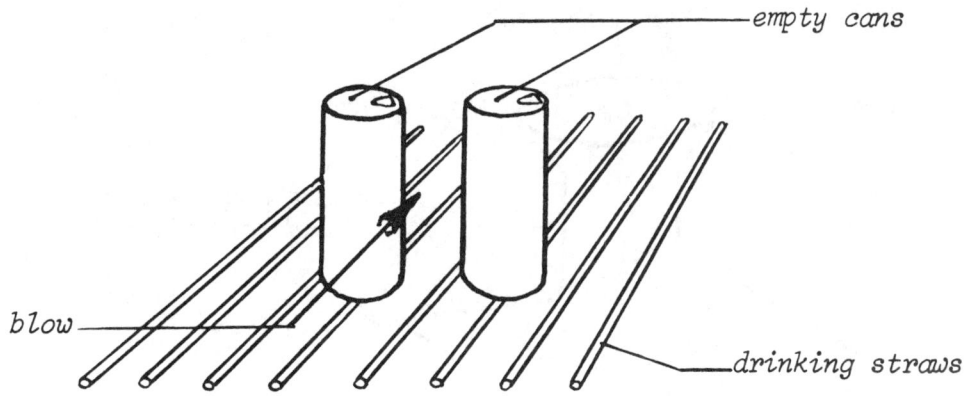

Procedure:
1. Spread the straws parallel to each other on the table and leave about 1/2 to 1 cm gap in between them.
2. Place the two cans upright about 2 cm from each other on the straws and show the students that they can easily move closer or further apart.
3. Ask: "What will happen to the cans if I blow in between them?"
4. Now spread the two cans about 5 cm apart. Ask: "Do I have to blow softer or harder to get the two cans clanging?" Blow harder!
5. Now place the cans about 20 cm apart. Ask: "Can I still get the two cans clanging against each other?" Take a deep breath and blow a constant stream of air on the right side of the left can and move your head towards the right, while constantly blowing (Cans will clang!).

Questions:
1. What made the cans move towards each other?
2. How far apart could the cans be placed and still be drawn together?
3. What does the flowing air create in between the cans?
4. Was a stronger flow of air necessary to bring the cans that were 20 cm apart together?

Explanation:
   Blowing in between the cans created a flow of air and thus a lower pressure compared to the stationary air on the other side of the cans. It is this lower pressure that drew them together. Theoretically, the cans could be placed an infinite distance away from each other and still be drawn together, as long as a constant flow of air on one side of one can moves along with it, to move it to the other can. Indeed, **the faster the flow of air, the lower the pressure it exerts**. But for the cans that were placed 20 cm apart, only a constant flow that could move with the can, was necessary.

FLOWING AIR — BERNOULLI'S PRINCIPLE

2.6. THE FLOATING CARD

Materials:
1. A paper card (3 x 5").
2. A thread spool and a pin.

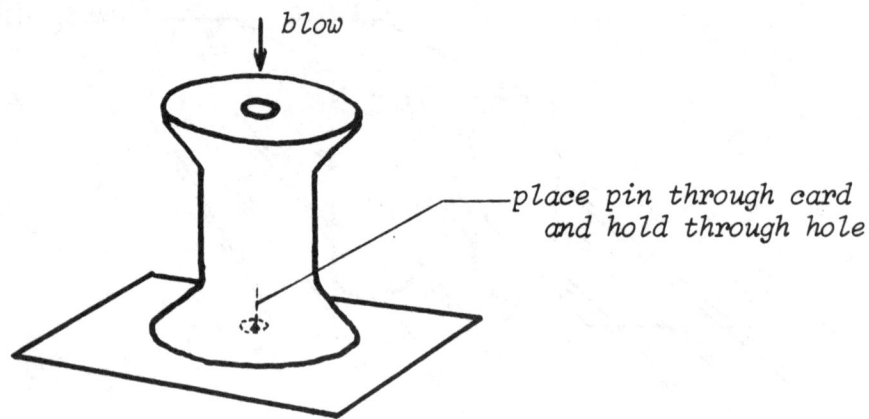

Procedure:
1. Hold the card up close to the mouth and blow against it. What do you observe? What did the card do?
2. Now push the pin through the center of the card.
3. Hold the card against the spool with the pin sticking in the hole of the spool. Ask: "What would you expect the card to do when I blow through the hole of the spool?" (Anticipated answer: 'blow away').
4. Now blow through the hole of the spool and let go of the card (card should stick against the spool).

Questions:
1. What did you observe when blowing against the card without the spool?
2. What did you observe when blowing through the hole of the spool against the card?
3. Where was the faster flow of air created?
4. What is different about the air above the card as compared to the air under the card (while blowing through the spool hole)?
5. What is keeping the card against the spool?

Explanation:
By blowing in the hole of the spool, we are creating a faster flow of air above the card, thus creating a partial vacuum at this spot: between the card and the spool. The relatively slower moving air, which is surrounding the card, exerts a higher pressure compared to the air between the card and the spool. This makes the card stay close to the spool. As soon as we stop blowing through the spool, the card drops, because the pressure above and below the card is equalized.

FLOWING AIR   BERNOULLI'S PRINCIPLE

2.7. THE FLOATING BALL

Materials: 1. A ping-pong ball or a beach ball.
            2. A 25 cm length of rubber tubing.
            3. A small compressor to produce a constant air flow (optional).
            4. A vacuum cleaner (when using the beach ball).

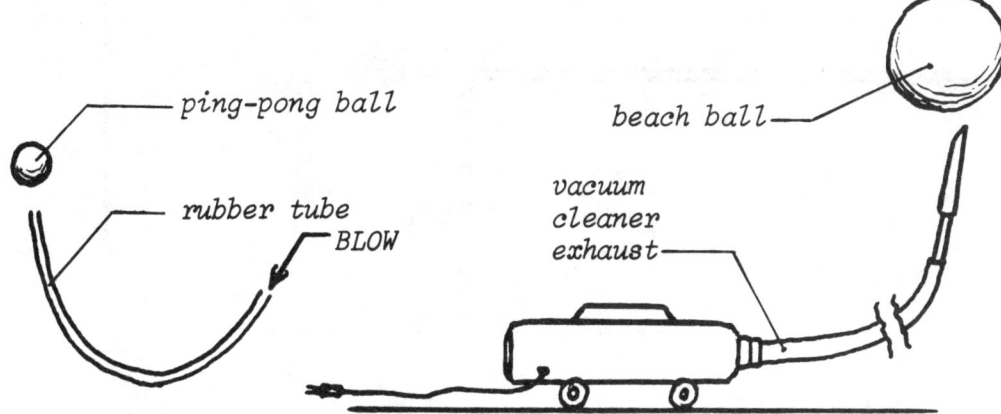

Procedure:
1. Take the rubber tubing with each hand on one end of the tube.
2. Place the ping-pong ball on one end of the tubing and blow through the other end, and balance the ball in the air flow.
3. When using a beach ball, you will need a vacuum cleaner. Detach the vacuum cleaner hose and attach it to the exhaust end, so that it blows through it rather than producing a suction.
4. Blow air into the beach ball and let the ball hang in the air stream that the vacuum hose is producing.
5. When a small compressor is available, attach a small orifice to the exhaust tube and hang a ping-pong ball in the air stream.

Questions:
1. Will we be able to do the same thing with other round objects?
2. What property of the round object is critical for it to stay afloat?
3. What does the blowing do to the movement of the air molecules?
4. How does the pressure of flowing air compare to that of stationary air?
5. Why does the ball stay afloat in the air stream?

Explanation:
   By blowing through the rubber tubing or by using the compressor or the exhaust of the vacuum cleaner, we are creating an air flow. This air flow causes the existence of a lower pressure area compared to the surrounding air, which is stationary or not moving as fast. **Bernoulli's Principle** states that: **the faster the flow of air, the lower the pressure.** This means that at the place of exhaust (end of the tube or hose), a cone of lower pressure is created. All the surrounding air is exerting a higher pressure and thus the ball is kept in the lower pressure area, which is the cone above the exhaust.

FLOWING AIR — TECHNOLOGICAL APPLICATION

2.8. LIFT WATER BY BLOWING

Materials: 1. Two drinking straws (transparent ones).
             2. A glass or beaker.
             3. Food coloring or ink & white sheet of paper.

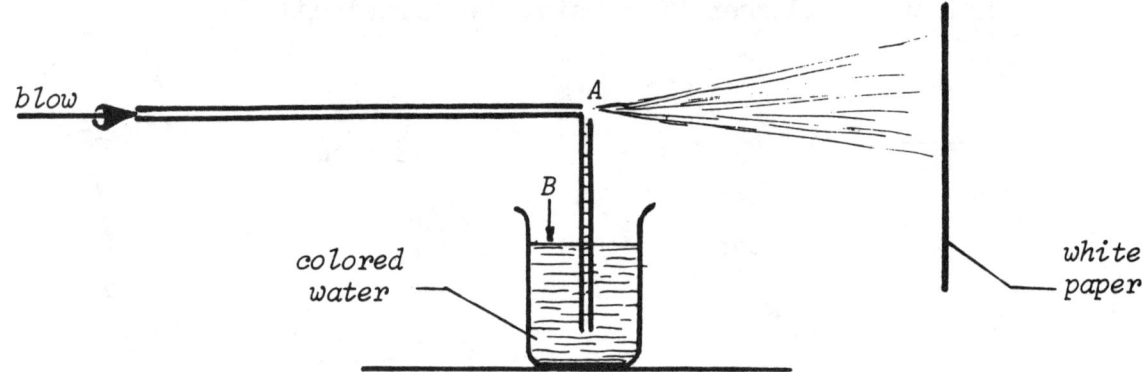

Procedure:
1. Fill the beaker about 3/4 full with water and add a few drops of food coloring or ink to it, and stir.
2. Cut one of the straws in half and dip this short straw vertically in the colored water (hold this with the left hand).
3. Place the long straw with the right hand horizontally against the top opening of the vertical straw (see Sketch above).
4. Ask a student to hold a white sheet of paper on the other side of the equipment set-up.
5. Blow through the horizontal straw until the colored water is sprayed against the paper. (The horizontal straw might need some adjusting up or down against the vertical one before the water comes out).

Questions:
1. What do you create by blowing the air with the straw?
2. How does the pressure at the meeting place of the two straws compare with that on the water surface of the beaker?
3. What is it that lifts the water in the vertical straw?
4. Would this also work with a longer vertical straw?
5. Where do we find this principle applied?

Explanation:
By blowing through the horizontal straw, we are creating a flow of air and **the faster the flow of air, the lower the pressure** (Bernoulli's Principle: see Event 2.1.). Thus the pressure at A is smaller than at B (see Sketch), and this pressure difference makes the water move upwards in the vertical straw. The longer the vertical straw, the harder we have to blow to create a larger pressure difference.

Applications of this principle are found in spray guns for painting, in spray bottles, spray pressure cans, etc.

FLOWING AIR   BERNOULLI'S PRINCIPLE

2.9. THE MYSTERIOUSLY MOVING FLAME

Materials: 1. A candle and matches & a drinking straw.

Sketch A

Sketch B

Procedure:
1. Light the candle and attach it to the table.
2. Blow through the straw directly against the flame, such that the students get the side view of it (Sketch A).
3. Now hold the straw to the side of the flame, blowing straight into the face of the observer, ask: "Where will the flame move to when I blow? Towards or away from the straw?" (Sketch B).
4. By moving the straw radially in a semi-circle and repeating point 3, make sure that every student has a good view of the moving flame.

Questions:
1. Why did the flame move away from the straw in set-up A?
2. Why did the flame move towards the straw in set-up B?
3. What variable is decreased by increasing the flow of air?
4. What principle is this phenomenon based on?
5. Where do we find this principle applied in daily life?

Explanation:
   This event again is based on Bernoulli's Principle, which states that **the faster the fluid flows, the lower the pressure the fluid exerts.** In this case the fluid is air. The air on the straw side of the flame is flowing and thus compared to the air on the other side of the flame, which is stationary, it exerts a lower pressure. For this reason the flame bends towards the straw. When blowing straight against the flame, this latter gets immediately into the air stream and can be compared to a flag waving in the wind.
   Applications of Bernoulli's Principle are encountered in our daily life in the lift that airplanes are getting when flying; the shape of the wings makes the air flow faster over the wing than under it (Event 2.11.). Other applications are found in throwing curved balls in baseball pitching and in using top spins in playing tennis.

FLOWING AIR — TURBULENCE

2.10. WHICH WAY WILL THE FLAME FLICKER?

Materials:
1. A candle and matches.
2. A paper card (3 x 5" or 10 x 15 cm).
3. A large empty jar (pickle jar).

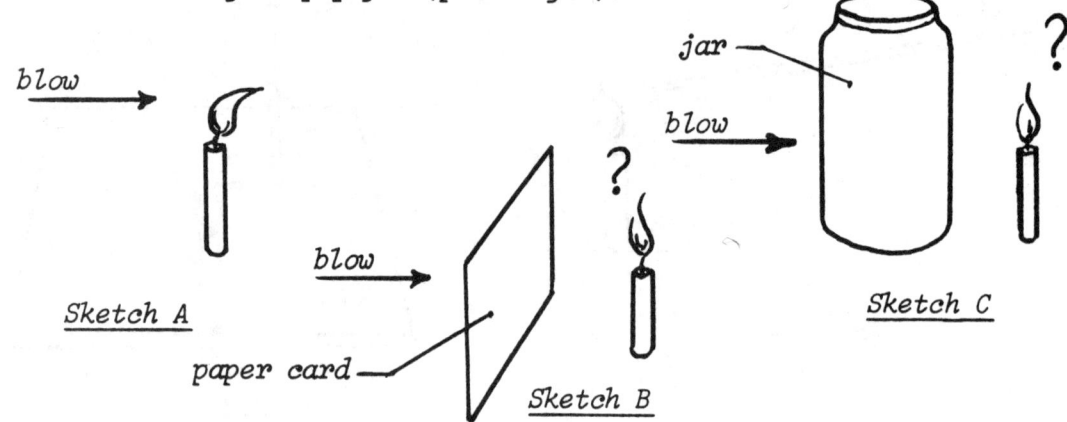

Procedure:
1. Light the candle and attach it to the table.
2. Stand to the side of the candle and ask: "Which way will the candle flame flicker when I blow from this side?"
3. Blow slowly against the flame: the flame moves away from you (Sketch A).
4. Place the card vertically between you and the candle, and ask again: "Which way will the candle flame flicker when I blow?"
5. Blow hard against the card and let students observe the flame. (B)
6. Place the empty jar between you and the burning candle. Repeat the question and blow hard against the jar (see Sketch C).

Questions:
1. What does flowing air against the card create?
2. From your observation of the candle flame movement with the card, on which side of the candle was the air pressure lower?
3. Why did the candle flame not behave in the same way with the jar as with the card when blown against?
4. What would happen with the candle flame behind the jar, if we blew really hard against the jar?
5. Why are fast moving vehicles almost never squared off at the rear?

Explanation:
Blowing air against the card creates a turbulence of air behind the card and thus a lower pressure. The flame is drawn to the spot of lower air pressure and therefore is bent towards the card. When blowing against the jar, the air can smoothly flow around the jar and when the blowing occurs hard enough, the flame can easily be extinguished. This is an example of a streamlined fast moving object. Vehicles that are squared off at the rear have problems with turbulence, which is creating a pocket of low pressure in the rear of the vehicle, forming a **drag**.

FLOWING AIR   FLIGHT

2.11. THE PAPER WING

Materials: 1. A piece of paper or paper card and cellotape.
             2. A round pencil or straw.

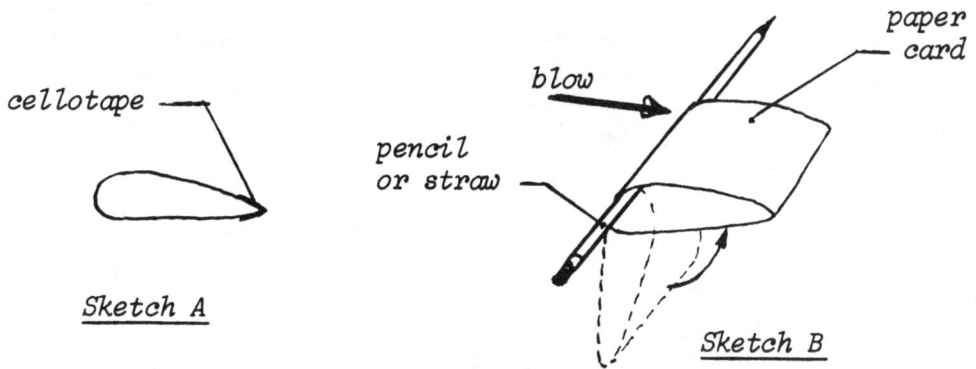

Sketch A

Sketch B

Procedure:
1. Cut a piece of paper of about 5 x 15 cm and bend it in the shape of an airplane wing (Sketch A) and tape the ends together.
2. Put a pencil or straw through the wide end of the wing and let the paper wing hang vertically down (see Sketch B).
3. Blow over the paper wing (it should move up horizontally).

Questions:
1. Why does the paper wing move up when blown over it?
2. What would happen if you blew under the paper wing?
3. What happens if you blow over the wing, but invert it (with the curved side facing down)?
4. What gives airplanes the air lift?
5. What do airplanes do with their wings when they take off or land?

Explanation:

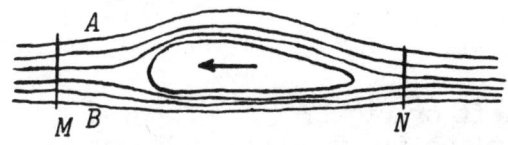

Bernoulli's Principle is underlying and causing the airlift that the airplane wing provides. The wing is built in such a way that the air over it flows faster than the air under it. It has a greater curvature on the top compared to the bottom side.

When we consider an air molecule A on the top side and molecule B below the wing, and they have to run from point M to N, it can be seen that A has to run faster than B, because of its longer path.

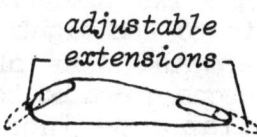

The wings of an airplane are provided with adjustable flaps that can be extended or retracted. When extended, it increases the curvature of the wing on the upper side and provides a greater lift for starting and landing.

FLOWING AIR — FLIGHT

2.12. THE PAPER AIRPLANE CONTEST

Materials:
1. Onion skin or tracing paper.
2. Small scissors (for each pair of students).

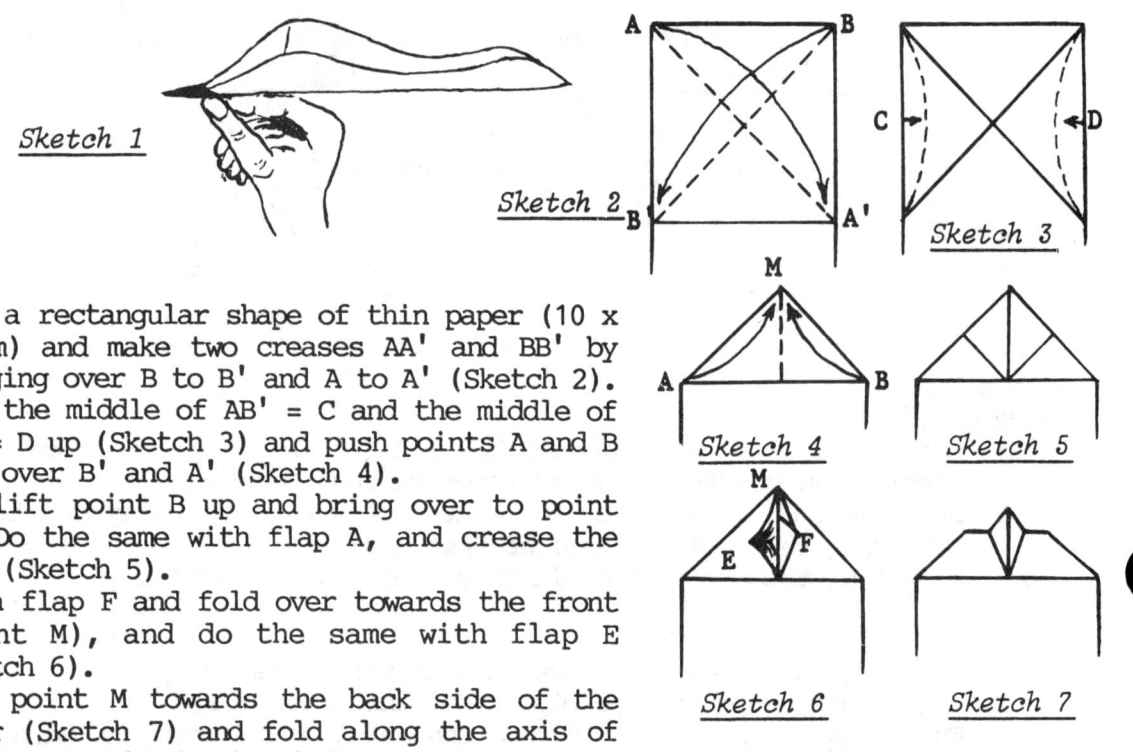

Procedure:
1. Take a rectangular shape of thin paper (10 x 20 cm) and make two creases AA' and BB' by bringing over B to B' and A to A' (Sketch 2).
2. Lift the middle of AB' = C and the middle of BA' = D up (Sketch 3) and push points A and B down over B' and A' (Sketch 4).
3. Now lift point B up and bring over to point M. Do the same with flap A, and crease the fold (Sketch 5).
4. Pinch flap F and fold over towards the front (point M), and do the same with flap E (Sketch 6).
5. Fold point M towards the back side of the paper (Sketch 7) and fold along the axis of symmetry to obtain Sketch 8.
6. Cut the plane in any shape you want (experiment with it) and bend wings up or down, and tail elevators up or down.
7. Fly the planes: the longest one to stay up wins (Sketch 1).

Questions:
1. What makes the paper plane fly farthest?
2. What makes the paper plane turn to the left or right?
3. Would thinner or thicker paper make the plane fly farther or longer?

Explanation:

When the elevators are angled up, the flowing air pushes it down and the plane climbs. The rudder works just like that of a boat. Turning it to the right moves the tail to the left and the whole plane turns to the right. Ailerons are used to roll the plane and the flaps to increase the curvature of the wings and increase the lift.

FLOWING AIR   WIND ENERGY

2.13. THE CORK RACE

Materials: 1. Different size corks (one per pair of students).
            2. A shallow tray (one per pair of students).

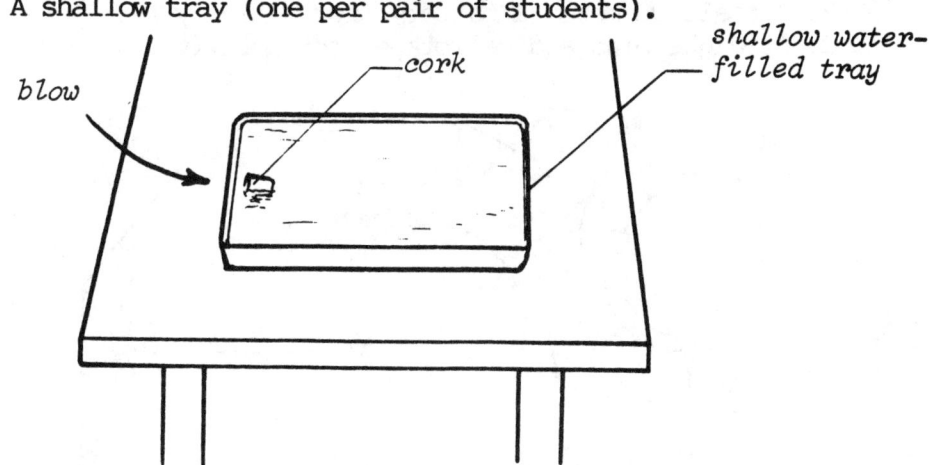

Procedure:
  1. Distribute one cork and one tray to each pair of students.
  2. Let the students fill the tray with water and float the cork on the water (making sure that the cork does not drag on the bottom of the tray; add more water if this is the case).
  3. Position the students facing each other with the tray in between them (the cork may not be touched during the race).
  4. On the count of three, one student blows the cork across the tray, and the other student indicates (raises hand) as soon as the cork has crossed over (reached the other side).

Questions:
  1. What made the cork move over the water?
  2. Which cork moved faster: the smaller or the larger one?
  3. What are other factors that influence the speed of the moving cork?
  4. What type of energy was used to move the cork?
  5. Which side of the cork, when blown against, results in a faster speed?

Explanation:
   The flowing air blowing against the cork consists of a stream of moving air molecules possessing kinetic energy. This energy is imparted upon the cork resulting in its movement. The speed of the cork depends on how hard one blows against the cork, whether the stream of air hits the upright ends or the round surface of the cork, and whether it is a larger or smaller cork. The larger cork has a larger surface to blow the air against and thus may move faster, even though it is slightly heavier than the smaller cork. Other variables influencing the speed might be investigated like: the type of liquid, the angle of air flow, etc.
   This demonstration shows that moving air is a form of energy: kinetic energy of moving air molecules.

FLOWING AIR   WIND ENERGY

2.14. THE PAPER FAN

Materials: 1. A paper card (5 x 8") or construction paper.
            2. A small DC motor & battery (for activity A).
            3. Straight pins and cellotape (for activity B).

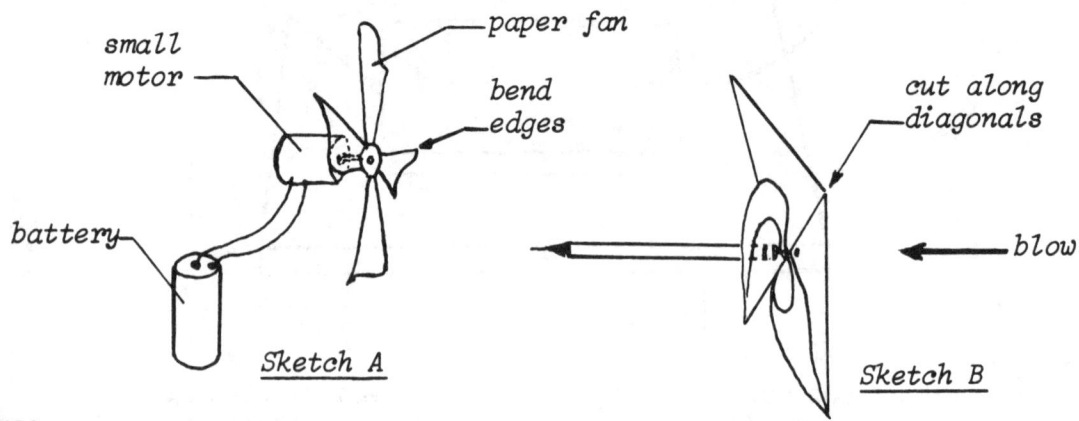

Procedure:
1. Cut a circle (12 cm/5" diameter) out of the paper card.
2. Draw four diameters perpendicular to each other on the circle and cut in with scissors till about 1 cm from the center.
3. Cut four sections off and bend the edges of the remaining sections (see Sketch A), and attach this fan to the small motor. Switch the motor on (connect to battery): a flow of air is created.
4. For activity B no motor is needed.
   Cut a square piece out of construction paper, draw the diagonals and cut along these lines till about 2 cm from the center.
5. Fold every other point back to the center and hold them in by pushing the straight pin through the points and the center of the paper.
6. Push the pin into the eraser part of a pencil and hold the paper mill in a stream of air, or move it back and forth in the air (Sketch B).

Questions:
1. What type of energy was transformed into the air flow in activity A?
2. Which way would the wind blow if the fan were turning in the opposite direction (in activity A)?
3. What type of energy did the air flow in activity B create?
4. What is needed to turn the paper mill in activity B?

Explanation:
   In activity A electric energy was transformed into mechanical energy: the moving of the paper fan. This in turn made the air flow, which is kinetic energy of the air molecules. If the fan would turn in the opposite direction, the air flow would also be reversed.
   With activity B, the kinetic energy of the air flow is transformed into mechanical energy: the turning of the paper mill. In order for this mill to turn, wind or a moving flow of air is needed.

FLOWING AIR   WIND ENERGY
                                                   BERNOULLI´S PRINCIPLE

2.15. THE LEAPING EGG

Materials: 1. Two identical wine glasses (or plastic cups).
             2. A hard-boiled egg (or ping-pong ball).

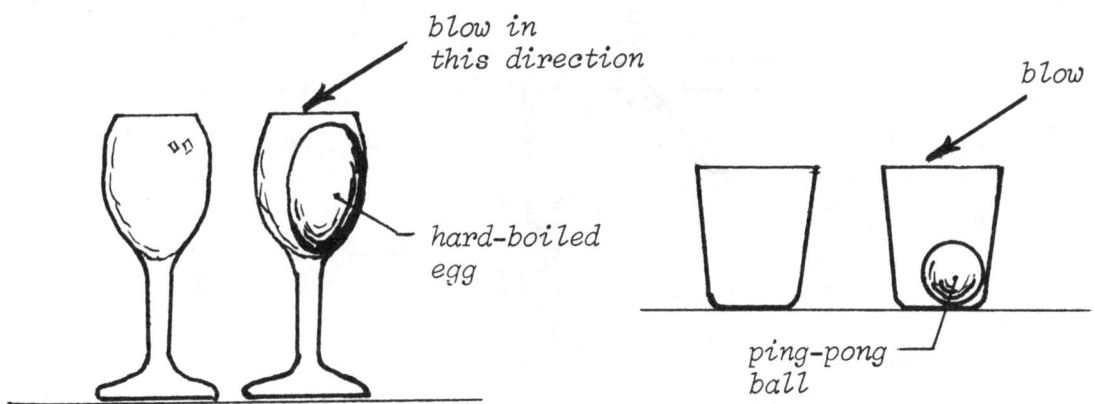

Procedure:
1. Place the two wine glasses or cups about 2-3 cm apart on the table and secure them down with tape or just hold them down.
2. Put the hard-boiled egg in one of the glasses (or the ping-pong ball in one of the cups) and ask the audience the question:
   "How can I move the egg (or ball) from one glass into the other without touching the egg and leaving the glasses as they are?"
3. Most people will respond with: "It´s impossible!"
   Now blow a short and hard puff obliquely into the far side of the wine glass that holds the egg or ball and watch the egg leap! (It may take a few practice blows to make the egg leap successfully).

Questions:
1. What makes the egg or ball jump out of the first glass?
2. Why would the egg or ball not go toward the one that blows?
3. What does flowing air create?
4. How far apart can we place the second glass in order for this one still to catch the leaping egg or ball?
5. What would happen if we blew on the near side of the egg or ball?

Explanation:
   Blowing obliquely into the far side of the glass builds the pressure on that side. It pushes the egg or ball out of the glass and the flowing air above both glasses guides the egg towards the second glass, because the flowing air actually creates a lower pressure. The harder we blow, the farther the second glass can be placed to catch the leaping egg. (Warning: <u>do not</u> try this activity with a raw egg!)

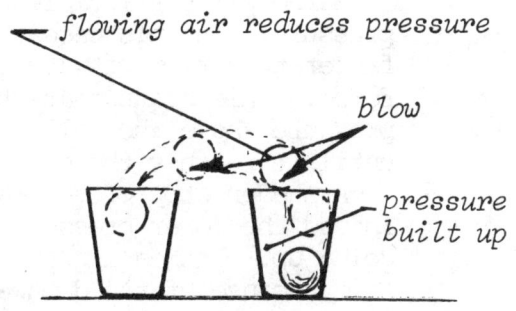

FLOWING AIR

FLOWING FLUID
BERNOULLI´S PRINCIPLE

2.16. LIFT A MOTH BALL WITH STREAMING WATER

Materials:
1. A small plastic vial (medicine vial or narrow jar).
2. A moth ball (or raw egg plus a small drink glass).
3. A water tap above a sink.

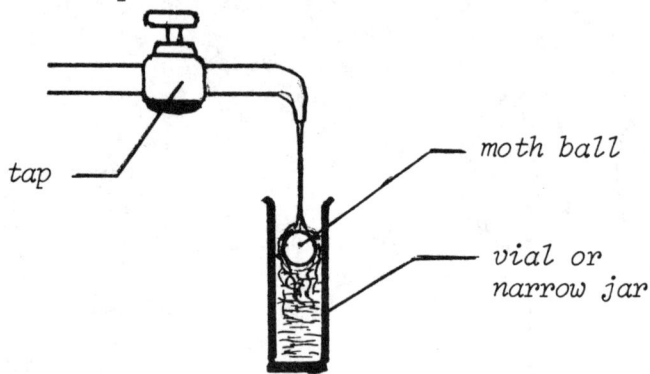

Procedure:
1. Place the moth ball in the vial (or egg in the small glass), then ask the students: "How can I lift the moth ball (or egg) up without turning the vial or glass upside down?"
2. Now open the tap and let the water run with a smooth medium size flow (this should be tried out beforehand so that you know what size stream you need for the lifting power).
3. Hold the vial with the moth ball or glass with the egg under the water stream, such that the water falls directly on top of the ball (or egg). Observe the ball (or egg) rise to the water surface!

Questions:
1. What made the moth ball rise up in the water?
2. Does a moth ball sink or float in water?
3. What happens if the water flow is suddenly stopped?
4. Would the moth ball keep floating if the water keeps running?
5. What would happen if the water stream was not falling head on?
6. What would happen to the ball if we used a larger/wider jar?
7. What happens if we used a stronger or weaker flow of water?
8. After the ball got lifted to the top, what would happen if the water stream were increased or decreased?
9. What happens to the ball if the vial was held in a slanted position?

Explanation:
   Bernoulli´s Principle states: **the faster the fluid flow, the lower the pressure.** In this case the fluid is the water streaming out of the tap. The faster this flow of water, the lower the pressure, but the larger the downward force. This means that there is an optimum flow where it creates a low enough pressure above the ball. This makes the ball rise up with the water flow until it reaches the surface.
   A sudden stop of the water flow will make the moth ball sink, showing that indeed the lower pressure created by the water flow was the cause for the ball going up. Increasing the water flow will also make the ball sink, but this time because of the downward force of the water.
   A technological application is the Venturi tube, a measuring instrument for fluid speed based on this same Bernoulli´s Principle.

FLOWING AIR   FLOWING FLUID
                                                                APPLICATIONS

2.17. THE SELF-PRIMING SIPHON

Materials: 1. A short wide glass tube, open on both ends (diam 2.5 cm,
                length about 8-10 cm long).
             2. A medicine dropper (or glass tube drawn at one end).
             3. One-hole stopper with a short glass tube.
             4. Two-hole stopper (both stoppers fitting in the wide tube).

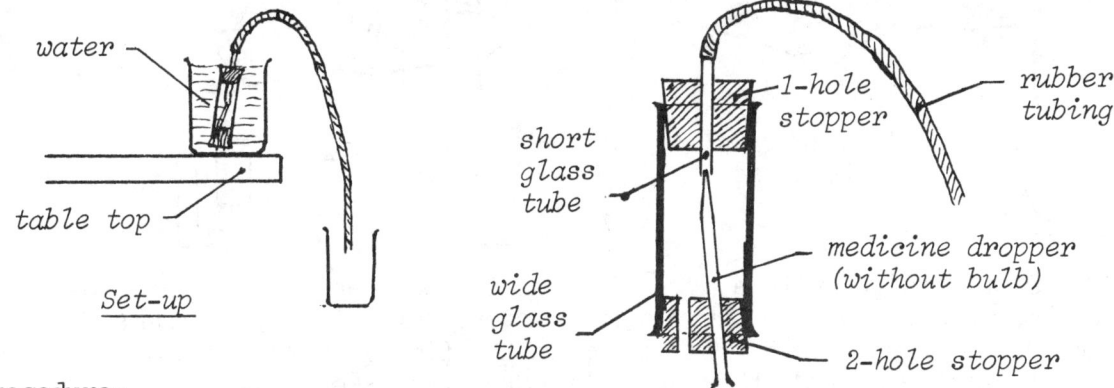

Procedure:
   1. Insert a short glass tube into the one-hole stopper and connect a long
      rubber tubing to the glass tube.
   2. Take the rubber off from the medicine dropper and insert the glass tube
      into one of the two holes of the two-hole stopper.
   3. Fit both stoppers into the wide glass tube, which is open on both ends,
      and make sure that the drawn end of the medicine dropper extends into
      the opening of the glass tube of the 1-hole stopper (see Sketch B).
   4. Plunge the wide tube with stoppers and all in a large water container
      on the table and bring the end of the rubber tubing below the table top.
      (the extent to which the drawn tube goes into the top glass tube may
      need some adjusting).

Questions:
   1. What is the cause for this siphon to start by itself?
   2. What happens to the air inside the wide tube at the moment that it is
      plunged into the water?
   3. What does the drawn tube create when water is pushed through it?
   4. How is a regular siphon started?

Explanation:
   At the moment that the wide tube is immersed into the water, water comes up
in the wide tube through the open hole in the 2-hole stopper. This causes the
air inside the wide tube to flow through the small tube in the 1-hole stopper.
Simultaneously, water is coming up through the drawn tube, and a small jet of
water is created. Both flows of air and water create a lower pressure and
thus an upward surge of the water, enough to overcome and fill the hump of the
rubber tubing. Once this hump is filled with water and the wide tube is also
filled with water, the siphon will continue running.
   A regular siphon has to be primed, either by sucking through the rubber
tubing, or by filling the tubing with water - by immersing the tubing in the
water, pinching the end and pulling it quickly over the edge of the container
and letting go of the tubing.

2.18. THE SELF-DIRECTING CARDS

Materials: 1. A deck of playing/bridge cards.

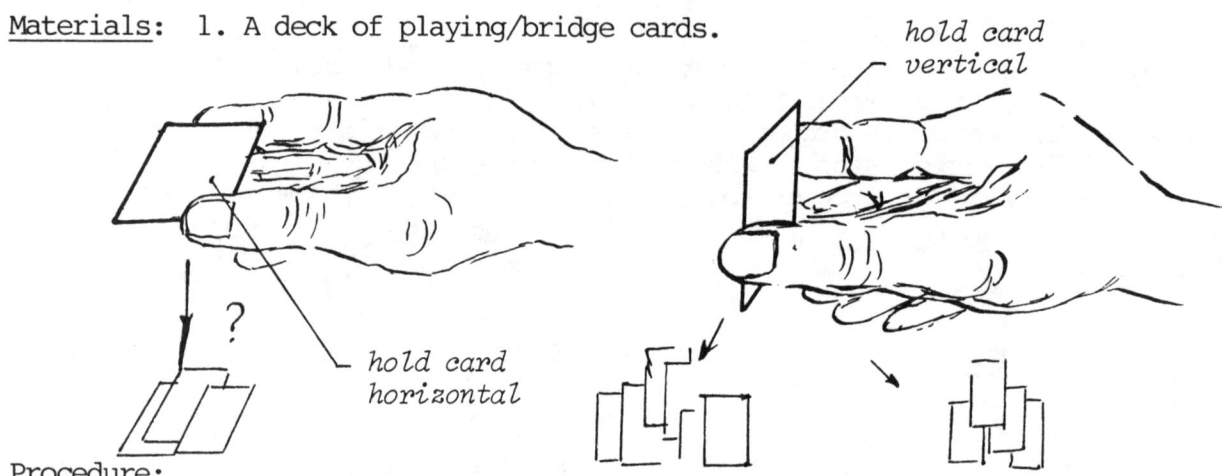

Procedure:
1. Stand up straight with a deck of cards in your hand. Hold one card at a time in front of you between two fingers horizontally and ask: "Where do you think this card will end up on the floor? Straight underneath it or off to the side?", then drop the card. (See Sketch A).
2. Hold the next card between thumb and index finger by the short sides, vertically with a slight slant to the left; ask the same question, then drop the card (See Sketch B).
3. Hold the next card the same way, but now slant it slightly to the right, ask: "Where will this card end up on the floor?", then drop the card.

Questions:
1. Where did the first, second and third card end up on the floor?
2. What made them fall so differently?
3. Which of the cards stayed steady while falling?
4. How could you separate the cards into three equal piles by dropping them to the floor? (This question could be used before doing the demo).
5. What do you think would happen if you held the cards by the long sides and let them fall from the three different positions?
6. What do you think the cards would do when dropping two at a time?
7. Where would the cards end up on the floor, if the cards were held in a random fashion before dropping?

Explanation:
When letting the card fall horizontally, it will stay horizontal and end up quite plumb underneath the position before falling. This is caused by the even flow of the air around the card. When holding the card vertically, the leading edge of the card cuts into the air, encounters a slightly higher pressure or resistance on one side, flips up, and the card starts rotating. Once it spins it will keep on spinning. The ones slanted to the left will rotate in a clockwise direction, and the ones slanted to the right will rotate counter-clockwise (from the point of view of the performer).

This demonstration must be done in a draft-free room. It will not have the same results outdoors. On a windy day, dropping the cards will have the same random result as holding the cards at random positions indoors before dropping them.

CHAPTER 3

WHAT FACTORS INFLUENCE THE WEATHER?

OBJECTIVES

After dealing with and studying the concepts and sub-concepts in this chapter, the students should be able to:

a. recognize the correct explanation of the influence of atmospheric pressure, wind speed and direction, and humidity on the weather;

b. explain correctly in their own words how atmospheric pressure, wind speed and direction, temperature and humidity are measured;

c. distinguish true from false statements concerning each one of the variables influencing the weather;

d. identify the correct explanation concerning variables that influence the climate, length of days, and the sun's heat.

3.1. A CLOUD IN A BOTTLE (I)

Materials:
1. A colorless transparent soda pop bottle.
2. A bicycle hand pump & bicycle tire valve.
3. A 1-hole stopper with short glass tube (fitting the bottle).
4. A short piece of thick walled rubber tubing.

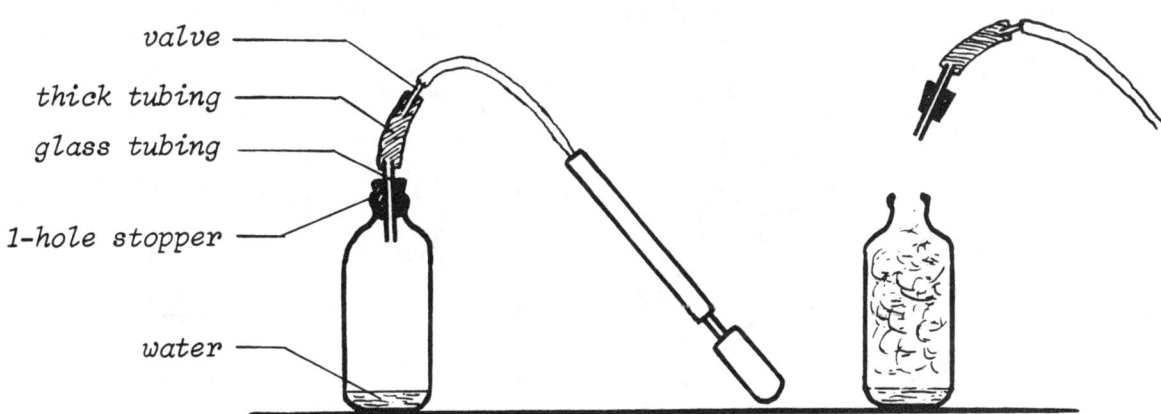

Procedure:
1. Place a few millilitres of water in the bottle.
2. Insert the 1-hole stopper tightly in the bottle opening and connect the bicycle pump as shown in the sketch.
3. Pump air into the bottle and let the students observe the bottle very carefully, especially the moment when the stopper pops outs.
4. When the cloud is not too visible, drop a lit match in the bottle or blow a little smoke in the bottle, then repeat point 3.

Questions:
1. What do we need the water in the bottle for?
2. At the moment that the stopper pops out, what does the air inside the bottle do?
3. At the moment that the stopper pops out, what is the temperature of the air inside the bottle doing?
4. What will water vapor do when it is cooled?
5. Why do we get a better cloud with a little smoke in the bottle?
6. Where do we see applications of this principle (with the smoke) in our daily life?

Explanation:
At the moment that the stopper pops out, we get a sudden low pressure in the bottle. The miniature atmosphere in the bottle had a high humidity because of the presence of the water. At the moment that the stopper pops, the air in the bottle rushes out, the temperature drops as air expands, and the water vapor condenses. This condensation process is enhanced by the presence of smoke, which is a suspension of solid particles in the air acting like nuclei for the water vapor to condense on. This is the reason why it is more likely that it will become foggy or rainy over polluted areas, such as larger cities with heavy industry, than over rural areas that have no industry.

3.2. A CLOUD IN A BOTTLE (II)

Materials:
1. A wide mouth gallon jar (pickle jar).
2. An uninflated rubber beachball.
3. Some heavy rubber bands or masking tape.

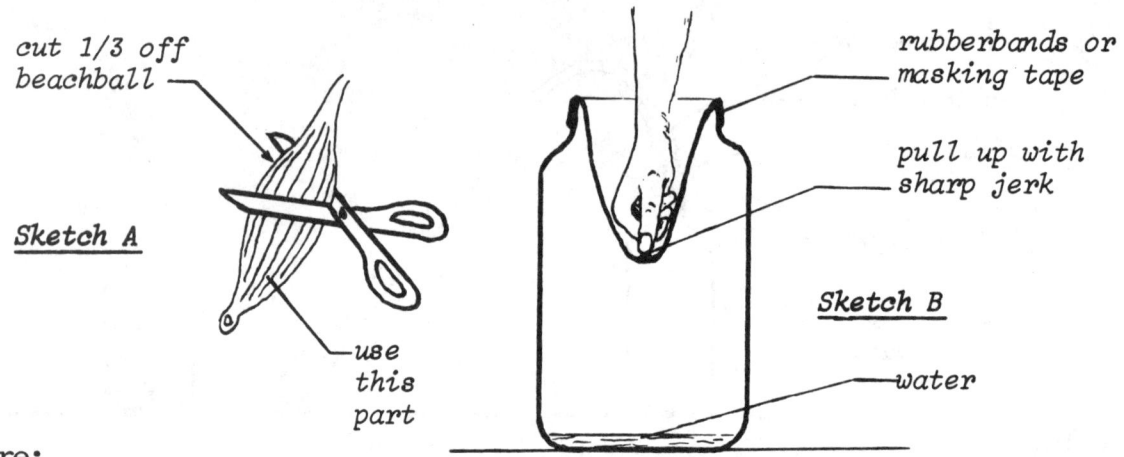

Procedure:
1. Place a few millilitres of water in the gallon jar.
2. Drop a lit match or blow some smoke in the jar.
3. Cut one third off of the beachball (Sketch A) and turn the larger part inside out. Place it over the jar, such that the cut edge slips over the rim of the jar, then fasten it with a few heavy rubber bands or masking tape around the jar mouth (make sure it is airtight).
4. Pull the rubber sheet up with a short sharp jerk and observe what is happening in the jar (see Sketch B).

Questions:
1. What function does the water in the jar have?
2. What do we need the smoke in the jar for?
3. What is suddenly increased by pulling the rubber sheet up?
4. How does the pulling of the rubber sheet affect the temperature?
5. Why is it usually more cloudy and foggy over industrial areas as compared to rural areas?
6. What would happen if the rubber sheet was not put on the jar airtight?
7. Would the cloud be more easily formed if the bottle were warm or cold?

Explanation:
The water in the jar is necessary to create a miniature atmosphere with a high humidity. The smoke provides the nuclei on which the water vapor can condense when the rubber sheet is suddenly pulled up. This pulling up of the rubber sheet causes the volume of the air inside the jar to suddenly increase, and thus the temperature to decrease. As the humidity is very high, the air contains the maximum capacity of water vapor. The slightest drop in temperature, therefore, is enough to cause the water vapor to condense and produce a mist or fog.

WEATHER   PRESSURE

3.3. THE JAR BAROMETER

Materials: 1. A wide-mouth gallon jar (pickle jar).
           2. A large balloon, 2 straws, a straight pin.
           3. A ruler, masking tape, a thermometer.

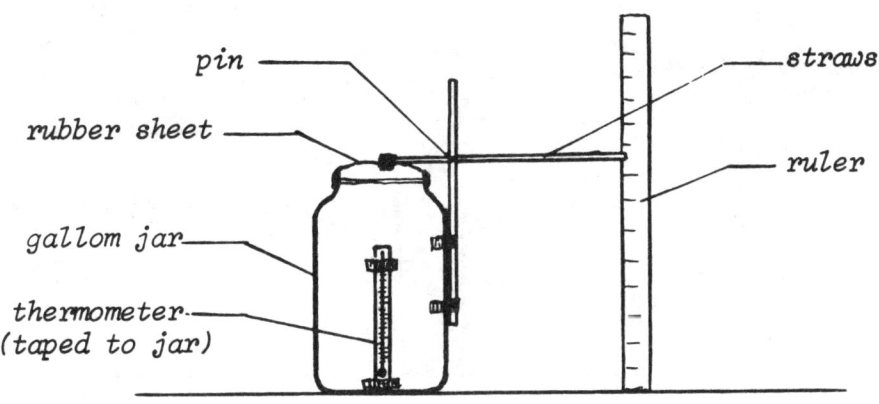

Procedure:
1. Blow the balloon up (preferably a day before use or a few hours before use) so that the rubber is stretched.
2. Cut a piece off the stretched balloon (large enough to cover the jar opening) and tape it airtight to the jar mouth.
3. Tape two straws pivoted with a pin to the center of the rubber sheet and to the side of the jar (see Sketch).
4. Place the jar such that the straw indicator points along a ruler.
5. Place a thermometer close to the jar or tape it to the jar, and read the pressure off the ruler at the same temperature, twice a day.

Questions:
1. When the atmospheric pressure gets higher, will the reading on the ruler be higher or lower?
2. Why is it necessary to stretch the rubber sheet beforehand?
3. Why does the reading of the pressure have to be done at the same temperature? What would a higher temperature do to the reading?
4. What does an increasing atmospheric pressure indicate for the weather forecasting?
5. What does a decreasing atmospheric pressure indicate?

Explanation:
   This simple barometer is only good for measuring relative atmospheric pressure changes. The ruler could be calibrated by listening to radio announcements of atmospheric/barometric pressures. The rubber diaphragm is more sensitive to pressure changes, when it is more flexible; that is why it needs prestretching. The reading of this simple instrument has to occur at the same temperature, as an increase in temperature will result in expansion of the air inside the jar, and thus a higher pressure in the jar and a lower reading. Cooling of the jar results in contraction of the air in the jar and thus a higher reading.

WEATHER — PRESSURE

3.4. THE BOTTLE BAROMETER

Materials:
1. A bottle with a long narrow neck.
2. A cylindrical tall jar & thermometer.
3. A grease marker (a felt marker will do).

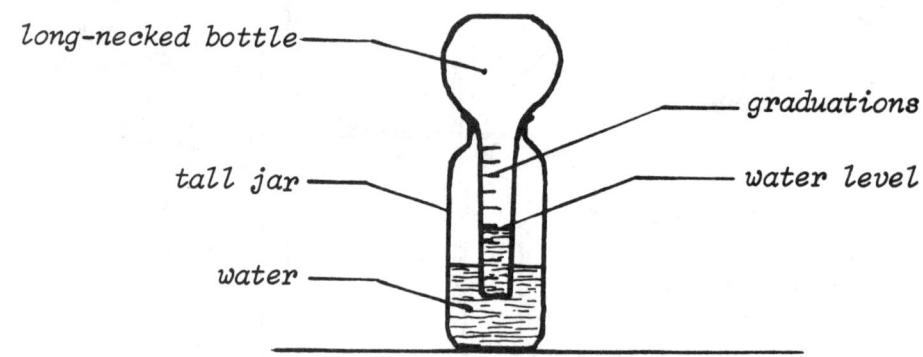

Procedure:
1. Find a bottle with a long narrow neck and a jar that can support the wider part of the bottle when inverted in the jar.
2. Draw horizontal even graduations on the bottle neck with a marker and assign numbers to the graduations (hold bottle upside down).
3. Fill the jar half full with water.
4. Pour some water in the bottle and invert it quickly into the slanted held jar, so that the water level in the bottle neck will be slightly higher than that in the jar.
5. Position the bottle barometer in a place where the temperature does not fluctuate, and read off the water level in the neck. Do this twice a day and compare readings.

Questions:
1. What should the water level in the bottle neck do if the atmospheric pressure increases? If it decreases?
2. Why does the barometer have to be put in a place where the temperature does not fluctuate?
3. At what point do these bottles function as a barometer?

Explanation:
When the atmospheric pressure increases, the water level in the jar is pushed down and thus pushed up in the bottle neck. When the pressure decreases, the water level in the jar goes up and thus the water level in the bottle neck goes down.

This simple barometer will only work when the surrounding temperature stays constant. When the surrounding temperature increases, the bottle and the air inside it warms up and expands, pushing and exerting pressure against the water resulting in a lower reading of the water level. A cooler surrounding temperature has the opposite reaction as a result.

Over a longer period of time, the water of the jar will evaporate and water has to be added to bring the water level back up to its original level. Otherwise it ceases to function.

WEATHER WIND

3.5. WHICH IS THE BEST WINDVANE?

Materials: 1. Drinking straws, straight pins, paperclips.
            2. Cellotape, paper cards, clay, scissors.
            3. An empty soda pop bottle.
            4. A small electric fan.

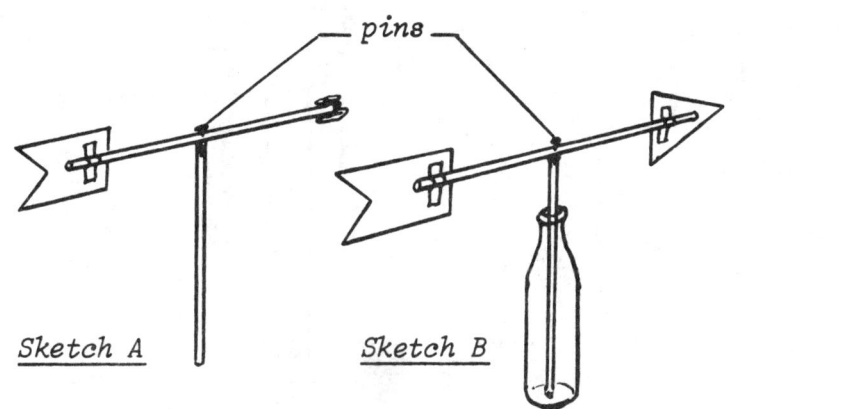 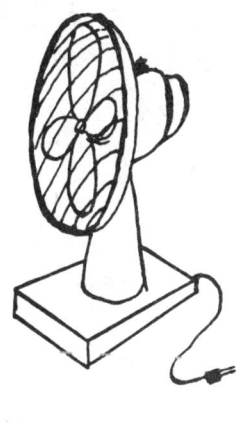

Procedure:
1. Provide students with straws, paper cards, pins, scissors, cellotape, paper clips (pop bottles if available).
2. Ask students to construct a good working windvane (the best one will turn fastest into the wind direction).
3. Ask students what the conditions are for the windvane to work.
4. If students have trouble making one, give them clues by first constructing windvane A (se Sketch A):
   Tape a piece of card to the end of a straw and insert some paper clips to the other end, find the balancing point (by letting it balance on your finger) and stick a pin through the straw at that point. Let this straw swivel in a vertically held straw. (set in a pop bottle: Sketch B)

Questions:
1. What is the difference between the front and the rear part of the windvane? Is there a difference in area?
2. What must the horizontal bar be able to do to turn in the direction of the wind?
3. Where must the pivot of the arrow (horizontal bar) be located?
4. What can be used as a base or stand for the windvane?
5. How would you construct a windvane that points where the wind is blowing to (opposite to where it comes from)?

Explanation:
   In order for a windvane to work, one end must be wider in surface area than the other end. By taping a cut-out card to one end of the straw, we increase the area at this end, but the weight at this end is thereby also increased. To counterbalance this, paper clips are inserted at the other end. This makes it possible to place the pivot in the center of the bar.

3.6. HOW FAST DOES THE WIND BLOW?

Materials:
1. Paper cards (3 x 5"), cellotape, scissors.
2. An empty plastic bottle with a narrow neck.
3. A wooden stick (round), knife, nail.
4. A small electric fan.

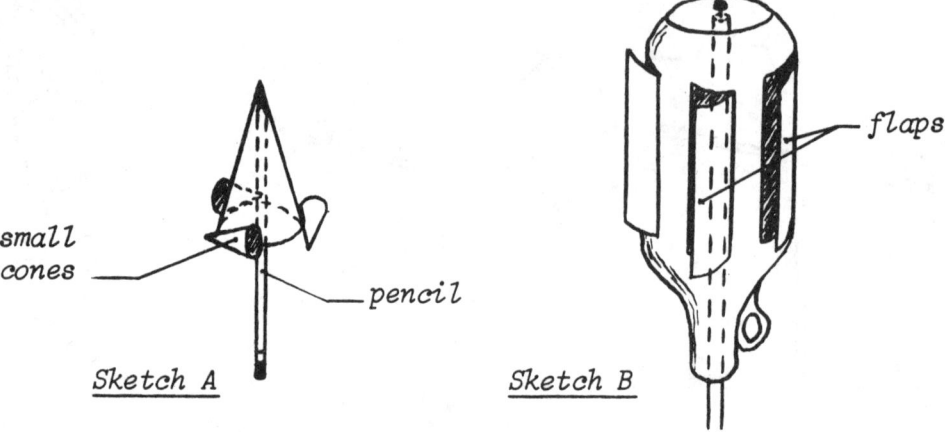

Sketch A   Sketch B

Procedure:
1. Fold four paper cards into cones and cut them into one large and three small cones. Tape the small cones to the edge of the larger cone, pointing in the same direction. Let the large cone swivel around a pencil or pen by holding it in the wind of the fan (see Sketch A).
2. Cut slits in the side of the plastic bottle and bend the flaps slightly outward. Make a small hole in the bottom of the bottle, turn it upside down on the stick and nail it loosely on the end of the stick so that the bottle can swivel easily (see Sketch B).
3. Color one of the small cones and one of the flaps of the plastic bottle red or black. To calibrate the instrument, hold it out a car window when the car moves with a speed of 20, 30, 40, 50 km/h and count the number of revolutions for each of the velocities (do this on a day when there is no wind blowing).

Questions:
1. Will anemometer A turn in the same direction as B (see Sketch)?
2. How can we measure the wind speed with this anemometer?
3. Does it matter where the wind comes from when using this anemometer?
4. In calibrating the instrument, why do we have to wait for a day when there is no wind?
5. What makes the anemometer turn only in one direction?

Explanation:
The anemometers as shown in the sketch will turn in opposite directions. The simple anemometer turns only in one direction when held in a stream of air, because the air resistance on one side of the cone or cylinder is less than on the opposite side. After calibrating the instrument it is just a matter of counting the number of revolutions to tell how fast the wind blows. A day with no wind should be chosen to calibrate it, as the existing wind will otherwise affect the reading.

WEATHER — WIND

3.7. FIND WIND SPEED AND DIRECTION TOGETHER

Materials:
1. A nylon stocking & a medium size wire (about 80 cm).
2. A narrow test tube & a small electric fan.

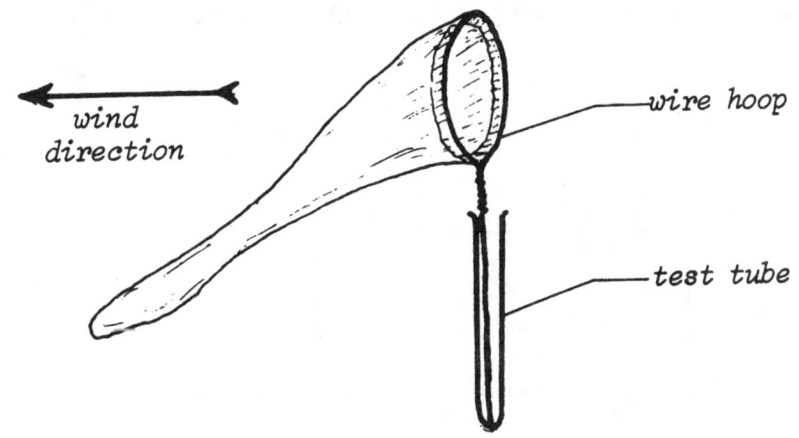

Procedure:
1. Make a hoop (circle) with a 10-15 cm diameter from one end of the wire, and let the other end protrude for at least 20 cm.
2. Fold the edge of the opening of the stocking around the hoop and tape it secure to the hoop.
3. Stick the protruding end of the wire into the test tube and make sure it can swivel freely in the tube.
4. Bring the "wind-anemo-vane" in the wind stream of the fan or place it outside in the ground on a windy day.

Questions:
1. What other material can be used instead of the stocking?
2. Why does the stocking turn away from the wind?
3. How can we tell the speed or velocity of the wind?
4. How would you calibrate this instrument?
5. What is a wind velocity meter called?
6. Why is the measuring of the wind speed important?
7. What do we need in order to forecast the weather?

Explanation:
   The stocking or strips of paper in the wind will act just like a flag in the wind. It will wave away from the wind, similarly the stocking will point in the direction where the wind is going to and not where it is coming from. The windvane, on the other hand, is pointing into the wind, opposite to what the stocking is doing.
   The stocking may be called a "wind-anemo-vane" because it does both: indicating the direction and the speed of the wind. The instrument measuring the speed of the wind is called: anemometer. The more horizontal the stocking, the stronger the wind is.
   Measuring the wind speed and direction is very necessary in the forecasting of the weather.

3.8. CREATE WIND CURRENTS

Materials:
1. An empty aquarium or any large transparent container.
2. Thin cardboard (bristol board or manila folders).
3. Black bristol board or black carbon paper.
4. A small candle, an old rag, matches, masking tape.

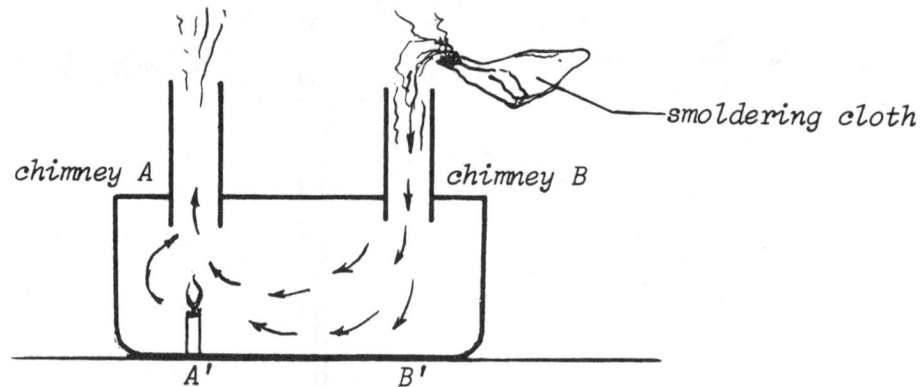

Procedure:
1. Make a cover for the aquarium with two holes, and also two fitting pipes, that will act as chimneys, from the thin cardboard.
2. Cover the backside of the container with black paper, so that the smoke currents can be more easily seen (use a spotlight if available).
3. Fix the candle under one of the holes in the bottom of the aquarium and light the candle.
4. Place the cover over the container and tape the edges airtight to it.
5. Make a smoke source from the rag (by burning it and letting it smolder) or light a cigarette, and hold the smoke close to chimney B (see Sketch).
6. Let the smoke fall into chimney B (do not hold the smoke source directly above the opening, but blow the smoke over the opening) and let the students observe the smoke currents.

Questions:
1. What makes the smoke fall in chimney B?
2. Why does the air rise through chimney A?
3. What is the purpose of the burning candle?
4. How can we create wind currents without the burning candle?
5. What places on earth can spot A' and B' be compared with?

Explanation:
The burning candle is a source of heat. It heats the air above and around the flame, and heated air rises. This makes the air rise through chimney A, and because air goes out of the container, air must replace it, and this comes in through chimney B, taking the smoke with it. Spot A' can be compared with a hot place on earth (f.i. the tropics) and B' with a cold place (the poles). The air currents or winds are created this way between such places on earth. Instead of the candle we can also use an ice cube above chimney B. Cold air will move down and create the same currents.

3.9. MAKE YOUR OWN THERMOMETER

Materials:
1. A small soda pop bottle & food coloring or ink.
2. A transparent plastic straw or glass tube.
3. A clump of clay or a one-hole stopper.
4. An alcohol burner & stand, or a candle.

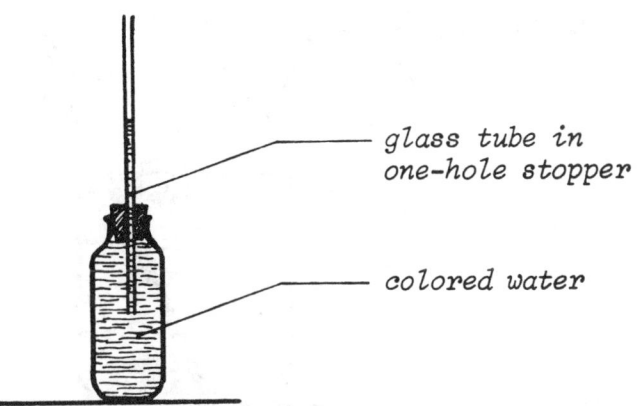

Procedure:
1. Fill the pop bottle full with cold water and color it with a few drops of food coloring or ink.
2. Put some clay around the straw or insert the glass tube in the one-hole stopper (make sure that only a short end will go in the bottle).
3. Immerse the straw or glass tube in the water-filled bottle, insert the stopper and mark off the water level in the straw or tube.
4. Heat the pop bottle or cool it down with an ice cube and let students observe the change in water level.

Questions:
1. Why does the straw or glass tube connection to the bottle have to be watertight?
2. What will heating or cooling do to the water level in the tube?
3. How similar is this instrument to a real thermometer?
4. How could we calibrate this simple instrument?
5. Could this instrument actually be used to measure temperatures?

Explanation:
The small soda pop bottle can be compared to the bulb of a regular thermometer. The larger this bulb, the more sensitive the thermometer is: thermometers for measuring body temperature have large bulbs and narrow tubes. Heating the bottle will result in heating the liquid and thus expanding it, causing the higher liquid level in the tube. Cooling the bottle causes the liquid to contract, resulting in a lower reading level. Calibrating of the bottle thermometer can be done by immersing the bottle in melting ice (0° C), marking the water level; then leaving the bottle at room temperature for some time (20° C), marking the water level in the tube, then dividing the space between the two marks by 20 for each degree Celsius.

WEATHER — DEW POINT TEMPERATURE

3.10. AT WHAT TEMPERATURE IS DEW FORMED?

Materials:
1. A beaker or regular drinking glass.
2. A thermometer (-5° to 50° or 100° C).
3. Ice cubes.

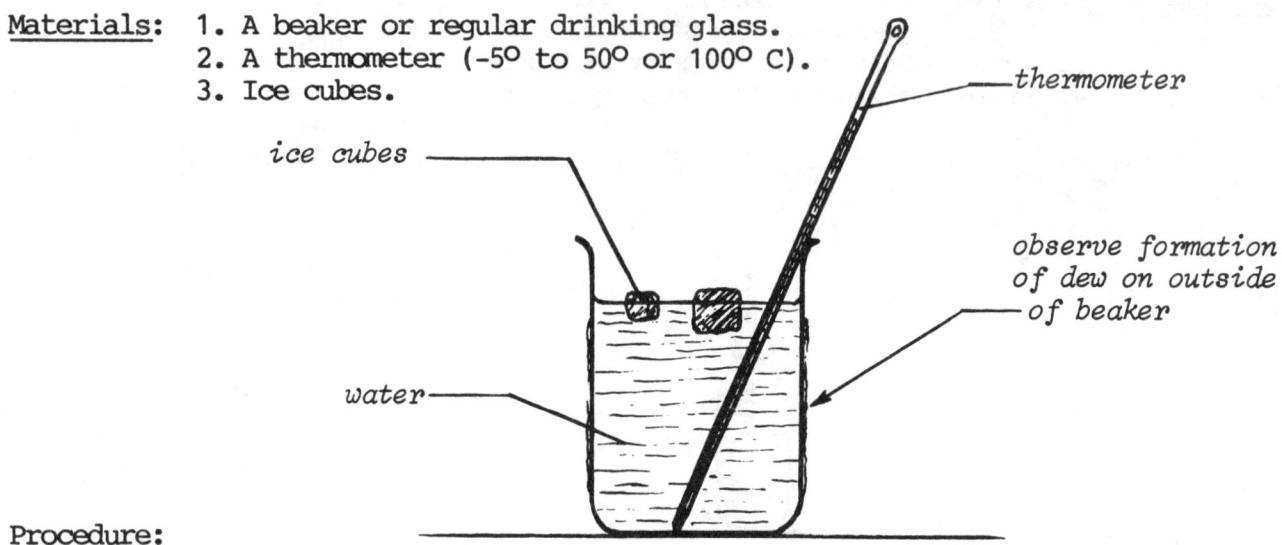

Procedure:
1. Fill the beaker almost full with water and add a few ice cubes.
2. Keep track of the temperature of the mixture and observe carefully the outside of the beaker or glass.
3. Stir the mixture once in a while with the thermometer (carefully) and note the temperature at which the dew is first formed on the outside of the glass (look for a dullness on the glass).

Questions:
1. Where does the dew on the outside of the glass come from?
2. At what temperature was the dew formed?
3. Is there always the same amount of water vapor in the air?
4. Will the dew point be higher or lower on humid days?
5. Where do we encounter a similar event inside the house in winter?

Explanation:
The ice cubes in the water are cooling the mixture slowly down and depending on how much moisture there is in the air, the sooner or later the water vapor will condense on the outside of the beaker; thus the higher or lower the dew point will be respectively.

The drier the air around the beaker is, the lower the temperature has to be, before any water vapor will condense on the beaker's surface, and thus the lower the dew point.

Just after a rainfall the air is very humid, meaning that this percentage of water vapor in the air is high, and thus the dew point is also high. On dry days or in areas of low humidity, the temperature has to drop quite low before any dew drops are formed on the grass in the morning.

WEATHER
WATER CYCLE
TEMPERATURE

3.11. RECYCLE THE WATER

Materials: 1. Two shallow trays or regular dinner plates.
2. Electric water kettle.

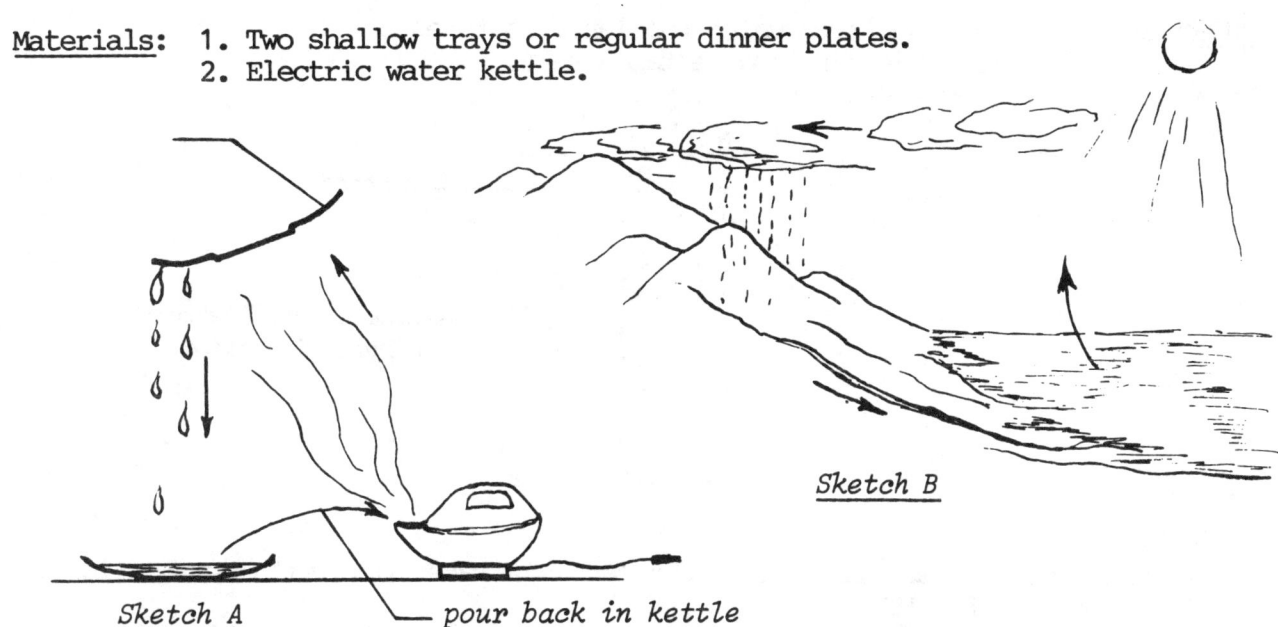

Sketch A — pour back in kettle

Sketch B

Procedure:
1. Heat water in a kettle until boiling.
2. Suspend a tray or plate in the steam or water vapor that comes from the kettle, and place the second plate under the first to catch the water drops (condensate). (Put cold water in the first plate).
3. Pour the water from the second plate back in the kettle to complete the cycle.

Questions:
1. What made the steam or water vapor turn back into water?
2. Would the steam also condense if the top plate was filled with hot water?
3. What are conditions for clouds to turn into rain?
4. What makes the water evaporate in nature?
5. What can the pouring back of water from the lower plate into the kettle be compared with in the natural water cycle?

Explanation:
The steam or water vapor is produced in the kettle, because the water in the kettle is heated. When the steam hits the cold plate, it condenses back into water. The kettle can be compared to the ocean water being heated by the sun rays. The water vapor accumulates in the higher layers of the atmosphere to form clouds. When these clouds move into colder regions, they saturate the air with water vapor and the cold temperatures turn it into rain--**condensation**. This may be compared to the steam hitting the cold surface of the first plate. The rain water in nature flows into rivers and these flow back into the ocean (Sketch B).

WEATHER

FROST FORMATION
TEMPERATURE

3.12. THE FROSTY CAN

Materials:
1. A tin can (from which the label is taken off).
2. Crushed ice and coarse table salt.
3. A thermometer and stirrer.

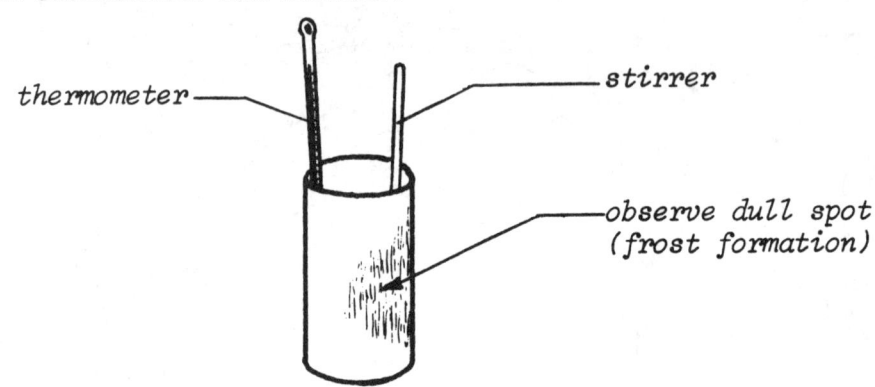

Procedure:
1. Fill the tin can, from which the label is taken off, with crushed ice and a handful of salt.
2. Place the thermometer in the can and stir with a separate stirrer.
3. Have students read off the temperature every half minute.
4. Have other students observe carefully the outside surface of the can and indicate to those observing the temperature, the moment that they observe frost formation (when the shiny surface gets dull).
5. A further extension of the activity might be the graphing of the data of observed temperature and the time elapsed.

Questions:
1. What is the purpose of the salt mixed in with the crushed ice?
2. How low does the temperature inside the can have to be in order to form frost on the outside?
3. What do you expect the temperature of the ice and salt will do after the moment that frost is formed?
4. After observing the shape of the graph after frost formation, what is the process of frost formation doing: absorbing or giving off heat?
5. Where does the frost come from; what is needed in the air?

Explanation:
The salt in the crushed ice makes the ice melt in the beginning, but brings the temperature down below the normal freezing point of water (0° C). The moisture in the air, hitting the cold surface of the can, turns into the solid state (frost) without passing the liquid state: **sublimation**. This process, just like condensation, is giving off heat, which is why the graph levels off after frost formation (see Graph on the right).

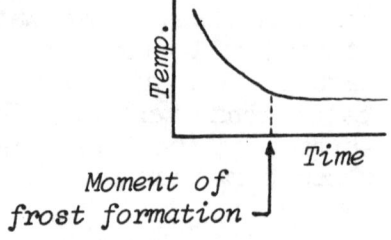

WEATHER   HUMIDITY

3.13. THE WET AND DRY THERMOMETER

Materials: 1. Two identical thermometers.
            2. A small swab of absorbent cotton, or piece of cotton cloth.
            3. A small beaker, two clamps and stand.

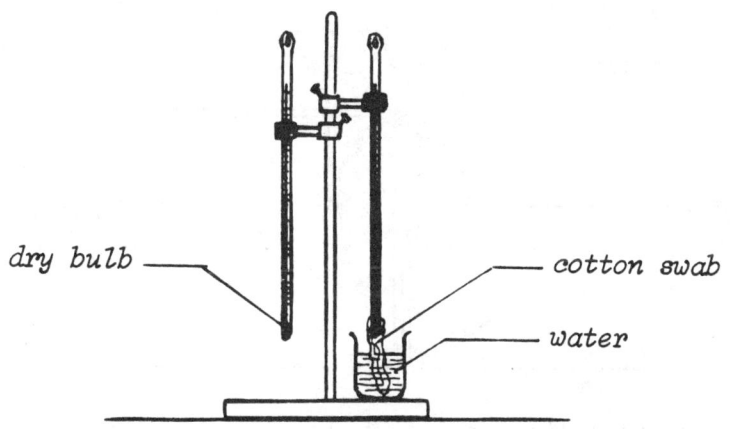

Procedure:
1. Clamp two thermometers close to each other on one stand.
2. Have students read off both temperatures.
3. Wrap one thermometer bulb with a small swab of absorbent cotton.
4. Place some water in the small beaker and keep the cotton around the bulb wet at all times.
5. Have students read off both temperatures again.
6. Determine the relative humidity from the humidity chart.

Questions:
1. Why is there a difference between the dry and the wet bulb temperature?
2. What causes the wet bulb temperature to decrease?
3. What effect does a higher relative humidity have on the difference between wet and dry bulb temperatures?
4. What can we say about the relative humidity when the difference between the two temperatures is large?
5. What can we say about the relative humidity when the difference between the two temperatures is small?

Explanation:
   The wet bulb temperature decreases, because the evaporating water withdraws heat from its environment. The drier the atmosphere is, the faster the water will evaporate, and thus the lower this wet bulb temperature gets, resulting in a greater difference between the two temperatures. When the relative humidity is high, the dry bulb can almost be considered to be a wet bulb; also the rate of evaporation of the water is much slower, because the air is almost saturated with water vapor. A smaller difference between the two temperatures therefore indicates a high relative humidity (see Humidity Chart in Appendix).

WEATHER   HUMIDITY

3.14. THE JAR HYGROMETER

Materials:
1. A wide-mouth gallon jar (pickle jar).
2. A large balloon, 2 drinking straws, a straight pin.
3. A ruler, adhesive tape, and a thermometer.

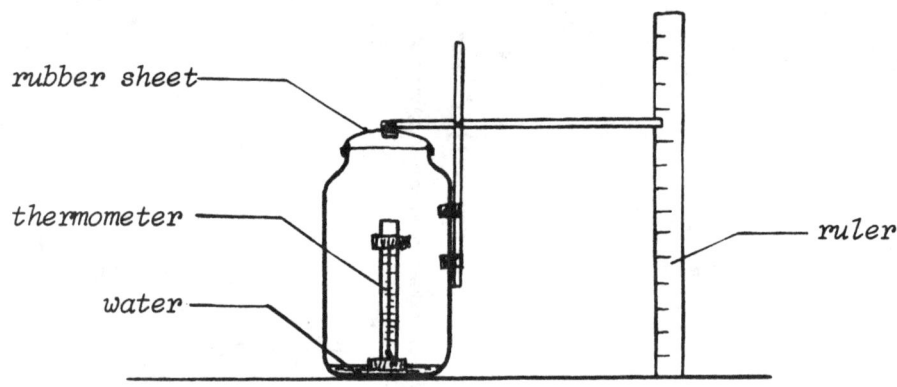

Procedure:
1. Place a few millilitres of water in the jar.
2. Blow up the balloon (preferably a day or a few hours before use), so that the rubber is stretched.
3. Cut a piece off the stretched balloon (large enough to cover the jar opening) and tape it airtight to the jar mouth.
4. Tape two straws pivoted with a pin to the center of the rubber sheet and to the side of the jar (see Sketch).
5. Place the jar such that the straw indicator points along a ruler.
6. Place a thermometer close to or tape it to the jar, and read the humidity off the ruler at the same temperature, twice a day.

Questions:
1. When the humidity gets lower, will the reading on the ruler be higher or lower?
2. What is the relative humidity inside the jar?
3. What is the absolute humidity as compared to the relative humidity?
4. How will an increase in temperature affect the relative humidity reading on this instrument?
5. What is a most comfortable relative humidity to maintain?
6. How can we increase the humidity in our homes in the winter?

Explanation:
The hygrometer only measures the relative humidity in the air. The absolute humidity is the actual amount of water vapor that the air holds at a particular temperature. The presence of the water in the jar makes the relative humidity 100%. This means that the air inside the jar cannot hold more water vapor: it is **saturated** with water vapor. This miniature atmosphere is being compared with the outside atmosphere. The less humidity in the atmosphere, the more the rubber diaphragm will bulge up, and thus the lower the reading. The term **relative** comes from this comparison with the maximum capacity of the air in holding water vapor.

A comfortable relative humidity is about 60 - 70%.

3.15. THE HAIR HYGROMETER

Materials:
1. A wooden cigar box, 2 drinking straws, pins.
2. A long human hair, string, and a weight (20 g).

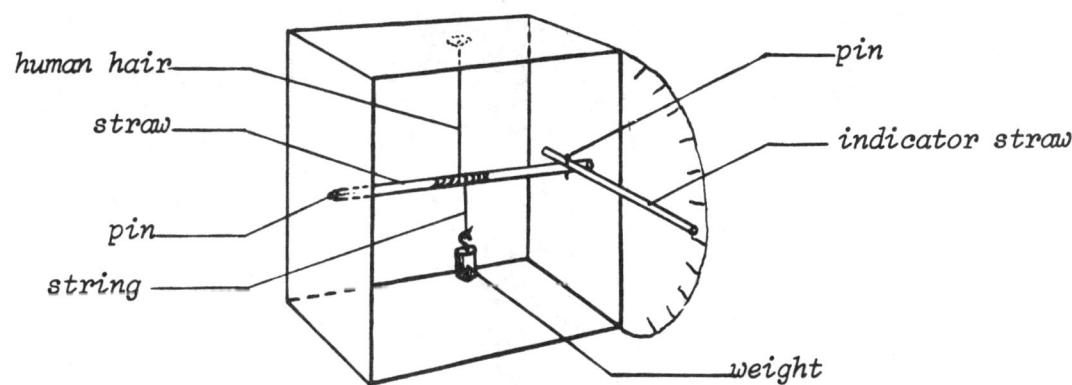

Procedure:
1. Attach a long human hair to the center of the topside of the box, and wind the other end around a horizontally rolling straw, turning freely around pins at each end (see Sketch) and tape the end of the hair to the straw.
2. Now tape a string or thread to the same straw and wind it such that the weight on the other end of the string will pull the hair straight.
3. Attach another straw perpendicular to the first one with a pin or tape. Attach a semi-circle cardboard to the side of the box closest to the indicator straw (draw some graduations on the edge of the semi-circle beforehand) (see Sketch).
4. Read off the relative humidity (position of the indicator straw) and compare with the readings announced on the radio.

Questions:
1. What does more or less humidity do to human hair?
2. Does higher humidity cause the indicator straw to go down or up?
3. What is the purpose of the weight?
4. How can we calibrate this instrument?

Explanation:
Human hair stretches when the air is humid and it shrinks when the air is dry. The weight just keeps the hair tight, thus when the hair stretches, the horizontal straw turns clockwise, and the indicator tip lowers. Lower humidity in the air causes it to move up higher.

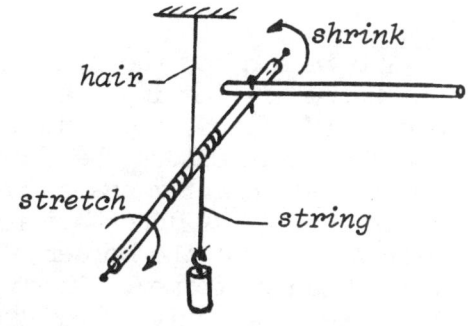

Calibrating of this instrument may be done by listening to radio weather reports and assigning the appropriate values to the graduations.

WEATHER   CYCLONE

3.16. MAKE A FIRE CYCLONE

Materials:
1. A old turntable (or a circular wooden board on marbles).
2. A metal mosquito wire screen (about 30 cm high to fit around the turntable (or the turning wood board).
3. A watch glass or evaporating dish, cloth, kerosene.

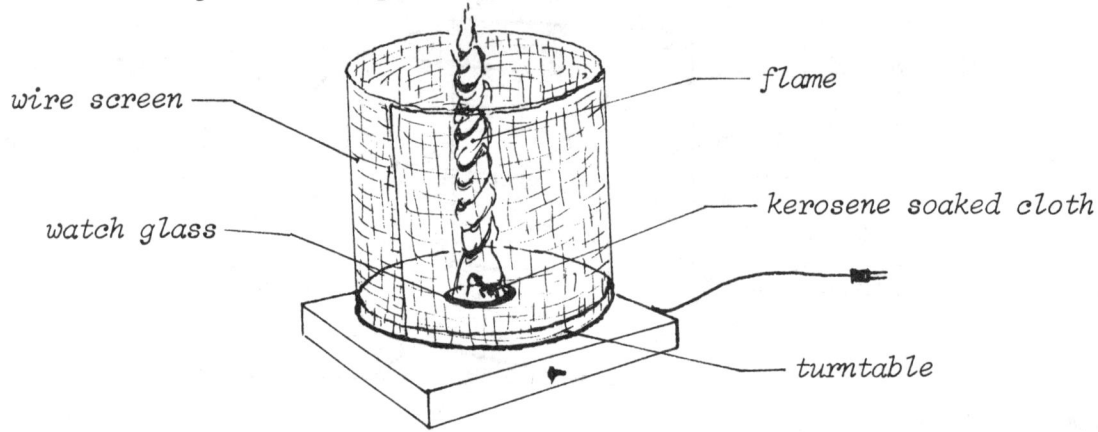

Procedure:
1. Roll the wire screen into a cylinder of about the diameter of the turntable and wrap it around the disk, fasten it with staples or with a wire.
2. Place the watch glass or evaporating dish in the center of the turntable.
3. Soak a small piece of cloth (cotton) in kerosene and place it on the dish.
4. Strike a match and set the cloth on fire, wait till the flame is a good size, then switch the turntable on (or start rotating the circular board).
5. Observe the flame shape!

Questions:
1. What hypotheses/inferences can you make to explain the event?
2. What does a flame need in order to sustain burning?
3. Where does the air come from to reach the flame?
4. In which direction do the air molecules move with respect to the rotation?
5. What shape did the fire take shortly after the rotation started?
6. What could we use instead of kerosene?
7. What event in nature does this resemble?
8. What is the reason for the flame to take that particular shape?

Explanation:
   The reason for the flame to be shaped like a **cyclone** is the direction from which the air feeding the flame has to come. Because of the rotating screen around the flame, the air molecules are given an initial **angular momentum** at the moment it enters the screen cylinder from outside. This angular or rotational momentum shapes the flame into this cyclone.
   In nature, cyclones are formed when two winds of opposing direction pass each other. At the border of the two passing masses of air, a **cyclone** or **vortex** can be formed. These cyclones are also called tornados or hurricanes. They are destructive wind storms that can occur over a large area, like hundreds of miles wide. The winds forming the vortex blow spirally inward, creating a low barometric pressure in its center. In the Northern hemisphere these cyclones rotate counter-clockwise and in the Southern hemisphere in a clockwise direction.

3.17. WHY LONGER OR SHORTER DAYS?

Materials:
1. A regular glowing light bulb or strong flashlight.
2. A medium size smooth rubber ball.
3. A knitting pin and marker.

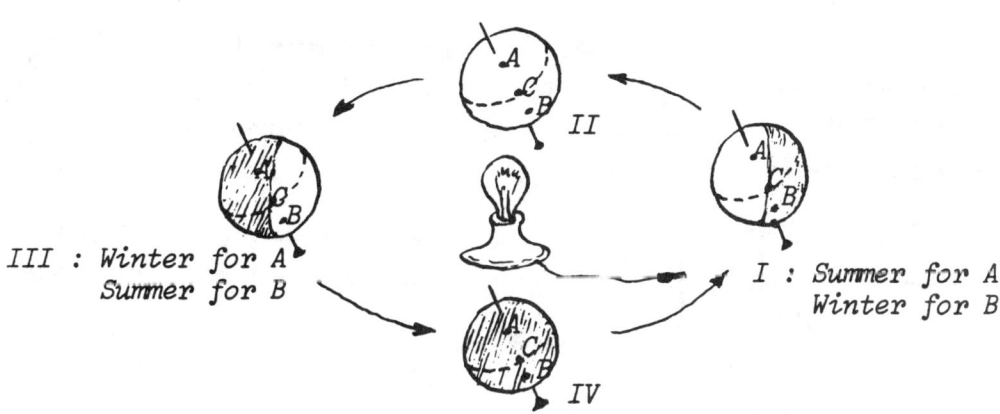

Procedure:
1. Stick the knitting pin through the center of the ball and light the bulb.
2. Hold the ball and knitting pin slanted to the left (top towards the light) at position I; and let the ball spin slowly around the pin.
3. Point out to the students spot A, B, and C (marked on the ball with a marker). Ask: "Which spot gets the most light?"
4. Sequentially, move the position of the ball globe from I to II, to III, and to position IV, and spin the ball around the pin as axis.

Questions:
1. At what position is it summer for place A on the globe?
2. At what position is it spring, winter, and fall for place A?
3. What makes the days longer in the summer for North Americans?
4. Why does the sun always set at 6 p.m. in a tropical country throughout the whole year?
5. Which point represents a tropical country on our globe?

Explanation:
The bulb represents the sun and the ball the earth with the knitting pin as the earth's axle (the poles). This axle, around which the earth is rotating, is not perpendicular to the plane in which the earth revolves around the sun. This slanted axle is the reason why we have four seasons in North America, Australia, and countries at about equal latitudes.

One turn around the axis represents one day and one revolution around the light represents one year. The difference between point A and B should especially be pointed out, at the moment that the shadow reaches one of them when the globe is being spun around its axis. Position I represents summer time, II autumn, III winter, and IV spring for spot A on earth; and winter, spring, summer, and fall respectively for spot B.

WEATHER — CLIMATE

3.18. THE COOL WINTER SUN

Materials:
1. A large smooth ball (without any designs).
2. A strong flashlight, a paper card or white cardboard.

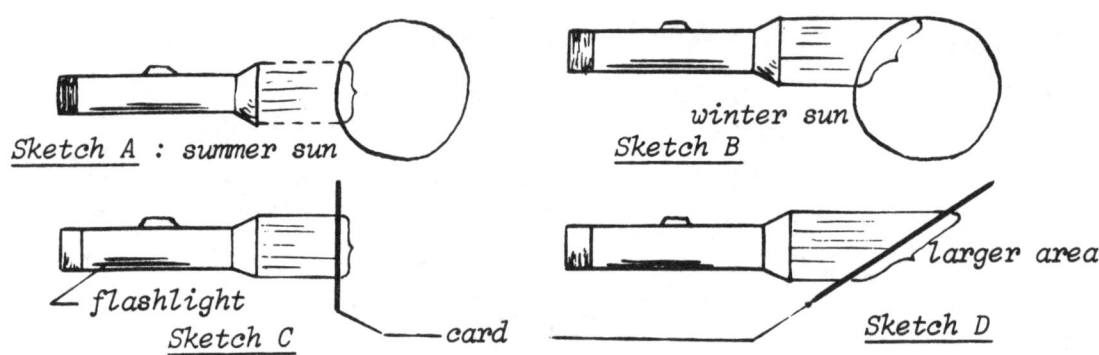

Procedure:
1. Shine the flashlight to the center of the ball (Sketch A), then shine it to the edge (Sketch B) and compare the brightness of the light spots.
2. The earth's surface in summer and winter may be compared with the positions of the card. Hold the card perpendicular to the light rays (C) and slant it away from the flashlight (Sketch D).
3. Let students observe the intensity of the light spot on the card.

Questions:
1. Which position of the ball or card shows the light less bright?
2. What is different between position A & B, or C & D?
3. Which of the positions has more surface area illuminated?
4. Was the amount of light changed to illuminate more surface area?

Explanation:

When the sun rays strike the earth perpendicular to the surface, they are hotter compared to when they strike the earth in a smaller angle. This is because the same amount of sun is spread out over a larger area. the intensity per square area is thus diminished.

The sun's rays in the summer are striking the earth's surface very similarly as the sun's rays strike the tropical areas on earth (perpendicular). The angle of the sun's rays striking the earth in winter may be compared with the angle of the sun's rays and earth at sunset. Although the sun's rays and their number do not change, their intensity is always much less at sunrise or during sunset.

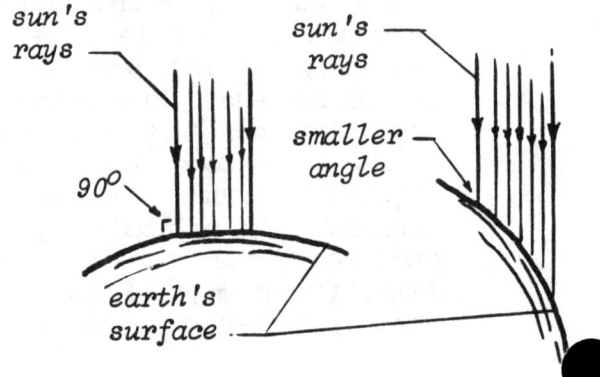

CHAPTER 4

WHAT ARE THE CHARACTERISTICS OF MATTER ?

OBJECTIVES

After dealing with and studying the concepts and sub-concepts of this chapter, the students should be able to:

a. recognize the correct explanation of an observed event based on each of the sub-concepts;
b. explain in their own words which of the sub-concepts is determining the course of an event;
c. distinguish true from false statements concerning each one of the sub-concepts;
d. identify the correct explanation of an event in daily life applying one of the sub-concepts;

all in relation to the following sub-concepts:

-- Between the molecules of matter there exists space.
-- Matter exists in three states: solid, liquid, and gas.
-- Melting and boiling points are specific for pure substances.
-- Cohesion is a force that keeps like molecules attracted to each other.
-- Adhesion is an attracting force between unlike molecules.
-- The forces in a soap film pull at the frame equally in all directions.
-- Surface tension exists at the surface of a liquid.
-- Surface tension is drastically lowered by soap molecules.
-- Density is characteristic for different materials.

CHARACTERISTICS OF MATTER   **MOLECULAR SPACING**

4.1. CAN THE CONTAINER HOLD MORE?

Materials: 1. A transparent container (glass or plastic).
2. Marbles, sand and water & a graduated beaker.

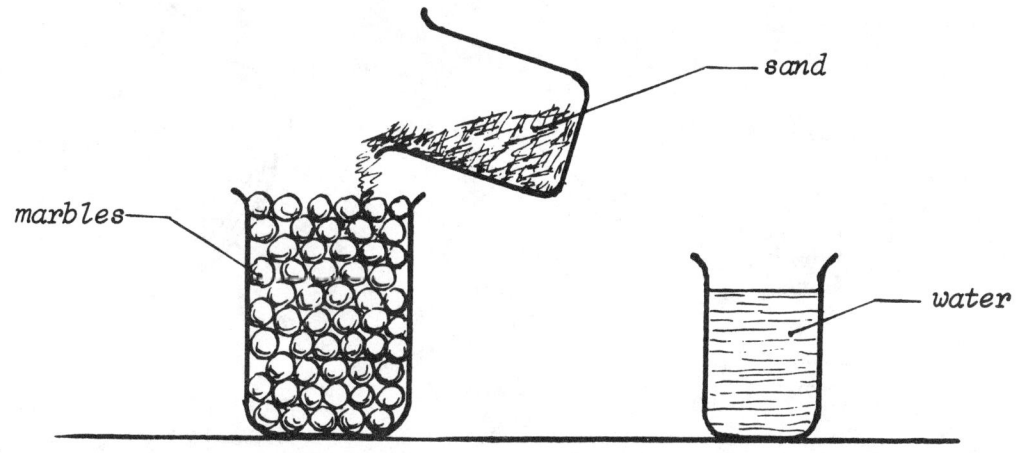

Procedure:
1. Fill the transparent container up to the brim with marbles.
2. Show the students that you still have sand and water; ask them: "Can I add any other material to this container?"
3. Add sand to the container (shake to settle the sand in between the marbles); ask the same question again.
4. Now add water to the mixture of water and sand.
5. Measure off the amount of sand and water added to the marbles (by measuring how much is left over in a graduated beaker).
6. Do not neglect to tell students that the marbles, sand, and water particles are only illustrating how molecules of matter are behaving and that they are not molecules themselves! It is only a model!

Questions:
1. Why could the container that was already filled with marbles still hold more sand and water?
2. Could we have started with water, then sand and marbles?
3. What would you infer that the sizes of molecules of different materials or substances would be?
4. What other materials could be used to do this documentation?

Explanation:
The marbles, sand and water particles are used only as an analogy of how molecules of different sizes would behave. The smaller sized molecules can slip in between the larger ones. Thus it is possible to slip the sand or the water in between the marbles, but not the other way around. It is especially important to point out to students, that the marbles or sand grains are not molecules, but that molecules are so small that we cannot see them, not even with a microscope.

CHARACTERISTICS OF MATTER — MOLECULAR SPACING

4.2. THE SHRINKING BALLOON

Materials:
1. A balloon for each of the students.
2. Flexible measuring tapes or string and meter sticks.

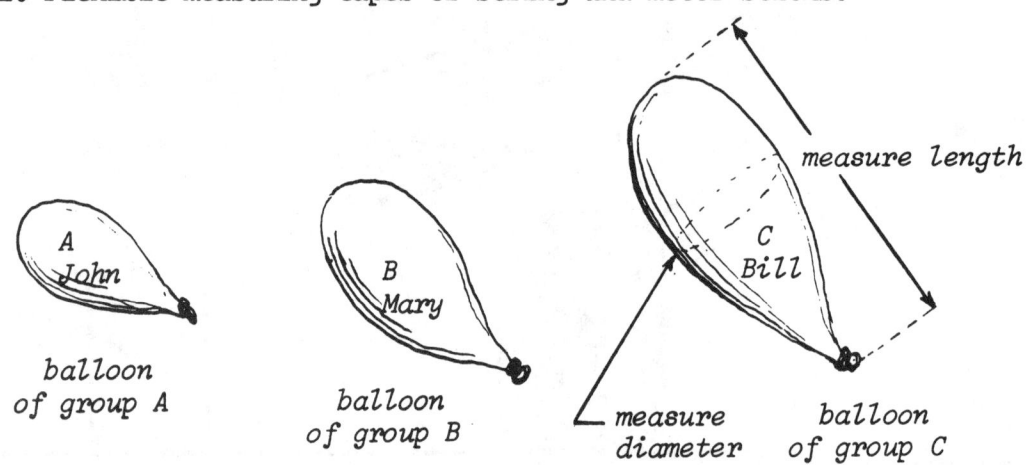

Procedure:
1. Divide the class into three groups A, B, and C; distribute the balloons.
2. Have group A blow the balloon up and let about half the amount of air out, group B blow the balloon up and then let about 1/4 of the air out, and have group C leave the balloon blown up fully, and tie a knot in it.
3. Have each student measure the diameter and length of his or her balloon with the measuring tape or place a length of string around the balloon and measure the string with a meter stick.
4. Let each student record these measurements and mark their balloons with their names, and put them away for the next day (they could be placed in cupboards or stuck to the walls).
5. The next day, have students measure the diameter and length of their balloons and compare these with those of the day before.

Questions:
1. What made the balloon shrink in size?
2. Which of the three groups had the fastest shrinking balloon?
3. Were there any leaks in the balloons?
4. What would the balloon do if kept for another day?
5. Where do we find applications of this principle in daily life?

Explanation:

The air molecules in the balloon are much smaller in size than the rubber molecules of the balloon itself. Although many layers of rubber molecules are contained in the balloon membrane, the tiny air molecules can slowly slip through these bigger ones that hold them inside the balloon. As the pressure inside is larger than outside, the air molecules move faster inside and push through the rubber. Over time, so many air molecules diffuse through the rubber that the pressure inside decreases, thus decreasing the size of the balloon. The rate of losing air molecules is greater for the larger balloon, because of the higher pressure.

CHARACTERISTICS OF MATTER MOLECULAR SPACING

4.3. THE SHRINKING MIXTURE OF LIQUIDS

Materials:
1. A test tube and small beaker.
2. Alcohol - ethyl-, methyl-, or isopropyl-alcohol (methyl hydrate or rubbing alcohol may be used).

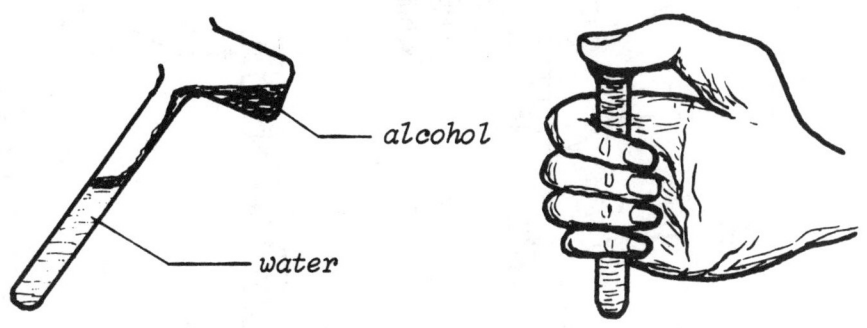

Procedure:
1. Fill the test tube half way with water.
2. Hold this test tube slanted and pour the alcohol slowly from a beaker until brim full.
3. Hold the test tube and place your thumb on the mouth of the tube, making sure that no air bubble is trapped.
4. Show the students that the tube is completely full.
5. Invert the tube several times (keep thumb on opening of test tube at all times; do not release the pressure).
6. Show to students that now the liquid level is lowered.

Questions:
1. Did the alcohol or water evaporate?
2. Was there any liquid spilled by inverting the tube?
3. Did the liquid shrink or contract?
4. Was there space between the water and alcohol molecules?

Explanation:
By closing the opening of the test tube tightly with the thumb, and not taking it off while inverting, no evaporation or spilling of the liquid could occur. By inverting the test tube, mixing of water and alcohol takes place, and because there are spaces between the molecules, the alcohol molecules slip in between the water molecules, thus making the total volume of the mixture become less. Although the spaces between the molecules cannot be seen by the naked eye, this demonstration shows that there must be room between the molecules. If ethyl alcohol is not available, methyl-hydrate (methyl alcohol) or rubbing alcohol (isopropyl alcohol) can be easily obtained from a drug store, and may replace it.

CHARACTERISTICS OF MATTER

MOLECULAR SPACING

4.4. IS THE CONTAINER LEAKING?

Materials:
1. Water softener salt pellets or table salt.
2. A graduated cylinder and a beaker.

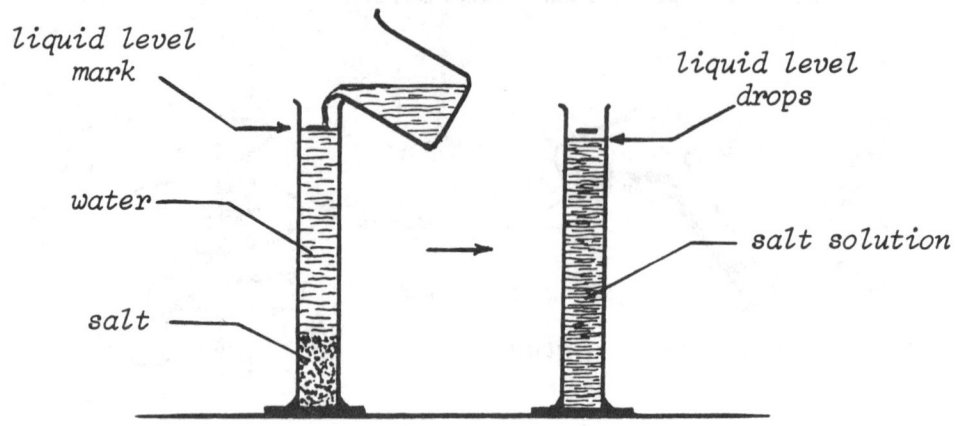

Procedure:
1. Fill the cylinder 1/3 full with water softener salt pellets or with table salt.
2. Add water until full and mark the liquid level with a grease pencil or tape or rubber band.
3. Let it stand for a few minutes and observe liquid level.

Questions:
1. Where did the water go?
2. Is the cylinder leaking any water?
3. What do the salt molecules do when they dissolve?
4. Can we use other salts to do this demonstration?
5. Can sugar be used to do this same demonstration?

Explanation:
   Students might incorrectly infer that the water is absorbed by the salt. As a follow-up to a response like this you might pour out the water before all the salt is dissolved and re-fill it with fresh water: the water level will drop again! The main cause of the water level dropping is, that the salt molecules break down into ions: NaCl molecules into Na^+ and Cl^- ions, which are much smaller than the molecules, and these ions can slip in between the water molecules. This makes the total volume decrease. Other salts that dissolve in water will exhibit the same property. Most inorganic water-soluble salts will do this. The sugar molecule is large in size and does not ionize when dissolved in water, so that the water level will stay at about the same spot when sugar is used.
   This demonstration shows students that there are spaces in existence between water molecules that are not detected by the naked eye.

CHARACTERISTICS OF MATTER

MOLECULAR NATURE

4.5 THE COOLING RUBBER BAND

Materials: 1. A wide rubber band, match and ruler.
             2. A stand and a 100 gram weight.

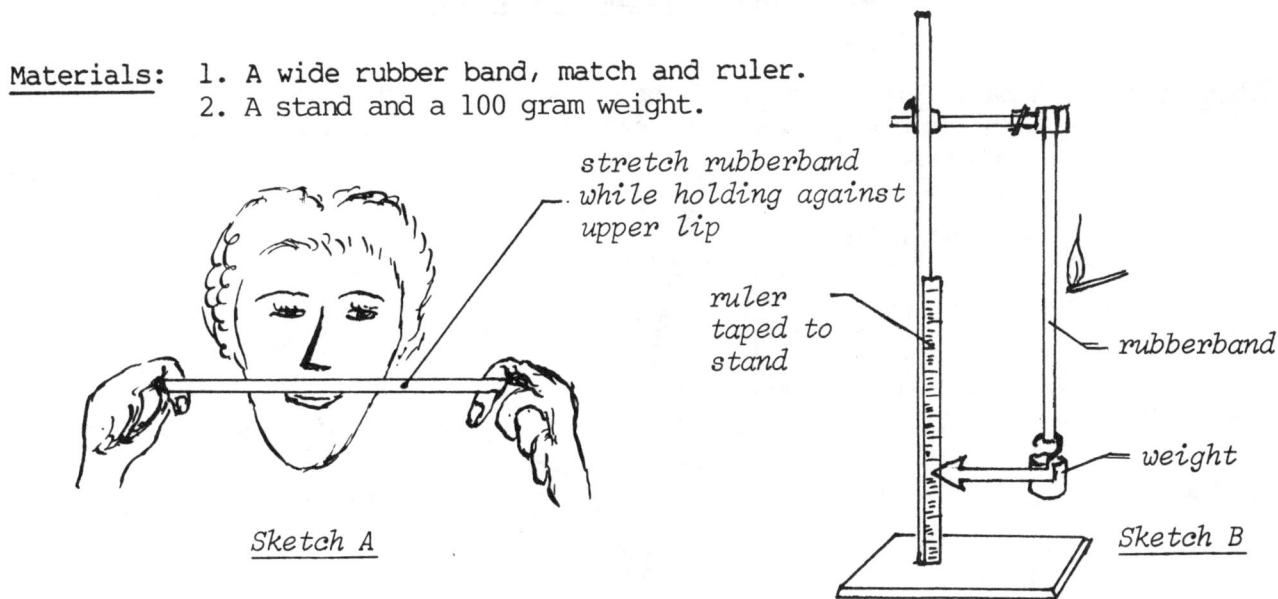

Sketch A   Sketch B

Procedure:
1. Hold the rubber band partly stretched between your forefingers, such that the band touches your upper lip.
2. While the rubber band touches your lip, stretch the band fully and release the tension slowly. What do you feel? (see Sketch A).
3. Hang the rubber band from a stand and hook a 100 gram weight to the band (tape a paper arrow to the weight pointing to the vertical ruler (see Sketch B).
4. Note the position of the paper arrow.
5. Now strike a match and hold the flame close to the rubber band. Observe the position (reading) of the paper arrow.

Questions:
1. How did the rubber band feel (against the upper lip) when it was stretched? When the tension was released?
2. What would stretching of the rubber do to the rubber molecules?
3. What did the rubber band do in demonstration B, when the flame was held near to it?
4. What do metals and other materials generally do when heated?
5. What other material behave the same way as rubber when heated?

Explanation:
Rubber is a high polymer, which means that it consists of large molecules with many side branches. When stretched, these molecules bump and slide against each other, creating heat. This makes the rubber band feel warm against the upper lip. When the tension is released, the molecules can relax back into their normal positions and cooling takes place in the rubber. This cooling process in the rubber means that it is withdrawing heat from its environment.

In demonstration B, this heat is supplied by the environment and forces the rubberband to contract. Other materials that are composed of polymers behave similarly, like: polyethylene, polystyrene, and other plastics.

CHARACTERISTICS OF MATTER — STATES OF MATTER

4.6. THE VANISHING ICE CUBES

Materials:
1. A bag of ice cubes (3-4 cubes per pair of students).
2. A thermometer, beaker, alcohol burner and stand (for each pair).
3. Graph paper for all students.

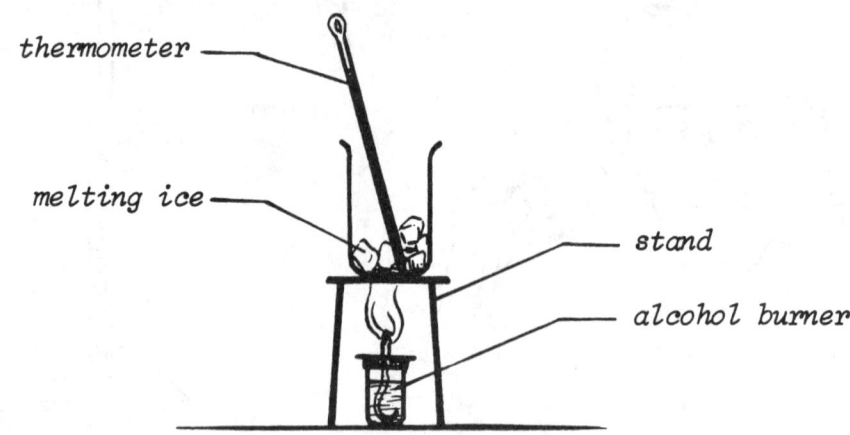

Procedure:
1. Distribute the materials to the students and have them work in pairs.
2. Have one student stir the ice-water mixture and another one observe the temperature every minute while it is being heated (a third student may record the temperature called out).
3. After all the ice is melted, have the students keep on heating the water and record the temperature every minute until the water boils for about ten minutes.
4. Have the students plot the observed temperatures (Y axis) against the time (X axis) on a graph.

Questions:
1. What was the temperature doing while there was still ice present?
2. Where did the heat from the alcohol flame go during that time?
3. What did the temperature do while the water was boiling?
4. Where did the heat from the flame go during this boiling time?
5. What will eventually be left over if the boiling was continued?
6. When heat is not added to the melting process, what will the melting ice do to the environment?
7. When heat is not added to the evaporating process, what will water that is evaporating do to the environment?

Explanation:
In the beginning of the heating process, the heat was used to melt the ice. This is why the temperature stayed on 0° C. This remains at that temperature as long as there is still ice present. As soon as all the ice is melted, the temperature rises steadily till 100° C, then it levels off and stays at the boiling temperature until all the water is evaporated (this temperature may be close to 98° C depending on the elevation of the place). The heat during the boiling process is used to transfer the liquid into the vapor state.

CHARACTERISTICS OF MATTER

STATES OF MATTER

4.7. THE INVISIBLE FLAME EXTINGUISHER

Materials:
1. Two beakers (300 ml) (transparent plastic cups will do).
2. Marble chips and hydrochloric acid (HCl) or carbon tetrachloride (CCl_4).
3. A candle and matches.

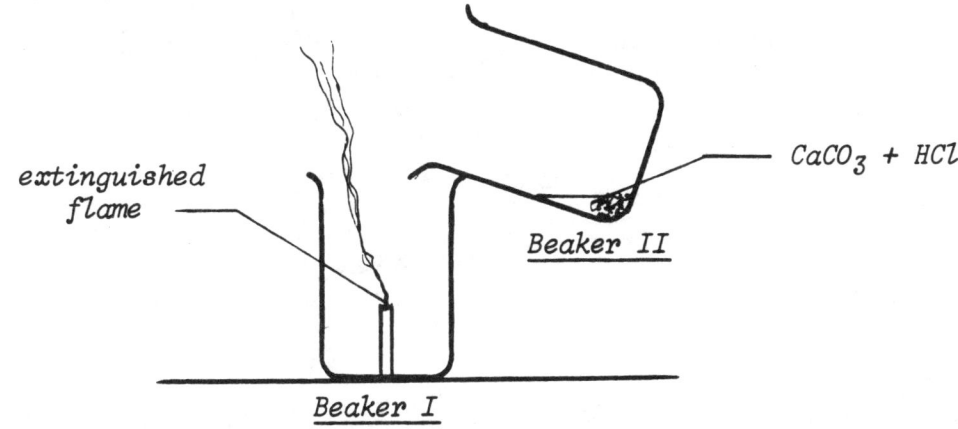

Procedure:
1. Secure the candle in the center of one of the beakers.
2. Put a teaspoonful of marble chips (calcium carbonate) in the other beaker and pour a few drops of dilute HCl on it, or pour a little carbon tetrachloride (CCl_4) in the beaker.
3. Light the candle in the first beaker and pour the gas that has developed in the second beaker in the first (see Sketch): the flame will be extinguished.
4. To repeat the lighting of the candle, the air has to be replenished by inverting the beaker and holding it upside down for a while, then the candle can be lit and the gas poured to extinguish it again.

Questions:
1. What was it that extinguished the candle flame?
2. What gas was formed in the second beaker?
3. Could we observe the gas pouring down into the first beaker?
4. Why does the first beaker have to be inverted to replenish the air?
5. What are other invisible gases that you can name?

Explanation:
   The gas formed in the second beaker is carbon dioxide (or CCl_4 vapor). By adding dilute hydrochloric acid to marble chips, we get the following reaction: $2 HCl + CaCO_3 \rightarrow CaCl_2 + H_2O + CO_2$. Carbon dioxide is an invisible gas (and so is carbon tetrachloride vapor), which is heavier than air, and which has the property of extinguishing flames. The fact that it is heavier than air makes it possible to pour the gas from one container into the other. This is also why it is necessary to invert the first container to replenish the air.
   By using carbon tetrachloride vapor we can make the illusion that an empty beaker is extinguishing the candle flame (by concealing the pouring of a few drops of the liquid in the second beaker from the students).

CHARACTERISTICS OF MATTER

STATES OF MATTER

4.8. THE DISAPPEARING LIQUID

Materials:
1. A medicine dropper & wall clock or watches.
2. Rubbing alcohol (isopropyl alcohol) or methyl hydrate.

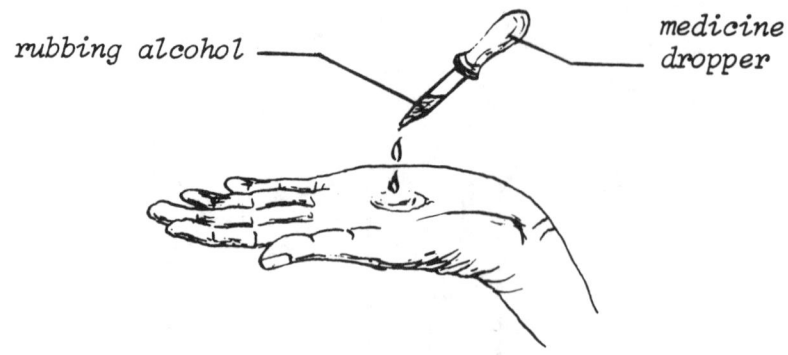

Procedure:
1. Have each of the students stretch out one of their hands with the palm facing up and held horizontally.
2. Come around with the 'colorless liquid' and the dropper; place 3-4 drops on each of the palms.
3. Have the students measure the time it takes for the liquid to evaporate completely.
4. You may repeat the activity, but this time have the students race to evaporate the liquid as fast as possible. They may do everything to the liquid except removing it from the palm or applying a match to it. Some may try blowing, spreading it, etc.

Questions:
1. Where did the liquid disappear to?
2. What did you observe during the evaporation process?
3. How long did it take for three drops to evaporate?
4. Which method proved to be most effective to speed up evaporation?
5. Where did the energy to change the liquid into a gas come from?

Explanation:
As alcohol is very **volatile**, it evaporates quickly. The three drops on the palm evaporate within a few minutes. The transfer of a liquid into its vapor state takes energy. This energy is usually supplied to the liquid in the form of heat. When this heat is not supplied to the liquid, and it evaporates by itself, it withdraws the heat from its environment. This is why the palm feels cool.

The evaporation process can be speeded up by blowing over the liquid, making the surface area larger by spreading it over the whole palm; doing both at the same time would probably by most efficient. Blowing warm air over the liquid would most likely increase the rate of evaporation.

CHARACTERISTICS OF MATTER

STATES OF MATTER

4.9. MAKE A PURPLE GAS

Materials:
1. An Erlenmeyer flask & small watch glass.
2. Alcohol burner and stand.

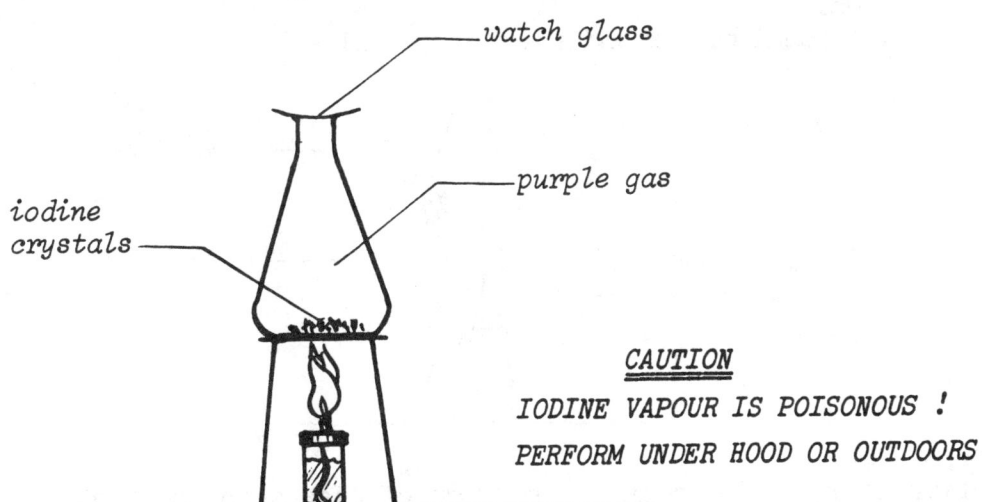

CAUTION
IODINE VAPOUR IS POISONOUS !
PERFORM UNDER HOOD OR OUTDOORS

Procedure:
1. Place a few iodine crystals in the Erlenmeyer flask and cover the opening of the flask with the watchglass.
2. Warm the crystals slowly over a small flame, and observe the flask.

Questions:
1. Did the solid iodine crystals melt by heating?
2. What is the process called when a solid changes directly into gas?
3. Is this process also reversible? Changing gas into solid?
4. What is needed to change gas into liquid or solid?
5. Where do we find gas to solid change in our daily life?

Explanation:
The iodine crystals, when heated, change directly into the gaseous state, and this process is called **sublimation**. When the purple iodine gas hits the cold surface of the flask or watchglass, it will sublimate back to the solid state. When we leave the alcohol flame on for a few minutes and then leave it to cool till the purple color starts to fade, we can see very tiny crystals forming on the cold glass surfaces of the upper part of the flask and the watchglass.

Do not lift the watchglass while the heating is taking place, as the iodine vapor is not fit to inhale. Keep the watchglass on the flask at all times to prevent the iodine fumes from escaping.

Sublimation from gas into solid in our daily life occurs when a freezer door is accidentally left open. Water vapor in the air hits the cold surface of the freezer and immediately turns into solid without passing the liquid state: the frost found in the freezer compartment.

CHARACTERISTICS OF MATTER

STATES OF MATTER

4.10. CHANGING GAS INTO CRYSTALS

Materials:
1. A medium size beaker.
2. Erlenmeyer flask or beaker with a larger base than beaker No. 1.
3. Alcohol burner and stand, moth balls.

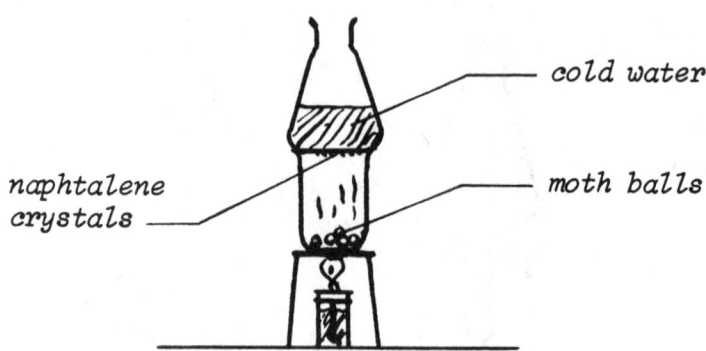

Procedure:
1. Place 4 or 5 moth balls or a scoop of flakes in the beaker and heat it slowly over a small flame.
2. Fill the flask halfway with cold water and stack the flask on top of the beaker holding the moth balls.
3. Leave the alcohol burner on until all the moth balls are melted, then blow the flame out and leave the flask on top of the beaker for another 5-10 minutes. (A longer time is even better.)
4. Lift the flask carefully and slant the bottom towards the students to show the crystals.

Questions:
1. What is the process of solid to liquid change called?
2. What is the process of solid to gas change called?
3. What is the process of gas to solid change called?
4. Where does the solid to gas change occur in daily life?
5. What purpose does the cold water have in the flask?
6. What other material behaves like the moth balls when heated?

Explanation:
By heating the moth balls, they change from solid into liquid state and also into the gaseous state directly. This first process is called **melting** or **liquifying** and the second is **sublimation**. Other processes are named in the following scheme:

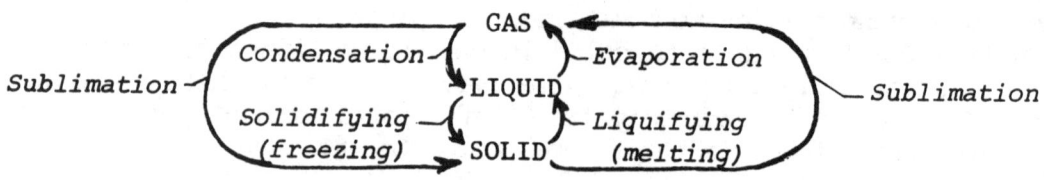

Other materials that sublimate like moth balls are sulfur (found in craters of volcanoes), iodine, and ice.

CHARACTERISTICS   COLLOIDS
OF MATTER   STATES OF MATTER

4.11. MAKE MILK FROM WATER AND OIL

Materials: 1. A glass/plastic clear jar (or beaker), stirrer.
2. Cooking oil, liquid detergent.

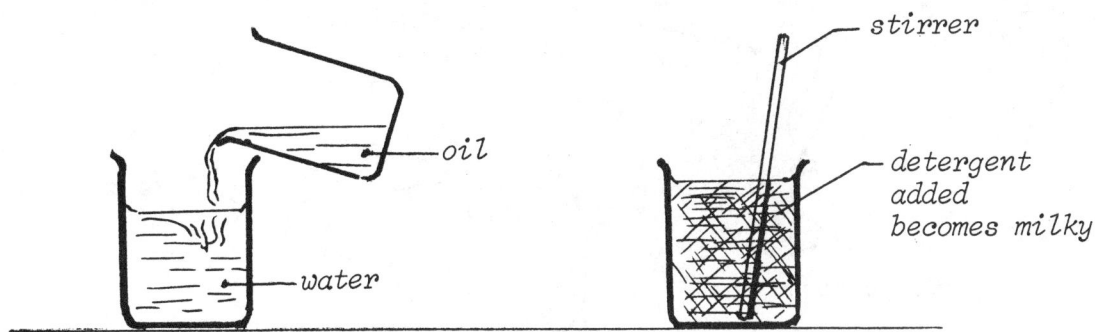

Procedure:
1. Fill the jar about half full with water and pour half of that volume (1/4 of the jar volume) of oil over the water.
2. Stir the two liquids with a spoon or glass stirrer and leave it for a while, and observe what is happening to the mixture.
3. Ask the students: "How can I make the two liquids stay mixed?" Now add a few squirts of liquid detergent and stir thoroughly.
4. After mixing thoroughly, leave the jar alone and observe (if the emulsion separates again you may need more detergent or more vigorous stirring or both).

Questions:
1. After stirring without detergent, what did you observe in the jar?
2. What made the two liquids stay mixed as an emulsion?
3. How would you define "emulsion"?
4. What is the term for a finely divided solid in a liquid?
5. Can you name some examples of finely divided solids in a liquid?
6. What is a finely divided solid in a gas? Examples?
7. What do we usually call fine droplets of liquid in the air?
8. What do we call finely divided gas bubbles in liquid?

Explanation:
   When mixing oil and water, the oil will break up in small droplets and be dispersed in the water very temporarily. After leaving the jar with the mixture alone for a while, the two clear liquids will separate: oil forming a layer above the water because of its smaller density. After adding some liquid detergent (the **emulsifyer**) and some vigorous stirring, the small droplets of oil will stay dispersed forming an **emulsion**. Examples of an emulsion: milk, mayonnaise, salad dressings, butter, etc.
   Finely divided droplets of liquid in gas is mist or fog, colloidal solid in liquid is called a **suspension**: muddy water; a solid **colloid** in gas is called **smoke**. An example of a colloidal dispersion of solid in solid is gold particles in ruby glass; gas bubbles in liquid is commonly called: **foam**.

CHARACTERISTICS OF MATTER   **STEAM & MOISTURE / STATES OF MATTER**

4.12. THE INVISIBLE STEAM

Materials: 1. An ordinary steam kettle (for boiling water).
          2. A large candle or an alcohol burner.

observe steam path

candle (flame in steam path)

steam kettle

Procedure:
1. Boil some water in the kettle.
2. As soon as the water boils, observe the steam leaving the kettle spout.
3. Light the candle/alcohol burner and place the flame slightly under the path of the steam. What do you observe?
4. Move the burning candle/alcohol burner to another spot in the steam path; what do you observe about the steam?

Questions:
1. Did you observe any steam above the burning candle flame?
2. What makes steam visible to the human eye?
3. How is fog formed in the atmosphere?
4. Does cold or warm air contain more moisture?
5. Why are clouds mostly formed at higher elevations?
6. What are conditions for fog to form?

Explanation:
    The steam is not visible in hot or warm air, because steam can only be seen when it is in the form of tiny condensed droplets. These droplets turn into vapour at higher temperatures. Fog and clouds are only formed when the atmosphere contains a lot of moisture and when the temperature is so low that the water vapour condenses. These lower temperatures are present usually at higher elevations.
    Moisture comes in the atmosphere from the soil, from plants, even from the breath that humans and animals exhale, but mostly from the ocean. Try exhaling against a mirror and you will see condensation of vapour.
    Hot air can contain much more water vapour than cold air. When the atmosphere is saturated with moisture, the decreasing of the temperature will precipitate out excess moisture. This precipitate is in the form of fog or rain or snow, depending on how low the temperature is.

CHARACTERISTICS OF MATTER

MELTING POINT

4.13. LIFT AN ICE CUBE WITH SALT

Materials:
1. An ice cube floating in cold water in a cup.
2. A sewing thread and salt shaker.

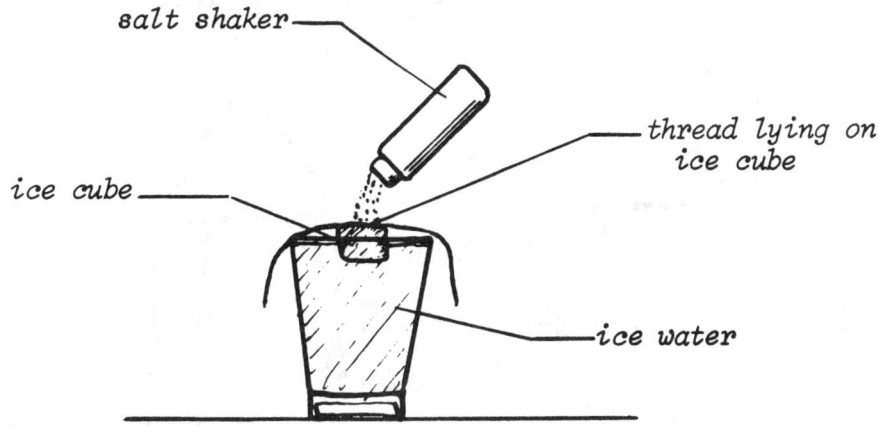

Procedure:
1. Let an ice cube float in a full cup with cold water.
2. Show students the salt shaker and the thread and ask: "How can I lift the ice cube out of the water without getting under the cube?"
3. Let the thread lie on the ice cube and sprinkle salt on it--especially close to the thread.
4. Wait a minute or so, lift the thread slowly to check whether the water above it has frozen; if so, lift both ends of the thread and lift the ice cube out of the water.

Questions:
1. What does salt do to ice on the road in the winter time?
2. At what temperature does ice melt?
3. What do you think salt does to the melting point of ice?
4. What must the melted ice above the string do in order to lift the cube?
5. Where else do we find this principle of lowering the melting point of ice applied in daily life?

Explanation:
As salt water has a lower melting or freezing point than pure water, the addition of salt to the ice cube makes it melt at the places where the salt is sprinkled. This means that the ice around the thread will melt, but the temperature of this water above the ice cube is still below 0° C. When the salt dissolves in more of the melted ice, the solution gets more dilute, increasing the freezing point close to 0° C. As this water still has a temperature of a few degrees below 0° C, it freezes again. This makes it possible to lift the ice cube out of the water. When making ice cream at home, crushed ice and salt are mixed in the mantle of the ice cream maker to lower the temperature below the freezing point of water.

CHARACTERISTICS OF MATTER — MELTING POINT

4.14. CUT THROUGH ICE WITH A WIRE

Materials:
1. A rather large ice block.
2. An iron or copper wire (about 1 m long).
3. Two 1 kg weights or large rocks, tripod or tall tin can.

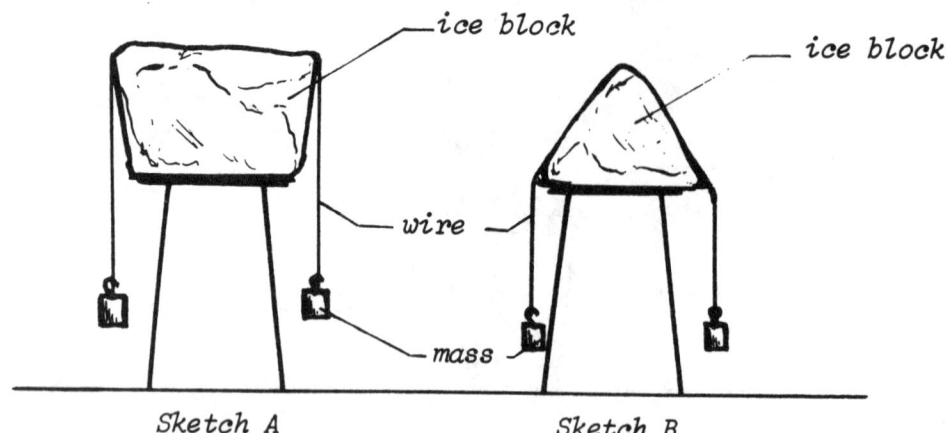

Sketch A Sketch B

Procedure:
1. Place the ice block on the stand or on a tall tin can such that the ice block overhangs the edge of the can (the ice block should be larger in diameter than the can or tripod).
2. Tie the wire ends to the weights and hang it over the ice block, such that the weights are hanging freely (see Sketch).
3. Leave the wire hanging for a while and observe the top part of the ice block on which the wire rests (hang heavier weights on the wire to speed up the process).

Questions:
1. Why does the wire cut into the ice?
2. What happened to the water above the wire?
3. What property is being lowered by increasing the temperature?
4. Would a smaller ice block be cut faster?
5. Would larger weights cut the ice block faster?
6. Will the wire cut faster through block A or B? (see Sketch)

Explanation:
Under high pressure the melting point of solids gets lowered. Under the wire, the ice molecules are pressed and thus they move faster. Immediately under the wire then, we get faster moving molecules, thus higher temperatures and melting of the ice. The wire moves through this water, but as soon as that water gets above the wire, the pressure is off and the temperature of the water gets below 0° C and it freezes again.

The larger the weights, the higher the pressure on the ice, and thus the faster the temperature under the wire is raised. This increases the rate at which the wire cuts through the ice. The same thing happens when the size of the ice block or the surface area between the wire and the ice is reduced (Sketch B).

CHARACTERISTICS OF MATTER   BOILING POINT

4.15. HEAT WATER ABOVE ITS BOILING POINT

Materials: 1. A small beaker (100 ml) & thermometer.
            2. A burner and stand.
            3. Table salt (sodium chloride).

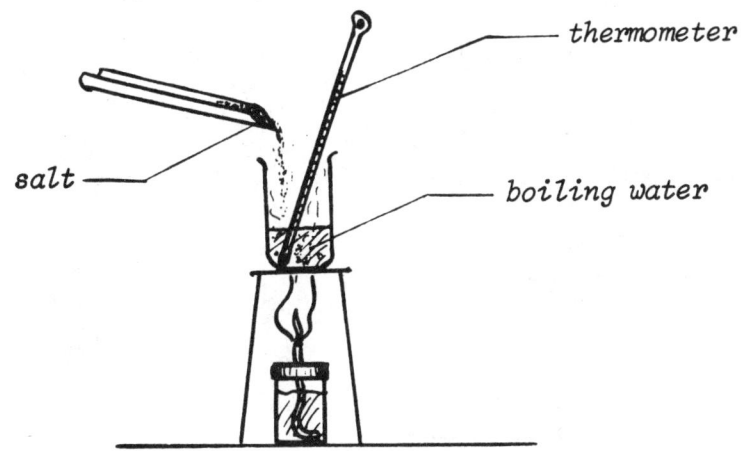

Procedure:
1. Let students work in pairs. Distribute the above materials to each pair of students.
2. Heat about 20 ml of water, place the thermometer in the beaker and record the temperature every half minute (one student stirs with the thermometer and reads off the temperature while the other records it).
3. As soon as the water boils, let students add different amounts of salt to the water (2, 4, and 6 g of salt could be given to three groups of student pairs).
4. Continue heating, stirring, observing the temperature, and recording it very half minute until the water boils again.

Questions:
1. What happened to the boiling when the salt was added?
2. What happened to the temperature after the salt was added?
3. What is the normal boiling point of water?
4. Which of the three groups had the highest boiling point?

Explanation:
   As soon as the salt is added to the boiling water, the temperature drops and the water stops boiling. This is because the salt has a lower temperature and dissolving it in the water absorbed some of the heat. With the salt in the water, the water molecules adhere not only to each other-- **adhesion**, but also to the salt ions, which makes it harder to transfer the water into the gaseous state. This is why the water has to have a higher temperature to boil. The more salt, the higher the boiling temperature of the solution, but the temperature of the water vapor stays at 100° C (this can be checked by lifting the thermometer above the liquid).
   This demonstration shows that impure water (water that contains minerals and salts) has a higher boiling point than pure (distilled) water.

CHARACTERISTICS OF MATTER

BOILING POINT

4.16. BOIL WATER WITH COLD WATER

Materials:
1. A round or flat bottom flask, clamp, boiling chips.
2. A one-hole stopper & thermometer.
3. A burner & stand.

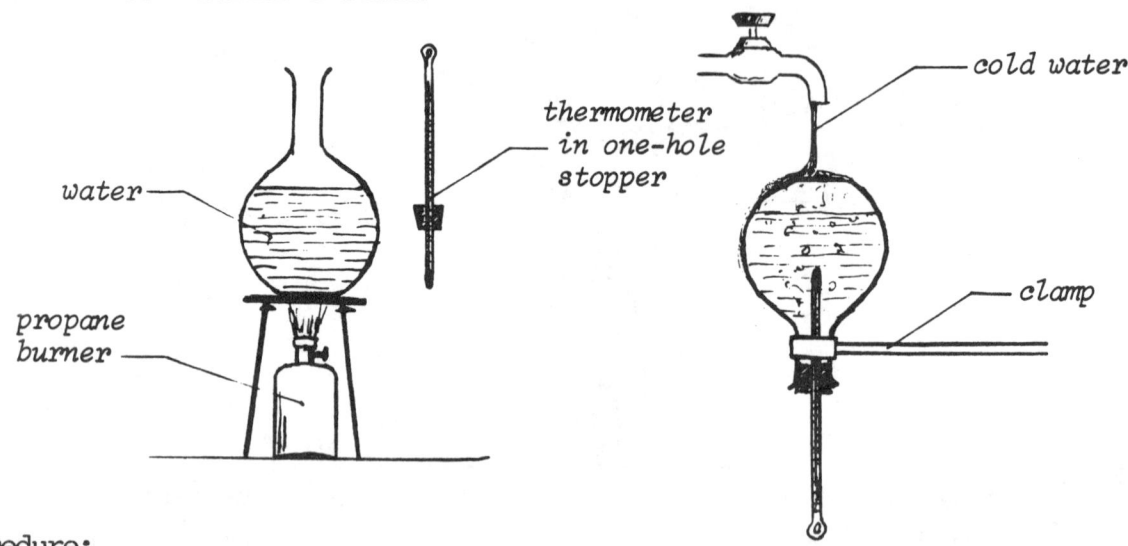

Procedure:
1. Fill the flask 3/4 full with water, put a few boiling chips in it, and heat over the burner and stand until boiling.
2. Attach clamp on flask's neck, shut flame off and immediately close flask with the one-hole stopper with the thermometer in it, and have a student read off the temperature (should be about 98°-100° C).
3. Invert the flask and hold it under a cold stream of water from the tap above a sink, or wipe the bottom of the flask with a towel soaked in cold water.
4. Observe boiling water and have students read off the temperature.

Questions:
1. Why does the water stop boiling after the heat is taken away?
2. At what temperature does water boil at normal pressure?
3. What is the cold water doing to the water vapor in the flask?
4. What do you think the pressure above the water in the cooled flask is?
5. What is the lowest temperature at which the water is still able to boil in the cooled flask?

Explanation:
   By boiling the water (the purpose of the boiling chips is to obtain an even boiling), steam is formed and this replaces almost all the air above the water in the flask. By closing it off immediately after boiling point is reached with the stopper and the thermometer, the air is trapped out of the flask. When the flask is cooled off with the cold water, the water vapor above the water condenses on the colder surface and a partial vacuum above the water is created. This makes the water boil at much lower temperatures than the **normal** boiling point of 100° C. **Normal** indicates the boiling point at **normal pressure**, which is 760 mm mercury or the barometric pressure at sea level.

CHARACTERISTICS　　　　　　　　　　　　　　　　　　　　　　　DISTILLATION
OF MATTER　　　　　　　　　　　　　　　　　　　　　　　　　　BOILING POINT

4.17. DRINK MUDDY WATER

Materials:
1. Distillation apparatus (flask, thermometer, cooler) (if not available revert to point 2).
2. Two beakers (200 ml) and a dinner plate.
3. Burner (propane) and stand.

Sketch A — flask, cooler, muddy water, water to sink, water from tap

Sketch B — dinner plate

Procedure:
1. Set the distillation apparatus up as in Sketch A. Place some muddy water in the flask and start heating it. (If the distillation apparatus is not available, proceed with the following points).
2. Place one beaker on the stand and burner, and fill it with some muddy water. Start heating this.
3. As soon as it starts to boil, hold a large dinner plate somewhat slanted above the beaker (in the steam) and let the condensed water drip into the other beaker (see Sketch B).

Questions:
1. At what temperature does water boil at normal pressure?
2. What temperature do you think would the boiling muddy water have?
3. Why do the impurities stay back in the flask (or 1st beaker)?
4. What is the purpose of the cooler (or dinner plate)?
5. What would be another way to purify the muddy water?
6. Would rain water be the same as distilled water?

Explanation:
The muddy water in the flask or first beaker would most likely reach a temperature above 100° C before it starts to boil, because of the impurities in the water. The thermometer, however, would indicate the boiling temperature of pure water, which would be a little below or close to 100° Celsius, depending on the elevation of the place. The impurities stay back in the flask, because it is only the water molecules that jump out of the liquid state into the vapor state, which are condensed back into the liquid state by the cooler or the cool plate. Another way of purifying water for cooler drinking would be to filter it through a thick layer of sand, although it would contain minerals, which distilled water would not. Rain water would pick up dust and chemicals in the air (especially in polluted areas).

CHARACTERISTICS OF MATTER   COHESION

4.18. THE CLINGING WATER STREAMS

Materials: 1. An empty milk carton or tin can.

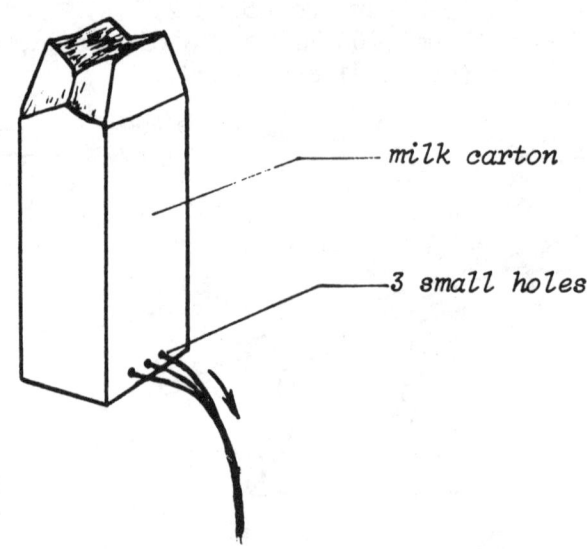

Procedure:
1. Make three small holes in one of the sides near the bottom of the milk carton--about 1/2 cm away from each other.
2. Fill the carton full with water and observe the water streams coming out of the holes.
3. Bring the streams together with the fingers to make it one big stream. Separate the streams by pushing one finger through the middle of the large stream (usually it requires several tries to separate them).

Questions:
1. Why does the water stay in one stream once they are brought together?
2. How far can the holes be placed apart for the water still to be able to cling together?
3. Is it easier to separate (or bring together) the streams with a full or almost empty carton?

Explanation:
   The closer the holes are placed in the carton, the easier to get a whole stream out of the carton, but the harder to separate it into three separate streams. The farther the holes are, the harder to bring the separate streams together. It is the **cohesive forces** between the water molecules that keep the streams together. The fuller the carton, the larger the pressure, and the easier the separate streams are obtained. With less water in the carton, we will get lower water pressure and thus an easier cohesive whole stream.
   We find this phenomenon in daily life in the shower head with the many holes in it. When the valve is turned wide open, separate streams are distinguished, but when the valve is only partially opened, the many small streams will cling together and form one whole stream.

CHARACTERISTICS OF MATTER   ADHESION

4.19. THE SMALLER, THE STRONGER

Materials: 1. Three or four capillary tubes of different diameter.
            2. A small beaker & food coloring.

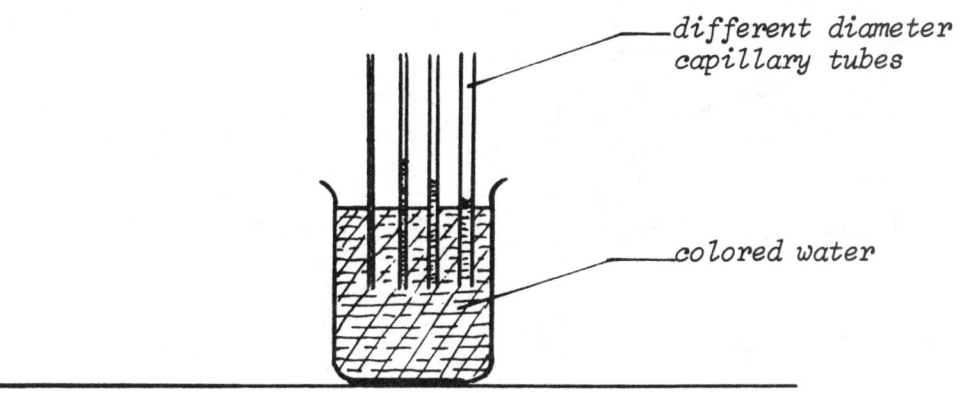

Procedure:
1. Fill the beaker with water and place a few drops of food coloring in it.
2. Hold the three or four capillaries close to each other and dip them in the water; observe the water level in each of the capillaries.
3. When a flat cell is available, the capillaries can be projected on the screen by slanting the overhead projector slightly.

Questions:
1. Will the water level in the capillaries change when the tubes are either moved higher or lower in the beaker?
2. What makes the water go up in the tubes in the first place?
3. Why does the water move up higher in the narrower capillary?
4. Where do we find an application of this capillary action in daily life?
5. What is the force between unlike or like molecules of matter called?

Explanation:
   When a capillary tube is dipped in water, the adhesive forces between the water and the glass of the tube are so large that the water is pulled upwards. The smaller the capillary, the more glass surface per unit number of water molecules exists, and thus the larger the adhesive force. Furthermore, the distance between the water molecules and the glass molecules is so much shorter in the narrower tube, that the adhesive attraction between these unlike molecules is greatly increased.
   This is how the fibers in plants and trees bring up the water from deep in the ground to high up in the leaves. **Capillary action** combined with **osmotic pressure** caused by the semi-permeable walls of the fibers and the plant juices, push up the water to the tree top. Another example of adhesion is the application of paint on any surface. The paint clings to the surface because of the adhesive forces between the paint and the surface molecules.

CHARACTERISTICS OF MATTER   COHESION & ADHESION

4.20. POUR WATER ALONG A STRING

Materials: 1. Two beakers or plastic cups.
2. A water-absorbent string (thin rope).

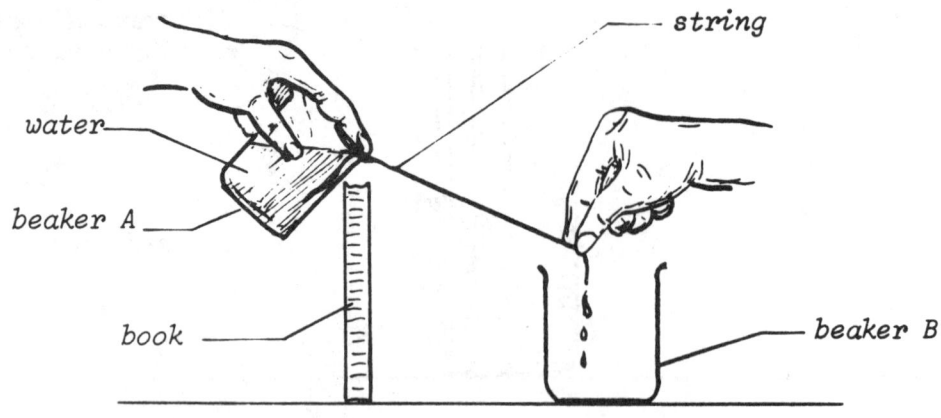

Procedure:
1. Fill one of the beakers about 3/4 full with water.
2. Stand a book about 20 cm away from the empty beaker.
3. Show students the string and ask: "How can I transfer the water from beaker A to B without moving beaker A over or around the book?
4. Wet the string thoroughly in the water.
5. Hold one end of the string in beaker A and the other end over beaker B (see Sketch) and pour the water slowly along the string.

Questions:
1. Why was it necessary for the string to be wet?
2. What forces were holding the water to the string?
3. What other materials can be used in place of the string?
4. Is it possible to pour other liquids along a string?

Explanation:
The string needs to be wetted so that water molecules would adhere to the molecules of the string. The water molecules are attracted to the string molecules by **adhesion.** Once the string is wet, the water can cling to the already present water molecules, because of the **cohesive forces** between like molecules of water. The transfer of water will not succeed with a dry string or any material which is not water absorbent.

Other materials that would have the same properties as the water-absorbent string would do the same job, like: cotton, cloth, paper, wood, etc. Materials that are not water-absorbent, like nylon or wool cannot be used for this purpose.

Liquids that have strong cohesive forces between their molecules, like oil, vinegar, syrup, etc. can be poured along a string as well, provided that we make sure that the string or whatever material is used to transfer the liquid can absorb it.

CHARACTERISTICS OF MATTER

COHESION & ADHESION

4.21. WHERE DOES THE CORK FLOAT?

Materials:
1. A regular drinking glass.
2. A small cork.

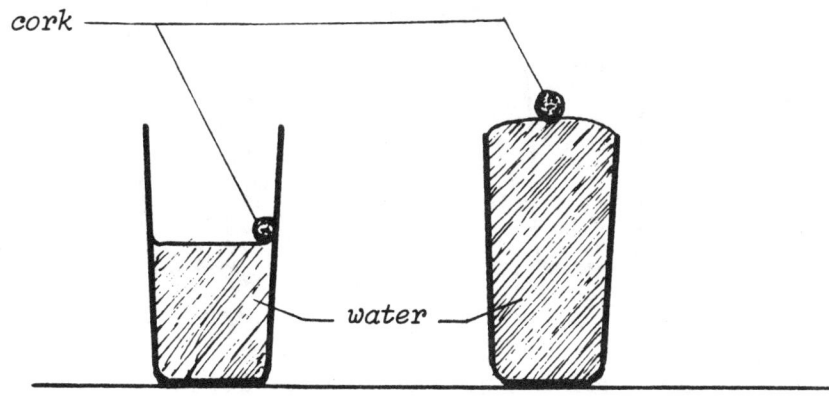

Procedure:
1. Fill the glass half way with water and float the cork on the water surface: observe!
2. Where does the cork float? (It should be attracted toward the side of the glass; move it somewhat if it stays in the middle.)
3. Now add more water to the glass and fill it brimful. Observe where the cork now floats.
4. Try to push the cork towards the edge: it will not stay there!

Questions:
1. Why is the cork moving towards and sticking to the side of the glass (with the glass half-filled)?
2. Where is the water level highest in the half-filled glass?
3. Why is the cork floating in the center of the full glass?
4. Where is the water level highest in the full glass?
5. Why can we fill the glass more than full without overflowing the water?

Explanation:
When the glass is half-filled with water, the circumference of the meniscus is the highest level. This is caused by the adhesive forces between the water and the glass molecules. As a result of this high level at the side of the glass, the cork moves to that place.

When the glass is brimful filled with water, the surface tension and the cohesive forces between the water molecules form a film on the surface of the water, making it possible to fill the glass more than full. The highest level of the meniscus is now in the center of the glass, which is where the cork will float. As floating objects tend to float at the highest spots of the meniscus, the center would be the spot where the cork would float when the glass is full.

CHARACTERISTICS OF MATTER — SURFACE TENSION

4.22. HOW MANY PENNIES CAN GO IN?

Materials: 1. Two regular drinking glasses, liquid detergent.
2. About 50 pennies (or other small coins).

Procedure:
1. Make sure that the drink glasses are very clean. Place one on the table and fill it to the brim (not too overfull, just full).
2. Now ask the students: "How many pennies can I put in the glass before it overflows?" Anticipated aanswer: "5, 10, 15, even 20 maybe".
3. Start putting the pennies in the glass of water, very carefully with its edge in first (vertically), and let the students count.
4. Surprised at the result? Now place the other glass on the table, in which you put a drop of detergent beforehand (without the students noticing it), and also fill this to the brim with water.
5. Now ask a student to put just as many pennies in this glass. What happened here?

Questions:
1. How many pennies could go in the first glass? In the second?
2. What made the water overflow so easily in the second glass?
3. What kept the water from overflowing in the first glass?
4. What shape did the meniscus take form in the first glass?
5. How would the number of coins compare if we used dimes instead of pennies? Nickels instead of pennies?

Explanation:
It is possible to put close to 50 pennies in the full glass of water, depending on how full the glass was filled. A thick-rimmed glass will look full even when the meniscus is a little below the rim. Then it is possible to put many more pennies in. A larger glass also increases the number of pennies that can be dropped. The water forms a **convex meniscus** due to the **surface tension,** which is nothing but a manifestation of cohesion of the surface molecules.

In the second glass, where detergent was present, this surface tension is broken. The cohesion between the surface water molecules is much smaller, and thus the water overflows much sooner. (Another way of inconspicuously putting detergent in the second glass, is wetting the second bunch of pennies with a little detergent before giving it to the student).

CHARACTERISTICS OF MATTER   SURFACE TENSION

4.23. FLOAT METAL ON WATER

Materials: 1. A needle, razor blade, or thin wire gauze.
            2. A beaker, and a candle.

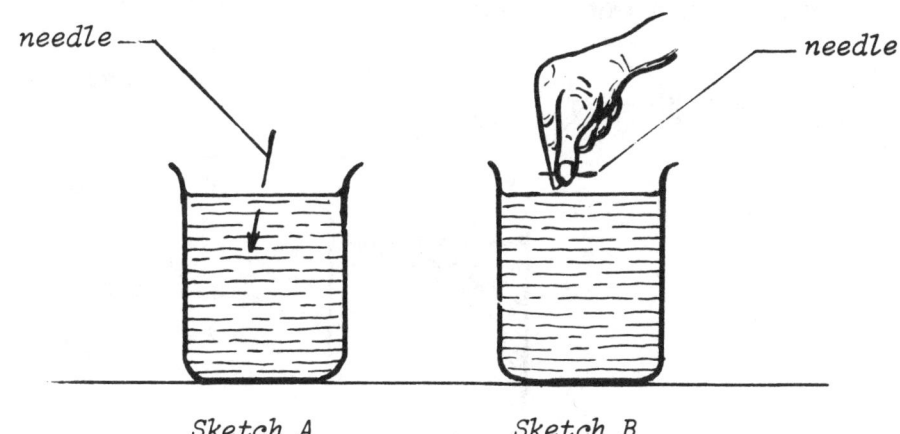

Sketch A   Sketch B

Procedure:
1. Fill the beaker with water.
2. Take the needle or razor blade and show that the density of the metal is higher than that of water by dropping the object vertically in the water (see Sketch A).
3. Fish the needle out of the water and dry it off, or use another needle.
4. Now hold the needle horizontally as close as possible to the surface of the water (without touching the water) and drop it (Sketch B).
5. If this does not succeed, rub the needle against the candle before dropping it on the surface of the water.

Questions:
1. Why is it not possible to float the needle on the water when it is dropped vertically?
2. Why does the needle have to be dry before floating it?
3. What purpose does the rubbing of the needle against the candle have?
4. What property of the water keeps the needle afloat?
5. How can we make a floating needle sink without touching it?

Explanation:
When the needle is dropped vertically in the water, it pierces the water surface. The surface tension at that very small area of the needle point is not large enough to hold the needle afloat. But when the needle is dropped horizontally, parallel with the surface of the water, a much larger area of the water surface is involved, and the combined cohesion forces between the water molecules under and around the needle, which forms the **surface tension**, is large enough to hold the needle afloat. The rubbing against the candle makes the object more water repellent and increases the chance the needle will lie afloat on the film of water molecules.

CHARACTERISTICS OF MATTER

SURFACE TENSION

4.24. SINK THE PEPPER

Materials:
1. A beaker or transparent container.
2. Whole and ground pepper.
3. Liquid detergent & dropper.

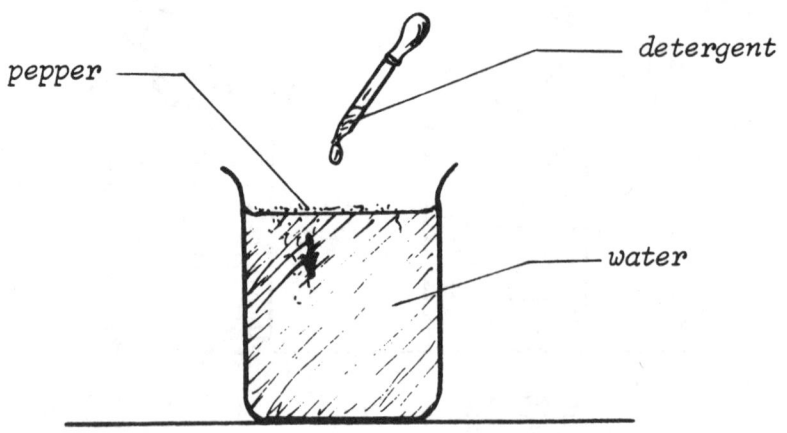

Procedure:
1. Fill the beaker with water.
2. Show the students a whole pepper kernel and ask: "Will the pepper sink or float in the water?" Then drop it in the water: it sinks!
3. Now shake some ground pepper on the water surface. Ask: "Why is the same pepper now not sinking?"
4. Place a drop of detergent in the water and observe the fine pepper now sink to the bottom.

Questions:
1. Why did the same pepper in fine state float on the water?
2. Do you think that pepper has a higher or lower density than water?
3. Did we change the density of the pepper by grinding it?
4. What did the detergent do to the water surface?
5. What forces were weakened between the water molecules when the detergent was added?

Explanation:
By showing the sinking of the whole pepper kernel in the water, the students should be able to conclude that the density of pepper is higher than that of water. The grinding of the pepper is not changing the density of it, yet it floats on the water because the pepper particles are light enough to be held up by the surface tension of the water. The water molecules and their cohesive forces on the surface form a film, which can hold up relatively light objects like a needle, razor blade, gauze, etc.

Detergent has the property of breaking the cohesive attraction forces between the water molecules, and thus weakening the surface tension of water. The film on the surface is removed, and the pepper sinks.

CHARACTERISTICS OF MATTER

SURFACE TENSION

4.25. THE DETERGENT PROPELLED BOAT

Materials:
1. A paper card (3x5"), a shallow tray or sink.
2. A piece of soap or a few drops of liquid detergent.

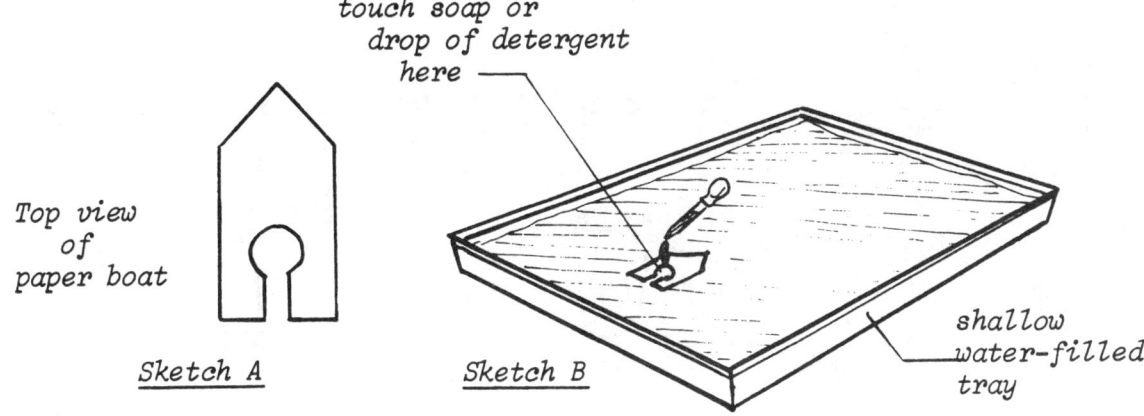

Procedure:
1. Fill the shallow tray (or sink) with water.
2. Cut out a boat from the paper card in the form of Sketch A.
3. Let the paper boat float on the water and touch a corner of the soap block to the water in the center opening of the boat (or place a drop of liquid detergent here).
4. Observe the movement of the boat!

Questions:
1. Why does the paper boat move forward only when the soap touches the water?
2. What made the paper boat be pulled forward?
3. What did the soap or detergent do to the cohesive forces between the water molecules?
4. What other shapes of boats would work the same way?
5. Would it work without a hole in the paper card?
6. What would happen if we touched the soap to the side of the boat?

Explanation:
The soap touching the water or the drop of detergent in the water is breaking the **surface tension**. There are **cohesive forces** between the water molecules in front and in the back of the boat, as well as on both sides. By touching the soap towards the rear of the boat, these cohesive forces are broken in the rear but not in the front of the boat. This makes the boat to be pulled towards the front. Similarly, before the detergent was dropped, the water molecules on the surface of the water were pulling the boat on all sides with an equal force. After the detergent was dropped, the attracting force in the rear of the boat was eliminated, so that the resultant force is to the front. The shape of the boat is irrelevant. Any shape will do the same thing. When the surface tension on the left is broken, the paper shape will move to the right. The propelling can continue until the detergent has lowered the surface tension considerably.

CHARACTERISTICS OF MATTER — SURFACE TENSION

4.26. THE WATER-FILLED CUPS

Materials:
1. Two regular identical drinking glasses.
2. A paper card and a dime (or razor blade).

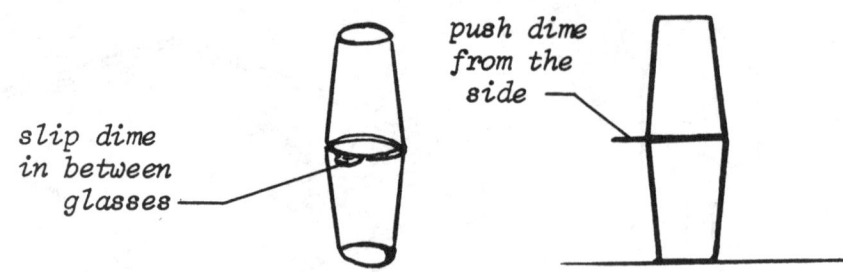

Procedure:
1. Fill the two glasses with water up to the brim.
2. Place a paper card on one of them and turn the glass upside down while holding the card against the glass.
3. Let go of the card, place the inverted glass on the other water-filled glass, and slip out the card carefully (keeping the top glass exactly over the bottom glass).
4. Place a paper towel around the bottom glass to guard eventual spillage.
5. Ask students: "Can anything be added to the water inside the cups?" (anticipated answer: 'No, not a thing').
6. Take the dime (or razor blade), push the dime in between the two glasses and slip it inside the glasses while holding the top glass.

Questions:
1. Why did the water not run out while the dime was slipped in?
2. Did any water spill while the coin was pushed in?
3. What indicated that some water must have run out?
4. What other objects could have been slipped inside the glasses?

Explanation:
In placing the top glass upside down on the bottom water-filled glass, it was the air pressure that held up the card and the water. While the dime (or razor blade) is pushed between the two glasses, some water is running out, which can be detected from the small air bubbles that float up in the top glass. The water in the top glass is held back mostly by air pressure pushing on the water surface between the two glasses. It is as if a film is covering the **surface of the water - surface tension.**

Other objects that are as thin as a dime or even thinner could be slipped inside the glasses.

CHARACTERISTICS OF MATTER   SURFACE TENSION

4.27. THE DETERGENT-FEARING POWDER

Materials:
1. A petri dish (bottom or top), a medicine dropper.
2. Liquid detergent or soap (solid or powder).
3. Lycopodium powder (or baking flour or fine pepper).

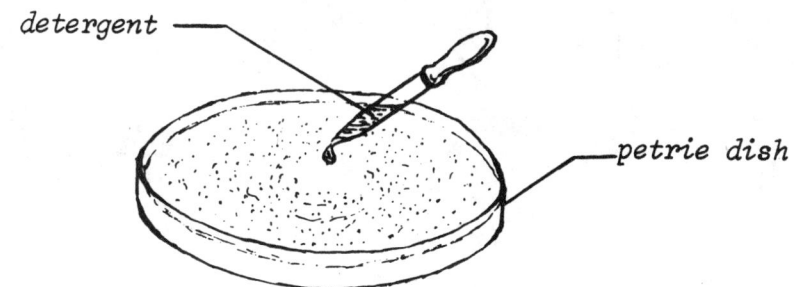

Procedure:
1. Fill a Petrie dish about half full with water (place it on an overhead projector and project the dish on a screen).
2. Sprinkle fine powder (face powder, baking flour, or fine pepper will do) on the water surface.
3. If you have liquid detergent: place a drop in the center of the Petrie dish. Observe the spreading of the powder.
4. If you use solid soap: touch the water surface with a corner of the soap block. Observe!
5. If you have powder soap: drop one or two flakes on the water surface. Observe!

Questions:
1. What made the powder move away from the detergent or soap?
2. What is the detergent doing to the surface tension of the water?
3. What forces are acting on the powder particles on the water surface?
4. What forces are acting between the water molecules on the water surface?

Explanation:
At the surface of the water, the molecules form a film-like layer, where **cohesive forces** are most prominent. The surface molecules have only three sides, where cohesive forces act upon them, as no water molecules are above them. When these cohesive attraction forces are weakened by the detergent at a particular spot, the molecules that are farther away from this spot will pull these weaker held molecules towards them. As the water molecules move, they bring the powder particles with them.

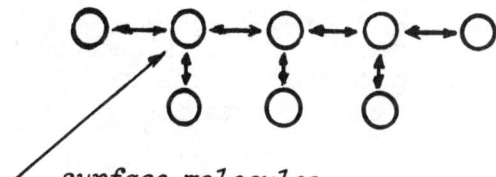

CHARACTERISTICS OF MATTER

SURFACE TENSION

4.28. WHICH BUBBLE IS STRONGER?

Materials:
1. A three-way glass valve or a T-tube with 3 rubber tube clamps.
2. Two (or three) short rubber tubes.
3. Two L-shaped glass tubes.
4. Liquid detergent, glycerine, a small beaker.

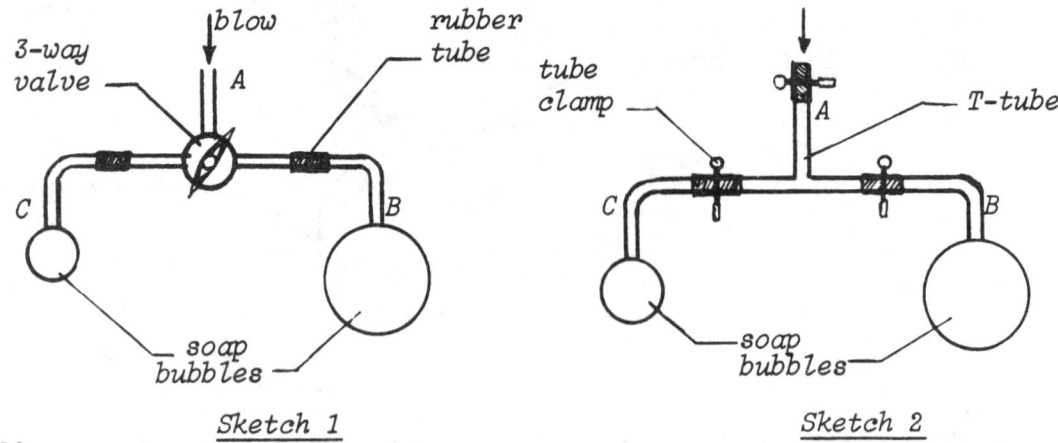

Sketch 1   Sketch 2

Procedure:
1. Clamp the three-way glass valve and connect the L-tubes to the short rubber tubes as in Sketch 1.
2. When no valve is available: use a T-tube with 3 rubber tubes and 3 tube clamps, and set the materials up as in Sketch 2.
3. Mix in a small beaker: 1 volume detergent, 2-1/2 volume glycerine and 3 volumes of water.
4. Dip the end of the L-tubes in the detergent mixture by bringing the beaker with the mixture to the set-up.
5. Set the valve such that tubes A & B are connected (when using a valve), or press open clamp A & B (close clamp B) and blow a small bubble on C.
7. Now freeze the air flow and ask: "Which one will blow which bubble up if they are connected to each other?"
8. Connect the two bubbles by turning the valve (when using a valve) or by pressing (opening) clamps B & C, (close clamp A) and observe.

Questions:
1. Which one is blowing which bubble up?
2. What can we conclude about the relationship between surface tension and radius of a soap bubble?

Explanation:
The smaller bubble will blow up the larger one, once the two bubbles are connected. This is because the surface tension in a soap bubble is inversely proportional to the radius of the bubble: the larger the bubble, the smaller the surface tension, thus the lower the pressure inside it.

CHARACTERISTICS OF MATTER   SURFACE TENSION

4.29. THE THREAD CIRCLE IN THE SOAP FILM

Materials: 1. A liquid detergent (preferably JOY), & glycerin.
             2. A square or rectangle wire frame and thread.
             3. A wide shallow container/tray (to fit the wire frame).

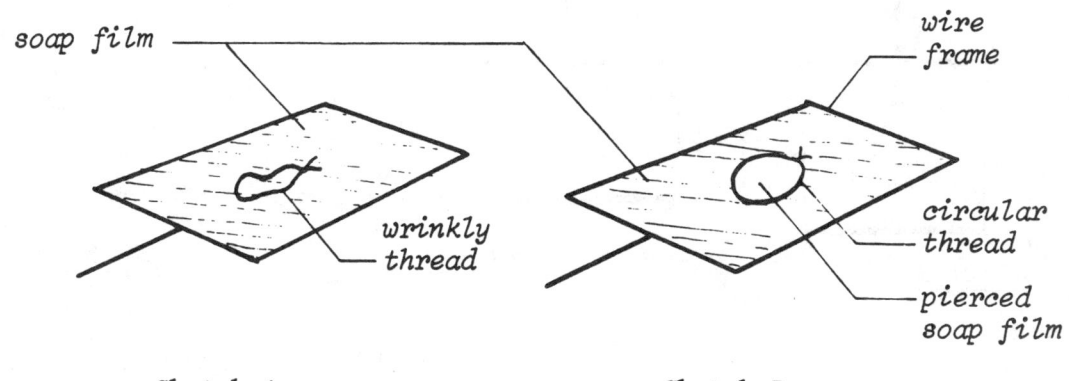

Sketch A   Sketch B

Procedure:
1. Make a soap solution by mixing the following liquids: 1 part JOY, 2-1/2 parts glycerin, and 3 parts water.
2. Place this soap solution in the shallow container, make a soap film by dipping the wire frame in the solution and taking it slowly out.
3. Make a small loop of thread (diam. 3-4 cm), wet it in the soap solution and lay it carefully in the soap film (see Sketch A).
4. Once the thread loop lies in the soap film, pierce the center of the loop with a dry object (pencil or dry finger).
5. Slant the wire frame, wiggle it, and observe the perfect thread circle move and travel throughout the whole frame!

Questions:
1. What made the wrinkly thread turn into a perfect circle?
2. What does the circular loop indicate about the forces working on it.
3. The thread circle stays the same shape no matter where it is in the wire frame, what does this indicate about the forces working on it in the soap film?

Explanation:
   This demonstration shows that there is indeed a tension working in the soap film. The wrinkly thread loop lying in the soap film suddenly turns into a perfect circle, after the inside of the loop is pierced, because the forces inside the loop are eliminated, and equal forces at all points on the outside of the loop are working and pulling on it.
   By tilting and wiggling the wire frame, the thread loop can move around the frame without changing shape, as a result of the cohesion forces in the soap film that are of equal magnitude throughout the whole film.

CHARACTERISTICS OF MATTER — CAPILLARITY COHESION & ADHESION SURFACE TENSION

4.30. THE STRONG SOAP FILM

Materials:
1. A piece of coat hanger or thick wire & thin sewing thread.
2. A medium sized glass or plastic funnel, a candle.
3. Liquid detergent (no additives e.g. JOY), glycerin.

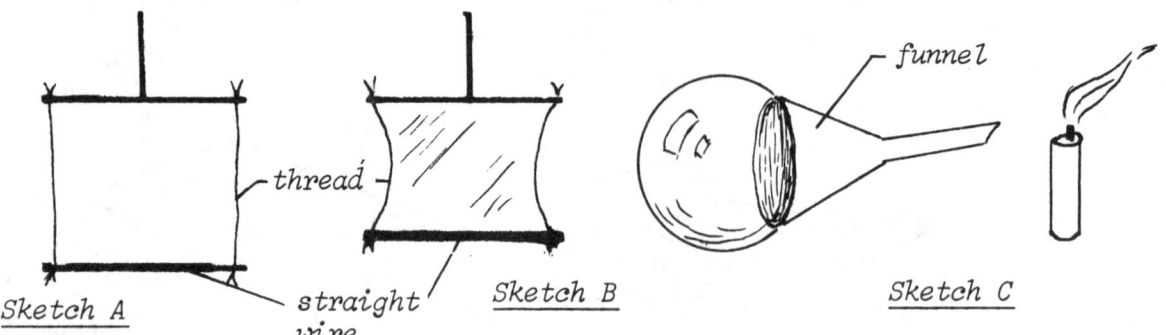

Sketch A   Sketch B   Sketch C

Procedure:
1. Make a soap solution by mixing 1 part of JOY, 2 1/2 parts of glycerin, and 3 parts of water.
2. Tie a short straight piece of wire (about 10 cm) with equal lengths of thread to the T-shaped wire (soldered together beforehand - see Sketch A).
3. Dip the wires into the soap solution and pull them slowly out, holding the T-shaped wire by the leg, and observe (see Sketch B).
4. Light the candle and secure it to the table top.
5. Hold the funnel by its long end and dip it into the soap solution. Blow through the funnel and form a medium sized bubble, but make sure that you let the bubble stick to the funnel.
6. Point the funnel opening (long end) to the candle flame and observe the candle flame (see Sketch C).

Questions:
1. What made the threads between the wires curve?
2. What would happen if the soap film between the wires is punctured?
3. What happened to the candle flame?
4. What happened to the soap bubble in the funnel?
5. What do both events indicate about the forces in a soap film?

Explanation:
In the soap film between the wires and thread there exists a force called: **surface tension**. It consists of adhesive forces between the liquid molecules and the solids (called **adhesion**), and cohesive forces between the molecules of the soap solution (called **cohesion**).

This same adhesive/cohesive force or sometimes also called **capillary force** exists in the bubble which is attached to the funnel. This force is pushing the air in the funnel out of the opening and is almost strong enough to blow the candle out.

Examples of capillary forces in action are: the sucking of water from the roots of plants travelling into the top leaves (also called **osmotic pressure**); the glueing of two pieces of plexiglass with a thin organic liquid (by placing a few drops of the liquid between the two solid surfaces).

CHARACTERISTICS OF MATTER

DENSITY
SURFACE TENSION

4.31. THE FLOATING OIL SPHERE

Materials:
1. A large, wide test tube or small beaker (100 ml).
2. A medicine dropper.
3. Ethyl or methyl alcohol, frying or light motor oil.

Procedure:
1. Pour about 50 ml of water in the beaker (or test tube).
2. Hold this beaker in a slant and slowly pour the alcohol on top of the water (about 50 ml - an equal amount as the water).
3. Let the beaker stand on the table.
4. Fill the medicine dropper with oil, bring the opening of the dropper in the beaker where the two liquids meet, and squeeze the oil out.
   Then withdraw the dropper carefully out of the liquid.

Questions:
1. Why does the alcohol float on top of the water?
2. What would happen if we poured the alcohol carelessly on top of the water?
3. Is alcohol completely immiscible with water?
4. Why does the oil take the shape of a sphere?
5. Why does the oil sphere stay between the two layers of liquid?
6. What would happen if the oil was squirted out of the dropper in the alcohol layer and not where the two liquids meet?
7. What would happen if the oil was squirted out in the water layer?
8. What would happen if the oil was squirted out in water only?

Explanation:
   As ethyl alcohol has a density of 0.794 and methyl alcohol even less than that, it can float on water and form a layer. It is not totally immiscible with water, but when poured slowly and carefully on the water it can form a layer and stay above it. Where the two liquids meet, however, the water and alcohol mix and form a liquid with a density very close to that of oil. This is why the oil forms a perfect sphere between the two liquid layers.
   A sphere is formed because it has the smallest surface area as compared to other three dimensional shapes. When the beaker is left standing, the alcohol evaporates slowly, and the oil sphere moves up slowly until it reaches the surface and then the sphere is knotted off at the top and slowly becomes a flat circle when all the alcohol has evaporated.

CHARACTERISTICS OF MATTER   **DENSITY**

4.32. THE FUNNY WATER

Materials:
1. Two identical beakers, two identical watch glasses.
2. One small and one larger piece of candle.
3. Alcohol (methyl, ethyl, or isopropyl).

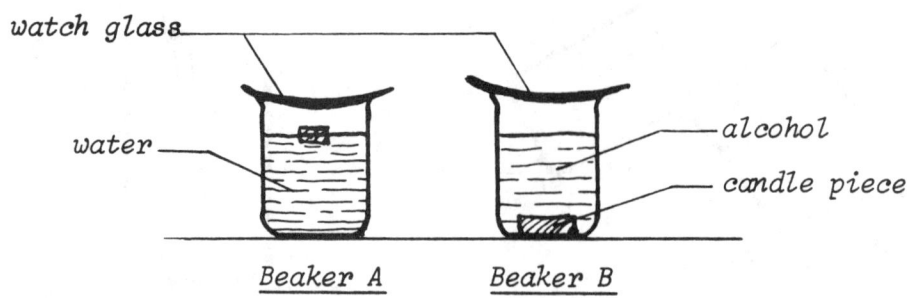

Procedure:
1. Fill beaker A about 3/4 full with water and beaker B with the same amount of alcohol and cover them with the watch glasses (do not reveal to students that the liquids are different).
2. Show the students the two pieces of candle: drop the short piece in beaker A and the longer candle in beaker B.
3. Ask: "Why is one candle floating and the other sinking?" "What will the candles do if I switch them around?"
4. Drop the long candle in beaker A (it still floats) and the short one in beaker B (it sinks).

Questions:
1. Why does the candle float in beaker A?
2. Why does the candle sink in beaker B?
3. What does it depend on whether something floats or sinks in water?

Explanation:
The alcohol has a lower density than water. This is why the candle sank in the alcohol and stayed afloat in water. Some students may suggest that it sank in beaker B because it is longer than the other piece that stayed afloat. But then, after switching the pieces of candle around, the longer candle stays afloat in the water and the shorter one sinks in beaker B. This most likely will give students a clue, that the liquid in beaker B is some kind of "funny water." This liquid may be methyl, ethyl or isopropyl alcohol, which all have a density of around .76. The density of water is 1, and that of wax in the vicinity of .85.

Density is defined as the ratio of mass and volume of a substance.
$D = \frac{M}{V}$ in which D = Density, M = Mass, and V = Volume.

Whether something sinks or floats depends on the relative density of the object compared to that of the liquid it is immersed in. Any object that has a density between that of water (1) and that of alcohol (.76) may be used for this demonstration.

CHARACTERISTICS OF MATTER   DENSITY

4.33. WHICH WILL FLOAT WHERE?

Materials: 1. Mercury, Carbontetrachloride, Kerosene (or gasoline).
            2. A piece of steel (bolt or nut), a piece of ebony wood, paraffin or candle), and a cork.
            3. A long thin measuring cylinder (100 ml), three small beakers.

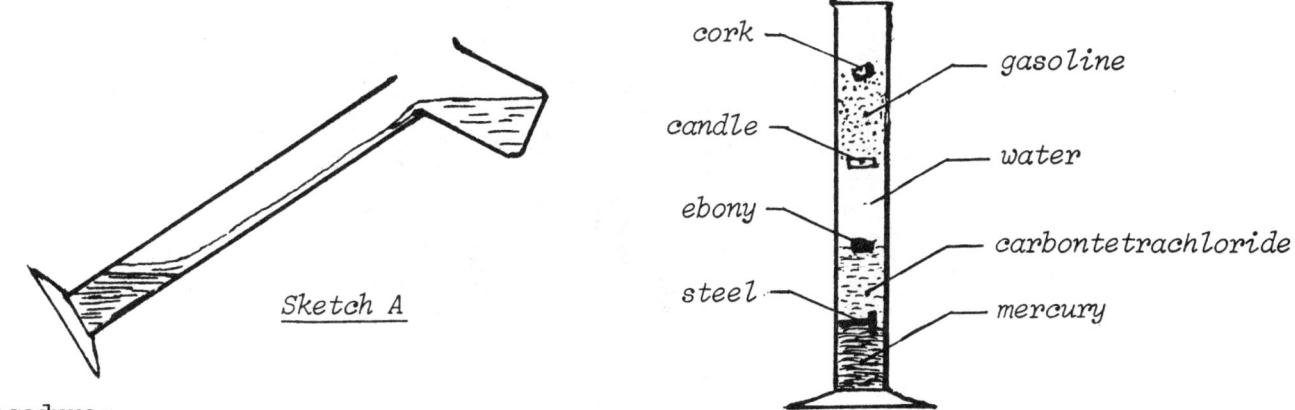

Procedure:
1. Pour about 20 ml of mercury in the cylinder, followed by 20 ml each of carbontetrachloride, water and gasoline. Do this pouring by holding the cylinder slanted so the liquids will slide on top of each other (Sketch A).
2. Now pick up the piece of ebony, and ask the students: "Where will this object end up, between which layers, or maybe on the bottom?"
3. Pick up the steel bolt and ask the same question (record students' answers) then drop it in the cylinder and observe.
4. Do the same with the piece of candle and the cork.

Questions:
1. Where did the piece of steel end up, was it on the bottom?
2. Where did the other objects end up floating?
3. How do the densities of steel and mercury compare?
4. How do the densities of the four liquids compare?
5. If we know that the density of water equals 1, what can we say about the densities of ebony, steel, carbontetrachloride and mercury?
6. What can we say about the densities of paraffin, cork, and gasoline?
7. On what property of the material does it depend, whether an object will float on a liquid or sink in it?

Explanation:
In ascending order the densities of the solids and liquids are as follows: cork, kerosene/gasoline, paraffin/candle, water (1), ebony, carbontetrachloride, steel, mercury. This demonstration shows that even steel can float on a liquid, as long as the liquid has a higher density than steel (see the Table of Densities in the Appendix).

The four layers of liquid can be used to roughly determine the density of an object. This can be done by just dropping the object in the cylinder and observing where the object ends up floating. In the order above, all those preceding water have a density of less than 1. Those that come after water have a density of greater than 1.

CHARACTERISTICS OF MATTER

DENSITY

4.34. THE BOBBING MOTH BALLS

Materials:
1. A graduated cylinder or narrow tall jar.
2. Four or five moth balls.
3. One or two Alka Seltzer tablets.

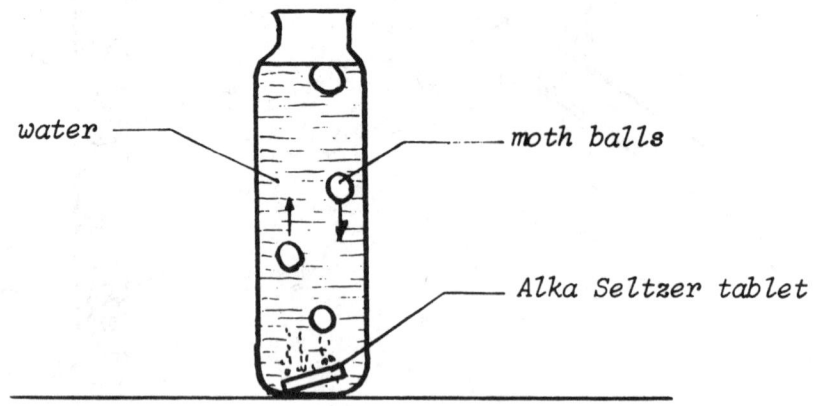

Procedure:
1. Fill the cylinder or jar with water.
2. Drop an Alka Seltzer tablet (or any tablet that produces carbon dioxide gas) in the water, and immediately afterwards the moth balls.
3. Let the students observe the moth balls (bobbing up and down).

Questions:
1. Why do the moth balls sink in the beginning (when they were dropped)?
2. What makes some moth balls bob up?
3. Why do the moth balls float only for a short while at the surface of the water?
4. What is the purpose of the Alka Seltzer tablet?
5. Would the moth balls bob up and down without the Alka Seltzer?
6. What do you observe the difference is between the rising and sinking moth balls? (Observe closely!)
7. What other liquid would give off carbon dioxide bubbles by itself?

Explanation:
   The moth balls have a density of a little over 1, so that they just sink in water. When the gas bubbles from the Alka Seltzer tablet adhere to their surfaces, they get lighter and rise up. As soon as they reach the surface of the water, the bubbles burst. This makes the moth ball become heavier again -- causing it to sink.
   If Alka Seltzer tablets are not available, other tablets that fizz in water may be used. Soda pop drinks, like 7-Up could be used as a replacement. These have the gas already dissolved in them and when they are poured into the jar, the bubbles escape, giving the same effect as the Alka Seltzer tablet's bubbles.

CHARACTERISTICS OF MATTER

DENSITY

4.35. WHICH GRAPE IS HEAVIER?

Materials:
1. Fresh grapes & 7-Up soda pop.
2. A clear drinking glass.

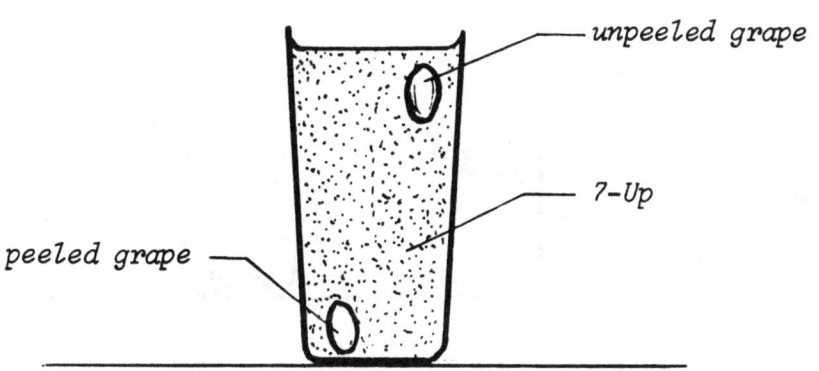

Procedure:
1. Pour and fill the glass with 7-Up soda pop.
2. Peel one grape (take the skin off and leave the grape whole).
3. Take another unpeeled grape in the other hand and pose the students the question: "Which of the two grapes is heavier?" (anticipated answer: 'the unpeeled one'--which is true!).
4. Drop both grapes (a peeled and unpeeled one) in the 7-Up and observe what they do.

Questions:
1. Why does the peeled grape sink to the bottom?
2. Which of the two grapes is lighter in weight?
3. What makes the unpeeled grape float?
4. Would an unpeeled grape also float in water?
5. Could we do this same demonstration with water and Alka Seltzer?
6. What property does the grape peel have?

Explanation:
   This demonstration is also very suitable in the teaching of adhesive and cohesive forces (bubbles adhering to the grape skin). The unpeeled grape has water repelling--**hydrophobe**--properties and thus the CO_2 bubbles from the 7-Up can adhere to this surface.
   The peeled grape lacks this **hydrophobe** skin (it is **hydrophyl**: water-attracting) and the bubbles of the 7-Up have no way of adhering to the grape. This causes the grape to stay at the bottom of the glass. The unpeeled grape becomes lighter in weight, because of the adhering bubbles, and rises to the surface of the liquid. There it loses some of the bubbles to become heavier again. It may sink a while, picking up more bubbles, to bob up to the surface again.

CHARACTERISTICS OF MATTER
DENSITY

4.36. THE DIFFERENT CLAY STICKS

Materials:
1. Molding clay and a short wooden stick.
2. A technical balance or equal arm balance.
3. Two beakers (glass or clear plastic).

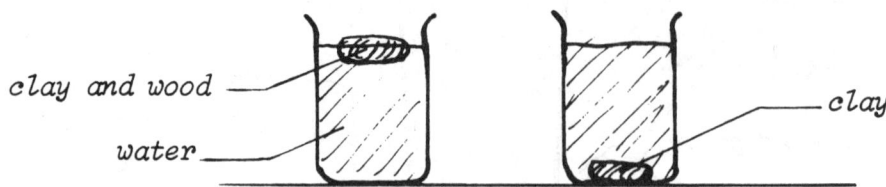

Procedure:
1. Wrap just enough clay around a short wooden stick, so that it would just barely float in water.
2. Place this on one pan of the equal arm balance and weigh off the same mass of clay on the other pan.
3. Roll a cylinder out of this mass of clay. You are now ready for the inquiry demonstration.
4. Show the students the two beakers filled with water and the two clay cylinders (to be placed in the water).
5. Show them that they weigh exactly the same, by placing them on each pan of the equal arm balance.
6. Now drop them in the two beakers with water (one sinks and one floats!).

Questions:
1. Why does one clay cylinder sink and why does the other float?
2. What property is the same for both sticks?
3. What property does it depend on whether an object sinks or floats in a certain liquid?
4. What force is holding the floating object afloat?

Explanation:
Although the mass of the two clay sticks are equal, the volume is definitely not the same, because the much lighter wooden stick is inside the one that floats. Since the density may be calculated from $D = M/V$, it can be seen that the density of the floating cylinder is smaller than the one that sinks. The force that holds up the floating cylinder is the buoyant force, which is the mass of the displaced liquid (water in this case). If the mass of the displaced water is the same as the mass of the object, it will float. If the former is less than the mass, the object sinks.

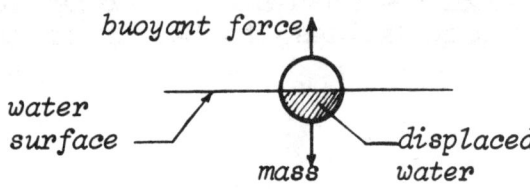

CHARACTERISTICS OF MATTER

DENSITY

4.37. IS THE CROWN MADE OF PURE CLAY?

Materials:
1. Molding clay, some sawdust or wood chips.
2. A graduated cylinder or graduated beaker.
3. An equal arm balance or technical balance.

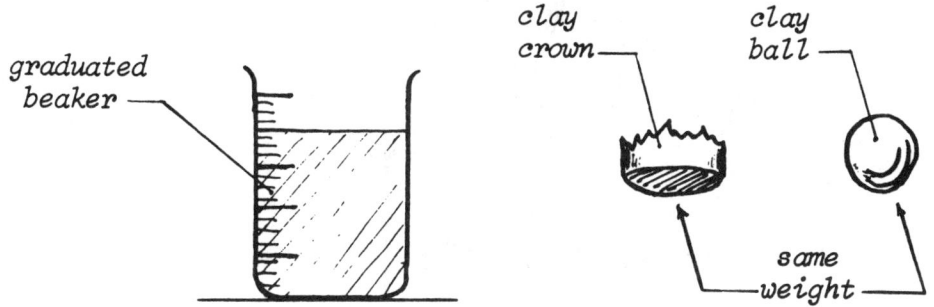

Procedure:
1. Make "impure" clay by mixing some sawdust or wood chips in the clay.
2. Now make an intricate shape (like a crown) out of the "impure" clay, and ask: "How can I find out whether the crown is made of pure clay or not?" (or tell the story of the King and Archimedes).
3. Weigh pure clay on the balance of the same mass as that of the crown.
4. Determine the volume of the crown and the pure clay by immersing them one by one in the graduated beaker. (Read off the water level before and after immersion and calculate the difference).

Questions:
1. If the crown and the clay ball were to displace the same amount of water, what may we conclude about the crown?
2. If the crown is made of impure clay, how would the two volumes differ?
3. Why does the ball need to be of the same mass as the crown?
4. What property are we actually determining of the crown?
5. What property is determined by immersing the object in water?
6. How did Archimedes determine the density of the King's crown?
7. Is density a specific characteristic of a substance?

Explanation:
The King suspected that his crown maker had cheated on him (by mixing in copper with the gold), so he summoned Archimedes to check whether his crown was made of pure gold or not. Archimedes made use of the definition of density ($D = M/V$); so he weighed in pure gold, the same mass as that of the crown, and then determined the volume of the crown and the chunk of gold by immersing it in water. If the crown were made of pure gold, the density of both objects should be the same, thus also the volume of the displaced water. If the volume of the crown came to be larger than that of the gold chunk, it would have been made of impure gold.

CHARACTERISTICS OF MATTER   DENSITY

4.38. FLOAT THE EGG WITH SALT

Materials: 1. A raw chicken egg & table salt.
            2. A large beaker or plastic container & stirrer.

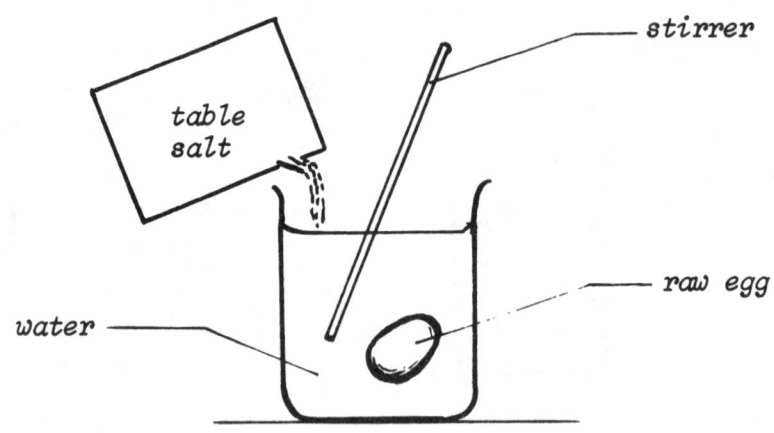

Procedure:
1. Fill the beaker about 2/3 full with water.
2. Ask students: "What will the egg do in the water, sink or float?" Immerse the egg carefully in the water.
3. Add salt to the water while stirring carefully (remember it is a raw egg), until the egg starts to float.
4. Try to dissolve other chemicals (instead of salt) in the water, like sugar, baking soda, corn syrup, etc.

Questions:
1. What would the density of the raw egg be compared to pure water?
2. By adding the salt to the water, what property was increased?
3. What two entities are we comparing, when we want to know whether an object will float in a particular liquid or not?
4. Would the egg float in alcohol? In corn syrup? In oil?
5. How can we determine the volume of an egg?
6. How can we calculate the density of an egg?

Explanation:
   From the behavior of the egg in water, it can be seen that the density of the raw egg is a little larger than that of pure water. It can actually be determined by weighing the egg and finding the volume of it my measuring the displaced water (the difference between the water levels), and by dividing this volume into the mass (weight) of the egg.
   By adding the salt to the water, the density of the water is increased. The mass of the displaced water is therefore also increased, and as this mass equals the buoyant force, the egg gets a larger force acting upwards upon it, and thus at a certain moment floats up. At this moment, the buoyant force is equal to the egg's mass. For this same reason it is much easier to swim in sea water compared to fresh water.

CHARACTERISTICS OF MATTER   DENSITY

4.39. CAN YOU MAKE CLAY FLOAT?

Materials: 1. A lump of molding clay.
2. A large beaker or transparent container.

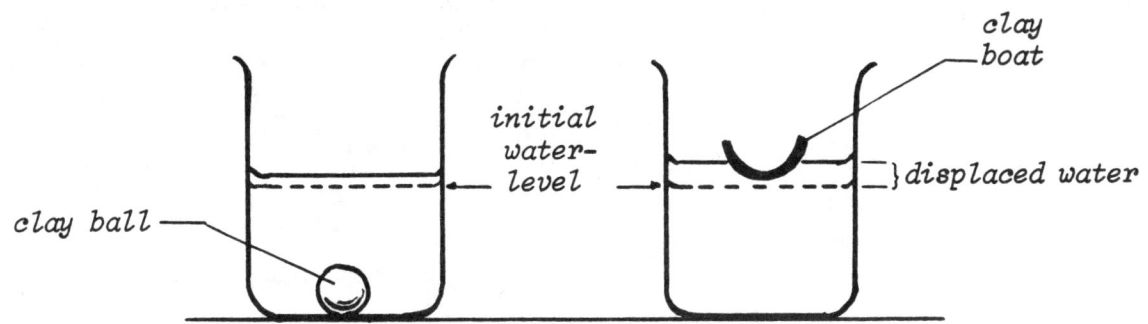

Procedure:
1. Make a ball of clay with a diameter of about 5 cm.
2. Fill the beaker half way with water and mark off the water level.
3. Ask students: "Will clay float or sink in water?" Plunge the clay ball in the water, mark off the water level, and take it back out.
4. Transform the same ball of clay into a small boat (make sure it is watertight), and let it float on the water.
5. Draw students' attention to the water level with the floating clay boat on the water.

Questions:
1. Was the clay ball heavier than the clay boat?
2. Which of the two displaced more water?
3. What is the volume of the displaced water equal to?
4. What is the weight of the displaced water equal to?
5. Would we be able to float iron or lead this same way?

Explanation:
The clay ball and clay boat were of the same weight, because the boat was made of the one and the same clay ball, and as mass is conserved and no clay has been taken away or added, the mass and thus the weight is the same. The boat, however, displaced more water because of its shape. The volume of this displaced water is equal to the volume of the submerged part of the boat or ball, whereas the weight of the displaced water is equal to the buoyant force, which is the force upward. This is equal to the mass of the clay boat when floating, but less than the mass of the clay ball (this is why the ball sinks).

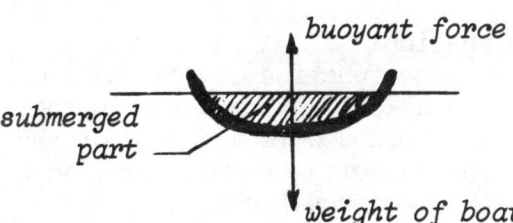

CHARACTERISTICS OF MATTER

DENSITY BUOYANCY

4.40. MAKE YOUR OWN LETTER SCALE!

Materials:
1. A measuring cylinder or tall slim jar.
2. A 20-30 cm piece of broom handle, or wooden dowel or pencil with an eraser end, and 10-20 pennies.
3. A 3x5 card, thumbtack, a small metal weight (steel nut).

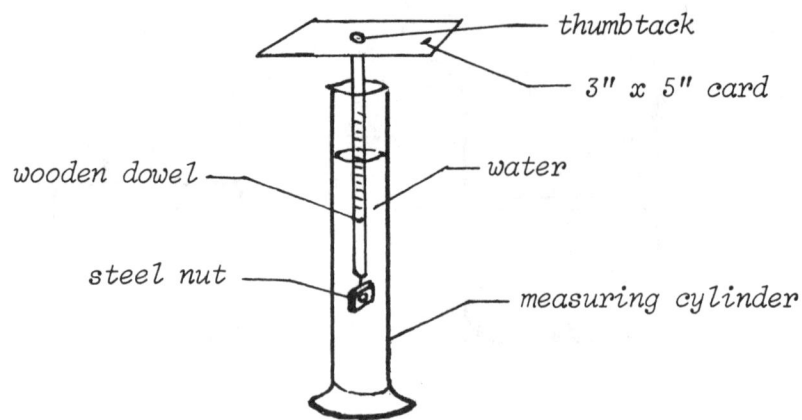

Procedure:
1. Tie a short thread to the weight and hook it to one end of the wooden dowel with a small nail or thumbtack.
2. Place it vertically in the water-filled cylinder to test whether it will float about half way in the water (if it floats too deeply, the weight is too big).
3. Attach the 3x5 card to the other end of the dowel with the thumbtack.
4. You are now ready to calibrate your home-made scale: a) Place a pencil mark at the water surface on the dowel. b) Put ten pennies on the card, and mark the dowel at the water surface (meniscus). c) Take the whole scale out of the water and divide the distance between the two marks in ten equal parts.
5. You are now ready to use the scale to weigh letters or small objects weighing less then 1 ounce or 30 grams (each line of the scale indicates .1 ounce or 3 grams: this depends on the weight of the coins used).

Questions:
1. What is the function of the small weight at the lower end of the dowel?
2. What would you say the density of wood is compared to water?
3. What other material could be used instead of a wooden dowel?
4. What variables influence the sinking distance of the dowel?
5. The capacity of the scale might be only 2 ounces or 60 grams; how can we increase its capacity?
6. How would using 2 or 3 dowels (bundled) influence the sinking distance?

Explanation:
The function of the small weight at the bottom end of the dowel is to make the dowel sink half way in the water and also to keep it floating vertically. The more dowels are used or the larger the diameter of the dowel, the greater the capacity of the scale, as the **displaced water** is increased and thus also the **buoyancy force.**

The placing of already known weights on the scale makes it possible to calibrate the wooden dowel against the sinking distance. In a less dense liquid the scale would float much lower, and in a denser liquid it would float much higher.

CHARACTERISTICS OF MATTER

BUOYANCY
MELTING POINT

4.41. THE WATER CANDLE?

Materials: 1. A drink glass or glass beaker.
2. A thick short candle, a small metal screw (or thumbtacks).

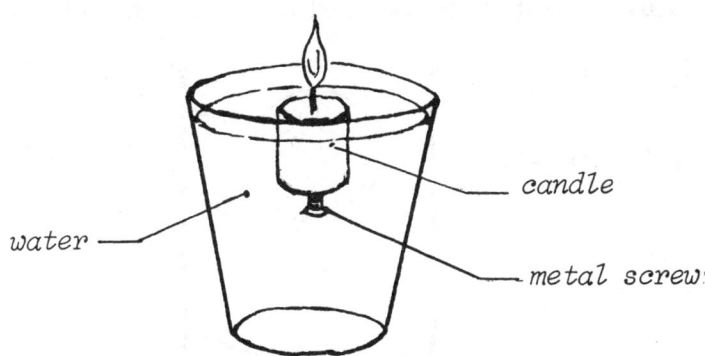

Procedure:
1. Attach the screw or thumbtacks to the bottom of the candle.
2. Fill the drink glass with water and let the candle float in it (if it sinks the screw size is too large or you used too many thumbtacks; use a smaller one or use less thumbtacks).
3. Light the candle and start asking questions.

Questions:
1. What is the function of the screw/thumbtacks in the bottom of the candle?
2. When the candle burns shorter, will the water extinguish the flame?
3. Will the candle float higher or lower in the water over time?
4. With time, will the density of the floating system change?
5. With time, will the buoyancy of the system change?
6. Will the candle actually become shorter over time?
7. When will the candle flame be extinguished?
8. What shape will the top of the candle be forming?
9. Will the candle tip over if the flame burns unevenly?
10. What would a long narrow candle do when floated in water and then lit?

Explanation:
The purpose of the metal screw or thumbtacks in the bottom of the candle is to make the candle heavier at the bottom, thus making it float in a vertical position. The top of the candle will burn leaving a concave cavity containing the molten wax. The outside of the candle which is in touch with the water stays cool and is therefore not melting. Slowly, however, the wax is burning off and the total weight/mass of the candle gets smaller.

As the mass as well as the volume of the candle become smaller, the density (mass over volume) stays the same - at least in the beginning of the burning process -, thus the part that is above water stays above water (just like 1/10th of an ice cube stays above water no matter what size of cube). But towards the end of the candle, the density becomes greater (because of the screw) and it will reach a point where the buoyancy force is not great enough to keep the candle floating.

CHARACTERISTICS OF MATTER
DENSITY OF GASES

4.42. THE FLOATING SOAP BUBBLE

Materials:
1. Soap solution (1 pt JOY, 2 1/2 pts glycerine, 3 pts water).
2. A large beaker (2 litre) or clear plastic/glass container).
3. Calcium carbonate or baking soda + dilute HCl or CCl_4.
4. A large plastic straw, and a small cup.

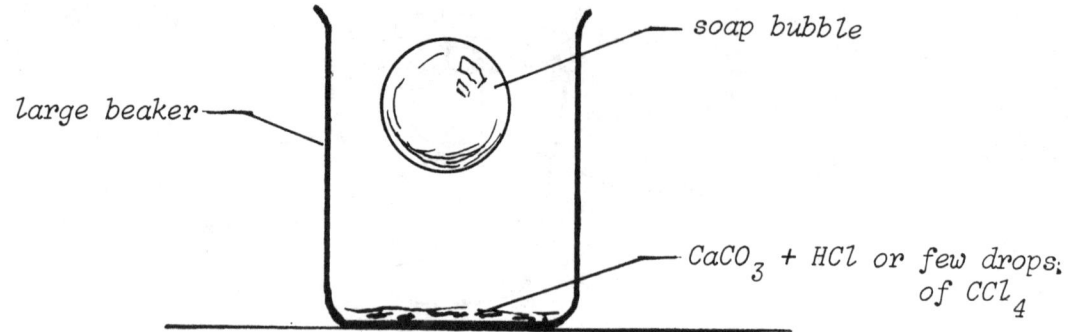

Procedure:
1. Pour the soap solution in the cup and get ready to blow a medium size bubble with the straw.
2. Put a scoop of calcium carbonate/baking soda in the large beaker and add about 50 ml dilute hydrochloric acid or a few drops of CCl_4.
3. Blow a medium size soap bubble (smaller than the beaker's diameter) and let it calmly fall in the large beaker (either guide it into the beaker or move the beaker to catch the bubble).
4. Observe the soap bubble (this demonstration/activity should be performed in a draft-free room).

Questions:
1. What is being produced in the large beaker? (when using $CaCO_3$ and HCl?
2. Why did the soap bubble float above the liquid in the large beaker and not descend all the way to the bottom?
3. How does the density of carbondioxide or carbontetrachloride gas compare with that of atmospheric air?
4. What other gas could be used instead of carbondioxide?

Explanation:
The chemicals, calcium carbonate or baking soda plus the dilute hydrochloric acid produce carbondioxide gas in the large beaker. This gas has a larger density (in other words is heavier) than air, and therefore stays in the beaker. The concentration of this gas is most likely higher closer to the surface of the liquid or bottom of the beaker.

The density of a soap bubble is just slightly greater than that of air. This is why bubbles ever so slowly descend in the air. In the large beaker the denser carbondioxide gas acts like a cushion, and this is why the bubble is not dropping all the way to the bottom.

Instead of carbondioxide gas a few drops of carbontetrachloride can be used in the large beaker. This liquid will evaporate and the gases are also heavier than air.

CHAPTER 5

HOW DO CHEMICALS BEHAVE IN OUR DAILY LIFE ?

OBJECTIVES

After dealing with and studying the concepts and sub-concepts of this chapter, the students should be able to:

a. recognize the correct explanation of an observed event based on each of the sub-concepts;
b. explain in their own words which of the sub-concepts is determining the course of an event;
c. distinguish true from false statements concerning each one of the sub-concepts;
d. identify the correct explanation of an event in daily life applying one of the sub-concepts;

all in relation to the following sub-concepts:

-- The existence of physical versus chemical changes.
-- Different indicators will give different colors in acids and in bases.
-- Some reactions produce gas, some produce precipitates.
-- The solubility of chemicals in water differ greatly.
-- Replacement and catalytic reactions.
-- Exothermic, endothermic, and spontaneous combustion reactions.
-- Addition reactions with unsaturated double bonds.
-- Polymerization or formation of large molecules.

CHEMISTRY   **PHYSICAL &**
                                                 CHEMICAL CHANGE

5.1. THE CHARCOAL SAUSAGE

Materials: 1. About 100 ml of sugar crystals.
             2. Two 100 ml beakers, two glass stirrers.
             3. Concentrated sulfuric acid (H_2SO_4).

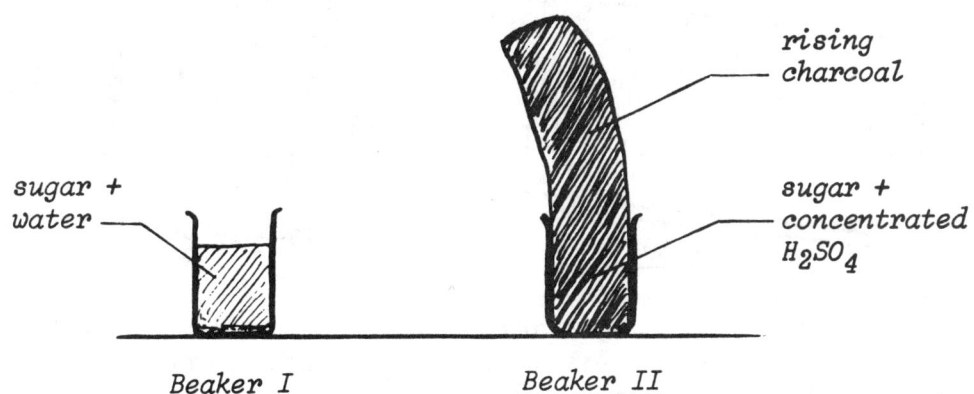

Procedure:
1. Fill each beaker half full with sugar.
2. Add about 40 ml of water to the first beaker and the same amount of concentrated sulfuric acid to the second beaker; stir and let stand.
3. Observe the difference between the two changes in the sugar. Ask: "In which beaker do the reactants still have the same properties?" (the reaction should preferably be carried out under a fume hood or close to an open window, or in the outdoors).

Questions:
1. What is the difference between the processes in beaker I and II?
2. How can we recognize or distinguish between a physical change and a chemical change?
3. In which of the two beakers could we get the sugar back as sugar?
4. What do you think happened in the second beaker?
5. What property do you think concentrated sulfuric acid has?
6. What do you think the black material in beaker two is?

Explanation:
   In beaker I, where the sugar was mixed with the water, a physical change was taking place. This means that the components of the mixture retained their properties; they could be separated and still have the exact properties as before the change. The water could be left to evaporate and the sugar would crystallize out of the solution.
   In beaker II a chemical change took place, leaving products, which have properties that are completely different from the original components of the mixture. A black charcoal mass is produced, which is expanding up because of the gases (SO_2) and water vapor being released. This is caused by the dehydrating properties of concentrated sulfuric acid.

CHEMISTRY — PHYSICAL CHANGE

5.2. WALK THROUGH A HOLE IN ORDINARY NOTEBOOK PAPER

Materials: 1. One sheet of ordinary notebook paper.
             2. A pair of large, sharp scissors.

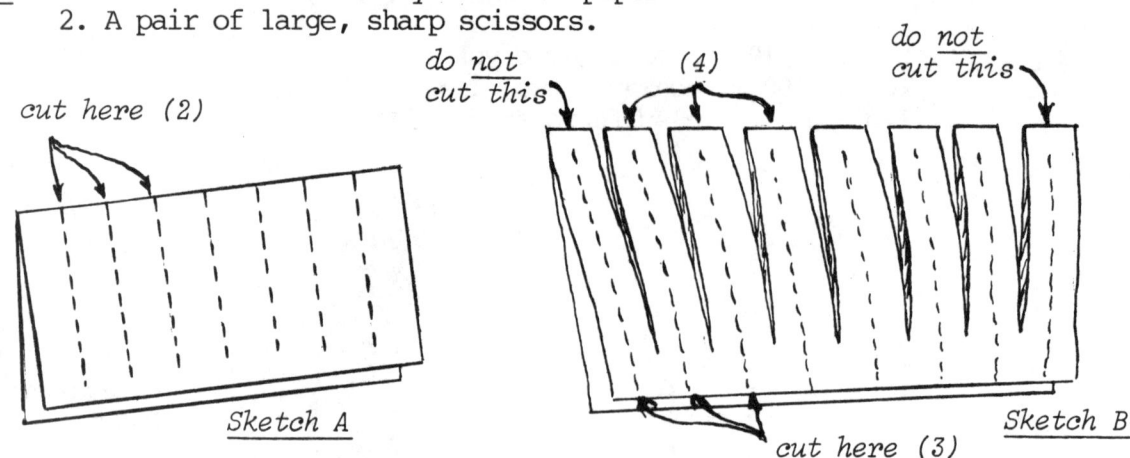

Procedure:
1. Tell the students that you are going to cut a hole in the ordinary piece of paper, and that you are going to walk right through the hole without tearing the paper!
2. Fold the paper in half, then make a series of straight cuts from the folded side, about 2 cm apart stopping about 1 cm from the edge of the opposite side (see Sketch A).
3. Turn the paper around and make cuts from the other side (see dotted lines in Sketch B), and also stop about 1 cm from the edge of the opposite side.
4. Except for the first and last strip at each end, now snip off the folded ends of the strips.
5. Then carefully open up the paper without tearing anything and walk through the hole!

Questions:
1. Do you think you could ride a bicycle through the hole?
2. We've changed the paper by cutting it; was it a physical or chemical change that the paper underwent?
3. Did we change any of the chemical properties of the paper?
4. What is the difference between a physical change and a chemical change in a sample of matter?
5. What are some other ways that you could make a physical change in this piece of paper?
6. What are some ways that you could make a chemical change in the paper?

Explanation:
Of course, the change that we made in the paper was a physical one. No chemical properties of the paper were changed, before or after the change. When a chemical change is occurring in a sample of matter, the chemical properties in the products after the change are completely different from those before the change. To make a chemical change in the paper, it could be burned or placed in sulfuric acid. The products of the burning process of paper would be CO_2 (carbondioxide) which is a gas, plus water vapour and carbon. All three products having totally different properties compared to those of paper (the sample of matter before the change).

CHEMISTRY
ACIDS & BASES

5.3. THE BLUE AND RED CABBAGE

Materials:
1. A few leaves of red cabbage.
2. A small beaker and four test tubes (test tube rack).
3. An alcohol burner and stand.
4. Vinegar, lemon juice, baking soda, lime.

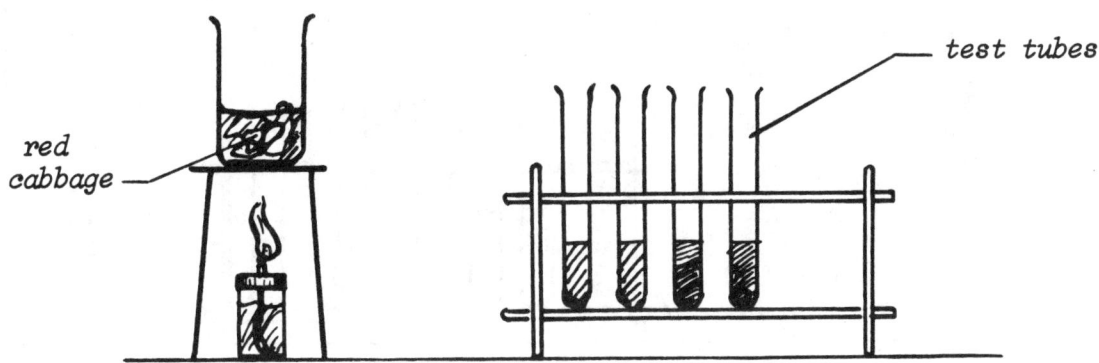

Procedure:
1. Cut a red cabbage leaf in small pieces; place it in the beaker and add about 10 ml of water to it.
2. Heat it above the burner until boiling, then pour equal amounts of the liquid in the four test tubes.
3. Place a few drops of vinegar in the first and a few drops of lemon juice in the second test tube: have students observe closely the color of the liquid.
4. Now add a pinch of baking soda to the third and a pinch of lime to the fourth test tube: observe the color change!

Questions:
1. Which two chemicals colored the cabbage juice red?
2. Which two colored the cabbage juice blue?
3. What would happen if we added baking soda to the cabbage leaf?
4. To which of the two groups of chemicals would orange juice belong? Pineapple juice? Grape juice? Soap water?
5. What would happen if we added vinegar to the blue liquid?

Explanation:
The red cabbage juice acts like litmus, an indicator that turns red in acids and blue in bases. Vinegar, lemon juice, and other sour tasting juices are all acids; whereas baking soda, lime, and soap water are bases. Chemists are talking in terms of pH (which will be dealt with in higher grades: the negative logarithm of the hydrogen ion concentration). A pH of 7 for a neutral liquid, less than 7 for acids, and higher than 7 (up to 12) for bases.

Other common indicators are: litmus paper, phenolphthalein, methyl orange, methyl red, universal indicator.

CHEMISTRY   ACIDS & BASES

5.4. THE FUNNY COLORS

Materials: 1. A test tube rack with six test tubes.
            2. Dilute hydrochloric acid (HCl). (1 Normal)
            3. Dilute sodium hydroxide solution (NaOH). (1 Normal)
            4. Phenolphthalein indicator, and a drinking straw.
            5. Medicine dropper, an opaque cup (styrofoam will do).

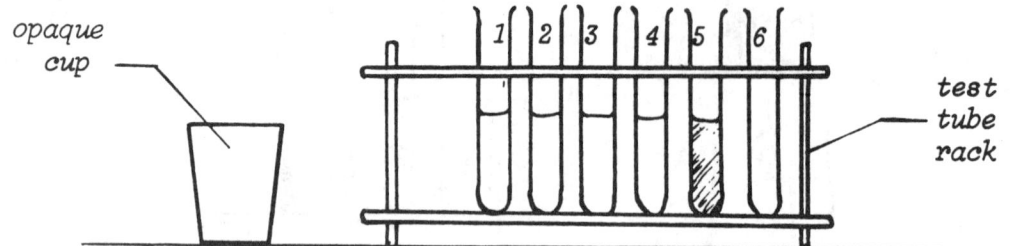

Procedure:
1. Number the six test tubes to avoid confusion.
2. Tape a white paper behind the tube rack for easier distinction of the pink color in the test tubes (a spot light will help).
3. Place 3 drops of phenolphthalein in test tube No. 2 and 4; and 1 drop of NaOH in test tube No. 6 and the cup each; and 3 drops of HCl in test tube No. 5.
4. Fill the opaque cup with drinking water from a pitcher, then halfway fill test tube 1, 2, 3, & 4 from the cup (2 & 4 appear pink).
5. Say: "There must be something wrong!" Pour the four test tube contents back into the cup. Now fill test tubes 1, 2, 3, 4, & 5. (All are pink except for No. 5).
6. Say: "What now?" Pour all 5 tubes back and fill all six tubes. No. 6 stays pink. Blow color away with the straw!

Questions:
1. What made the pink color in test tube No. 2 and 4?
2. What made the color disappear in test tube No. 5?
3. What is in our breath that can neutralize the basic solution?
4. What else could we use to take away that pink color of tube 6?

Explanation:
Phenolphthalein is an indicator for basic solutions. It turns pinkish red in dilute sodium hydroxide solution or any other basic solution (with a pH higher than 7). Test tube 5 contains hydrochloric acid, which will neutralize the base being poured in it. The pH is brought back to lower than 7 and thus turning the liquid colorless.

The air that human beings are exhaling contains more carbon dioxide than atmospheric air. This gas is slightly acid in water and neutralizes the weak basic solution. Other substances that can be used to neutralize bases are: vinegar (acetic acid), lemon juice (citric acid), etc.

CHEMISTRY   OXIDATION REACTION

5.5. DRAW WITH FIRE

Materials: 1. Potassium or sodium nitrate (saltpeter).
            2. A small beaker and a glass stirring rod.

paper

marked spot.
touch with
glowing match

Procedure:
1. Dissolve a full scoop of potassium nitrate (or sodium nitrate) in about 20 ml of water in the beaker, and stir with the glass rod until all the solid material dissolved.
2. Keep adding another scoop of KNO_3 or $NaNO_3$ and stir until the solid material stays as a solid on the bottom of the beaker.
3. Take a blank sheet of paper and draw an animal or other object on the paper with the glass rod dipped in the saturated solution. (make the lines rather thick and use the liquid rather liberally).
4. Leave the paper to dry at room temperature. Before it is completely dry, mark one spot where the drawing was started with a pencil.
5. Show the students the completely blank sheet of paper, and tell them that you will start a drawing on the paper with fire! Strike a match, blow it out, and with the glowing tip touch the marked spot on the paper; observe!

Questions:
1. Why does the pencil mark have to be made just before the drawing was completely dry?
2. What made the glowing continue further on the paper?
3. Was it actually a flame that made the drawing?
4. What kind of a reaction was it: endothermic or exothermic?
5. Is this type of a reaction a spontaneous reaction?
6. What was the glowing wood tip initially needed for?

Explanation:
   Potassium or sodium nitrate is a strong oxidant. After the water of the saturated solution evaporated, there was only potassium nitrate left on the paper in a very finely divided state. The glowing tip of the match supplied the initial **activation energy** needed to release very active oxygen from the potassium nitrate. This active oxygen is further oxydizing the paper (seen in the form of a burning glow), which is exothermic (giving off heat). This energy is enough to keep further releasing the oxygen from the KNO_3. Since this is only found on the line of the drawing, the glow of the paper burn is only following the path of the drawing.

CHEMISTRY — REDUCTION REACTION

5.6. TURN A RED ROSE INTO A WHITE ONE

Materials:
1. A small red rose, and sulfur powder (S).
2. A short metal strip, and a short metal wire.
3. A glass beaker and a watchglass to cover it.

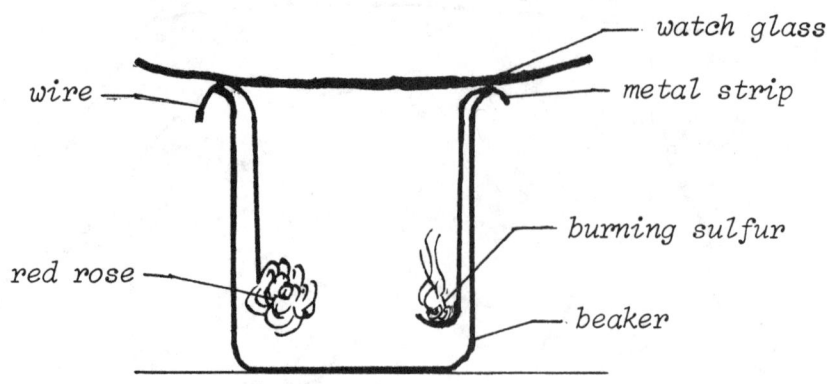

Procedure:
1. Attach the wire to the stem of the rose and let it hang inside the beaker from the rim (make a hook in the wire).
2. Bend the short metal strip in an S-shape so that it can hang from the rim of the beaker (hang this on the opposite side: see Sketch above).
3. Take this metal strip out of the beaker and place a small heap of sulfur powder on the bent part of the metal strip, hold a lit match close to the sulfur until it catches fire. Immediately lower the metal strip into the beaker and cover it with the watchglass. Observe!
(Keep the beaker covered; it would be best to perform this under the fumehood or outdoors).

Questions:
1. What color did the red rose turn into?
2. What part of the rose started to change color first?
3. What type of reaction is actually taking place?
4. What gas was formed by burning the sulfur?
5. What other bleaching agents do you know?
6. What is the function of the watchglass over the beaker?

Explanation:

By burning the sulfur, sulfurdioxide gas is formed (SO_2) which is a strong reductor. The red coloring in the rose is being reduced and turns white. Sulfurdioxide is a very pungent gas and also poisonous when inhaled in large doses. This is why it is very important to dispose of the fumes under an exhaust hood or outdoors.

Other bleaching agents are chlorine compounds, like: chloride of lime, sodium chlorite, sodium hypochlorite (used in laundry bleaches) and calcium hypochlorite. In solution, these agents release chlorine gas, which reacts with colour molecules and thus also remove colour.

The sulfurdioxide gas plus the water in the rose petals form sulfurous acid, which works as a reducing agent. The reaction with the water is as follows: $SO_2 + H_2O \longrightarrow H_2SO_3$

CHEMISTRY   GAS PRODUCTION

5.7. GRIND A CRACKER

Materials: 1. A mortar and pestle, & scoop.
            2. Potassium chlorate (KClO$_3$), and sulfur (S) powder.

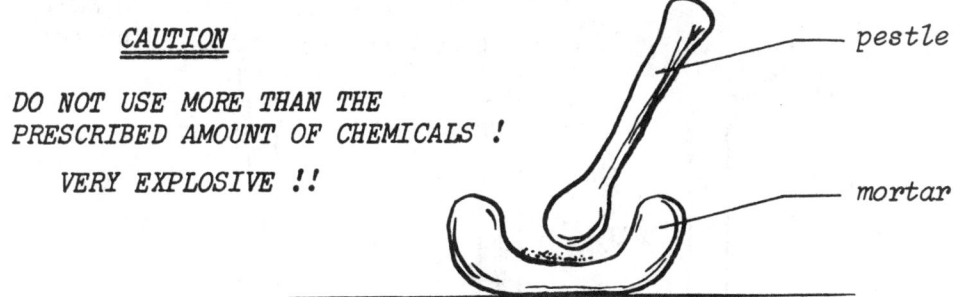

CAUTION

*DO NOT USE MORE THAN THE
PRESCRIBED AMOUNT OF CHEMICALS !*

VERY EXPLOSIVE !!

Procedure:
1. Take a tip of a scoop of potassium chlorate and the same amount of sulfur and place it in the mortar. (less than .5 g of each!)
2. Now show the students as if you were just casually mixing the two powders. Put a little pressure on the pestle (hold the mortar and pestle with outstretched arms), and watch the students jump out of their seats!
3. Place cotton balls in your ears if you can't stand loud noises!

Questions:
1. What are the products of the reaction?
2. Why do we hear an explosion in general?
3. What happens to the air in a bursting balloon?
4. What made the explosions in the mortar?
5. What do you think the rate of gas production is in this reaction?

Explanation:
We can hear an explosion, because a sudden large wave in the air hits our ear drums. This large wave is usually caused by a sudden expansion or production of gas. In this case a sudden production of sulfur dioxide (SO$_2$) is the cause of the explosions. The reaction is as follows:

$$KClO_3 \rightarrow KCl + 3 O_n \quad \ldots \ldots (1)$$
$$S + 2 O_n \rightarrow SO_2 \quad \ldots \ldots \ldots (2)$$

The grinding of the powder causes heat to develop between the mortar and the pestle, which triggers reaction (1) to proceed to the right. The oxygen released in **status nacendi** (O_n) is in atomic form and very reactive, thus immediately attacking the surrounding sulfur (reaction 2).

The sudden production of SO$_2$ gas causes a sudden movement of the surrounding air, thus a large wave is created and this reaches our eardrum, with the result of hearing the explosion. Potassium chlorate is very much used in the firecracker industry, together with charcoal and sulfur.

CHEMISTRY   GAS PRODUCTION

5.8. WHICH GAS IS WHICH?

Materials: 1. Zinc powder or chips, calcium carbonate (marble chips), potassium chlorate, and hydrochloric acid (dilute).
2. An alcohol burner and wood splints.
3. Three test tubes.

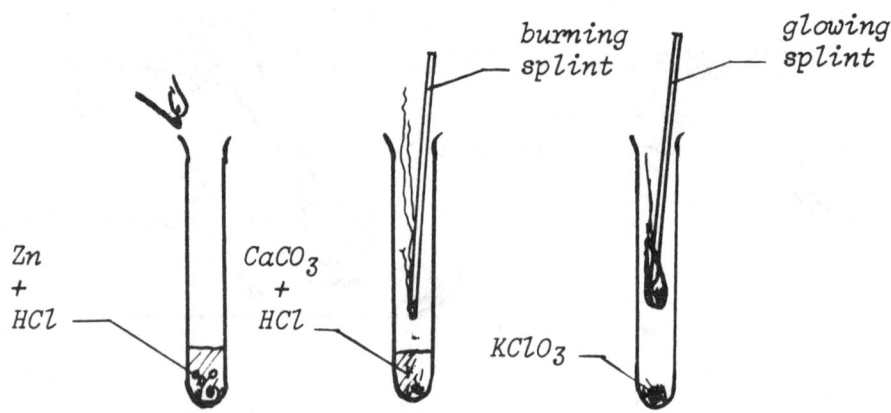

Procedure:
1. Place a few zinc chips in the first test tube, add about 3 ml of dilute HCl, and place your thumb over the tube. Light a match and bring it close to the test tube's mouth: release thumb!
2. In the second test tube, put a few marble chips and add about 3 ml of dilute HCl. Burn a wood splint and lower the flame into the test tube. (What happens to the flame?)
3. Put a little $KClO_3$ plus a little of the burnt ashes from the wood splint into the third test tube. Mix them by shaking the tube and heat it slowly over the alcohol flame. Test the gas with a glowing wood splint (not a burning one; flame should be blown off before inserting in tube).

Questions:
1. What gas is produced in the first test tube?
2. What unique characteristics does this gas possess?
3. What gas is produced in the second test tube?
4. How can we recognize this second gas from the others?
5. What gas is produced in the third test tube?
6. How different is this third gas compared to the other two?

Explanation:
In test tube I: $Zn + 2\ HCl \rightarrow ZnCl_2 + H_2$; Zinc chloride and Hydrogen gas is produced. This gas is highly flammable and gives off a characteristic small explosion, when ignited in the test tube.

In test tube II: $CaCO_3 + 2\ HCl \rightarrow CaCl_2 + H_2O + CO_2$; calcium chloride, water, and carbon dioxide are formed. This gas is also invisible, but it has the property of extinguishing a flame.

In test tube III: $2\ KClO_3 \rightarrow 2\ HCl + 3\ O_2$; potassium chlorate is decomposed by heat into potassium chloride and oxygen gas, which makes a glowing splint flare up. Oxygen is present in the air and makes up 20% of the atmospheric air: a necessary component for all living things.

CHEMISTRY   PRECIPITATES
                                                       GAS PRODUCTION

5.9. TURN WATER INTO WINE, MILK, AND BEER!

Materials:
1. Sodium bicarbonate ($NaHCO_3$), sodium carbonate (Na_2CO_3).
2. Phenolphtalein, saturated barium chloride solution ($BaCl_2$)
3. Two drink glasses, a wine glass, a beer mug.
4. Conc. HCl, and Bromothymol Blue.

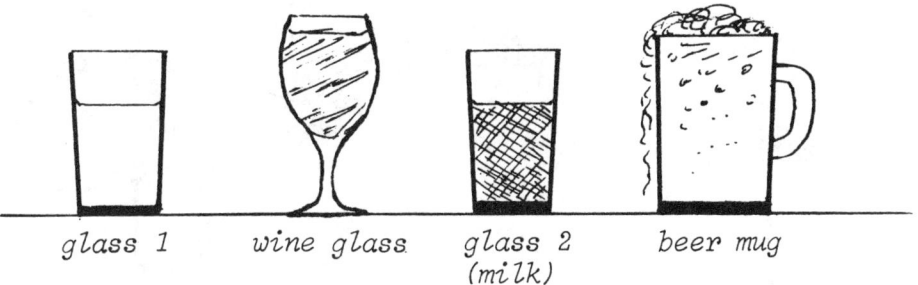

glass 1   wine glass   glass 2   beer mug
                           (milk)

Procedure:
1. Fill glass No. 1 three-quarters full of water and add 10 ml of saturated $NaHCO_3$ and 20% Na_2CO_3 solution (pH 9).
2. To the wine glass add a few drops of phenolphtalein.
3. To glass No. 2 add about 25 ml of saturated $BaCl_2$ solution.
4. To the beer mug add 5 ml of conc. HCl and about 3 ml of bromothymol blue. (Points 1-4 should be done without the students knowing).
5. Now you are ready for the demonstration:
   Starting with glass No. 1, tell the class/audience that you have water in the glass; then pour contents into the wine glass while saying: "Now I am turning it into wine!"
   Now pick the wine glass up and pour contents into glass No. 2, saying: "Then I am turning it into milk!"
   Pick this glass with "milk" up and pour its contents into the beer mug, while saying: "Now the milk turns into beer!"

Questions:
1. What made the colourless solution turn red in the wine glass?
2. What chemicals would produce a white precipitate?
3. What reaction would produce a gas?
4. Could we turn the beer back into milk?
5. What is a one-way reaction like the above one called?
6. What reversible reactions do you know?

Explanation:
   The change of the colourless liquid in glass No. 1 into the red coloured solution in the wine glass is caused by the phenolphtalein, which turns red in basic solutions. The barium ions in glass No.2 precipitated out in the presence of carbonate ions from the Na_2CO_3, and thus gave the milky white appearance. This milky mixture is actually a suspension of $BaCO_3$ solid particles in water. Pouring this suspension into the concentrated hydrochloric acid and bromothymol blue produces CO_2 gas throughout the liquid, thus forming the foam like in a beer. The bromothymol blue gave it the yellow colour at that particular pH.
   This demonstration would be most effective if steps 1 through 4 were not shown to the students!

CHEMISTRY — SALT FORMATION

5.10. MAKE A SMOKE SCREEN

Materials:
1. Two wide (large size) test tubes.
2. Four L-shaped glass tubes in two 2-hole stoppers.
3. A T-tube and two pieces of rubber tubing.
4. Concentrated HCl and ammonia (NH_4OH).

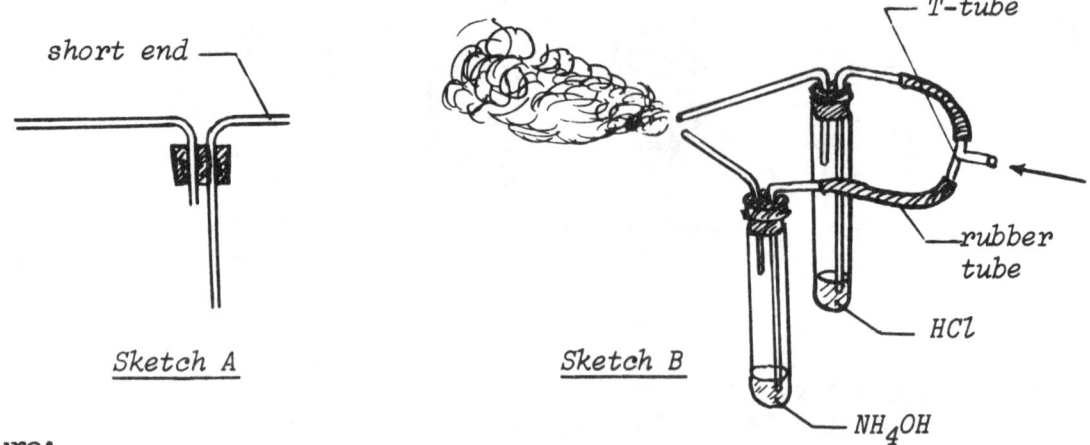

Procedure:
1. Fit the four L-tubes in the two 2-hole stoppers as in Sketch A.
2. Attach two rubber tubes to the short ends of the L-tubes above the rubber stopper (see Sketch A).
3. Attach the other end of the two rubber tubes to the glass T-tube (see Sketch B).
4. Place about 10 ml of concentrated hydrochloric acid (HCl) in one test tube and 10 ml of concentrated ammonia (NH_4OH) in the other.
5. Close the test tubes off with the stoppers (and tubes in them).
6. Hold the two test tubes in one hand (making sure that the two free ends of the glass tubes are close together), hold the T-tube in the other hand, and blow through the free end of the T-tube.

Questions:
1. What reaction is taking place?
2. What is making up the smoke particles?
3. How could we do this reaction without the glass and rubber tubes?
4. What does smoke in general consist of?

Explanation:
The reaction that takes place where the smoke is formed, is as follows: $NH_4OH + HCl \rightarrow NH_4Cl + H_2O$. The NH_4Cl (ammonium chloride) is a solid and as the reactants are gases or actually fine mists, the ammonium chloride formation results in very fine solid particles. Smoke is exactly this: a fine suspension of solid particles in a gas. Smoke that is used for sky-writing by planes, is produced by spraying special oils and paraffins on the hot exhaust of the airplane.

If glass and rubber tubes are not available, the reaction can also be accomplished by holding the mouths of the two test tubes with HCl and NH_4OH close to each other.

CHEMISTRY   RATE OF REACTION

5.11. THE RETURNING COLOR

Materials:
1. One Erlenmeyer flask (medium size), & two beakers.
2. Potassium iodide (KI), hydrogen peroxide (H_2O_2), sodium thiosulfate ($Na_2S_2O_3$), starch solution, and dilute hydrochloric acid (HCl).

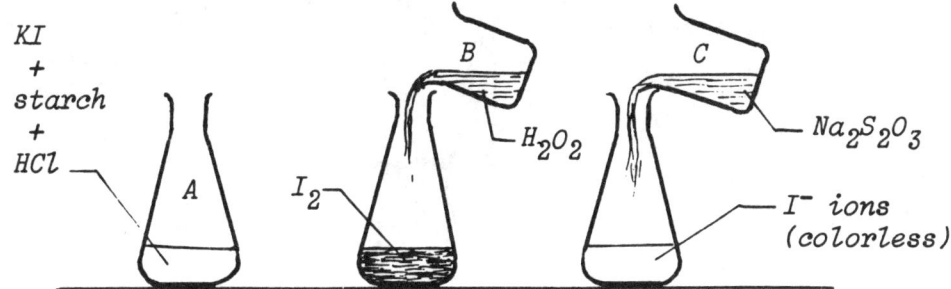

Procedure:
1. Dissolve a little KI in water, add some starch solution and a little dilute HCl (Sketch A).
2. Prepare solutions in beaker B and C. Place a 3% H_2O_2 solution in beaker B; dissolve some $Na_2S_2O_3$ in about 100 ml of water in beaker C.
3. Add to flask A some of liquid B (turns blue-black) and immediately afterwards a little of liquid C (turns colorless).
4. Place flask on table and observe: sudden color change!
5. The addition of liquid C can be repeated several times. (To make it more interesting, after addition of liquid C, hold the flask behind your back for a few moments and bring it back on the table as a blue-black solution).

Questions:
1. How is the blue-black color created?
2. What was it that took away the color?
3. Which reaction is slower? The color-producing or the color-removing one?
4. How long can we keep repeating the second reaction?
5. What reactions are usually slower in rate?

Explanation:
   The reactions that are taking place are the following:

Flask A + Beaker B : $2 KI + H_2O_2 \xrightarrow[\text{Starch}]{\text{HCl}} 2 KOH + I_2$ (blue-black)

Addition of Beaker C : $I_2 + Na_2S_2O_3 \longrightarrow NaI + Na_2S_2O_5$ (colorless)

   Reaction (1) is an oxidizing reaction, where the iodide ions are oxidized into iodine, which turns starch into a blue-black color. The addition of $Na_2S_2O_3$ reduces the iodine back into iodide ions, which are colorless. The fact that the blue-black color keeps on returning, even after the $Na_2S_2O_3$ is added, shows that the first reaction is still proceeding and thus is a much slower reaction than the second.

5.12. DISSOLVE MORE SALT BY COOLING?

Materials:
1. Sodium nitrate, sodium chloride, potassium nitrate, potassium chloride, calcium acetate.
2. Three beakers (250 ml), stirrers, & scoop.
3. Alcohol burner & stand.

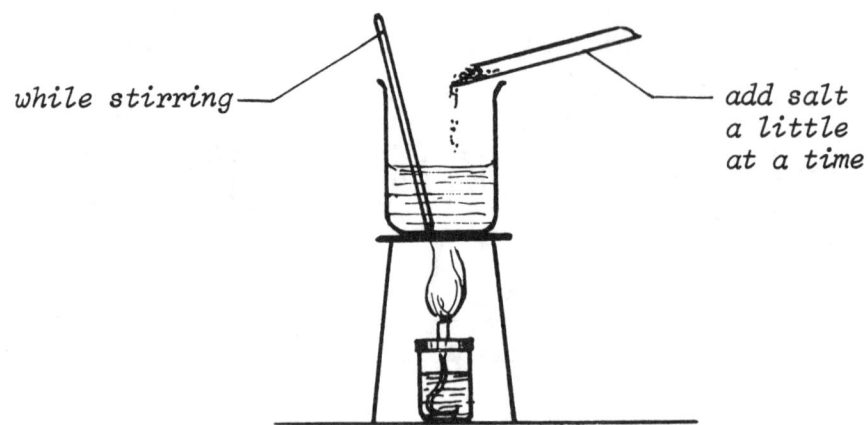

Procedure:
1. Make a saturated solution of each of the following salts: potassium nitrate (or sodium nitrate or potassium chloride), sodium chloride, and calcium acetate, at room temperature, by adding half a teaspoon at a time to 100 ml of water, until some solid is left behind after stirring for one minute. (Stir after every addition of solid).
2. Now heat beaker I (with KNO_3) and as soon as the solid particles dissolve, add half teaspoon at a time until some solid stays undissolved.
3. Do the same with beaker II (NaCl): not as much can be dissolved!
4. Heat beaker III (with Ca acetate): more solid precipitates out!

Questions:
1. How could we dissolve more Ca acetate in Beaker III?
2. Most salts behave like the first two salts; what variable goes up with increasing temperature?
3. Which of the salts could dissolve much more with increasing temperatures?
4. What happened to the three salt solutions after being cooled?

Explanation:
The solubility of a substance is generally increasing with increasing temperatures (see solubility curves I and II on the right). Solubility curve III is very seldom encountered. Students can keep track of the amount of salt added to the water and the temperature, in order to plot a solubility curve.

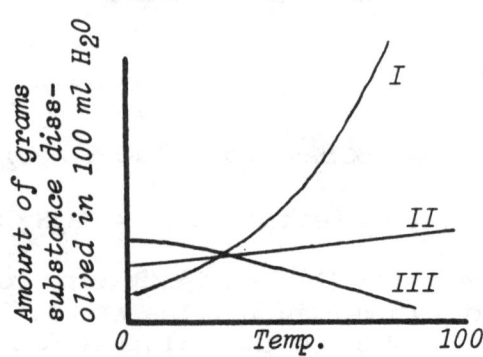

CHEMISTRY   SOLUBILITY

5.13. THE THREE-LAYERED LIQUID

Materials: 1. Carbon tetrachloride (CCl_4).
             2. Ether (di-ethyl), iodine crystals.
             3. A tall (graduated) cylinder (100 ml).

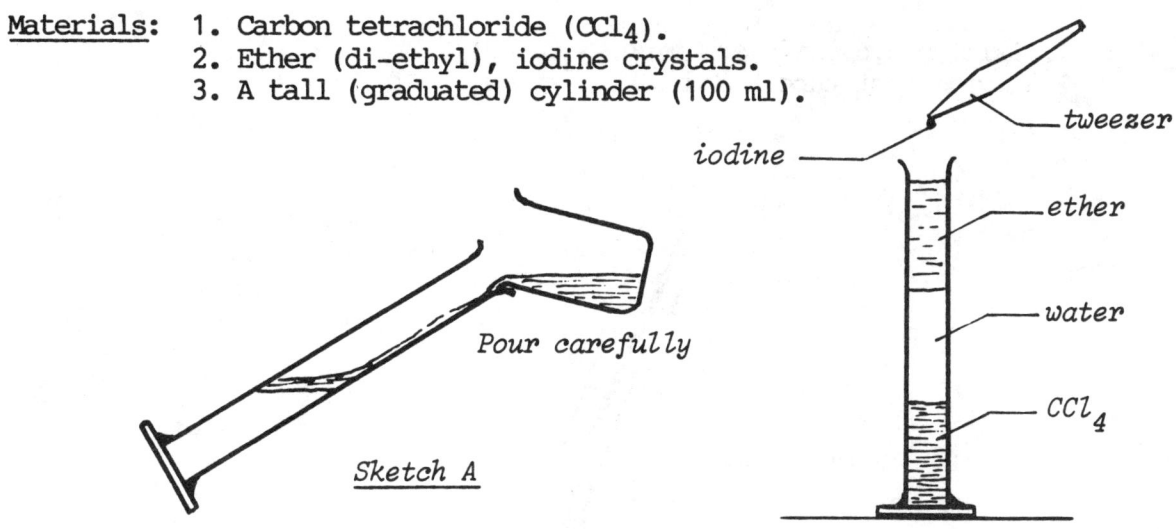

Sketch A

Sketch B

Procedure:
1. Pour about 30 ml of carbon tetrachloride in the cylinder.
2. Carefully pour 30 ml of water on top of the carbon tet, by slanting the cylinder when pouring (see Sketch A).
3. Do the same thing with 30 ml of ether on top of the water.
4. Stand the cylinder on the table and drop several crystals of iodine in the layered liquid. Observe the colors.

Questions:
1. Why did the water float on top of the carbon tet?
2. Would ether be able to float on top of carbon tetrachloride?
3. What does it depend on whether we can make layered liquids or not?
4. Why did we get different colors in the liquid?
5. What does a dark color indicate about the solubility?
6. In what liquid did iodine dissolve the least?
7. With what other liquids could we make layers?

Explanation:
   The liquids could stay layered on top of each other, firstly because one is lighter than the other (smaller in density), and secondly because they are insoluble in each other: the carbon tetrachloride is **immiscible** with water, and the ether is also insoluble in water. But ether can dissolve, or is **miscible** in carbon tetrachloride and thus floating the first into the other would not work. Most organic liquids, like oils, gasoline, benzene, etc. except alcohols, are **not soluble** in water or **immiscible** with water.
   The iodine crystal, when dropped through the layered liquid, dissolves in different degrees in each of the solvents. This is why different color intensities appear in the liquid layers.

CHEMISTRY   SOLUBILITY &
                                         PRECIPITOUS REACTION

5.14. THE DISAPPEARING PRECIPITATE

Materials: 1. Calcium hydroxide solution -- $Ca(OH)_2$.
            2. A wide test tube & a drinking straw.

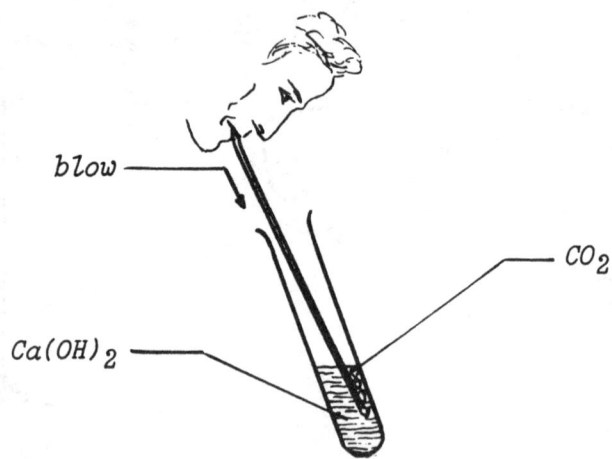

Procedure:
1. Fill the test tube about one-third with calcium hydroxide solution. If in powder form, make a saturated solution by dissolving some in warm water and filtering excess powder; use filtrate.
2. Blow through the straw into the calcium hydroxide solution until a precipitate is formed (cloudy solution).
3. Now continue blowing into the cloudy liquid until the liquid clears up again: the cloud is blown away!

Questions:
1. What is the reaction that forms the precipitate?
2. In general, why and when is a precipitate formed?
3. What made the formed precipitate disappear?
4. When is a precipitate or solid powder in a liquid invisible?
5. Can you name other precipitous reactions?

Explanation:
By blowing into the solution, we were adding carbon dioxide gas to it. The reaction is as follows: $CO_2 + Ca(OH)_2 \rightarrow CaCO_3 + H_2O$, where the calcium carbonate formed is the precipitate. Although this precipitate is in the form of a milky or cloudy substance, it is still a deposit of solid material. This is apparent when the blowing is ended at this stage and the test tube left standing for a while. A precipitate will then collect at the bottom of the test tube.

By continuing the blowing, more carbon dioxide is added to the precipitate, resulting in the reaction: $CaCO_3 + H_2O + CO_2 = Ca(HCO_3)_2$, which is calcium bicarbonate, and this substance is soluble in water.

In general, a precipitate is formed when the compound formed has a low solubility in water. When this solubility is increased, either by raising the temperature or by changing the compound itself, it goes back into solution and becomes invisible (clear liquid).

CHEMISTRY   SOLUBILITY OF GASES

5.15. THE AMMONIA FOUNTAIN

Materials:
1. One round- or flat-bottomed flask (400 ml).
2. One test tube and a 1-hole stopper and bent glass tube.
3. A 2-hole stopper with a long glass tube drawn at one end in one hole and a dropper in the other hole.
4. A 500 ml beaker, a propane or alcohol burner.
5. Ammonium chloride (NH_4Cl) and calcium carbonate ($CaCO_3$).

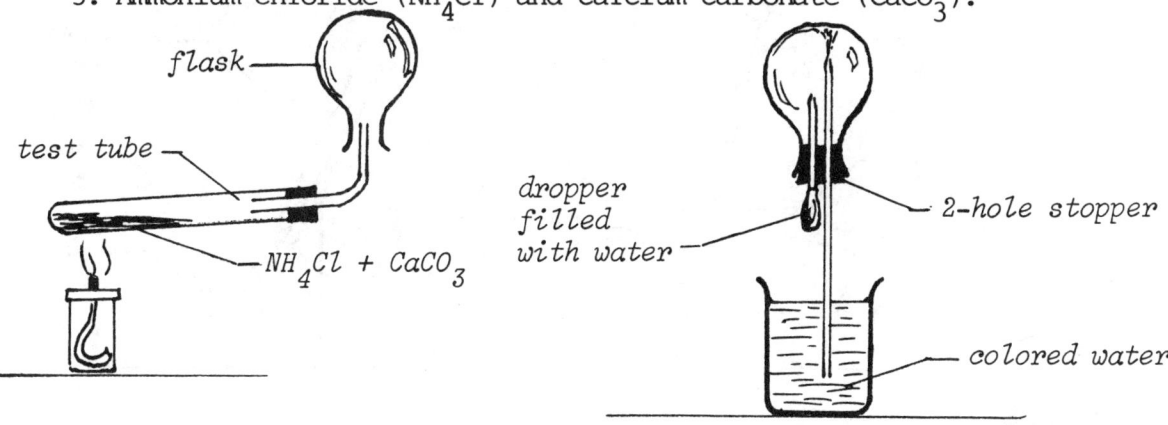

Procedure:
1. Mix one scoop of NH_4Cl and a scoop of $CaCO_3$ in the test tube and clamp the test tube slanted over the burner (see Sketch A).
2. Insert the 1-hole stopper with the bent tube in the test tube.
3. Fill the dropper (in the 2-hole stopper) with water.
4. Light the burner and heat the mixture of powders in the test tube while holding the flask upside down over the bent tubing (see Sketch A again).
5. As soon as ammonia vapours are detected (strong pungent smell), continue a few seconds longer with the heating, then close the flask off with the 2-hole stopper (with the long tube and dropper).
6. Immerse the end of the long tube in the beaker of water, then squirt the water from the dropper into the flask by pinching the rubber bulb (see Sketch B). Observe what is happening!

Questions:
1. What was produced by heating the two powders?
2. Why does the gas have to be collected in an upside down flask?
3. What was the function of the water in the dropper?
4. At what moment did the water rush into the flask?
5. What made the water rush up into the flask like a fountain?

Explanation:
The following reaction takes place in the test tube by heating the two powders: $2 NH_4Cl + CaCO_3 \rightarrow 2 NH_3 + CaCl_2 + H_2O$.

The ammonia gas is lighter than air and therefore it is collected in an upside down flask. As soon as the flask is full of ammonia, it is closed off and since ammonia is very soluble in water, the squirt of water from the dropper was enough to dissolve all the ammonia. This made a sudden partial vacuum in the flask and the water from the beaker is therefore sucked into the flask. This demonstrates the solubility of ammonia gas in water.

To make it more interesting we may put a few drops of an indicator like methyl red or bromothymol blue in the water of the beaker.

CHEMISTRY   **REPLACEMENT REACTION**

5.16. MAKE A SILVER TREE

Materials: 1. A thick solid copper wire (about 10 cm long).
               2. Silver nitrate solution—$AgNO_3$.
               3. A test tube or when available: petri dish & overhead projector.

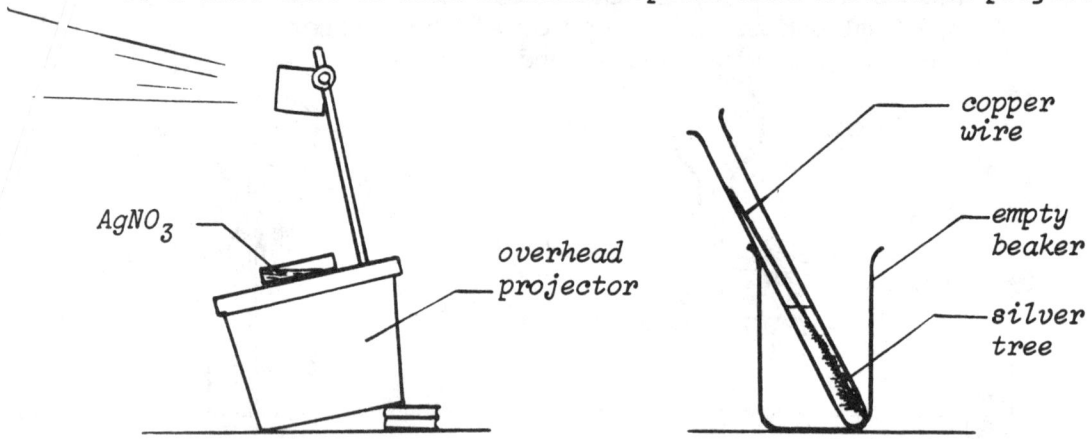

Procedure:
1. Strip the insulation off a heavy electric wire.
2. Place about 10 ml of $AgNO_3$ (silver nitrate) in a petri dish, or in the test tube (when dish & overhead projector are not available).
3. Put the dish on the overhead projector and slant the project towards the screen by placing some books under the far end.
4. Place the copper wire in the dish, such that the projection shows up vertically on the screen (or insert the wire in the test tube).
5. Let students observe closely when doing the reaction in a test tube, but they may not jar or touch the test tube.

Questions:
1. What did you observe on the screen or in the test tube?
2. What is the 'tree' composed of?
3. What specific reaction took place?
4. What would happen if the test tube were jarred or touched?
5. What caused the liquid to turn blue?

Explanation:
    The reaction that takes place in the petri dish or test tube is as follows: $Cu + 2\ AgNO_3 \rightarrow Cu(NO_3)_2 + 2\ Ag$. The silver precipitates out of the solution but clings to be copper wire, which gives the illusion of a growing tree. The reaction occurs because copper has a larger tendency to go into solution than silver. As the copper goes into solution, it forms copper ions, which are responsible for the blue color of the liquid. The longer the copper is left in the liquid, the more copper is going into solution, thus the more copper ions are formed and the deeper blue the liquid gets.
    The silver, which is deposited along the wire, is just hanging loosely on it and a very slight disturbance (jarring of the container) will send the silver crashing down.

CHEMISTRY   REPLACEMENT REACTION

5.17. THE CONFUSED BLUE SOLUTION

Materials: 1. Dilute copper sulfate solution & dilute HCl.
            2. Coarse iron filings (steel wool will also do).
            3. Two small Erlenmeyer flasks.
            4. A funnel and filter paper.

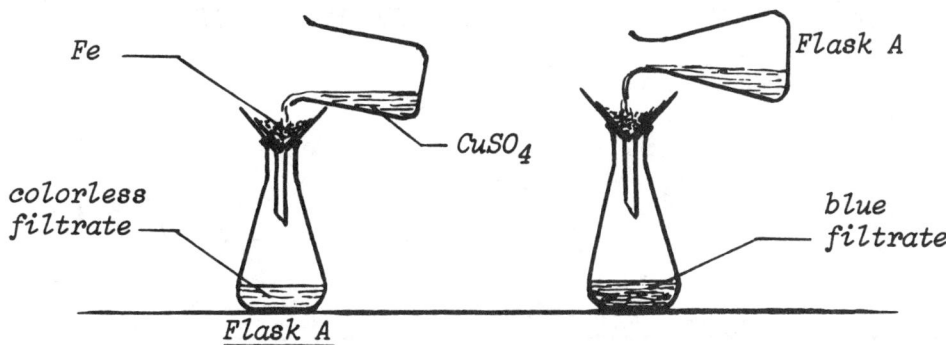

Procedure:
1. Place a filter paper in the funnel and about half a teaspoon of iron filings or a wad of steelwool on it (do not show this to students).
2. Put about 1 ml of dilute acid in flask A.
3. Place the funnel on flask A and pour slowly the copper sulfate solution through the filter (the filtrate should come through colorless).
4. Move the funnel now to flask B and pour the filtrate through the filter paper again (the filtrate should become blue again).

Questions:
1. What made the blue color disappear in the first filtrate?
2. What do you infer would be in the funnel?
3. Can we filter out the blue color of copper sulfate?
4. What made the blue color reappear in the second filtrate?
5. What is actually responsible for the blue coloring?

Explanation:
   The replacement reaction during the first pouring of the copper sulfate solution through the filter is as follows: $CuSO_4 + Fe \rightarrow FeSO_4 + Cu$. Copper sulfate + iron → ferrous sulfate + copper. It is the replacement of the copper ion in the solution with ferrous ions that makes the solution colorless. With the acid in the first filtrate, and pouring this the second time through the filter, we get the following reaction: $Cu + 2 HCl \rightarrow CuCl_2 + H_2$. The copper is transformed into the copper ion again and this ion is responsible for the blue color in the liquid.
   The first reaction occurs because iron has a greater tendency to go into solution or to form ions compared to copper, thus iron replaces the copper, and the copper ion is transformed into the copper metal, which is retained by the filter paper. Ions will pass through filter paper, whereas solid metal particles will not.

CHEMISTRY — CATALYTIC REACTION

5.18. THE FIERY WATER

Materials:
1. An asbestos plate or evaporating dish.
2. A medicine dropper, mortar and pestle.
3. Aluminum powder and iodine crystals.

CAUTION !

PERFORM UNDER FUMEHOOD OR OUTDOORS !

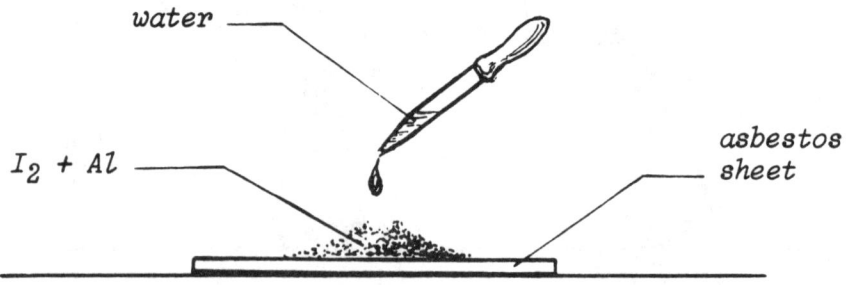

Procedure:
1. Grind the iodine crystals (about a teaspoonful) in the mortar.
2. Take about the same amount of aluminum powder (make sure that the aluminum powder is dry, otherwise heat in oven before using).
3. Mix the two dry powders on the asbestos sheet or in the evaporating dish with a spatula and make a small indentation in the heap.
4. Place the asbestos plate or evaporating dish under an exhaust hood or outdoors, and add just one drop of water to the mixture of powders (use the medicine dropper).
5. Stand back and observe!

Questions:
1. Why do the iodine crystals have to be finely ground?
2. Why do the reagents have to be dry?
3. What function does the water have in this reaction?
4. Would the two substances react without the water?
5. Where in daily life does water speed up a reaction like this?

Explanation:
   The iodine crystals have to be ground first to increase the contact surface between the two reagents. Smaller particles have a much larger surface area than the relatively larger crystals. If the aluminum powder is damp, the reaction may not succeed too well, because the moisture might take away the released heat needed to initiate the reaction. The water's function is that of a **catalyst**, which brings the molecules of the reagents even closer together. Heat is evolved from the first few molecules reacting, and this heat sets off the rest of the reaction.
   In our daily life, water speeds up the rusting process of metals when left in the damp outdoors, like bicycles, nails, etc.

CHEMISTRY   CATALYTIC REACTION

5.19. THE GLOWING PENNY

Materials:
1. A penny or other copper coin.
2. Copper or other thin metal wire.
3. A glass stirrer, small beaker, alcohol or propane burner.
4. Acetone (about 20 ml).

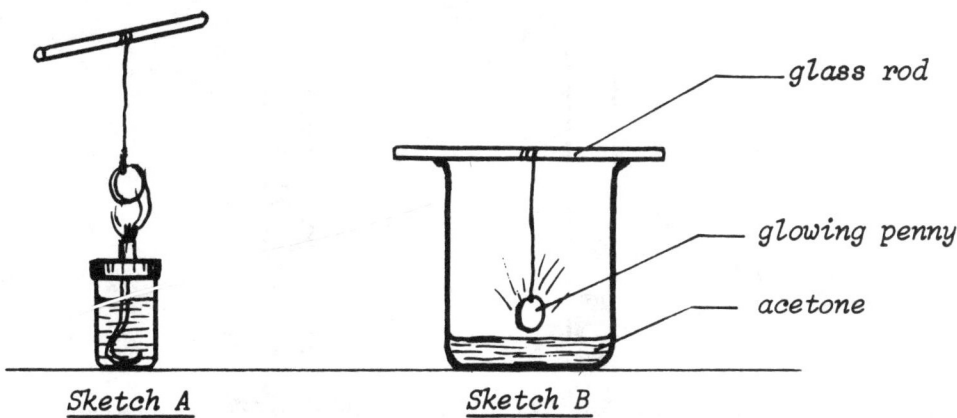

Sketch A   Sketch B

Procedure:
1. Make a small hole in a penny (punch a small nail through) and tie a short length of wire (about 10 cm) through the hole.
2. Wind the other end of the wire around the middle of a glass stirrer and heat the coin in an alcohol or propane flame until red hot (Sketch A).
3. Immediately after heating, hang the coin about 3 mm above the acetone level in a beaker (check this level before heating the coin, by rolling more or less windings around the glass stirrer--see Sketch B).
4. Observe the glowing penny in the dark.

Questions:
1. What made the penny glow in the flame?
2. What made the penny glow in the beaker?
3. Would the penny glow if it was hanging higher above the acetone level?
4. How high can the coin hang and still have it glowing?
5. How long will the penny stay glowing?
6. Would other metals do the same when hung above the acetone?

Explanation:
Holding the penny in the flame makes it so hot that it gives off a red glow. While the temperature of the penny is still high, it can speed up or facilitate the reaction of:

$$CH_3 - CO - CH_3 \xrightarrow{Cu} CH_3COH$$
$$\text{acetone} \qquad\qquad\qquad \text{acetaldehyde}$$

without participating in the reaction itself. Copper functions as a **catalyst** for the above reaction. The reaction is exothermic, thus it produces heat, which keeps the penny glowing. The glowing will stop only when all the acetone has evaporated. No other metal is as efficient a catalyst as copper for this particular reaction.

CHEMISTRY — CATALYTIC REACTION

5.20. THE RISING SUDS

Materials:
1. Two tall cylinders or narrow jars.
2. Liquid detergent, glycerine, hydrogen peroxide (H_2O_2), manganese dioxide (MnO_2).
3. Two long stirrers (long straws will do).

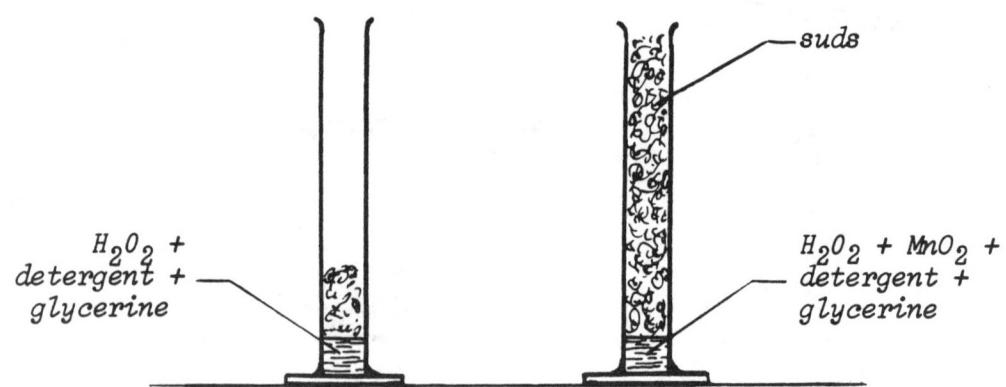

Procedure:
1. Place about 10 ml of hydrogen peroxide and 5 ml of detergent, and about 5 ml of glycerine in each cylinder.
2. Add a pinch of MnO_2 to one of the cylinders and stir the mixture in both cylinders with the long stirrers.
3. Observe the difference in height of the produced suds in the two tall cylinders.

Questions:
1. In which of the two cylinders did the suds rise higher?
2. What causes the suds to go up higher in one of the cylinders?
3. What do we need to make soap bubbles?
4. What function does the peroxide have in this reaction?
5. What do the MnO_2 and the glycerine do?
6. Would the suds rise just as high without the stirring?
7. What other substances might be used in place of MnO_2?

Explanation:
The detergent functions like the soap when blowing soap bubbles. By adding some glycerine to the detergent, the surface tension of the liquid is increased, and the bubbles will stay longer before collapsing. In order to get suds in a soap solution when blowing bubbles, we need to blow through a straw into the detergent mixture. In this case we do not need to blow. Where are we getting the gas to blow the bubbles from? The gas is supplied by the hydrogen peroxide, which decomposes into water and oxygen. The MnO_2 acts as a catalyst for this oxygen producing reaction. It does not partake in the reations but only facilitates it. This means that the properties of MnO_2 before and after the reaction are unchanged. Dust and dirt particles or ashes could replace the MnO_2.

CHEMISTRY   SPONTANEOUS COMBUSTION

5.21. THE CHEMICAL FLAG

Materials: 1. A pair of tongs, a filter paper.
2. A test tube and tube holder.
3. White phosphorus and carbon disulfide.

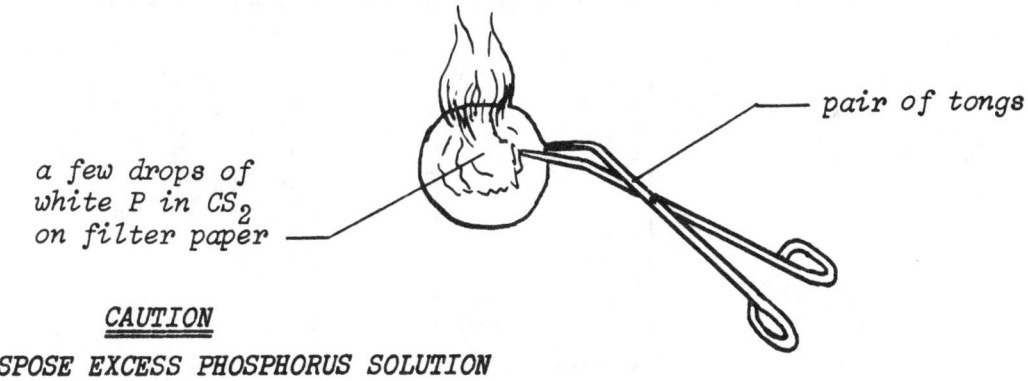

CAUTION
DISPOSE EXCESS PHOSPHORUS SOLUTION ON BARE SOIL, OR BURN OFF !

Procedure:
1. Cut a small piece of white phosphorus off a stick (do not touch the phosphorus with your fingers).
2. Place the piece of phosphorus in about 5 ml of carbon disulfide (CS_2) in a test tube.
3. After all the phosphorus has dissolved, pour a few drops of the solution on a filter paper which is held with a pair of tongs.
4. Wave the filter paper slowly in the air (like waving a flag): let students observe the 'chemical flag' catching fire!

Questions:
1. Why does white phosphorus have to be kept under water when it is stored?
2. Why is it dangerous to touch white phosphorus with the fingers?
3. At what moment did the flame burst?
4. In what state is the phosphorus spread on the filter paper?
5. What is the phosphorus reacting with?
6. Is the reaction endothermic or exothermic?

Explanation:
White phosphorus is kept under water when stored, because it would otherwise react with the oxygen in the air. By dissolving the phosphorus in carbon disulfide, it is dispersed in atomic particles (**status nacendi**). This makes the phosphorus much more reactive, and as soon as the solvent (carbon disulfide--CS_2) is evaporated, the finely dispersed phosphorus that is left on the filter paper reacts with the oxygen in the air. As this reaction: $4P + 5O_2 \rightarrow 2P_2O_5$ is exothermic, the evolved heat kindles the paper and bursts into flame.

This is the reason why it is dangerous to get the solution on the skin or any other combustible material. When accidentally dropped on the skin, it should be washed off immediately with soap and water.

CHEMISTRY — SPONTANEOUS COMBUSTION

5.22. THE BARKING DOG

Materials:
1. Three or four different size cylinders.
2. A test tube and tube holder, & medicine dropper.
3. Filter paper or paper towel pieces.
4. White phosphorus (P), and carbon disulfide (CS_2).

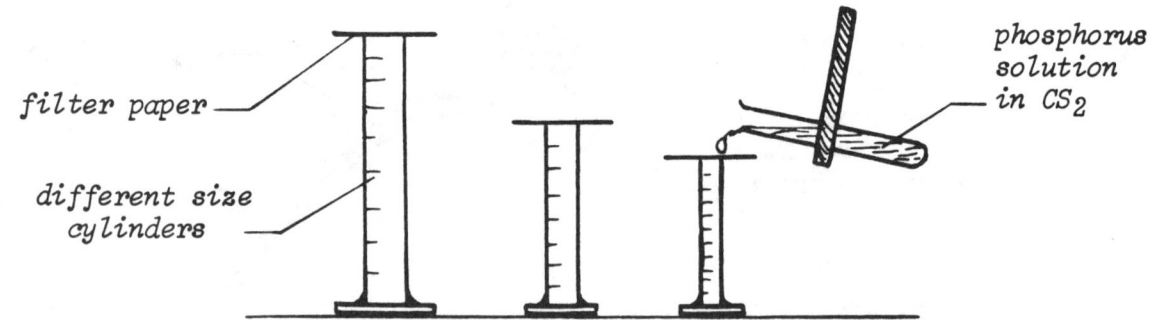

Procedure:
1. Cut a piece of white phosphorus (hold phosphorus with a pair of tongs) and dissolve it in about 10 ml of CS_2 in the test tube.
2. Hold the test tube with the test tube holder and do not touch or spill the liquid (wash off immediately when spilled accidentally).
3. Place a filter paper on each of the cylinders, then carefully place five drops of the phosphorus solution on each of the filter papers (use a long-stemmed medicine dropper).
4. Move the original phosphorus solution away from the cylinder, stand back and observe the cylinders. **CAUTION: PERFORM UNDER HOOD OR OUTDOORS!**

Questions:
1. Where did we get the lowest "bark"? Where the highest?
2. What was causing the small explosions?
3. What did we need to have in the cylinders for it to cause an explosion?
4. Would CS_2 gas be lighter or heavier than air?
5. What initiated the first spark for the explosions?
6. What should you do to the cylinders when repeating the demonstration?

Explanation:
   The white phosphorus dissolves in the carbon disulfide and is dispersed throughout the liquid in atomic form. When a few drops of the solution are poured on the filter paper covering the cylinders, the solvent (CS_2) evaporates and leaves the phosphorus behind in **status nacendi**, which means in atomic state. This state is much more reactive than when it is in the molecular state. Thus the phosphorus atoms react immediately with oxygen from the air. This is an exothermic reaction and the temperature exceeds the kindling point of the paper. This spontaneous combustion forms the first spark, which sets off the explosion of the CS_2 and air mixture in the cylinders. The different heights in the cylinders produce the high and low pitches in the "barks," the tallest giving the lowest "bark."

CHEMISTRY

STRONG OXIDANTS
SPONTANEOUS COMBUSTION

5.23. COLOR THE FLAME

Materials:
1. A mortar and pestle, medicine dropper.
2. Four asbestos sheets or evaporating dishes.
3. Potassium permanganate ($KMnO_4$), potassium chlorate ($KClO_3$), fine sugars, concentrated sulfuric acid (H_2SO_4), barium-(Ba), strontium-(Sr), copper- (Cu), and sodium- (Na) nitrates or chlorides.

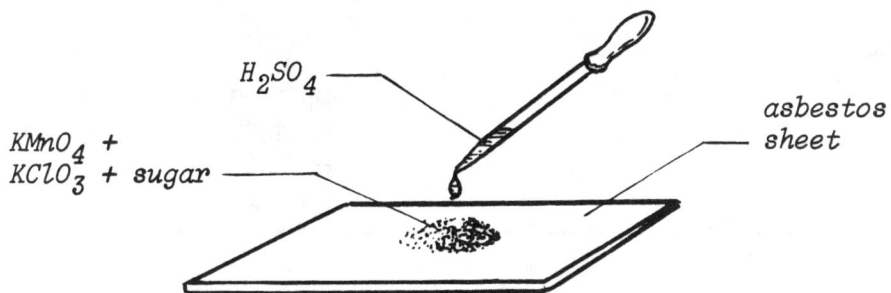

Procedure:
1. Grind about half a teaspoonful of $KMnO_4$ in the mortar and put it aside on a piece of paper.
2. Do the same with half a teaspoonful of $KClO_3$, put it aside and grind a full teaspoon of fine sugar (or sugar crystals).
3. Mix all three powders very carefully in the mortar (do not apply pressure on the pestle; heat of friction might set off the reaction!) and divide the mixture into four small heaps on the four separate asbestos sheets (or evaporating dishes).
4. Put a pinch of each of the salts containing the element Ba, Sr, Cu, and Na on the heap and mix it in well with the powder.
5. Place with a stretched-out arm (using the dropper) one drop of concentrated sulfuric acid on one heap of mixture at a time (this is most dramatic in a darkened room). **CAUTION: PERFORM UNDER HOOD OR OUTDOORS!**

Questions:
1. What reaction is taking place?
2. What color did each of the flames give?
3. What caused the different colors in the flames?
4. What purpose did the drop of sulfuric acid have?

Explanation:
The drop of concentrated sulfuric acid reacts mainly with the sugar in the mixture and releases heat. This heat triggers the decomposition of $KClO_3$ and $KMnO_4$, which produce oxygen in **status nacendi** (atomic form). This very active oxygen combines with the sugar and supplies the needed agent for the burning process.

The different ions of Barium, Strontium, Copper and Sodium produce the following colors respectively in the flame: yellowish green, red, bluish green, and yellow. These salts are very much used in the fire cracker

CHEMISTRY — EXOTHERMIC REACTION / COMBUSTION

5.24. THE MAGIC WAND

Materials:
1. Potassium permanganate & concentrated sulfuric acid.
2. Mortar & pestle, alcohol burner.
3. A small evaporating dish & glass stirrer.

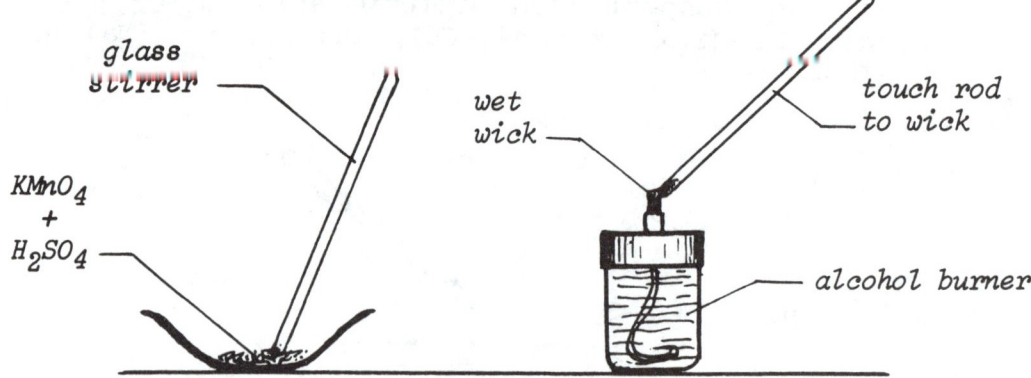

Procedure:
1. Grind about half a teaspoon of $KMnO_4$ (Potassium permanganate) in the mortar with the pestle, and transfer it to the evaporating dish.
2. Make sure that the alcohol burner's wick is soaked and wet with the alcohol by turning the burner upside down.
3. Mix a few drops of concentrated sulfuric acid with the $KMnO_4$ in the evaporating dish.
4. Turn the glass rod into the 'magic wand': touch the wick of the alcohol burner with the rod end that was dipped in the mixture.
5. The spontaneous lighting of the alcohol flame may be repeated as many times as needed.

Questions:
1. Why did the potassium permanganate have to be finely ground?
2. What made the alcohol burner's wick burst into flame?
3. What is the reaction that occurred between the mixture of potassium permanganate + H_2SO_4 and the alcohol?
4. Why did the wick have to be soaking wet with alcohol?
5. What other substances would also burst into flame when touched with the mixture of $KMnO_4$ and H_2SO_4?

Explanation:
The potassium permanganate had to be ground to increase its surface area. This compound is a very strong oxidant, just like concentrated sulfuric acid is. When the two are mixed, heat is produced and oxygen atoms are released, which attack any material in its neighborhood that can be oxidized. The reaction is as follows: $2\ KMnO_4 + H_2SO_4 \rightarrow K_2SO_4 + 2\ MnO_2 + H_2O + 3\ O_n$; this oxygen in **status nacendi** is very reactive, and attacks easy oxidizable substances like: alcohols, sugars, and other carbohydrates. In the process of oxidation, much heat is being released--**exothermic**, reaching a temperature above the kindling point of alcohol. This causes the alcohol to burst into flame.

CHEMISTRY   EXOTHERMIC REACTION
                                                SPONTANEOUS COMBUSTION

5.25. THE ROCKET EXHAUST

Materials: 1. An alcohol burner, test tube and holder.
            2. A medicine dropper, a scopula.
            3. Potassium chlorate ($KClO_3$), glycerine, and MnO_2.

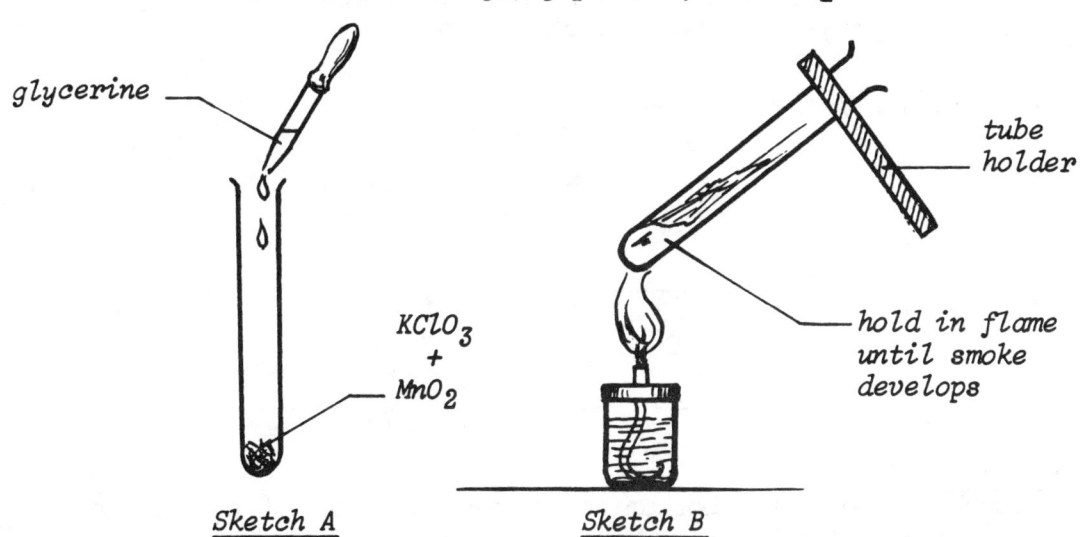

Sketch A   Sketch B

Procedure:
  1. Place a small amount of potassium chlorate ($KClO_3$) and a scopula tip of manganese dioxide (MnO_2) in the test tube and mix by shaking.
  2. Hold the test tube vertical and let 3 - 4 drops of glycerine fall on the powder (use a medicine dropper--see Sketch A).
  3. Hold the test tube (with the holder) above an alcohol or small Bunsen flame, just until the moment that smoke develops, then remove from flame.
  4. Observe the test tube and the sudden burst of flame.

Questions:
  1. What reaction takes place in the test tube?
  2. What purpose or function do each of the reagents have?
  3. Why was it necessary to heat the mixture?
  4. Why was it heated only until it started to smoke?
  5. If the test tube were suspended freely hanging horizontally from two strings, would the burst of flame move it? Which way?

Explanation:
  The potassium chlorate ($KClO_3$) is the source of oxygen, the manganese dioxide (MnO_2) functions as the catalyst for the decomposition of $KClO_3$ into KCl and O (potassium chloride and oxygen). This atomic oxygen immediately attacks the glycerine, which can easily be oxidized into alcohols, aldehydes, acids, and other gases. This reaction needs heat only in the beginning to supply the initial energy (to produce the oxygen). Once the oxygen producing reaction is going (it is extremely **exothermic**), the heat produced is enough to set the rest of the glycerine into flame.

CHEMISTRY   EXOTHERMIC REACTION
                                                   COMBUSTION

5.26. THE ANGRY BUCKET

Materials: 1. A syrup bucket or any can with a tightly fitting friction lid.
           2. A small funnel & rubber tubing (1 m).
           3. A 15 cm candle & small piece of tissue paper.
           4. Lycopodium powder or corn starch.

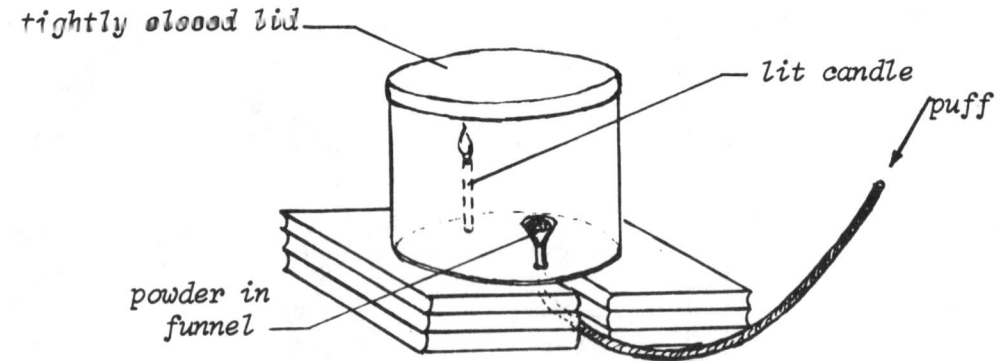

Procedure:
1. Punch a hole in the bottom of the bucket (somewhat to the side) to fit the funnel stem (with the rubber tubing attached).
2. Set the bucket on two stacks of books to allow the funnel stem to protrude below the bottom (see Sketch).
3. Place a single thickness of tissue paper in the funnel mouth to support a teaspoon of lycopodium powder or corn starch.
4. Attach the candle on the opposite side of the bucket from the funnel and light it.
5. Close the bucket lid firmly and blow a sharp puff through the rubber tube (make sure you are demonstrating it under a sturdy ceiling and away from the lights!).

Questions:
1. What made the bucket blow its top?
2. What is the purpose of the burning candle?
3. What reaction occurred inside the bucket?
4. Did all of the powder react? How do we know?
5. What other powders could we use in place of lycopodium?
6. Where do we find examples of this phenomenon in daily life?

Explanation:
   At the moment that the powder was blown into the bucket, it was spread finely throughout the space and mixed with the existing air. As the powder consists of combustible hydrocarbons, it forms an explosive mixture with the oxygen in the air, and the lit candle is there to trigger the explosion. Most of the powder reacted with the oxygen to form CO_2 and water vapor. These are gases, and their sudden production suddenly increased the pressure inside the bucket, resulting in blowing the bucket's top.
   This phenomenon is encountered in coal mines full of coal dust, and when someone carelessly strikes a match an explosion is triggered.

CHEMISTRY   SPONTANEOUS COMBUSTION

5.27. BURN PAPER WITH ICE?

Materials: 1. Sodium peroxide - Na_2O_2.
             2. Finely chopped tissue paper, or sawdust, or starch.
             3. A small chip of ice.

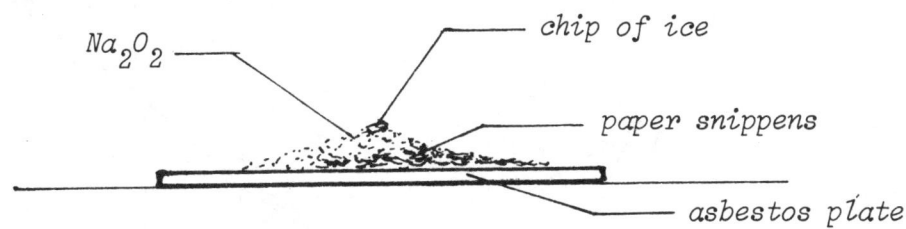

Procedure:
1. Before doing anything, show students a piece of tissue paper and ask: "Would I be able to burn this piece of paper with a chip of ice?" Anticipated answer: "Impossible!"
2. Tear or cut the tissue paper into very tiny pieces and place them on a heap on an asbestos/tile plate, and build it up to a cone which is about 5 cm high in the center.
3. On top of this cone, place a half teaspoon of sodium peroxide.
4. Now show the students the small chip of ice and put it on top of the heap stand back and observe!

Questions:
1. What reaction took place? What made the paper burn?
2. What does the burning process actually need in terms of chemicals?
3. What was the function of the sodium peroxide?
4. **Why was it necessary to divide the paper in such small pieces?**
5. What else beside sawdust or starch can be used to replace the paper?
6. What does the ice do when left at room temperature?
7. What is the reaction between sodium peroxide and water?
8. Would this reaction be endothermic or exothermic?
9. Would regular writing paper work better or worse than tissue paper? Why?

Explanation:
· The chip of ice at room temperature will melt and turn into water. The reaction between water and sodium peroxide is as follows:·

$$Na_2O_2 + H_2O \longrightarrow 2NaOH + O_n + \text{energy (heat)}$$

The oxygen released from the above reaction is in **status nacendi**, this means that it is in atomic form, and thus very reactive. The very reactive oxygen immediately attacks the combustible paper snippers and set it into flame. By chopping the paper into fine small snippers, we actually decrease the **kindling temperature.**

In place of the paper we can use fine sugar, lycopodium powder, fine coal dust, or any other easily combustible material. The decomposition reaction above is very **exothermic,** and the released heat is enough to decompose more of the sodium peroxide, which releases more active oxygen.

CHEMISTRY

SPONTANEOUS COMBUSTION
GAS PRODUCTION

5.28. THE POTASSIUM CHLORATE BOMBS

Materials:
1. Potassium chlorate, white phosphorus (kept under water).
2. Carbon disulfide, Kleenex tissue.
3. Asbestos plate (or tile).
4. Small beaker and medicine dropper.

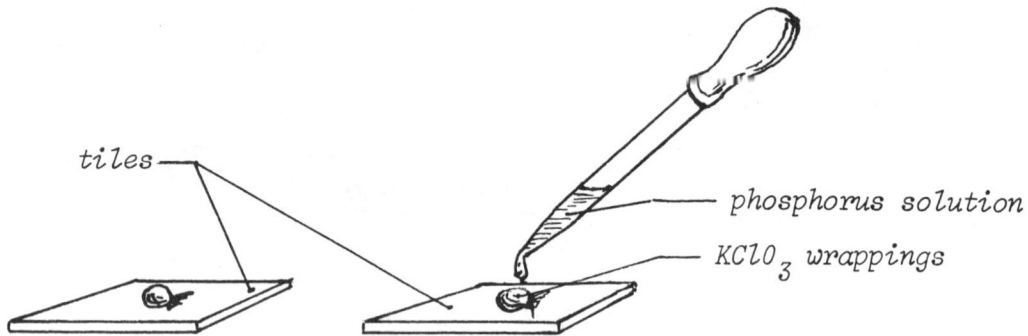

Procedure:
1. Make a solution of white phosphorus in carbon disulfide, by cutting a small chunk off the white phosphorus stick (hold with tongs) and placing it in some carbon disulfide in the small beaker (a half cm cube of phosphorus in about 10 ml of carbon disulfide).
2. Wrap a very small amount of potassium chlorate in tissue paper and tie it off with a thread, such that it makes pea-size wrappings (cut the excess tissue paper off).
3. Place these little wrappings on the tile or asbestos plate, one on each tile and place the tiles about 10 cm apart from each other.
4. Now, with the dropper place two or three drops of the phosphorus solution on each of the wrappings, and stand back!

Questions:
1. What triggered the sudden explosions?
2. What does the phosphorus do when dissolving in the carbon disulfide?
3. What kind of liquid is carbon disulfide?
4. What do we call a liquid which evaporates quickly?
5. Why is white phosphorus kept under water when stored?
6. What happens to the P-solution after dropping it on the wrappings?
7. What gas is released by the potassium chlorate?
8. What is it that makes us hear an explosion?

Explanation:
White phosphorus is very reactive in the presence of oxygen. This is the reason why it is kept under water when stored. Dissolving it in carbon disulfide, which is a very **volatile** and **combustible** liquid, breaks it up in very small, atomic particles. After dropping the solution on the wrappings of potassium chlorate, the volatile (easily evaporating) carbon disulfide evaporates and atomic phosphorus is left on the tissue paper. The phosphorus reacts with the oxygen in the air, releasing heat, and heat causes the breakdown of the potassium chlorate into potassium chloride and atomic oxygen: $KClO_3 \longrightarrow KCl + 3O_n$. Thus a sudden formation of gases is the result, causing a sudden large wave in the air, which our eardrum senses in the form of a BANG!

CHEMISTRY   SPONTANEOUS COMBUSTION
                                                         REACTIVITY OF SODIUM

5.29. BURN A PIECE OF METAL IN WATER

Materials: 1. Sodium metal (kept under liquid paraffin or oil).
             2. A petri dish, and phenolphtalein.
             3. An overhead projector (for large audiences).

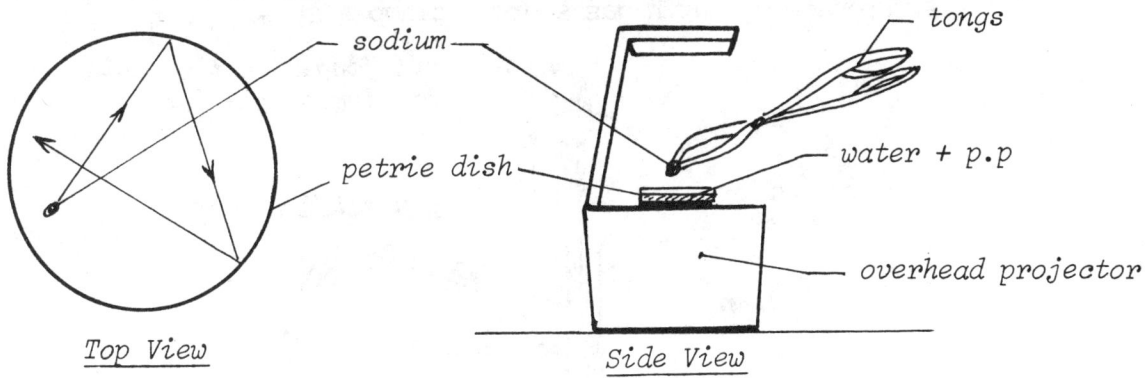

Procedure:
1. Fill the petri dish about half way with water and place it on the overhead projector. Add a few drops of phenolphtalein and stir.
2. Cut a *small* chunk of sodium metal (use tongs to take the sodium out of the liquid, dry the sodium off on a paper towel and cut a piece off with a knife) and place the rest of the sodium back in the liquid.
3. Pick the piece of sodium up with the tongs and drop it in the water in the petri dish on the overhead projector (with an outstretched arm). Let students observe the screen as well as the dish from the side!

Questions:
1. What did the sodium do after it was dropped in the water?
2. What did you notice about the colour of the water after the dropping?
3. In what environment does phenolphtalein turn reddish pink?
4. What do you expect the pH of the water to be after the dropping?
5. What are the products of the reaction of sodium and water?
6. What was the cause of the flame? (Not every time is a flame created).
7. Why did the sodium piece make such eratic motions over the water?
8. Why would it be dangerous to use too large a chunk of sodium?

Explanation:
   Sodium metal is so reactive that it has to be kept under liquid paraffin. If it is kept in the open air, it will react with the water vapour in the air. Dropping a piece of sodium in water will give the following reaction:
   $2Na + 2H_2O \longrightarrow 2NaOH + H_2$ (gas). This is an **exothermic** reaction, and since hydrogen gas is flammable, the heat produced is high enough to ignite the hydrogen. The sodium hydroxide produced gives the water a basic reaction. The pH gets to be higher than 7 and the phenolphtalein is turning red.
   Since the hydrogen gas is produced more on the contact point with water on one side than on the opposite side of the piece of sodium, a movement of the piece is caused in the water. This movement is eratic and random because the sodium piece is irregular.
   Too large a chunk of sodium in water might produce sudden large amounts of hydrogen gas, which may explode when brought in contact with the water. Therefore this is very dangerous!

CHEMISTRY   COMBUSTION

5.30. THE AIR-BURNING FLASK

Materials:
1. A distillation flask (with a small hole in the bottom).
2. Glass tubing (25 cm long) & a 1-hole stopper.
3. A glass tube (10 cm long) & 1-hole stopper (fits on side arm).
4. Propane or earth gas source, clamp & stand.

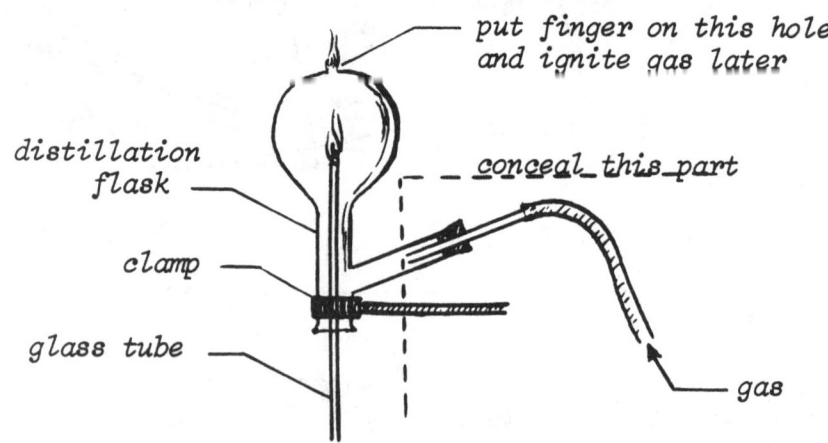

Procedure:
1. Make a small hole in the bottom of the distillation flask by heating the flask with a torch and pushing the back end of a file or other sharp tool through the red hot glass.
2. Set the flask up as pictured in the sketch with the gas valve closed (conceal the gas connection by covering it with cardboard).
3. Open the gas valve, keeping a finger over the hole in the bottom of the flask until the gas flows out through the glass tube (check smell).
4. Ignite the gas coming from the glass tube, then release the finger holding the hold at the top of the apparatus (flame will travel up and continue to burn inside the flask).
5. Ignite the unburned gases coming out of the top hole.

Questions:
1. Is it air that burns in the flask?
2. If the top flame is a gas flame burning in an air atmosphere, how can you describe the flame in the flask?
3. What would happen if we turned the whole apparatus upside down?
4. Why is the gas only burning at the end of the tube in the flask?

Explanation:
   This demonstration shows students that gas fuel cannot burn without the presence of air. The flame inside the flask is only burning at the end of the glass tube, because that is where air is present, and that is the only place where air is continuously being supplied. There is not enough air supplied to burn all the gas issued to the flask, and whatever is in excess can be burned off at the top like a regular flame coming directly from the gas supply (source).
   When the whole apparatus is turned upside down, the flames are switched around: the bottom hole supplies air for the flame inside the flask and the excess gas will burn at the end of the tube.

CHEMISTRY

FLAME AND BURNING

5.31. THE HUMAN FLAME THROWER

Materials:
1. A sheet of paper (towel or writing paper).
2. A candle and match.
3. A container of water.

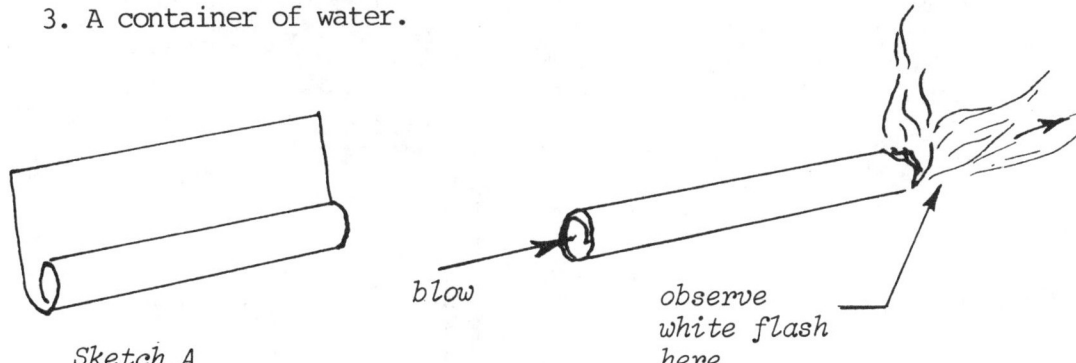

Sketch A

Procedure:
1. Light the candle and attach it to the table top
2. Roll up the sheet of paper rather tightly in a cylinder of about 3 cm diameter (see Sketch A).
3. Light one end of the paper roll with the burning candle and hold the roll horizontally with the burning end slightly upwards.
4. Without touching your lips to the other end of the paper roll, blow a short puff through the roll (the blowing can be repeated).
5. Observe a brighter flash of light and listen to the sound it makes!
6. Dump the burning paper in the container of water.

Questions:
1. What makes the flame move outwards?
2. What causes the brighter flash of light?
3. What kind of sound did the flame make during the blowing?
4. Why must we hold the roll with the burning end slanted upwards?
5. What would happen if the burning end was held lower than the other end?
6. Why is it better to hold your lips a little away from the roll of paper when blowing through it?
7. What are the products of the burning process?
8. What are the elements that paper molecules are made up of?

Explanation:
After burning one end of the paper roll, it is better to hold the burning end slightly higher than the other end, so that the smoke formed by the burning is flowing up and not through the roll to your mouth side. This demonstration works best if the roll was held perfectly horizontal. This way the smoke will be kept inside the roll and at the time that we blow, the smoke comes out the other end. This smoke mixed with the air burst into flame and gives a bright flash of light. At this moment we hear a ...pllaap...sound, indicating the sudden formation of gases.

Paper is composed of plant fibers and these consist of cellulose molecules: $C_{12}H_{22}O_{11}$ and when burned gives off CO_2 (carbondioxide) gas, and water vapour, and leaves behind the black material, which is carbon (C).

5.32. THE MAGIC CANDLE

Materials:
1. A medium size candle.
2. A book of matches.

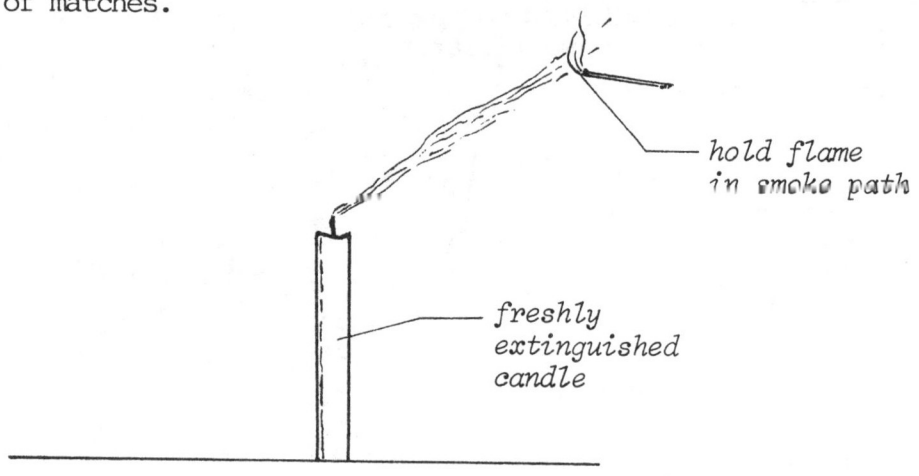

hold flame in smoke path

freshly extinguished candle

Procedure:
1. Light the candle and let it burn until the flame becomes a good size.
2. Strike another match. While holding the burning match in your hand, blow out the candle flame with a short puff.
3. Immediately after the flame is out, hold the flame of the burning match in the smoke trail coming from the candlewick (about 3-4 cm from the wick).
4. Observe the travelling flame! (If smoke trail is not thick enough, repeat points 1, 2, and 3 again. A draft-free room is needed to carry this out).

Questions:
1. What makes the flame travel to the candlewick?
2. Why does this demo have to be carried out in a draft-free room?
3. What is actually burning of the candle?
4. Would wax without a wick sustain a flame?
5. What else beside a wick can we use to make wax sustain a flame?
6. What creates the smoke after a candle is blown out?

Explanation:
　　What is actually burning of a candle is the vapour which comes from the heated candlewax. When a candle is being lit, it is the wick that burns first. The heat of the flame melts the wax, the wick absorbs the molten wax and the hot wax vapours burn around the wick. When blowing out the candle flame, the hot wick keeps heating the molten wax. This is the reason why a smoke trail is formed. This smoke actually consists of very small **dispersed particles of wax** in the air, making it highly combustible. Holding a flame within this smoke trail sets it aflame and the flame "travels" to the wick.
　　The "Magic Candles" that can be bought in confectionary stores are candles which cannot be blown out. After blowing the flames out of these candles, they light themselves again. This is because the wax is treated with red phosphorus. The glowing wick will make the little particles of phosphorus spark up and make the candle burst back into flame.

CHEMISTRY FLAME AND BURNING

5.31. THE HUMAN FLAME THROWER

Materials:
1. A sheet of paper (towel or writing paper).
2. A candle and match.
3. A container of water.

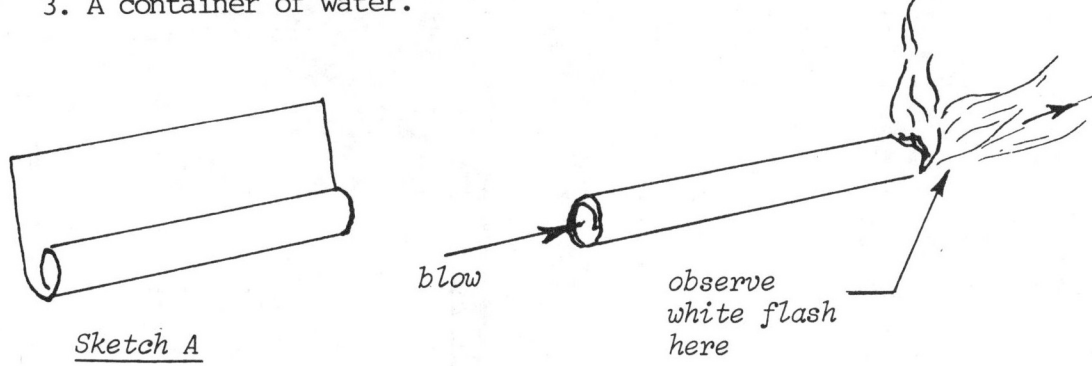

Sketch A

Procedure:
1. Light the candle and attach it to the table top
2. Roll up the sheet of paper rather tightly in a cylinder of about 3 cm diameter (see Sketch A).
3. Light one end of the paper roll with the burning candle and hold the roll horizontally with the burning end slightly upwards.
4. Without touching your lips to the other end of the paper roll, blow a short puff through the roll (the blowing can be repeated).
5. Observe a brighter flash of light and listen to the sound it makes!
6. Dump the burning paper in the container of water.

Questions:
1. What makes the flame move outwards?
2. What causes the brighter flash of light?
3. What kind of sound did the flame make during the blowing?
4. Why must we hold the roll with the burning end slanted upwards?
5. What would happen if the burning end was held lower than the other end?
6. Why is it better to hold your lips a little away from the roll of paper when blowing through it?
7. What are the products of the burning process?
8. What are the elements that paper molecules are made up of?

Explanation:
After burning one end of the paper roll, it is better to hold the burning end slightly higher than the other end, so that the smoke formed by the burning is flowing up and not through the roll to your mouth side. This demonstration works best if the roll was held perfectly horizontal. This way the smoke will be kept inside the roll and at the time that we blow, the smoke comes out the other end. This smoke mixed with the air burst into flame and gives a bright flash of light. At this moment we hear a ...pllaap...sound, indicating the sudden formation of gases.

Paper is composed of plant fibers and these consist of cellulose molecules: $C_{12}H_{22}O_{11}$ and when burned gives off CO_2 (carbondioxide) gas, and water vapour, and leaves behind the black material, which is carbon (C).

CHEMISTRY   FLAME AND BURNING

5.32. THE MAGIC CANDLE

Materials: 1. A medium size candle.
             2. A book of matches.

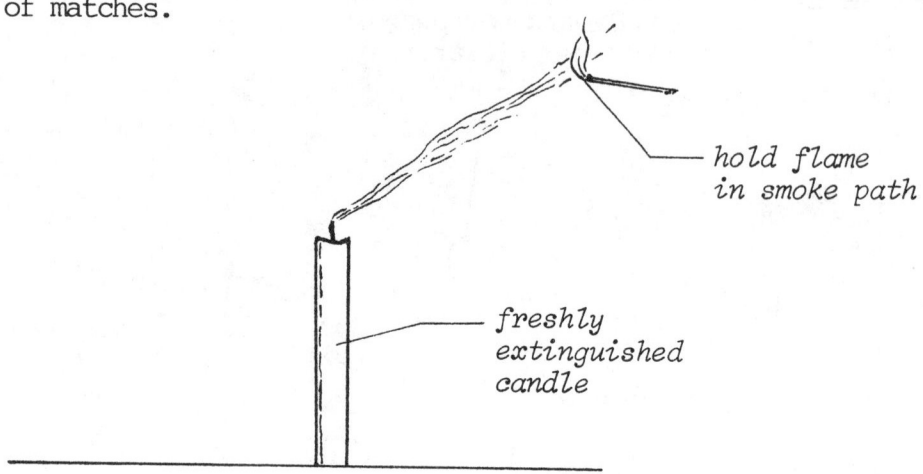

Procedure:
1. Light the candle and let it burn until the flame becomes a good size.
2. Strike another match. While holding the burning match in your hand, blow out the candle flame with a short puff.
3. Immediately after the flame is out, hold the flame of the burning match in the smoke trail coming from the candlewick (about 3-4 cm from the wick).
4. Observe the travelling flame! (If smoke trail is not thick enough, repeat points 1, 2, and 3 again. A draft-free room is needed to carry this out).

Questions:
1. What makes the flame travel to the candlewick?
2. Why does this demo have to be carried out in a draft-free room?
3. What is actually burning of the candle?
4. Would wax without a wick sustain a flame?
5. What else beside a wick can we use to make wax sustain a flame?
6. What creates the smoke after a candle is blown out?

Explanation:
   What is actually burning of a candle is the vapour which comes from the heated candlewax. When a candle is being lit, it is the wick that burns first. The heat of the flame melts the wax, the wick absorbs the molten wax and the hot wax vapours burn around the wick. When blowing out the candle flame, the hot wick keeps heating the molten wax. This is the reason why a smoke trail is formed. This smoke actually consists of very small **dispersed particles of wax** in the air, making it highly combustible. Holding a flame within this smoke trail sets it aflame and the flame "travels" to the wick.
   The "Magic Candles" that can be bought in confectionary stores are candles which cannot be blown out. After blowing the flames out of these candles, they light themselves again. This is because the wax is treated with red phosphorus. The glowing wick will make the little particles of phosphorus spark up and make the candle burst back into flame.

CHEMISTRY

REACTION ENERGY

5.33. THE GLOWING ALUMINUM

Materials:
1. Aluminum chunks (small pea size), and bromine (poisonous vapours).
2. A 300-400 ml Erlenmeyer flask + cork, wide masking tape.

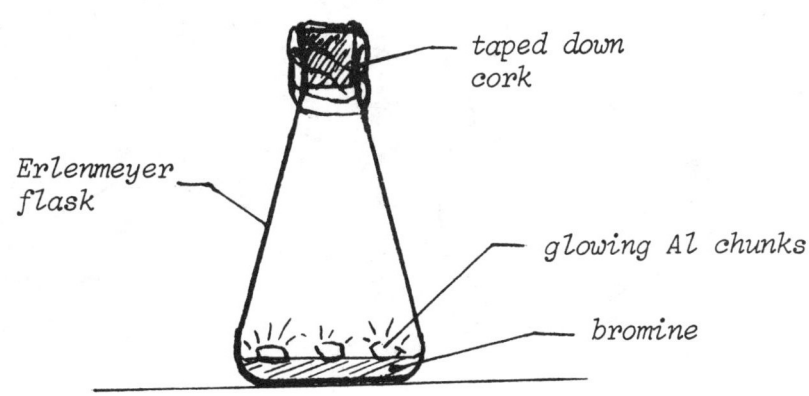

Procedure:
1. Pour some bromine into the Erlenmeyer flask (cover the bottom with about 1 cm layer. Use gloves when working with bromine and work under a fume hood or outdoors!)
2. Open the cork and drop two or three aluminum chunks in the bromine. Stopper the flask immediately and tape the cork down securely.
3. Darken the room and observe the glowing aluminum!

Questions:
1. What made the aluminum glow?
2. What particular reaction was taking place?
3. Did the reaction give off or take in energy?
4. Why did the stopper have to be taped down to the flask?
5. Would other halogens also react the same way with aluminum?
6. Which of the halogens would react faster or slower with aluminum?
7. How can we promote contact between two substances that are both solids?

Explanation:
The reaction between aluminum and bromine is exothermic. This means that it is giving off energy in the form of heat. This heat makes the aluminum glow in the dark. The reaction between aluminum and bromine is as follows:
$$2\ Al\ (s) + 3\ Br_2\ (l) \longrightarrow 2\ AlBr_3\ (s) + energy$$

Other halogens are chlorine and iodine, and both will react with aluminum. Chlorine is a gas and will react much faster with aluminum than bromine. Iodine reacts much less readily with aluminum. In order to let them react with each other, both solids have to be divided in very small particles, in other words they have to be very finely ground. When mixed with each other, they still do not react, but a tiny drop of water will start the reaction (see THE FIERY WATER).

CHEMISTRY   REACTION ENERGY

5.34. TOUCH A CRACKER!

Materials: 1. Concentrated ammonia - $NH_{3(aq)}$ Conc.
            2. Iodine (solid crystals).
            3. A meter stick, filter paper, a balloon, a pin.

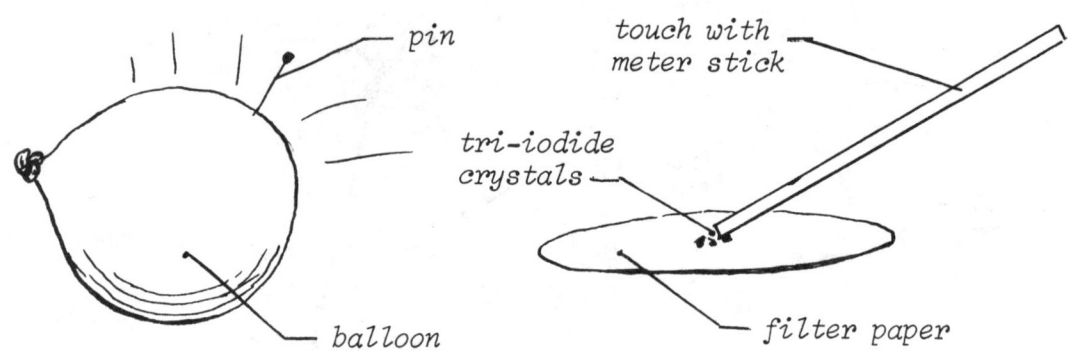

Procedure:
1. Dissolve some iodine in concentrated ammonia; make sure that all solid material disappears (use a glass stirrer and a small beaker).
2. Let the solution stand in the air for a while until some solid brown material is formed on the bottom of the beaker.
3. While this solid material is still wet, transfer it with the stirrer to a filter paper and let it dry in the air (this will take about an hour to completely dry; the drier the better).
4. Place these dry crystals about a meter away from you, let students stand back, hold the meter stick at the end and tap the other end of it on the crystals. (**CAUTION: TAP ONLY A FEW CRYSTALS AT A TIME: VERY EXPLOSIVE!**)

Questions:
1. What compound was formed by dissolving the iodine in ammonia?
2. What is it that makes a chemical so explosive?
3. What are other chemicals or chemical mixtures that are explosive?
4. What makes us hear the loud noise in an explosion?
5. What is always suddenly formed in an explosion?
6. Does an explosion always have to be accompanied with a fire?
7. Does oxygen always have to be present for an explosion to occur?
8. What can we compare the trapped energy in the nitrogen tri-iodide with?
9. How does a bursting air balloon compare with this explosion?

Explanation:
  The wet nitrogen tri-iodide is still safe to handle, but as soon as all liquid is evaporated (usually excess ammonia), the dry solid is very unstable and explodes the moment it is touched or walked on. It is harmless when very small quantities at a time are used.
  Nitrogen tri-iodide can be compared to a set mousetrap, and the touching of it to the releasing of the trap. It is an unstable compound that suddenly forms nitrogen gas when it is mechanically upset. This sudden formation of gas is the cause for us to hear the explosion. Explosions are always accompanied with the sudden formation of gases, whether it is in the formation of water vapour from its elements, or the igniting of explosive mixtures (See Events 5.7; 5.26; and 5.28).

CHEMISTRY   ENDOTHERMIC REACTION

5.35. THE STICKY BOARD

Materials:
1. A small beaker (250 ml), glass or plastic stirrer.
2. A small thin wooden board (10 cm x 10 cm)
3. Barium hydroxide - Ba(OH)$_2$ and ammonium thiocyanide - NH$_4$SCN

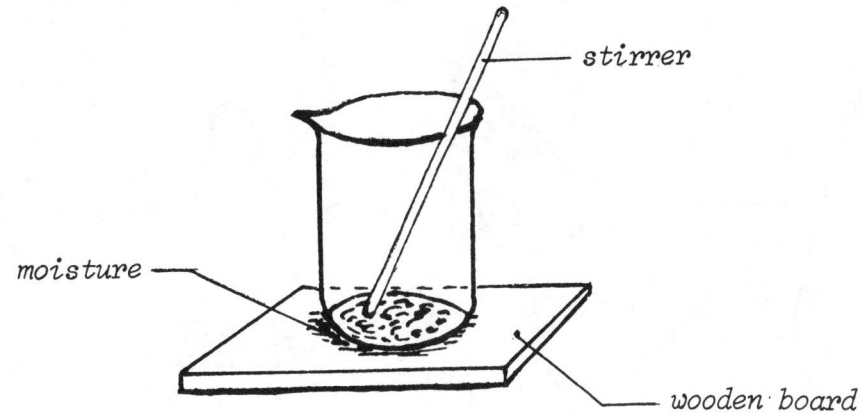

Procedure:
1. Wet one side of the wooden board thoroughly with water.
2. Place the beaker on the wet side of the board.
3. Add one teaspoon of each of the chemicals to the beaker and stir while holding the beaker down tight to the board.
4. After the two chemicals are thoroughly mixed, lift the beaker slowly up (formation of frost on the outside of the beaker is an indication that it is ready to be lifted).

Questions:
1. Why did the wooden board have to be wetted?
2. What kind of reaction occurred between the two chemicals?
3. Why did the wooden board stick to the beaker?
4. How else can we show that the reaction took away heat?
5. What other types of reactions do you know?

Explanation:
Most reactions are exothermic, which means that the reaction gives off heat (See Events 5.23 - 5.29), but this particular one is endothermic. This means that the reaction takes away heat, in other words the reaction requires calories in order for it to occur and if this heat is not supplied, it withdraws the heat from its environment. This cools down the environment and thus freezes the moisture that is between the beaker and the wooden board. This is the reason why the wooden board sticks to the beaker after the two chemicals are mixed in it.

The reaction between the two chemicals is as follows:

$$Ba(OH)_2 + 2NH_4SCN \cdot (H_2O)_4 \longrightarrow Ba(SCN)_2 + 2NH_4OH + 8H_2O - energy$$

CHEMISTRY — REVERSIBLE REACTION

5.36. SHAKING THE "BLUES"

Materials:
1. An Erlenmeyer flask (250/400 ml) and stopper.
2. Potassium hydroxide (5g), Glucose or dextrose (3g).
3. Methylene blue.

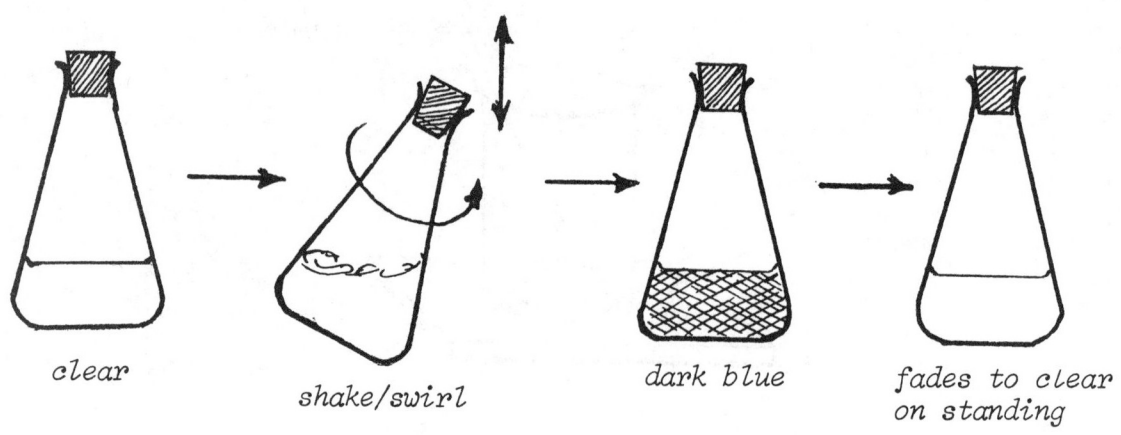

clear → shake/swirl → dark blue → fades to clear on standing

Procedure:
1. Dissolve the above chemicals in the Erlenmeyer flask with 250 ml of water.
2. Show to the students that it is a clear colourless liquid.
3. Stopper the flask and shake the solution vigorously and show the change of colour.
4. Place the flask on the table and leave it standing.
5. Draw students attention to the colour of the solution.
   (The shaking may be repeated to show the colour change for several hours).

Questions:
1. What type of reaction is this? One that can be repeated seemingly almost endlessly?
2. When will the turning into the blue colour stop? Or will it ever stop?
3. What makes the clear, colourless liquid turn blue?
4. What can we do to make the liquid turn blue again? (after it has stopped turning colour?

Explanation:
   Methylene blue is reduced to a colourless compound by an alkaline solution of a reducing sugar: in this case the glucose or dextrose. When we shake the flask, the colourless solution is reoxydized by the oxygen above the liquid into the blue dye: methylene blue. When it is left standing, it is being reduced again, thus the gradual fading of the blue colour.
   Slowly, however, the oxygen in the flask above the liquid is seeping through the cork and is also reacting with the reducing sugar. This is why it eventually (after a few hours) stops changing into the blue colour. To make the solution turn colour again, the oxygen has to be replenished. This can be done by leaving the flask open to the air for a few moments.
   After a few days standing, the solution will turn yellow and to brown. A freshly prepared solution should be used for this demonstration.

CHEMISTRY

ORGANIC CHEMISTRY
UNSATURATED BONDS

5.37. THE COLOR ABSORBING BACON

Materials: 1. Three or four strips of crisply fried bacon.
2. Bromine in an Erlenmeyer flask (250/400 ml), stoppered with cork.

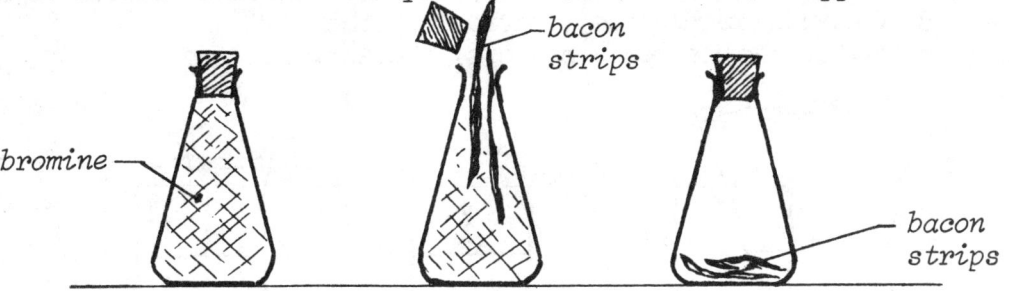

Procedure:
1. Place a few drops of bromine in an Erlenmeyer flask and quickly stopper it with a cork (make sure that the cork fits the flask prior to adding the bromine to the flask!).
2. Show students the brown colour of the bromine vapours, and tell them that in compound form, that is if they react with other chemicals, that there is no characteristic of bromine vapour left.
3. Open the stopper of the flask and quickly drop the bacon strips into the flask, close again and shake.
4. Hold a white sheet of paper behind the flask and let students observe the colour change!

Questions:
1. What colour did the brown vapour turn into?
2. What made the brown colour disappear?
3. What do double bonds do when a halogen is added to it?
4. What other halogens would behave the same way as the bromine?
5. How do we know whether all the bromine has reacted?
6. How do we know whether all the unsaturated bonds have reacted?
7. What are unsaturated bonds in organic compounds?

Explanation:
Crisply fried bacon has a high content of unsaturated fats. These are organic compounds containing double or triple bonds between the carbon atoms. When halogens are added to these compounds, the halogens are active enough to open these bonds and attach themselves to the organic molecule. See reaction below:

$$Br_2 + -\overset{H}{\underset{|}{C}} = \overset{H}{\underset{|}{C}}- \longrightarrow -\overset{H}{\underset{Br}{\overset{|}{C}}} - \overset{H}{\underset{Br}{\overset{|}{C}}}- \quad \text{or} \quad Cl_2 + -\overset{}{\underset{}{C}} \equiv \overset{}{\underset{}{C}}- \longrightarrow -\overset{Cl}{\underset{Cl}{\overset{|}{C}}} - \overset{Cl}{\underset{Cl}{\overset{|}{C}}}-$$

brown   colourless   green   colourless
gas   gas

These reactions are called **addition reactions**. Iodine vapours would most likely also react with the bacon strips, except that it would take a little longer, as iodine is the least reactive of the three halogens.

CHEMISTRY

ORGANIC CHEMISTRY
HIGH POLYMERS

5.38. MAKE A NYLON THREAD OUT OF TWO LIQUIDS

Materials:
1. Hexamethylene diamine (1,6 hexanediamine) $NH_2(CH_2)_6NH_2$
2. Sebacyl chloride (1,10 decanedioxyl chloride) $COCl(CH_2)_8COCl$.
3. Carbontetrachloride (tetrachloroethane).
4. Sodium carbonate, 2 beakers, tweezer, stirrer.

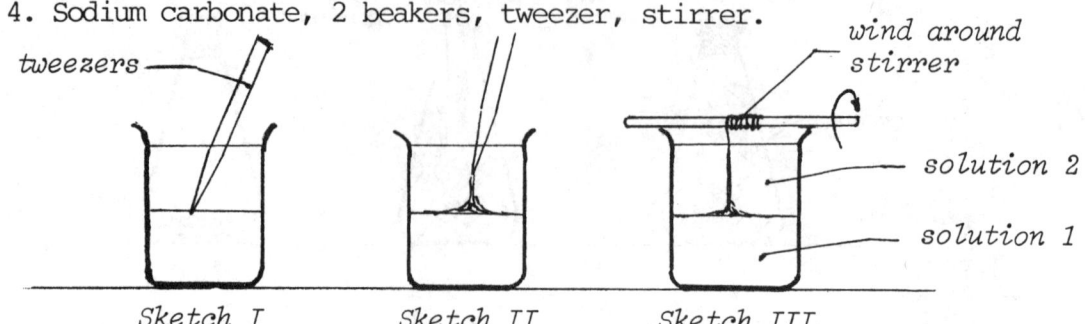

Procedure:
1. Make in a 100 ml beaker a solution of 2 ml sebacyl chloride in 50 ml of carbontetrachloride.
2. Make in another beaker a solution of 2 g of 1,6 hexanediamine and 4.0 g of sodium carbonate in 50 ml of water.
3. Pour this second solution very carefully on top of the first solution by slanting the first beaker and slowly pouring the second solution over the first one.
4. With the pair of tweezers, pick up the center interface between the two liquids and pull it slowly out of the liquid (see Sketch I and II).
5. Then wind this thread around a glass stirrer and keep turning the glass rod to continue pulling the nylon thread out.

Questions:
1. What compound is formed at the interface of the two liquids?
2. What would happen if the top liquid was just poured carelessly into the first solution?
3. What would happen if the two liquids were stirred together?
4. Is the formed compound identical to the industrial Nylon 66?
5. How long do you think the thread can be pulled?
6. At what point would the thread stop forming?

Explanation:
The liquid containing sebacyl (or adipyl) chloride and the other liquid containing hexamethylene diamine react at the interface to form Nylon-610.

$$NH_2(CH_2)_6NH_2 + COCl(CH_2)_8COCl \xrightarrow{-HCl} -\overset{H}{\underset{|}{N}}-(CH_2)_6-\overset{H}{\underset{|}{N}}-C-(CH_2)_8-C-\overset{H}{\underset{|}{N}}-(CH_2)_6-\overset{H}{\underset{|}{N}}-C-(CH_2)_8-$$

hexamethylene   sebacyl chloride   Nylon-610

The above reaction is a polymerization reaction. Large and extremely long molecules are formed. This same polymerization occurs in the production of high polymers like, polyethylene, polystyrene, polyurethane, and other plastics.

CHEMISTRY   TECHNOLOGICAL
                                                   APPLICATION

5.39. THE STICKY MATCHES

Materials: 1. A box of wooden safety matches.

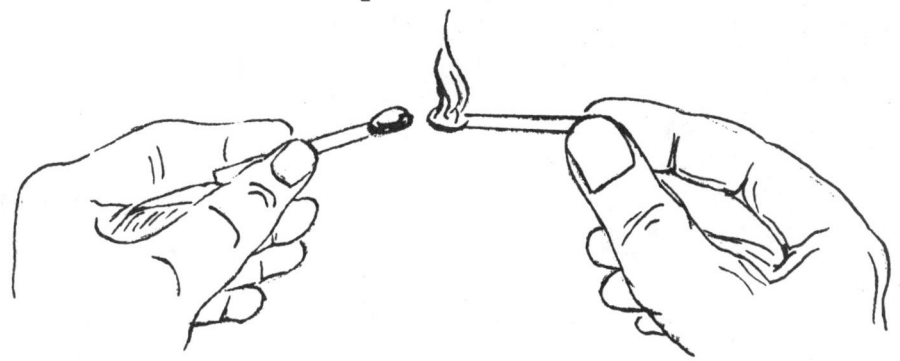

Procedure:
1. Take two match sticks out of the box and hold one in each of your hands.
2. Strike the one in your right hand against the box (if you are right handed) to light the match.
3. Immediately hold the lit match head against the match head in your left hand and wait till the left match gets lit.
4. As soon as the hissing sound of the ignition stops, blow out the flame. Make sure that you are holding the two matches with a steady hand against each other (see Sketch above).
5. Slowly let go of the match in your right hand (without moving/jarring the match - if you don't succeed the first time, try again!)

Questions:
1. What made the two matches stick together?
2. What would happen if the first match was put against the second match not immediately after it was lit?
3. How are matches produced?
4. What are matches made of?
5. What is the match head made of?
6. What makes a match head burst into flame when rubbed against the box?
7. What chemical is on the striking surface of the box?

Explanation:
   The red tips of the matches burn and fuse together, making the matches stick. The burning of the red tips form chemicals that are porous and rough in texture. These porous and rough tips "grab" each other and thus making the matches stick together.
   The match stick itself is made of softwood veneer. The wood splints are soaked into a bath of sodium silicate, ammonium phosphate or sodium phosphate and then dried. This impregnation prevents the afterglow. The wood splint is then dipped in a paraffin bath, which sustains burning. The match head consists of an oxygen carrier (potassium chlorate, chromate or lead oxide), sulfur and abrasives like powdered glass, and binding agents (dextrin or gums).
   The side of the box contains red phosphorus and powdered glass. Safety matches can only be lit by striking against the side of the box. The friction produces heat, which releases the oxygen from the potassium chlorate and this reacts with the sulfur to produce the flame.

SECTION II

ENERGY

This section consists of seven chapters dealing with the different types or forms of energy and the transformation from one into the other. Each of the chapters contains demonstrations that can be used to start off lessons dealing with the sources of energy, heat, magnetism, static and current electricity, light and sound.

Chapter 6 deals with how heat compares with temperature, heat of fusion and vaporization, the sources of energy: chemical, mechanical, heat energy, and nuclear energy, and how they can be transformed from one into the other. It also deals with potential versus kinetic energy.

Chapter 7 deals with heat and how it travels. It moves from one place to the other by conduction in solids, by convection in fluids, and by radiation through fluids and vacuums. Heat makes solids, liquids, and gases expand.

Chapter 8 contains demonstrations to show the characteristics of magnets, the nature of magnetism, and the rules of magnetism. The construction of simple compasses, the showing of magnetic lines, and magnetic induction are dealt with in this chapter.

Chapter 9 should be used when dealing with static electricity. It deals with the attraction of uncharged objects, the positive and negative charges and how they attract or repel each other, the induction of static charges, and the storage of static charges in capacitors.

Chapter 10 deals with current electricity. The discrepant events can be used to start lessons in conductivity, circuits, sources of electricity, and electromagnetism.

Chapter 11 contains discrepant events to initiate lessons dealing with light, its reflection, refraction, and defraction properties.

Chapter 12 deals with sound and its properties: the medium of travel, how loudness and pitch are influenced, the nature of resonance, and its velocity in air.

In Science Teaching......

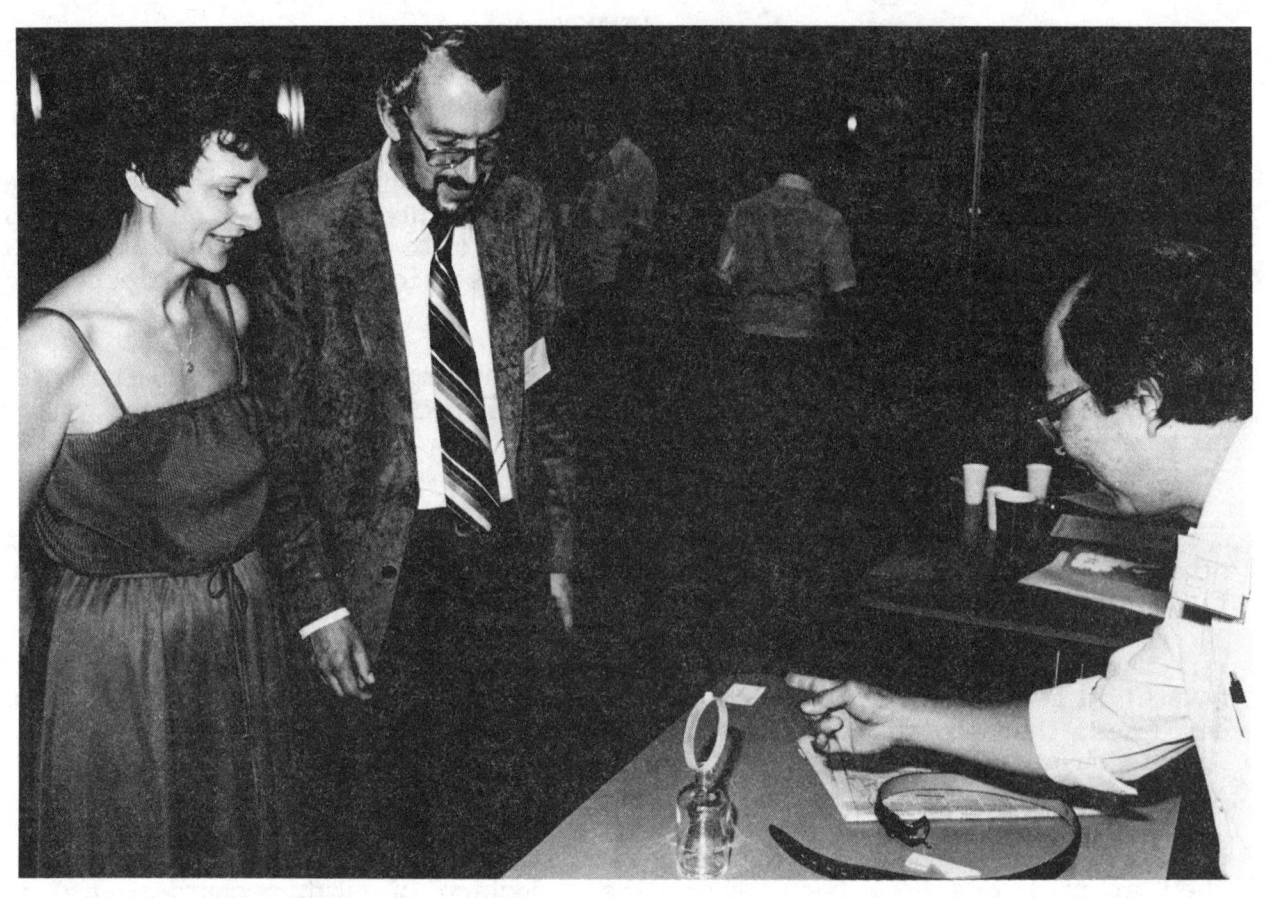

a Demonstration is Ten Thousand Words !

CHAPTER 6

WHAT FORMS OF ENERGY ARE THERE ?

OBJECTIVES

After dealing with and studying the concepts and sub-concepts of this chapter, the students should be able to:

a. recognize the correct explanation of an observed event based on each of the sub-concepts;
b. explain in their own words which of the sub-concepts is determining the course of an event;
c. distinguish true from false statements concerning each one of the sub-concepts;
d. identify the correct explanation of an event in daily life applying one of the sub-concepts;

all in relation to the following sub-concepts:

-- The ability to hold heat depends on the mass and material of the object.
-- Temperature is a measure of how hot or cold an object or substance is.
-- Heat is required to change a solid into a liquid, a liquid into a gas, or a solid into a gas.
-- Energy can be developed from chemical, mechanical, magnetic, or nuclear sources.
-- All forms of energy can be transformed from one form into the other.
-- All forms of energy can be traced back to solar energy.

ENERGY HEAT vs TEMPERATURE

6.1. THE HOT BOLT

Materials:
1. A heavy steel bolt, a small iron nail.
2. Two identical beakers, two thermometers.
3. One beaker or tin can (to boil water in).
4. A pair of tweezers or tongs.
5. Burner & stand, or hot plate.

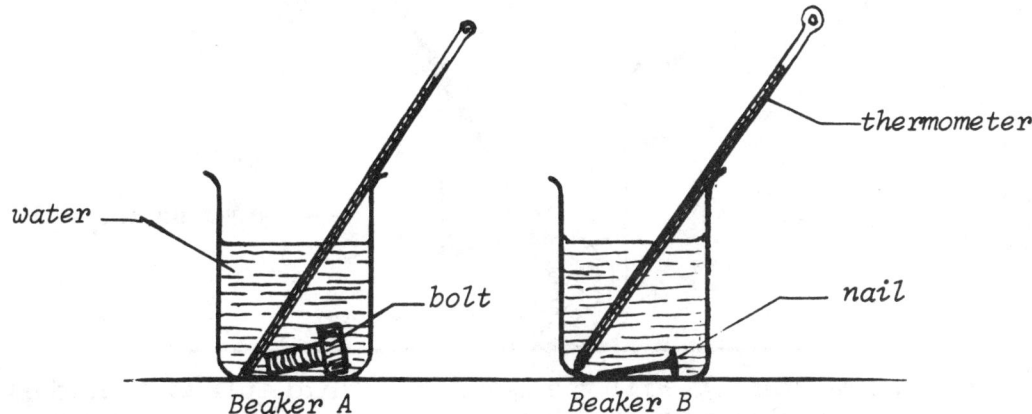

Procedure:
1. Place the bolt and nail together in a beaker or tin can in about 100 ml of water, heat over burner or hot plate and bring to a boil.
2. Pour 100 ml of water in each of the two identical beakers and read their temperature off with the thermometer.
3. Transfer the bolt to beaker A and the nail to beaker B with the tongs.
4. Read off the temperature in each of the beakers every 15 seconds until the temperature starts to level off (have a student assist you).

Questions:
1. What is the temperature of the boiling water?
2. What is the temperature of the bolt and nail before the transfer?
3. In which of the two beakers did the temperature rise higher?
4. Why did the water in the two beakers rise in temperature?
5. Which has a higher heat capacity, the bolt or the nail?
6. What determines the ability for an object to hold heat (heat capacity)?

Explanation:
The water is boiling at 100° C, at normal pressure. The bolt and the nail that were in the same water have also the same temperature of 100° C before they were transferred to Beaker A & B, but they did not absorb the same amount of heat. The large bolt holds more heat or its **heat capacity** is higher, because it is larger in mass compared to the nail. The extra heat is enough to raise the temperature of the water in beaker A more than that in beaker B. The ability to hold heat or the heat capacity of an object is determined by its mass and the material it is made of. Temperature is a measure of how hot or cold an object or substance is in degrees Celsius or Fahrenheit. Heat refers to the total energy of the moving molecules of a substance, measured in calories or BTU's (British Thermal Units).

ENERGY HEAT OF FUSION

6.2. MELTING ICE BELOW FREEZING?

Materials:
1. A medium size beaker & thermometer.
2. Common table salt, crushed ice.

thermometer

crushed ice + salt

Procedure:
1. (Before class starts) Fill the beaker one-third full with crushed ice.
2. Pour 3 or 4 tablespoons of table salt in the center of the beaker on the crushed ice.
3. Cover this with another one-third beaker full of crushed ice.
4. Now show the students the beaker with crushed ice and ask: "At what temperature will ice melt?" (Anticipated answer: 'at 0° C').
5. Insert a thermometer in the ice and let a student read it off.

Questions:
1. What is the thermometer reading of the melting ice?
2. At what temperature does ice usually melt?
3. What do you think made the ice melt below freezing point?
4. Does the melting process need heat or does it give off heat?
5. Was any heat supplied or withdrawn from the system?

Explanation:
   The freezing or melting point of salt water is much lower than that of pure water. The addition of salt to the ice at 0° C causes it to melt, because salt water cannot freeze at 0° C, just as pure water does not freeze at 5° C.
   The melting process decreases the temperature because melting needs heat, and as heat was not supplied to the system of melting ice, it withdraws heat from its own environment. Just as the forced evaporation of liquid without any heat supply will have a cooling effect, this decrease in temperature is a result of a forced melting process.
   In colder climates salt is used to melt ice on slippery roads in winter.

6.3. WARMING BREEZES

Materials:
1. A small electric fan.
2. Two identical thermometers.

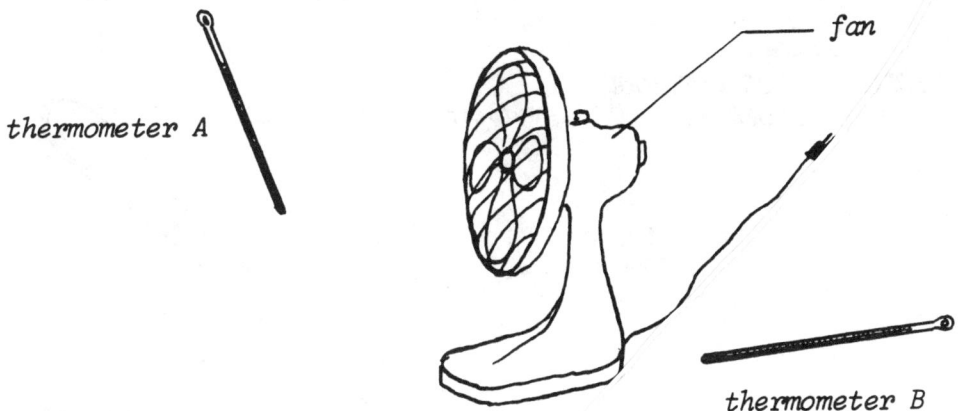

Procedure:
1. Show the students the two identical thermometers A and B, and let two students read the temperature of each of the thermometers.
2. Place thermometer A in the breeze of the fan and thermometer B near the fan, but not in the breeze.
3. Ask the students: "Which of the two thermometers will show a lower temperature?" (Anticipated answer: Thermometer A).
4. After leaving the fan on for several minutes and leaving the two thermometers in the two designated places, have two other students read off the two temperatures.

Questions:
1. Which of the two thermometers do you predict will show a lower temperature? The one in the breeze or the one not in the breeze?
2. What temperature did the two thermometers show?
3. Why was there no difference in temperature?
4. What causes the cool feeling on our skin in a breeze?
5. What would you predict would happen to temperatures A and B if the bulbs of the thermometers were moistened?

Explanation:
Most of the students will predict that thermometer A will show a lower temperature. Actually, both temperatures stay the same. A very sensitive thermometer might even show that spot A (the one in the breeze) is slightly warmer than spot B, since the motor of the fan gets a little warmer and this heat is carried in the wind stream. An additional reason for it to get a little warmer is the friction of air molecules colliding with the bulb.

The decrease in temperature in a breeze is only caused by the evaporation of moisture. For evaporation to occur, heat is needed. When heat is not supplied, this heat of evaporation is withdrawn from the liquid's surroundings, resulting in the decrease of temperature. If the thermometer bulbs were moistened, temperature A would show up much lower.

6.4. THE HOT ACID

Materials:
1. Potassium chlorate ($KClO_3$), powdered sugar, concentrated sulfuric acid (H_2SO_4).
2. Asbestos plate or evaporating dish, medicine dropper.

CAUTION

PERFORM UNDER FUMEHOOD OR OUTDOORS !!

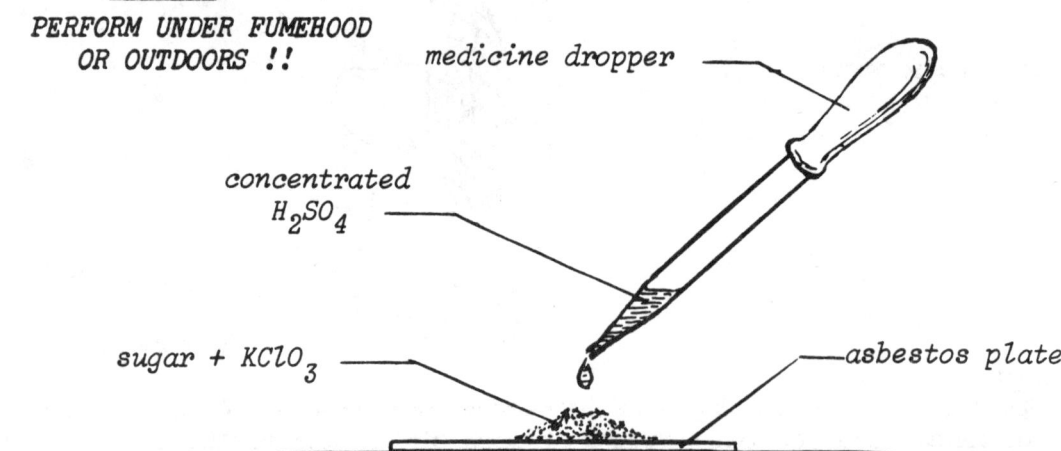

Procedure:
1. Mix about one teaspoon of powdered sugar and one teaspoon of potassium chlorate on the asbestos plate or evaporating dish.
2. Leave the powder in a heap and make a small dent on top of the heap.
3. Suck some of the concentrated sulfuric acid in the medicine dropper (be very careful not to get any of the acid on the skin or clothing).
4. Place one drop of the acid on the heap of powder (stretch your arm as far out as possible in doing this)--do this outdoors or under the exhaust hood.

Questions:
1. What type of energy source was creating the heat?
2. Which of the two powders was actually burning?
3. What do you think was the oxygen supply for the burning?
4. What created the heat for this spontaneous combustion?

Explanation:
This is an example of a chemical source for heat: a spontaneous combustion. The initial heat needed for the oxygen to be released from the potassium chlorate was created by the drop of concentrated sulfuric acid and the sugar. This heat activated the release of very active oxygen in atomic form from the potassium chlorate ($KClO_3$). This oxygen further oxidized the sugar that was present. This further oxidation released so much heat that the sugar burst into flame. Of the two powders then, it is the sugar that was burning and the potassium chlorate provided the needed oxygen for the burning process. Most chemical reactions are releasing heat: **exothermic**. Some withdraw heat: **endothermic** reactions.

ENERGY   CHEMICAL SOURCES

6.5. THE LIGHTER-FUEL CANNON

Materials: 1. Three empty soda pop cans.
            2. Two styrofoam cups.
            3. Lighter fuel (or gasoline) and a match.

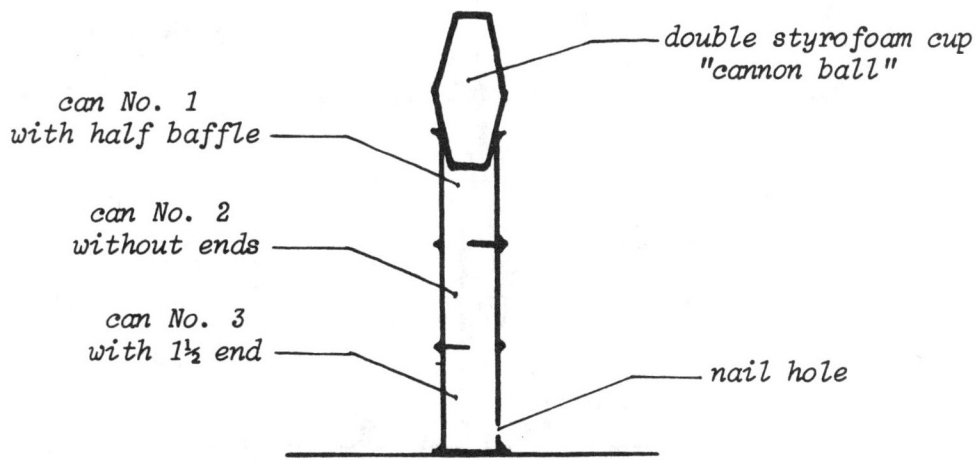

can No. 1 with half baffle

can No. 2 without ends

can No. 3 with 1½ end

double styrofoam cup "cannon ball"

nail hole

Procedure:
1. With a can opener cut open the ends of the soda pop cans, except for one half-end of one can (top), and the bottom and one half of the other end of the third can (bottom can).
2. Connect the three cans on top of each other with masking tape.
3. Punch a hole about 2 cm from the bottom, in the side of the bottom can about 1/2 cm in diameter (use a large nail).
4. Tape the two styrofoam cups together rim to rim, and place it tightly in the top opening (see Sketch).
5. Place two or three squirts of lighter fuel in the bottom hole and shake the stack of cans (let the fuel drip along the walls), and let it stand for a few seconds. You are now ready for ignition!
6. Strike a match and hold the flame close to the bottom opening (Watch out for the double cup cannon ball!).

Questions:
1. What are the two baffles in the 'cannon' for?
2. What purpose did the shaking of the stack of cans have?
3. What kind of energy resulted from the chemical explosion?
4. What other kinds of liquids do you think could be used in place of the lighter fuel?

Explanation:
   The baffles were left in the soda pop can 'cannon' to enhance the mixing of the fuel with the air in the cylinder. The shaking of the cylinder was done immediately after the fuel was injected for exactly the same reason. The better the mixture of fuel vapors and the air, the better the explosion. The chemical energy stored in the lighter fuel is transformed by the combustion into kinetic energy of the moving cannon ball. Gasoline or alcohol may be used instead of lighter fuel.

ENERGY — MECHANICAL SOURCES

6.6. WARM A BOTTLE BY SHAKING

Materials:
1. A medium size jar with lid.
2. Dry sand (beach sand).
3. A thermometer.

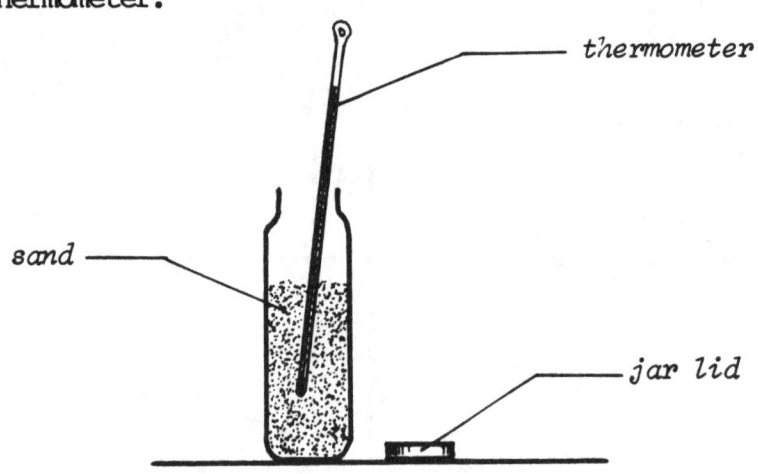

Procedure:
1. Fill the jar about two-thirds full with dry sand.
2. Push a thermometer in the sand and let one of the students read off the temperature.
3. Take the thermometer out of the jar, close the jar tightly and let the students take turns shaking the jar for 5 to 6 minutes in total (about 1/2 minute each for 10 to 12 students).
4. Open the jar after the shaking, push the thermometer in the sand and let a student read off the temperature again.

Questions:
1. What were the temperature readings before and after the shaking?
2. What is the cause for the temperature to rise?
3. What type of energy was turned into heat?
4. To what type of energy as the original source can this produced heat be traced back?

Explanation:
The shaking of the bottle or jar is mechanical energy, which was turned into heat. The mechanical shaking of the sand gave the sand particles kinetic energy, which caused friction between the sand particles, thus creating heat. The energy needed for the shaking of the jar was supplied by muscle power of the students. To give this energy, the students needed to eat (chemical energy), and the food the students ate came from plants and animals. The animals also needed to eat plants, and the plants in their turn obtained the energy from the sun to grow. All forms of energy can thus be traced back to **solar energy**.

ENERGY   MECHANICAL SOURCES

6.7. THE WIRE HEATER

Materials: 1. A piece of heavy wire (coat hanger).
            2. A wire cutter.

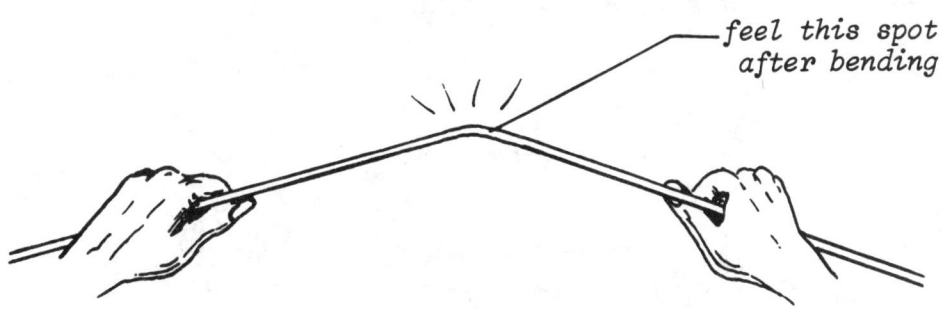

feel this spot after bending

Procedure:
1. Take a coat hanger and cut about a length of 30 cm piece off with the wire cutter.
2. Hold the two ends in each hand and bend the wire several times back and forth at the same spot.
3. Let students touch the spot that was bent, with their fingers.

Questions:
1. Did the spot you touched feel warm or cold?
2. What type of energy was used to bend the wire?
3. What made the wire heat up (explain in terms of moving molecules)?
4. Would the same happen with a plastic or wooden rod?
5. What would eventually happen if we keep bending the wire?

Explanation:
   Bending the metal wire back and forth is actually putting mechanical energy from our muscles into the wire. This energy forces the molecules of metal to move faster at the bent spot and the friction of the faster moving molecules produces heat, thus causing the bent spot of the wire to feel warm to the touch. When this bending of the wire is continued, it will eventually break: this is called **metal fatigue.**
   When using a plastic or wooden rod instead of a metal one, the bending can be carried on for a longer period before it breaks, because of the flexibility and elasticity of the material, unless it is bent beyond its flexibility point. The molecules here are not sliding past each other as much as in the metal wire, so that heat is not likely generated so readily.

ENERGY   HEAT SOURCES

6.8. USE HEAT TO TURN A WHEEL?

Materials:
1. A hot water kettle.
2. A fitting one-hole stopper for the kettle spout.
3. A bent glass tube fitting in the stopper.
4. A cork, 6 old ink pens, 4 paper clips.
5. A piece of soft wood or board.

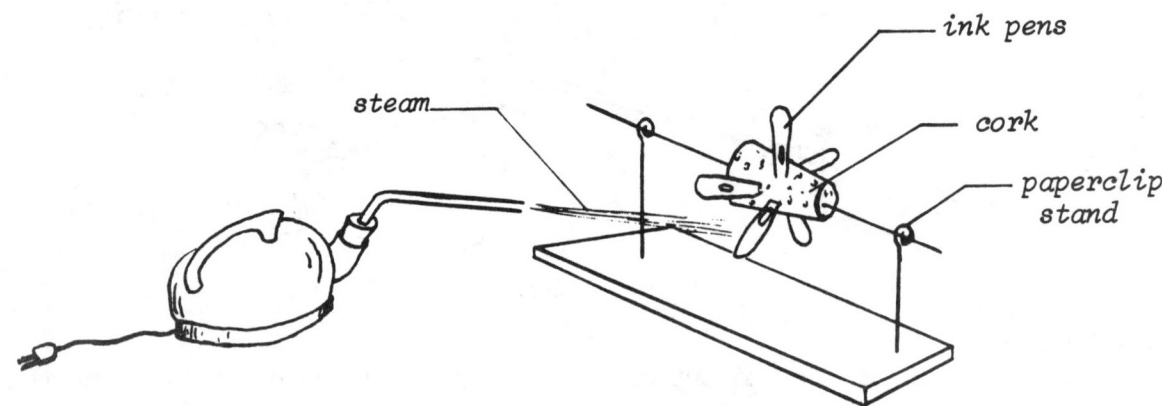

Procedure:
1. Fill the electric kettle with water and plug it in.
2. Insert the bent glass tube in the one-hole stopper and cover the kettle spout with the stopper.
3. Stick the six old ink pens (or any other small metal narrow plates) in the side of the cork and set up the cork to move freely as in the sketch, with modified paper clips as a stand on the wood board.
4. As soon as the water boils, direct the steam towards the lower or upper side of the cork against the baffles.

Questions:
1. What is the original source of energy used to turn the wheel?
2. What forms of energy do we encounter in this event?
3. Where do we find this principle applied?
4. How is electricity produced and what could be the original form of energy that is transformed into electricity?

Explanation:
The original source of energy used is electricity for the electric kettle, which may be traced back to chemical sources or water power. The electricity is turned into heat in the heating coils of the kettle, which causes the water to boil and thus forming the steam. By leading this steam through a narrow tube, the vapor particles are pushed out with a velocity and thus gain in kinetic energy. In directing this steam flow to the scoops of the wheel, the kinetic energy is converted into mechanical energy. This finds application in the production of electricity in turning the turbo engines and in other steam engines.

6.9. MAGNETIC PERPETUAL MOTION?

Materials:
1. A small steel ball (from a ball bearing).
2. A strong bar magnet.
3. A plastic ruler (with a ridge on top side).
4. Two lengths of heavy wire (coat hanger wire).

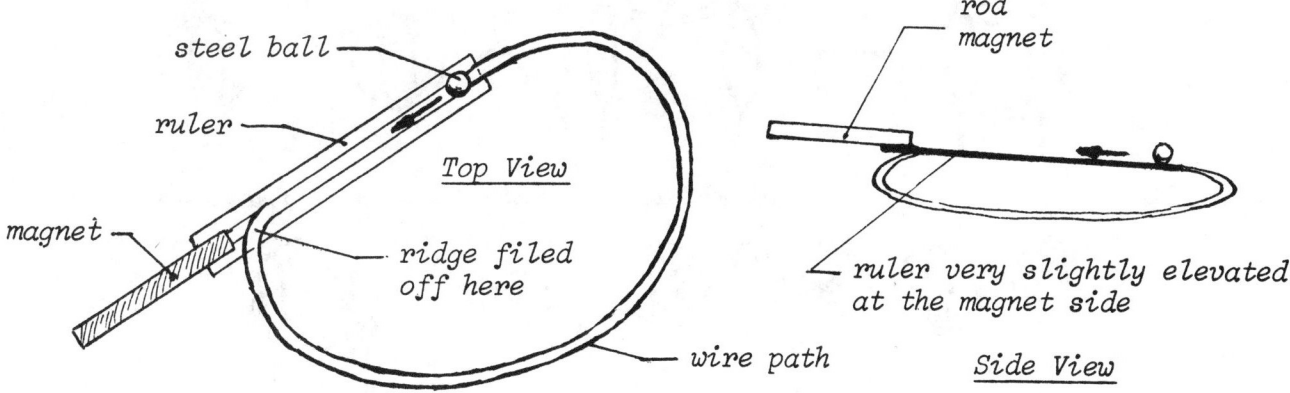

Procedure:
1. Make a wire path out of two heavy wires in the shape of three quarters of a segment of a circle.
2. File one side of the ridge of the ruler off about one third of one end about the width of the ridge.
3. Mount the wires by glueing them to the ruler in the set-up as in the sketch above.
4. Experiment with the slant of the ruler (this should be almost horizontal, slightly higher at the magnet side).
5. The steel ball should roll smoothly over the ridge of the ruler and over the wire rail, gaining speed as soon as it comes off the ruler on the wire rail.

Questions:
1. Will this motion of the steel ball be perpetual?
2. Where is the energy coming from? Where is it lost?
3. If the magnet is too strong, what would we need to keep the ball from going towards the magnet?
4. In order for the ball to be able to shoot up the ruler ramp, what is the minimum height required for the initial entrance of the wire rail?
5. Which force or component force have to be overcome to pull the ball up?
6. Where and in what form is energy being lost?

Explanation:
The force to overcome the gravitational component becomes smaller when the slant of the ruler is less steep. The energy to keep the ball moving comes from the magnet, which eventually will be depleted, at which moment the ball would stop rolling.

Energy is lost on the track in the form of heat. Thus the smoother the track the more efficient the ball will roll.

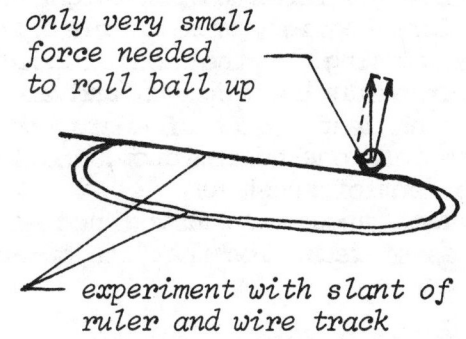

6.10. FISSION AND FUSION

Materials:
1. Soap solution (1 pt Joy, 2-1/2 pt glycerine, 3 pt water).
2. Two circular wire frames.
3. A wide shallow tray.

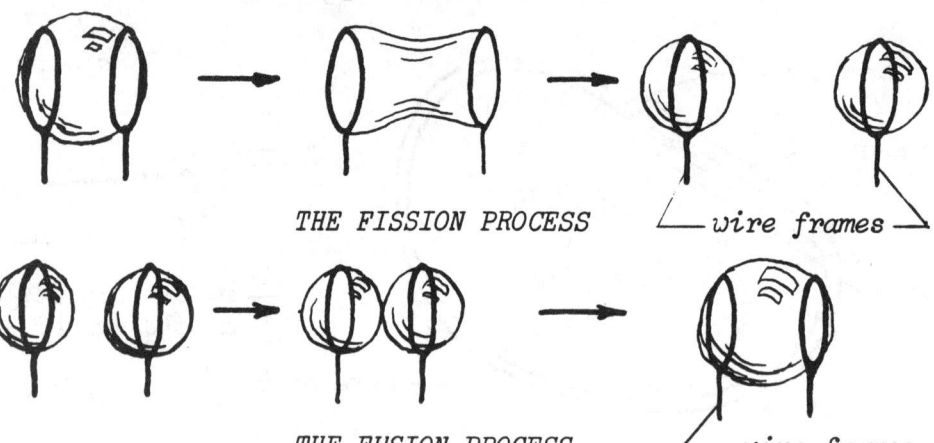

THE FISSION PROCESS — wire frames

THE FUSION PROCESS — wire frames

Procedure:
1. Place the soap solution in the shallow tray and dip the two circular wire frames in the solution.
2. Blow through the wire frame and blow a bubble with a diameter a little larger than the frame, and catch the bubble between the two frames.
3. Now stretch the bubble by pulling the two frames farther apart until the bubble separates into two bubbles in each frame (fission). When this is done a litle faster a small bubble is released, illustrating the released neutron.
4. Bring the frames and the bubbles in it together; let the bubbles press against each other until they form one large bubble (fusion).

Questions:
1. Why are bubbles always spherical in shape?
2. What is missing from this model of fission or fusion?
3. What examples can you give on nuclear energy sources in nature?
4. What are the dangers in the use of nuclear energy?

Explanation:
This demonstration with the soap bubbles is only a model for the actual process of fission or fusion. In the case of **fission**, a very heavy nucleus splits and forms medium-weight nuclei. When two light-weight nuclei combine to form heavier, more stable nuclei, it is called **fusion**. The elements that are missing in the soap bubble model, are the neutrons and the energy released during fission, and the protons and energy released during fusion.

A nuclear source of energy in nature is the sun, where four hydrogen atoms fuse into one helium atom, and in the process of fusion it releases energy in the form of sunshine.

The dangers in using nuclear energy lie in radiation leakages and the doses of radiation that the human being can withstand without being harmed.

ENERGY   SOLAR SOURCES

6.11. THE TEST TUBE GREENHOUSE

Materials:
1. Two thermometers (0-100°C).
2. A large, wide test tube or long jar with a narrow mouth.
3. One-hole stopper to fit the thermometer in the test tube.
4. A white or infra red spot light.

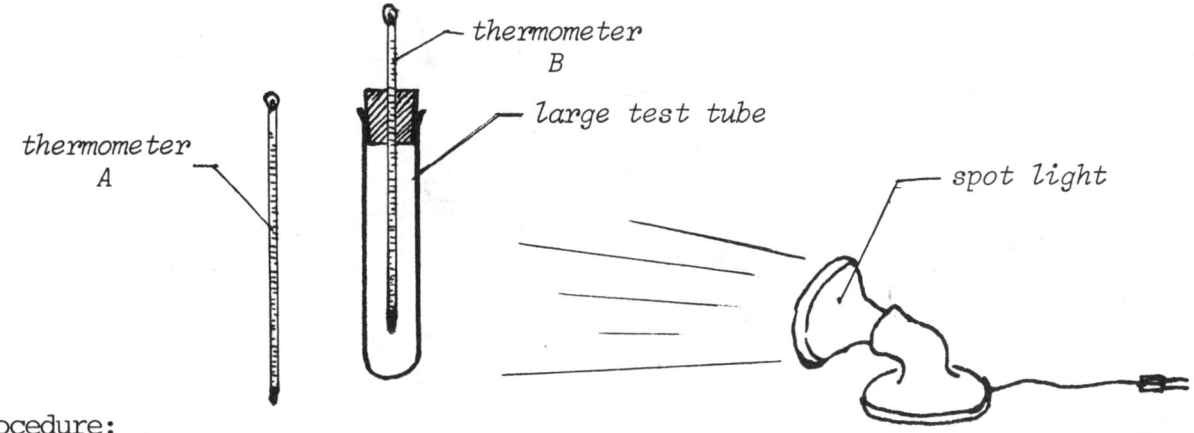

Procedure:
1. Insert one thermometer in the one-hole stopper and place this over the large test tube or jar (if a loose thermometer is not available, a paper covered jar may be used instead of the large test tube and the stopper).
2. Place both thermometers in the sunlight or shine the spot light on them from about 50 cm away (see Sketch above).
3. Have the students read and record the temperatures of both thermometers every minute for about 15 minutes.

Questions:
1. Which of the thermometers shows a higher temperature?
2. What is the light heating around thermometer A, and what is it heating around thermometer B? (see Sketch above)
3. What do these results tell us when applied to a sun room or greenhouse?
4. Which direction should these sun rooms face when your home is located in the Northern hemisphere? In the Southern hemisphere?
5. Would the temperature on thermometer B be higher or lower if the tube or jar were made of opaque glass? If the tube or jar were covered with white paper? If it were covered with black paper?

Explanation:
   Like the atmosphere of the earth, glass around the thermometer can trap heat energy. The light rays heat up the air in the tube, which cannot move around, contrary to the air around thermometer A.
   Similarly, sunlight energy is trapped in the atmosphere, as the sun's energy is absorbed by the earth. Part of this energy is reradiated into the atmosphere but cannot pass out of the atmosphere.
   A greenhouse works on the same principle. If the glass walls of a greenhouse were covered with black material, it would make the room much hotter, but since plants need light, glass is needed. Other examples of energy traps are: closed cars left in the sun, especially the dark colored ones; attics of homes with black roofs, etc.

ENERGY   SOLAR SOURCES

6.12. START A FIRE WITH A MAGNIFYING GLASS

Materials:
1. A large magnifying glass (or strong reading glasses).
2. A bright sunny day, over-exposed film or dark glasses.
3. A piece of tissue paper (or any easily combustible material).

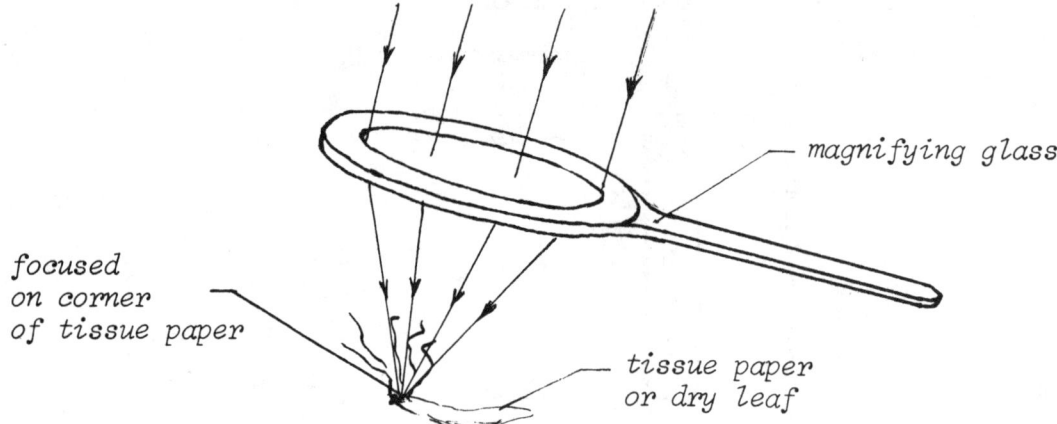

Procedure:
1. Take the class outdoors on a bright sunny day, and bring the large magnifying glass or a pair of strong reading glasses with you. You also should have a piece of tissue paper or any thin paper in your pocket.
2. Place the tissue paper on a spot where everybody can see it. Let the students use the darkened film or dark glasses to observe the paper, while you hold the magnifying glass about 20-30 cm above the paper. This distance has to be varied by moving the lens up or down, such that the sun rays are focussed on the paper. (The small bright circle has to become a small point if it is well focussed).
3. Hold the magnifying glass at that position (focussed position) for awhile (about 15-20 sec) until suddenly the paper starts to smolder and burn.

Questions:
1. What made the paper burn?
2. Why would the paper not burn without the magnifying glass?
3. What would it depend on how high we have to hold the lens to focus it?
4. What else beside thin paper can we use to burn?
5. How are forest fires sometimes started by nature?
6. What did the magnifying glass actually do with the sun rays?

Explanation:
What the magnifying glass or lens does, is concentrating all the sun rays that were falling on it, into one point. This increased the intensity of the heat 100 or may be 200 fold, depending on how large and how strong the lens is. You will find that on clear sunny days it is much easier to start the fire, than it is on hazy or cloudy days. The stronger the lens (more convex) the closer we have to hold the lens to the piece of paper to focus the sun rays.
  Dry leaves can easily be substituted for the tissue paper. In nature, forest fires are sometimes caused by concentrated sun rays. Lingering dew drops between plant leaves can act as lenses and the concentrated sun rays may fall on very dry leaves, and a forest fire is in the making!

ENERGY   POTENTIAL vs KINETIC

6.13. DOES A BOOK HAVE ENERGY?

Materials: 1. A heavy book (encyclopedia or dictionary).
            2. A meter stick, pencil and eraser.
            3. A high stool.

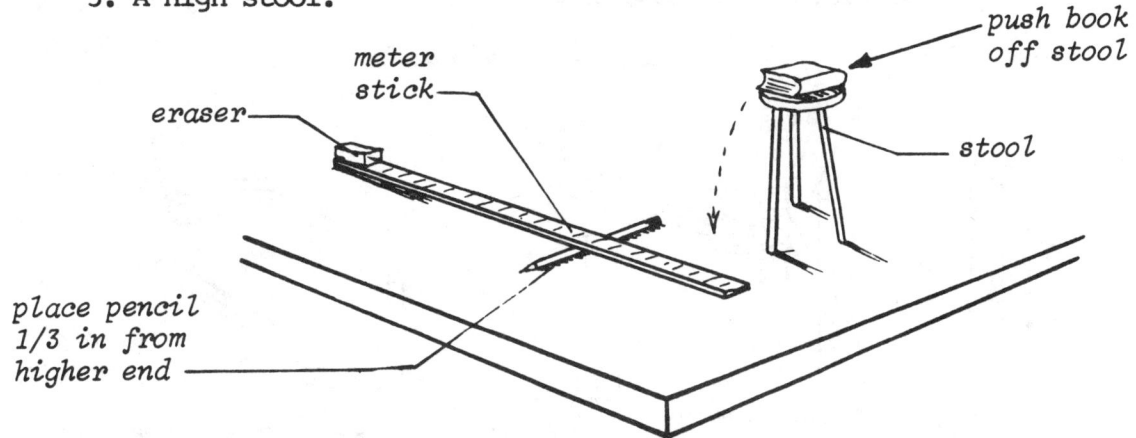

Procedure:
1. Place a stool on the table top and put a heavy book on the stool. Ask students: "Does the book have any energy at this position?"
2. Place the meter stick on the table top about 20-30 cm in front of the stool, such that when the book is pushed off the stool, it will fall on the end of the ruler.
3. Insert a thick pencil under the meter stick, about one-third from the end where the book will be falling on.
4. Place an eraser on the other end of the meter stick.
5. Push the book off the stool (and let it fall on the meter stick). Observe the eraser fly up!

Questions:
1. What kind of energy did the book have when it was lying on the stool? When it was falling on the ruler?
2. Does the book have any energy lying on the table top?
3. What type of energy did the book impart on the meter stick?
4. What kind of energy did the eraser obtain?
5. What was the original source of energy that triggered the flying eraser?

Explanation:
   When the book was lying on top of the stool, it had potential energy. As long as the book has the potential of falling towards the center of the earth, it has potential energy. Thus when it was lying on the table top after falling off the stool, it still possesses potential energy (to fall off the table and send another eraser flying). At the moment that the book was falling, this potential energy was transformed into kinetic energy, which was turned into mechanical energy of the moving meter stick, and this was imparted to the eraser, which obtained kinetic energy.
   The original source of energy of this whole chain reaction was human muscle energy, which can be traced back to solar energy.

ENERGY POTENTIAL vs KINETIC

6.14. DOES WATER IN A LAKE HAVE ENERGY?

Materials:
1. A large beaker, two short glass tubes.
2. A cork, 6 old ink pens, 2 paper clips.
3. Two clamps and two stands.

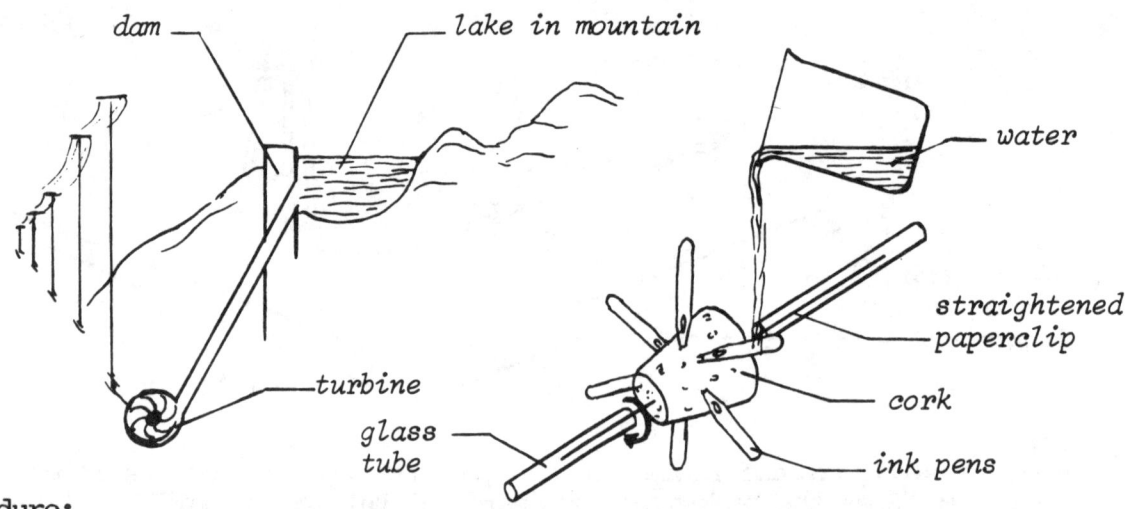

Procedure:
1. Push the six ink pens radially in the cork and set the materials up like in the sketch, so that the cork can rotate freely.
2. Hold the water-filled beaker above the cork and tell students: "This represents water in a lake up in the mountains."
3. Ask the students: "Does the water have energy? What kind?" Pour the water on one side of the cork on the protruding pens (do this over a sink or a container to catch the water).

Questions:
1. What kind of energy does the water in the beaker have?
2. What kind of energy was moving the cork?
3. What is the resultant type of energy produced?
4. If the water in the beaker can be compared to the water in a lake, what can we compare the rotating cork with?
5. What kind of energy do we need to bring the water up in the beaker?
6. What type of energy is needed to fill up a lake with water?

Explanation:
The water in the beaker, held up high, is comparable to the water in a lake up high in the mountains, possessing potential energy. The energy may be released and transformed into kinetic energy by building a dam and controlling the flow of the water. By letting the flow of water pass the turbines, represented by the cork and protruding ink pens, the kinetic energy of the water is transformed into mechanical energy, which in turn is transformed into electrical energy. This last step takes place in the generators, being rotated by the turbines.

ENERGY   POTENTIAL vs KINETIC

6.15. THE SPINNING RINGS

Materials: 1. A 1 cm diameter wooden dowel (about 1 m long).
2. Large rubber washers (about 2-1/2 - 3 cm in diameter).

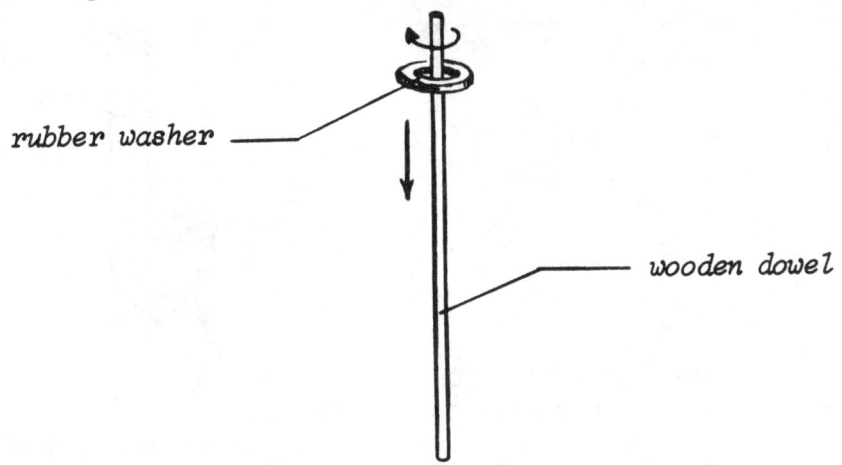

Procedure:
1. Hold the dowel vertically on the table top and the rubber washer at the top of the dowel. Ask the students: "How long will it take for the ring to drop the length of the wooden dowel?" (anticipated answer: 'A fraction of a second').
2. Give the ring a spin at the top of the dowel and time its fall. Ask: "Why is the fall not occurring in a uniform motion?" Observe the spinning and the falling speed carefully (when the falling slows down, the spinning or rotation of the washer increases).

Questions:
1. How long did it take for the spinning ring to fall?
2. What made the ring keep on spinning?
3. What kind of energy did the ring at the top of the dowel have?
4. Why did the ring spin faster after a sudden drop in height?

Explanation:
The friction between the dowel and the ring prevents the ring from dropping straight down. The faster the ring is spinning, the larger the force perpendicular to the dowel, and thus the stabler its plane of rotation. By dropping, potential energy is transferred into rotational kinetic energy. It can be shown that the rotation of the ring requires energy by spinning it on a horizontal dowel (one spin will probably spin it for a few seconds). The potential energy that the ring possessed at the top of the dowel is not completely transformed into the rotational kinetic energy, but some heat of friction has no doubt developed between the dowel and the ring.

ENERGY

POTENTIAL VS KINETIC
THE PENDULUM

6.16. WILL THE HEAVY BRICK HIT YOUR NOSE?

Materials: 1. A heavy masonry brick or large bowling ball.
2. A strong nylon rope.

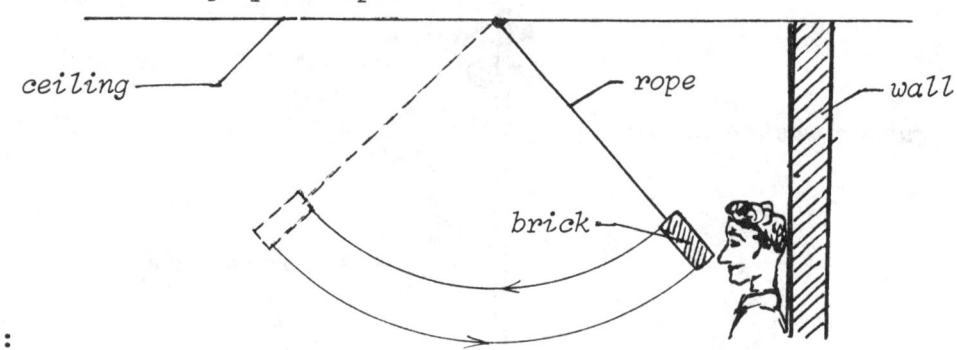

Procedure:
1. Tie the brick or bowling ball to one end of the rope and suspend the pendulum about 3 m from the wall to the ceiling.
2. Before securing the knot in the rope, adjust the length of the pendulum such that the highest swing just reaches your nose when you stand with your back against the wall (see Sketch above).
3. Make sure that the knots are tight and secure and cannot slip.
4. Now bring the pendulum as close as possible to your nose and let go. (**CAUTION: DO NOT PUSH THE PENDULUM AWAY!**).
5. Stay perfectly still against the wall and let the pendulum swing back. (Ask the students: "Who dares to do this?")

Questions:
1. Why is it perfectly safe to stand against the wall and not get hit?
2. Will the pendulum ever swing higher than its first position?
3. What would give the brick a higher swing?
4. What forms of energy are involved in a swinging pendulum?
5. Where does the energy go? What does it turn into?
6. In which positions of the swing does the pendulum have the greatest kinetic energy? The greatest potential energy?
7. What would happen if the brick was swung in a vacuum?
8. What makes a clock pendulum swing? When does it stop?

Explanation:
   When the brick pendulum is released for the first time, it loses some of its energy due to friction at the suspension point and mostly in air resistance. This energy is turned into heat or actually faster moving rope and air molecules. This loss occurs with every swing of the brick and thus each swing becomes lower and lower (**amplitude** gets smaller). A higher swing would only be obtained if we add energy to the pendulum by giving it a push (do not do this while you are still standing against the wall!).
   If this brick pendulum was swung in a vacuum, energy loss would only occur in the rope and the suspension point, and the swinging would last for a much longer time before it stops.
   In a clock pendulum the swinging is kept going by the spring of the clock. A small amount of energy is added to each swing by the potential energy of the wound up spring. When the spring is completely unwound the clock and the swing of its pendulum stops.

ENERGY

ENERGY TRANSFER
THE PENDULUM

6.17. THE TWIN PENDULUM

Materials:
1. Two identical washers (or any other object to make a bob).
2. Thin string or strong thread.
3. Two vertical stands.

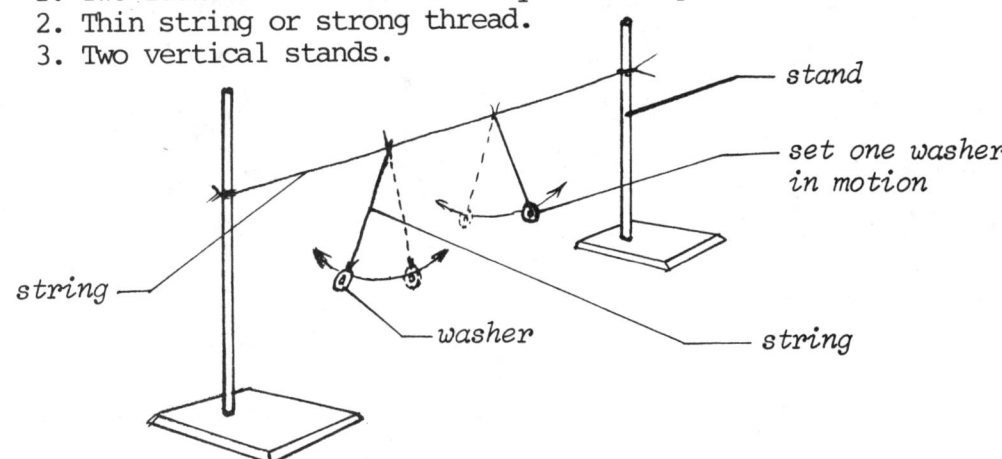

Procedure:
1. Place the two vertical stands about 30 cm apart from each other and tie a piece of string horizontally from one stand to the other.
2. Tie a short piece of string (about 15 cm) to each of the washers (or pendulum bobs) and tie the other end to the horizontal string, about 5 cm apart from each other - make sure that the bobs hang down from the horizontal string at the same height!
3. Let the two pendulums hang perfectly still: now you are ready for the demonstration. Ask: "What will happen to the second pendulum if I start swinging the first one?" Anticipated answer: "Nothing!"
4. Pull out the first pendulum carefully by the bob (hold the string tight) and let go of the bob: OBSERVE!

Questions:
1. What happened to the second pendulum?
2. Where did all the energy of the first pendulum go to?
3. What made the second pendulum pick up the energy?
4. Would a pendulum that was a little longer or shorter than the first one also be able to pick up the energy?
5. Would the second pendulum also pick up energy from the first if they were both tied to a solid metal rod instead of a string?
6. Did the first pendulum completely stop after the energy was transferred?
7. If compared to just one pendulum swinging from a similar support, which of the two systems would swing the longest? Which are the variables?

Explanation:
Only when the two pendulums are perfectly identical in length, will the energy of the first pendulum be transferred totally to the second. The first pendulum then stops and in turn picks up the energy from the second, etc. until the energy is dissipated into heat, and both pendulums stop swinging.

The energy from the first pendulum is transferred to the second only because of the flexible horizontal support, which moves in phase with the swing of the second pendulum. These movements keep on strengthening the swing just like the push that we give to a child on a real swing in the park (see also Events 12.8 & 12.7).

CHAPTER 7

HOW DOES HEAT AFFECT THINGS?

OBJECTIVES

After dealing with and studying the concepts and sub-concepts in this chapter, the students should be able to:

a. recognize the correct explanation of an observed event based on each of the sub-concepts;
b. explain in their own words which of the sub-concepts is determining the course of an event;
c. distinguish true from false statements concerning each one of the sub-concepts;
d. identify the correct explanation of an event in daily life applying one of the sub-concepts;

all in relation to the following sub-concepts:

-- Heat travels by conduction in solids.
-- Heat travels by convection in fluids - liquids and gases.
-- Heat travels by radiation in fluids and in vacuum.
-- Heat makes solids, liquids, and gases expand.
-- Cooling makes solids, liquids, and gases contract.
-- A substance does not burn unless its kindling point is surpassed.

HEAT — CONDUCTION

7.1. THE COIL CANDLE-SNUFFER

Materials:
1. Insulated solid copper wire (about 30 cm).
2. A wire stripper or knife.
3. A birthday candle & matches.

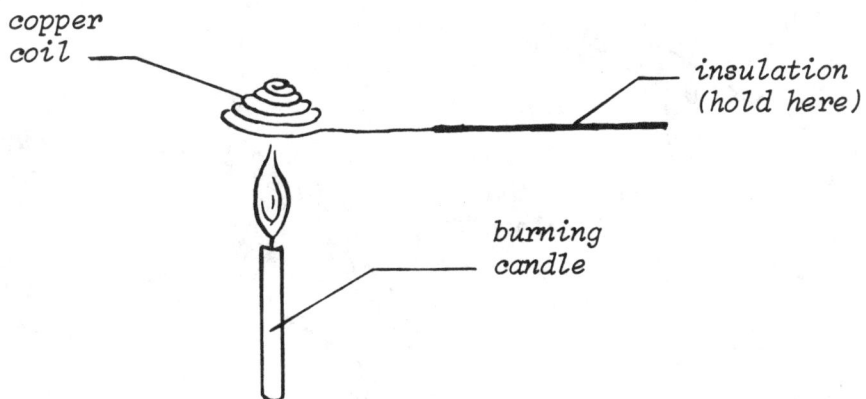

Procedure:
1. Take the copper wire and strip 3/4 of the insulation off one end (use the wire stripper or knife).
2. Make a spiralling coil of the stripped end of the copper wire.
3. Light the candle and fasten it to the table with a drop of molten wax.
4. Lower the coil over the candle flame quickly; flame snuffed!
5. Relight the candle; and now lower the coil slowly over the flame (hold it in the flame for a while), then lower the coil over the candle: flame stays on!

Questions:
1. What made the flame go out when the coil was lowered quickly?
2. Was the flame cut off from the air?
3. Why did the flame stay on after the coil was held in the flame for a longer period?
4. What was the temperature of the coil the first time compared to the second time it was lowered over the flame?
5. Why was the insulation not completely stripped off the wire?

Explanation:
In lowering the copper spiral over the flame, it was conducting the heat away from the flame. This made the surrounding of the flame suddenly drop in temperature, which made the flame go out. In other words, the temperature dropped below the **kindling temperature** of the candle wax.

When holding the coil for a longer period of time in the flame, it is heated to a higher temperature. When this hot coil is now lowered over the flame, it will not extinguish it, because the surroundings of the flame have a temperature which is higher than the **kindling point** of candle wax.

The insulation of the wire was left on at one end of the wire in order to prevent conduction of the heat to the fingers. Without insulation it would be too hot to hold the coil with the bare fingers.

HEAT — CONDUCTION

7.2. THE CHARLESS COTTON

Materials:
1. A quarter coin or any large coin (copper is best).
2. A piece of old cotton cloth.
3. A cigarette and matches.

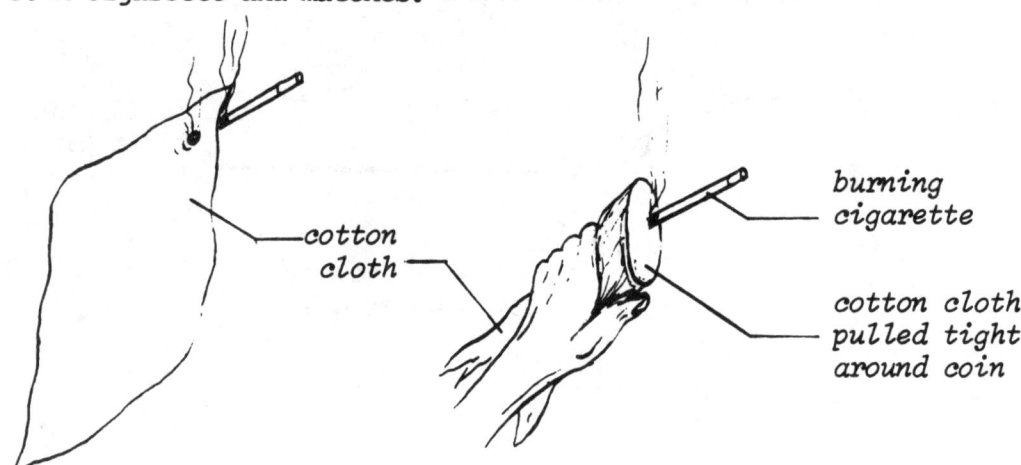

Procedure:
1. Take the piece of cotton cloth, light the cigarette and show students how easy it is to burn a hole through the cloth (in the corner of it).
2. Wrap the cotton cloth over the coin tightly and hold it in one hand.
3. With the other hand, push the lit cigarette against the cloth on the coin -- cotton cannot be charred or burnt!
4. Take the cigarette off the cloth and show no charred spot on the cloth.

Questions:
1. Why was it not possible to char the cotton on the coin?
2. Why is it so easy to burn a hole in the cotton without the coin?
3. After the attempt to char the cotton, how does the coin feel?
4. Could the quarter be replaced with a penny? A nickel? An iron disc? A wooden disc? A plastic disc?
5. What temperature does the cotton have to reach in order for it to burn?
6. What was the purpose of the coin?
7. How did the heat travel away from the cotton?

Explanation:
   The heat from the lit cigarette is enough to easily char or burn a hole in the cotton cloth. With the cloth drawn over a coin, however, this heat was absorbed by the coin and conducted away from the cloth. The heat of the burning cigarette disappeared into the coin, lowering the temperature of the cloth, with the result that the cloth could not be charred.
   The cloth has to be held very tight over the coin in order for this demonstration to succeed. The closer the contact between cloth and coin, the better the **conduction** of the heat away from the cloth.

7.3. THE SCORCHING PAPER

Materials:
1. A short wooden dowel (about 10 cm).
2. A piece of copper rod or tubing (with the same diameter as the wooden dowel).
3. Paper, cellotape, alcohol burner.

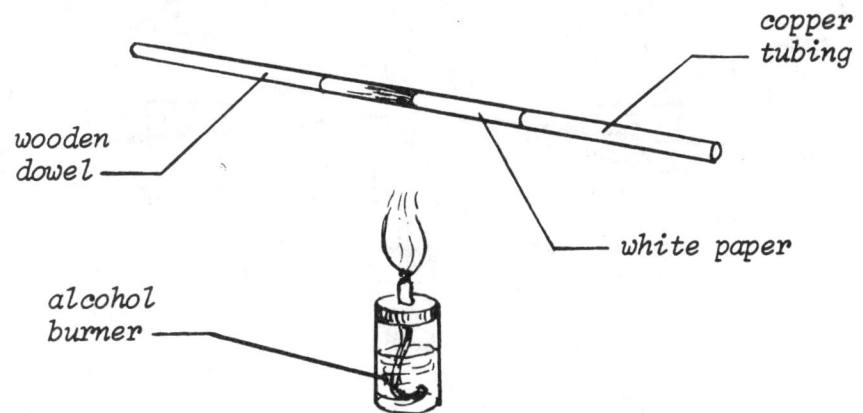

Procedure:
1. Take the wooden dowel and the copper rod (or tubing) of the same diameter, and join them together by wrapping one layer of white paper around them (use cellotape only at the end of the paper--make sure that the paper is tight against the rods).
2. Hold the metal rod and heat the piece of paper evenly over the colorless flame of the alcohol burner, by moving it back and forth above it (do not let the paper touch the flame). Continue this heating until some scorching of the paper occurs.

Questions:
1. Where did the scorching occur?
2. Why did the paper only scorch on one side?
3. Could a candle flame replace the alcohol burner?
4. Why do we need to heat the paper evenly?
5. What would happen to the paper if the copper rod were replaced by a piece of rubber tubing? Glass tubing?
6. What would the paper do if we had more than one layer of paper wrapped around the rods?

Explanation:

The copper tubing, compared to the wooden dowel, is a much better heat conductor. The heat on the copper side was therefore conducted away from the joint which was heated, whereas the heat on the wood side stayed where it was. The temperature at the wood side got much higher than that at the copper side, thus scorching occurred on the wood side.

When the metal rod is replaced with less good conductors of heat, like rubber or glass, the scorching will most likely occur on both sides of the paper, as the heat build-up will take place on both sides.

HEAT — CONDUCTION

7.4. THE HEAT RACE (I)

Materials:
1. Two small corks.
2. One long and 2 short pieces of copper wire.
3. An alcohol burner.

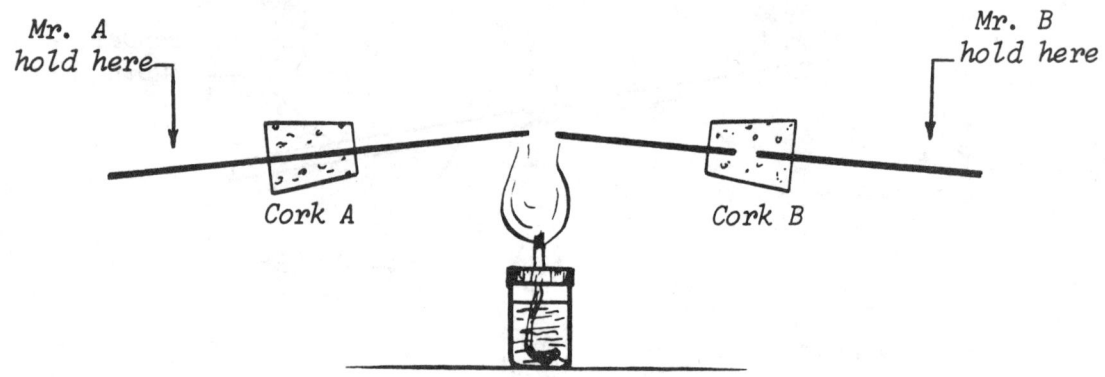

Procedure:
1. Cut the following lengths of copper wire: 10 cm, two 4-1/2 cm.
2. Push the 10 cm wire through one of the small corks (see Cork A).
3. Push the two short wires through each end of the other cork (Cork B).
4. Let two students hold one end of the wire and keep the other end in the alcohol burner flame. (Do not reveal to the students that the wire in cork B is broken--soon Mr. A will let go of the wire!).

Questions:
1. Why did Mr. A have to let go of the wire?
2. Why can Mr. B keep on holding the wire in cork B?
3. What can you infer about the wire in cork B?
4. By what means was the heat transferred to Mr. A's fingers?
5. What would Mr. A have to do in order to be able to further heat cork A?

Explanation:
   Mr. A had to let go of the wire containing cork A, because the heat from the alcohol flame was conducted to the end where the wire was held. This end eventually became too hot to hold. The students could infer about the wire in cork B, that it was made of poor-conducting metal or some other non-conductive material, or that it was broken in the cork and thus not further conducting the heat to the fingers.
   In order for Mr. A to keep on holding the wire with cork A, he has to put some non-conductor of heat between the wire and his fingers. This will stop the heat conduction to his fingers and he will be able to keep holding the wire. In Mr. B's case the heat conduction was stopped in the cork, because the wire was broken in cork B.

7.5. THE HEAT RACE (II)

Materials:
1. A glass rod or tube (about 20 cm long).
2. A copper, iron, and aluminum wire (15 cm each).
3. An alcohol burner.
4. A candle and three paper clips.

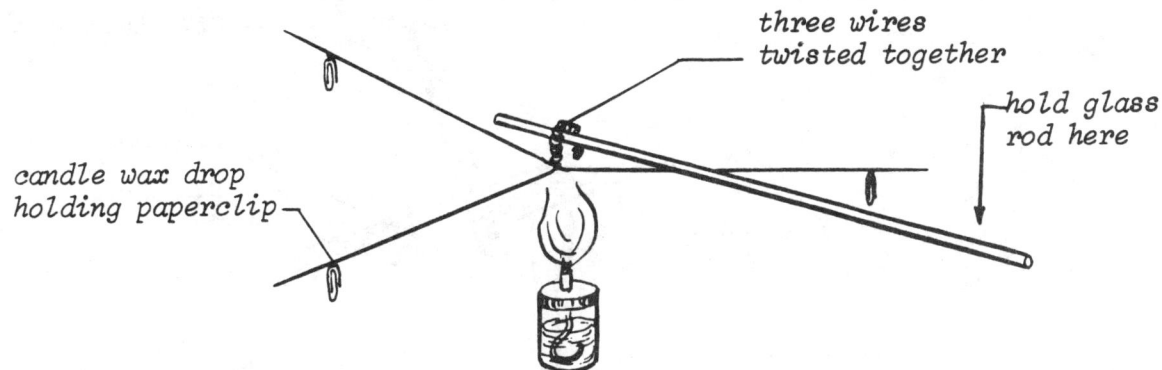

Procedure:
1. Take equal lengths (15 cm) of a copper, iron, and aluminum wire and twist them together at one end.
2. Spread the other end wide out with about 120° angle between them.
3. Attach a paper clip at equal distances (10 cm) from the twisted end with a drop of melted wax (from the candle).
4. Bend the twisted end so it can hang over the glass rod.
5. Heat the twisted end over the alcohol flame and measure the time it takes for the paper clips to drop.

Questions:
1. Which of the paper clips dropped first, second, and last?
2. Which of the metals is the best heat conductor? Which the worst?
3. Which variables do we have to keep constant for this race?
4. Which are the manipulated and responding variables?
5. Would more or less wax holding the clip have any influence in the race?

Explanation:
Although the paper clips are attached at the same distances from the twisted end and thus from the alcohol flame, they do not fall off at the same time. This is because the **thermal conductivity** of each of the metals is different. Copper has the highest thermal conductivity, then comes aluminum, and last is iron. The variables that are influencing the race are: the position of the flame, the amount of wax used to attach the paper clip, the dimensions of the wire, the distance of the paper clip from the flame, the size of the paper clip, and many more, which have to be held constant for this race. The manipulated variable is the kind of metal (this is different in the three wires) and the responding variable is the time the clip could hang on to the wire after heating was started, or the thermal conductivity of the metal. The heat is conducted from the twisted end towards the open end in each wire. The heat melts the drop of wax to which the paper clip is attached, and this latter falls off.

HEAT

CONDUCTION
CONVECTION

7.6. CAN ICE-WATER BOIL?

Materials:
1. Small ice cubes (crushed ice).
2. A wide test tube and holder.
3. A small weight or rock.
4. An alcohol or Bunsen burner.

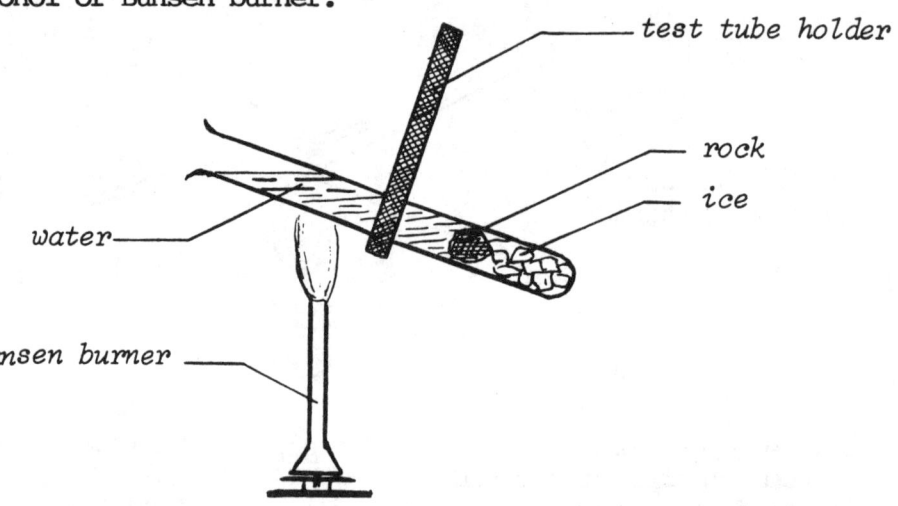

Procedure:
1. Place some ice cubes in the test tube (about 1/3 full) and put a rock or small weight on it in the tube.
2. Pour water in the tube until almost full.
3. Clamp the test tube in the tube holder and hold the upper part of the test tube in the flame of the alcohol or Bunsen burner (see Sketch), until the water boils.
4. Show students that the ice cubes are still present beside the boiling water in the top of the tube.

Questions:
1. Why did the ice not melt by the heating?
2. What is the purpose of the rock?
3. Would this demonstration work without the rock?
4. Was the water conducting the heat well?
5. Were there convection currents taking place in the test tube?

Explanation:
   The ice is kept in the lower part of the test tube by the rock or small weight. Without the weight, the ice would float in the test tube and heating this would melt the ice first before the water can boil. By keeping the ice down in the test tube, the water in the upper part can be heated and boiled without melting the ice, because hot water is lighter than cold water. Very little convection takes place: only in the upper part of the test tube. Another reason for the ice not melting is, that water is a **poor conductor** of heat. The heat from the upper part of the test tube is therefore not conducted to the ice in the lower part of the tube.

HEAT   CONVECTION

7.7. THE CONVECTION TESTER

Materials: 1. Candles (for each pair of students).
            2. Matches (a book for each pair of students).

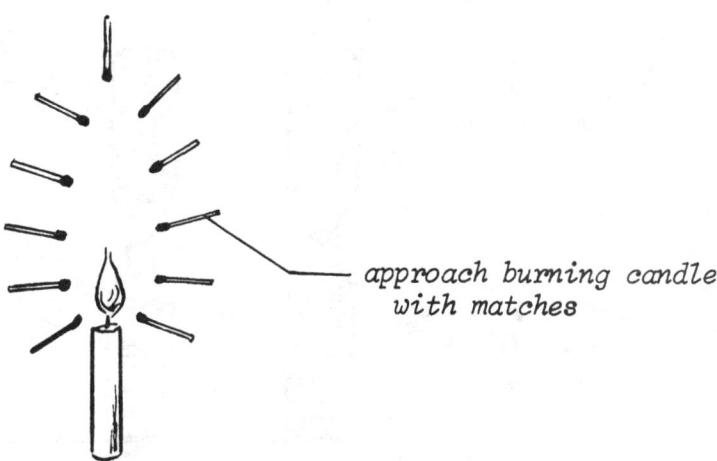
approach burning candle with matches

Procedure:
1. Distribute a candle and a book of matches to each pair of students in the class, and instruct them to do the following points.
2. Light the candle and secure it to the desk.
3. Approach the candle flame with the match head from different angles until it bursts into flame (make sure that the flame is straight up).
4. Estimate (or measure) and record the nearest distance between the unburnt match and the flame at the different locations (see Sketch).

Questions:
1. Why can we approach the flame closer at the bottom of the flame?
2. What does the air do above the flame?
3. Why can't we come closer to the flame if we approach it with the match head from above without burning it?
4. What temperature is exceeded when the match head bursts into flame?
5. Do the match heads burst into flame at exactly the same temperature?
6. What are the variables influencing the ignition of the match heads?
7. Would a propane or Bunsen burner flame behave in the same manner?

Explanation:
   The heat from the flame creates a convection in the air, making the spot have the flame hotter than anywhere else around the flame. The reason for this is that hot air is lighter than cool air. As the spot above the flame is hotter, the match heads can approach the flame at the bottom of it much closer than at the top of the candle flame before bursting into flame. The heat reaching the match heads at the lower part of the candle flame is mostly by **radiation** and not **convection** of heat.

7.8. THE CONFUSED BOTTLES

Materials:
1. Four empty identical soda pop bottles.
2. Food coloring.
3. A 3x5" card.

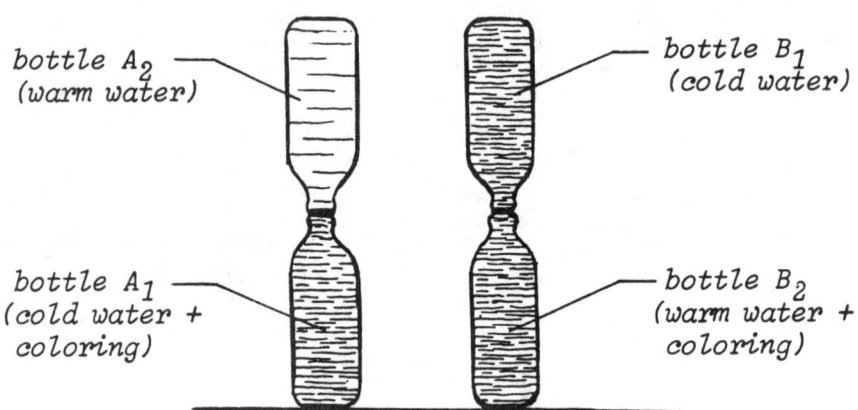

Procedure:
1. Fill two bottles (A_1 and B_1) with cold and two bottles (A_2 and B_2) with warm water (do not reveal to students the temperature difference).
2. Color the water in bottles A_1 and B_2 with a few drops of food coloring and mix the color evenly (cover the bottle with your thumb and turn upside down).
3. Cover the bottles A_2 and B_1 with a small piece of the paper card, and place them upside down on the colored bottles (one finger on the card will keep the water from spilling while turning it upside down; center the bottles A_2 and B_1 carefully over A_1 and B_2 and slip out the piece of card by holding the top bottle).
4. Let the students observe what is happening to the color

Questions:
1. Why did the top bottle B get colored and not Bottle A?
2. Do you think the temperature of all four bottles of water was the same?
3. Which of the four bottles were warmer?
4. Does bottle A_2 ever get colored; if so, when?
5. What would happen to the color if the temperature of all four bottles were the same?

Explanation:
The water in bottles A_1 and B_1 was cold and that in A_2 and B_2 was warm. Warm water is lighter in weight or less dense than cold water and thus rises. Since the warm water in B_2 was colored this water rises into the top bottle and the cold water sinks bringing with it **convection currents**. As the water in A_2 is warm and already above the cold water in A_1, no convection is occurring in this set of bottles, and thus no coloring of the top bottle.

When the water temperature of this top bottle gets to be equal to that of the lower bottle, **diffusion** of the color will occur, but no **convection**. This process is much slower than convection and is caused by the constant vibration of molecules.

HEAT   CONVECTION

7.9. INTERESTING CURRENTS

Materials:
1. A large beaker (800 - 1000 ml).
2. An alcohol or propane burner, and tripod.
3. Potassium permanganate ($KMnO_4$) or food coloring or ink.

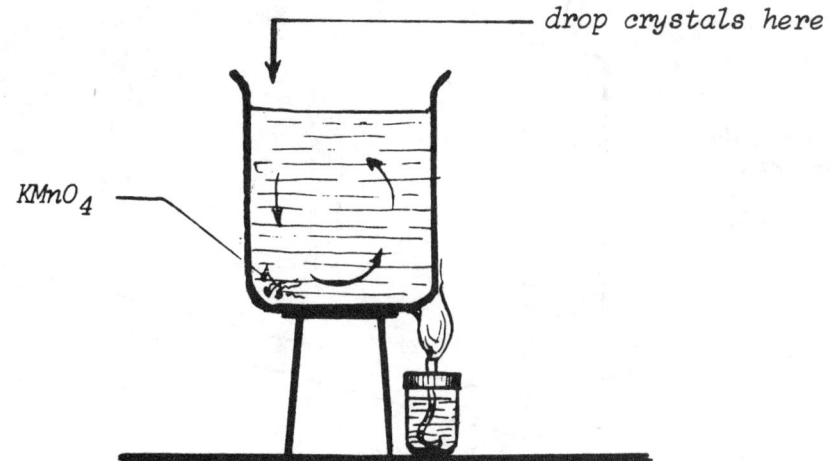

Procedure:
1. Fill the large beaker with water and place it on the stand.
2. Light the alcohol or propane burner, but keep it away from the beaker.
3. Drop a few crystals of potassium permanganate ($KMnO_4$) or a few drops of food coloring close to one side of the beaker.
4. Quickly place the flame under the opposite side of the beaker, and let students observe the color movements in the water.

Questions:
1. What makes the $KMnO_4$ crystals sink to the bottom of the beaker?
2. What does the flame do to the water?
3. When the fluid expands, does it get lighter or heavier per unit volume?
4. When the fluid is heated, what characteristic is decreased?
5. Instead of using the flame, what could we use to achieve the same convection currents?
6. Where do we find these interesting currents in our daily life?

Explanation:
   The flame of the burner is heating the water on one side of the beaker, which makes the water expand at this spot, making the water less dense. As the density here decreases, the water gets lighter per unit volume and rises. The crystals of $KMnO_4$ are greater in density than water and thus they sink. As the hot water on one side of the beaker rises, the cold water on the opposite side of the beaker moves down, and a **convection current** is created in the water. This convection current is made visible by the color in the water, which is moved along with it.
   In our daily life, warm currents can occur in oceans, like the warm Gulf Stream moving up north along the American Eastern Seaboard.

7.10. THE MYSTERIOUSLY RISING NAPKIN

Materials:
1. A thin paper napkin (the cheapest quality, rather crisp to the touch, used to replenish refillable dispensers).
2. Matches.

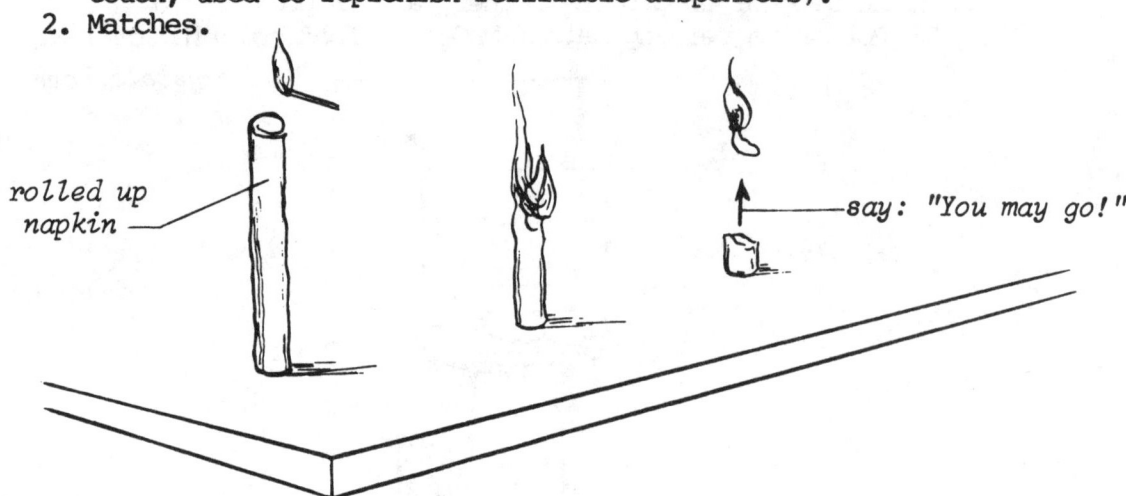

Procedure:
1. Test the napkin before the actual demonstration:
   a. Roll the napkin in a cylinder with about 3 cm diameter and let it stand upright on the table.
   b. Burn the napkin at the top and see whether the ashes say together and lift off the table at the end of the burning.
   c. If the ashes stay together and it does not rise up towards the end of the burning, try it with 3/4 or 1/2 of the napkin, by tearing 1/4 or 1/2 of it off.
   d. Once the right kind of paper is found, you are ready for the actual demonstration.
2. Roll the napkin in a cylinder and let it stand on the table top.
3. Start to burn the napkin at the top of the cylinder. At this moment you may tell a story (about three guys who passed away and who were waiting to go into heaven at the pearly gate: St. Peter looking at the three stacked-up guys, was judging them for what they have done on earth; the first was not good enough, the second did not do enough good but the third was OK and can go).
4. Just before the flame hits the table, say: "You may go!" (or whatever comment suits your own story).

Questions:
1. What made the ash shoot up in the air?
2. Why did the burning paper not rise up in the air before the flame reached the bottom?
3. Where can we see this phenomenon in our daily life?

Explanation:
   The ash content of the paper is critical in this demonstration. If the ash stays together and is light enough, it will reach the proper weight to be carried up by the convection current created by the flame.

HEAT — RADIATION

7.11. WHICH IS THE WARMER COLOR?

Materials:
1. Two identical beakers (100 ml).
2. Two identical thermometers.
3. A flood light or spot light.
4. A piece of white and a piece of black paper.

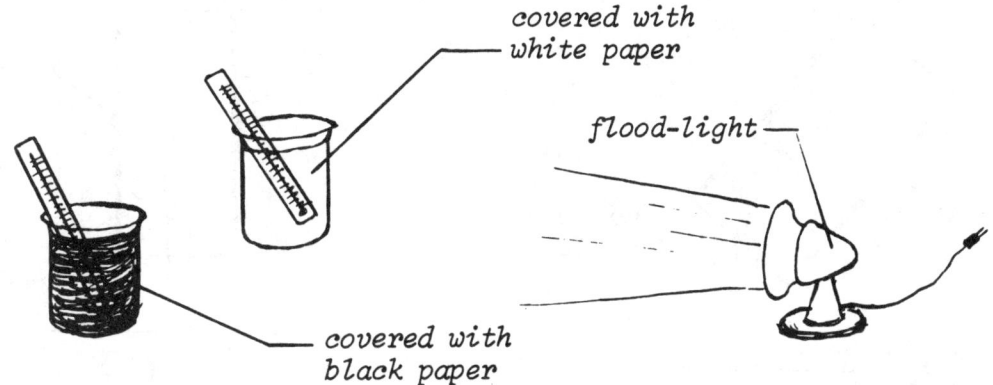

Procedure:
1. Wrap 1 of the beakers with white and the other with black paper on the outside (tight against the glass).
2. Fill both beakers with the same amount of water and immerse a thermometer in each and read off the temperature.
3. Place both beakers at the same distance (about 50 cm) from the spot light and switch the light on.
4. Read the temperature of each beaker after a 5 minute exposure to the light or radiated heat.

Questions:
1. Which thermometer is indicating a higher temperature?
2. Why would the two beakers show different temperatures?
3. Which are the warmer colors in general?
4. What happens to the radiated energy hitting warm colored surfaces? And cold colored surfaces?

Explanation:
The black surface absorbs the radiated heat much more than the white surface, which reflects all colors back. The black color is actually the absence of all other colors, and thus no emission of any other color occurs. The white color emits or reflects back all colors and thus does not absorb the **radiated** heat on the surface. In terms of molecules then, we can visualize that the molecules of the black surface vibrate much stronger than those of the white surface. The **radiated energy** is therefore imparted much easier to the water by the black surface than by the white one.

The warmer colors are red, orange, brown, and yellow in contrast to the cold colors of green, blue, purple, violet. An experiment could be set up to test the absorption of radiated heat on these colored surfaces.

HEAT — RADIATION ABSORPTION

7.12. WHICH COIN WILL STAY ON LONGER?

Materials:
1. An empty juice can (or any other large tin can) of which the top and bottom has been taken off (remove with can opener).
2. A medium size candle, two identical coins.

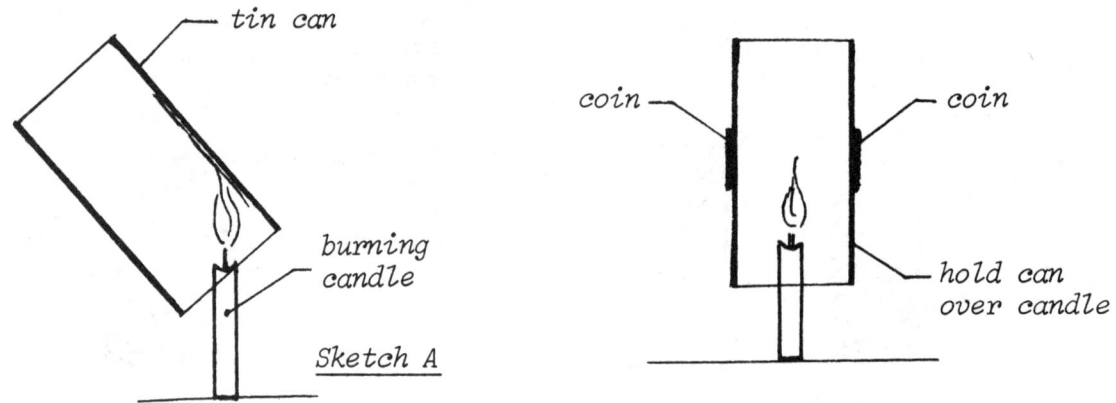

Sketch A

Procedure:
1. Light the candle and blacken one half of the inside of the can with soot from the candle, by holding it sideways close to the candle flame (see Sketch A). Once this is done, let stand and cool.
2. Against the outside of the tin can, attach one coin on each side of the can (the blank side and the soot-blackened side) with one drop of molten wax from the candle.
3. Now place the tin can (with the two coins attached) over the burning candle, and ask the students: "Which of the two coins will stay sticking to the can longer?" - Make sure that you place the can such that the candle is exactly in the center of the can. (see Sketch B).

Questions:
1. What is the cause of one coin dropping first?
2. Which of the coins dropped first?
3. Will the other coin also drop off eventually or not?
4. Would blackening the outside of the can have the same effect?
5. How did the heat travel from the flame to the can?

Explanation:
The coin attached to the side of the tin can that was blackened with the candle soot would fall off first. This is because the heat that was **radiated** from the candle flame was **absorbed** more by the black surface, as compared to the shiny metallic surface. Because of the higher degree of absorption of heat by the black half, the temperature increased more rapidly and thus the wax melted sooner and the coin would drop off.

If the outside of the can was blackened with soot instead of the inside surface, there would be no difference in the absorption of heat from the candle flame, as the inside surface would reflect the heat rays away in the same fashion. Absorption of heat from outside the can might then have an influence on making the difference in temperature, and thus in the length of adherance of the coins to the tin can.

HEAT EXPANSION & CONTRACTION

7.13. THE RISING JUICES

Materials:
1. A flask or bottle with a narrow neck.
2. A glass tube (30 cm) in a 1-hole stopper (for flask).
3. One large and one small container for water.
4. Food coloring or ink.

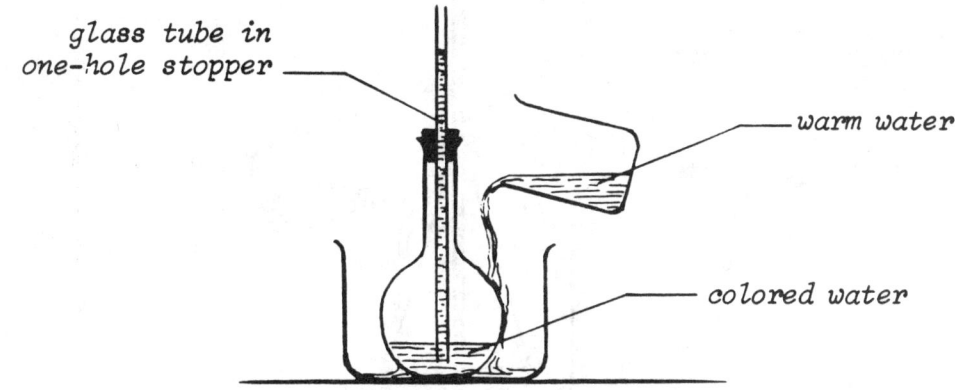

Procedure:
1. Fill the flask or bottle 1/4 full with water and put a few drops of food coloring in the water.
2. Insert the one-hole stopper with the glass tube in the flask opening such that the lower end is immersed in the water.
3. Push on the stopper: the water level should rise in the tube.
4. Place the closed off flask in the larger container.
5. Pour some warm water from the smaller container over the flask: observe the water level inside the glass tube!

Questions:
1. What made the water level in the tube rise?
2. When the pouring is stopped, what made the water level in the tube go down? Where will the water level stop?
3. Was the water that was poured on the flask cold or warm?
4. How different will the rising and falling of the water level be, if we poured oil instead of water on the flask?

Explanation:
The water poured on the flask was warm, and this caused the air inside the flask to expand, pushing the water up in the tube. The higher the temperature of the poured water, the more the air inside the flask will expand and the higher the water inside the small tube is pushed up. When the pouring is discontinued, the water outside the flask is evaporating and cooling. This cooling contracts the air inside the flask and the water column descends. If we use cold water instead of warm water to pour over the flask, the air inside it will contract and bring down the water level in the tube.

When using oil instead of water of the same temperature to pour over the flask, the rising of the water level inside the tube will be the same, but the falling will most likely be at a slower rate, because oil does not evaporate as quickly as water, and is thus not cooling as much.

HEAT — EXPANSION & CONTRACTION

7.14. WITHDRAWING JUICES

Materials:
1. An Erlenmeyer flask (250 ml).
2. A glass tube in a one-hole stopper (fitting in flask).
3. A large transparent container, small beaker.
4. Food coloring.

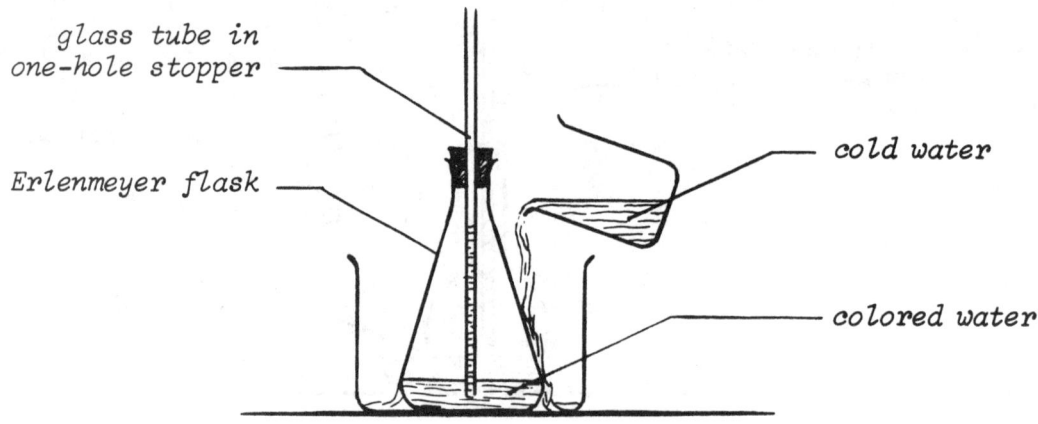

Procedure:
1. Fill the Erlenmeyer flask 1/4 full with colored water.
2. Insert the tube with the one-hole stopper over the flask (make sure that the tube dips in the water before pushing down on the stopper--the water level in the tube should stand higher than in the flask).
3. Pour cold colorless water over the Erlenmeyer flask and have the students observe the water level in the tube.

Questions:
1. Why did the water level in the tube go down?
2. Why did the water level in the tube rise again after the pouring was discontinued?
3. Was the poured water of the same temperature as the water inside?
4. What would happen to the water inside the tube if we would blow on the wet flask?
5. How can we make the water level in the tube rise in the first place?

Explanation:
The cold water that was poured over the Erlenmeyer flask made the air inside the flask contract, thus the pressure on the water inside the flask was lowered. This made the water column inside the tube go down. When the pouring of the cold water is discontinued, the cooling action is not completely stopped because the flask is still wet, and evaporation of the water takes place. Not until the flask is completely dry, will the water column inside the tube rise again.

By blowing on the wet flask, the cooling action--by evaporation of the water--is even more increased and a quick descent of the water results.

The water level in the tube can be raised by pouring warm water over the flask or simply by holding the flask in both hands. Heating the flask over a flame will have the same results.

HEAT   EXPANSION & CONTRACTION

7.15. THE CURVING TAPE

Materials: 1. Masking tape and cellotape.
            2. An alcohol burner or other burner with colorless flame.

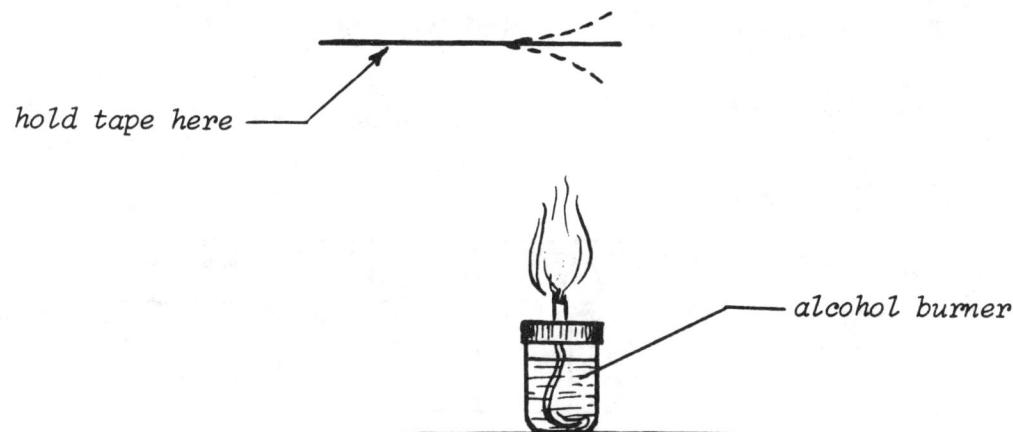

Procedure:
1. Take a 10 cm piece of masking tape and an equal length of cellotape.
2. Stick the two pieces of tape with their sticky side together, such that one covers the other completely (cut any excess off).
3. Hold one end of the double tape above the alcohol flame and let students observe what happens with the tape.
4. Take the tape away from the heat, straighten it out, flip it upside down (conceal this from students) and bring it back over the flame: which way does tape bend now?

Questions:
1. Why did the tape curve up or down when heated?
2. Would a regular piece of masking tape behave the same way?
3. What would two pieces of masking tape stuck together do over the flame?
4. Which of the two tapes, masking or cello, expands more when heated?

Explanation:
    A regular strip (single or double) of masking tape would not curve up or down when held above the flame. The curving occurs, because the cellotape expands more than the masking tape, and because one cannot slide past the other, it makes the double tape bend towards the masking tape.
    This same principle is in operation in the **bimetallic strip,** which is a strip of two different metals with different **expansion coefficients** (they expand at different rates). These bimetallic strips are used for thermal switches, thermostats, on-off lights, etc. When the temperature reaches a certain point, the strip bends enough to switch the current off; when it gets cooler, the strip bends back and switches the current on again.

7.16. THE MOVING ROD

Materials:
1. A metal rod (about 50 cm long).
2. Two small pieces of window glass.
3. A straw and pin.
4. Four or five candles or alcohol burners.

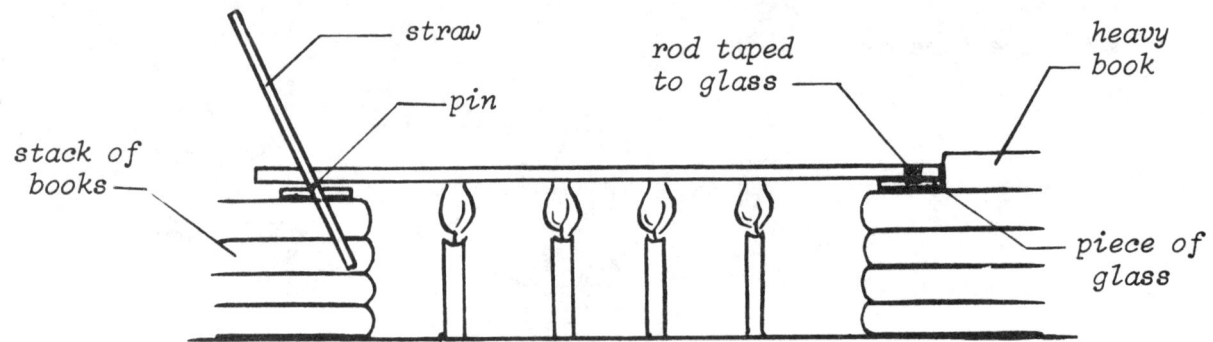

Procedure:
1. Place the metal rod on the two small pieces of window glass on two stacks of books, so that it lies horizontally (see Sketch).
2. Tape one end of the rod against the glass and place a heavy book against the end of the rod.
3. Push a pin through a straw and place the pin between the other end of the rod and the glass piece, such that it can rotate freely.
4. Place four or five candle flames or alcohol flames under the rod.
5. Let students observe the straw.

Questions:
1. How long after the candles were lit did the straw move?
2. In which direction did the straw move?
3. What was the cause for the straw to move?
4. Would the straw move the same way if the heavy book was removed before the heating of the rod?
5. What function do the glass pieces have?
6. What difference would we see in the movement of the straw, if we were to use more candles to heat the rod?
7. How would a change in the length of the rod affect the straw movement?

Explanation:
The pin is tight in the straw and thus when the rod is heated, it **expands** and turns the pin. This is magnified by the movement of the straw. The straw does not move immediately after the candles are lit, but probably a few minutes afterwards. This is because the rod needs to absorb the heat and impart this energy to its molecules.

The two glass pieces under the rod function as an insulator of heat, as well as provide a smooth surface for the pin to roll evenly. The heavy book at one end of the rod keeps this end from sliding. More candles will heat the rod to a higher temperature and thus expand it more. The longer the rod, the larger the increase in length.

HEAT   EXPANSION & CONTRACTION

7.17. THE HEAVY HEAT

Materials: 1. A brass rod or tube with screw in one end.
             2. A razor blade, or small wooden block or molding clay.
             3. An alcohol burner.

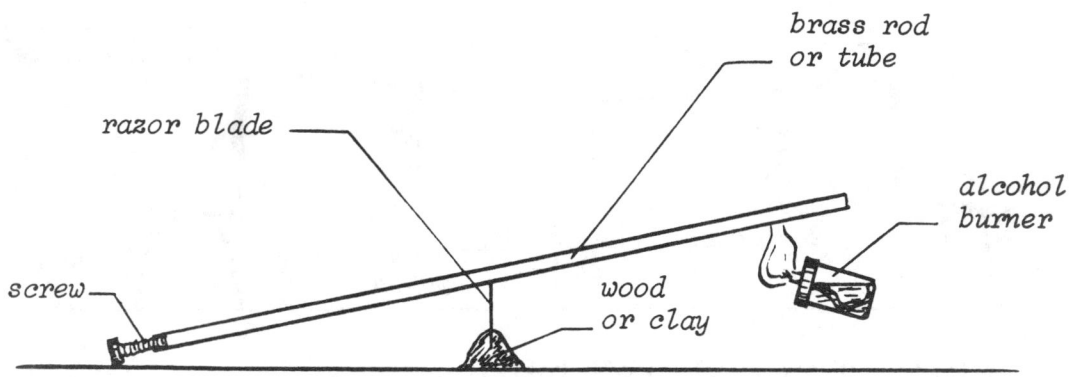

Procedure:
   1. Turn the screw into the brass rod or tube about half way.
   2. Stand the razor blade vertically in the block of wood or clump of clay (this functions as the pivot for the balance).
   3. Place the rod or tube on the pivot and balance it, by moving it more to the left or to the right.
   4. Turn the screw out so that the rod goes off balance (towards the screw).
   5. Heat the other end of the tube with the alcohol burner (see Sketch).

Questions:
   1. What made the rod go off balance before the heating?
   2. What happened to the rod when it was heated?
   3. What made the rod tip towards the heated end?
   4. What happened to the rod after the heat was taken off (after the rod tipped towards the heated end)?

Explanation:
   Turning the screw out shifted the center of gravity of the rod a little towards the screw's end, causing the rod to tip over to the right, because there is more weight at that end. When the other end of the rod is heated, it **expands** and actually gets a tiny bit longer, causing the center of gravity to shift towards the heated side. As a result the rod is coming back to a level position. When the heating is continued, the rod would expand even more, and the rod tips over to the heated end. After the rod has tipped over towards the heated end and the heating has ceased, the rod cools off, **contracts**, gets lighter on this end, and tips back to the screw end.
   It might seem that the heating caused the mass of the heated end to increase, but this is not so. It only increased in length.

HEAT — EXPANSION & CONTRACTION

7.18. THE WEAKENING WIRE

Materials:
1. A length of wire (copper or iron) about 3 m long.
2. A 1/2 - 1 kg weight (or rock of the same weight).
3. Eight candles or alcohol burners, a meter stick.
4. Two straight-back chairs.

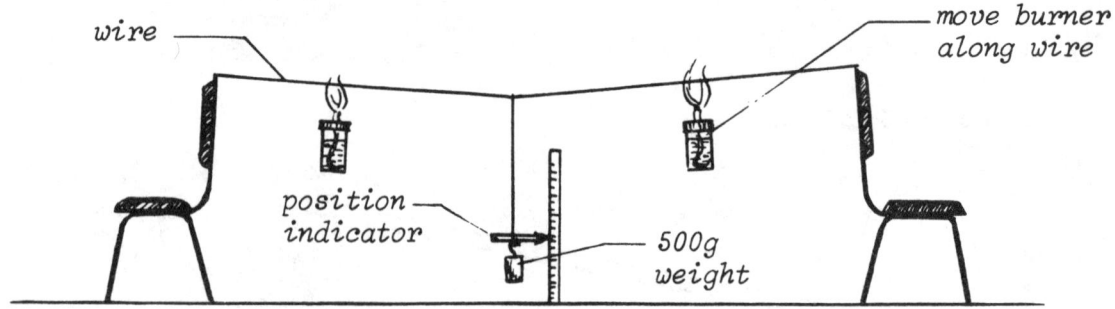

Procedure:
1. Cut a length of about 50 cm off the wire.
2. Attach the remaining wire to two straight-back chairs and keep the wire as straight as possible by moving the chairs out.
3. Attach the 50 cm long wire to the center of the stretched wire, and hang the 1/2 kg weight at the end of this wire. (You might need some weight on the chairs to keep them from moving--books or students sitting).
4. Tape a small paper arrow to the weight and point it to a ruler, held vertically next to the vertical wire (taped against wall or table).
5. Heat the stretched wire with at least two alcohol burners (have a student assist you in heating one side of the wire--move the burner slowly under the wire along the whole length of it). Observe arrow!

Questions:
1. Why did the weight sag down when the wire was heated?
2. How much did the weight come down?
3. Would the number of burners or candles affect the amount of sagging?
4. Would wires of different materials sag the same distance?
5. How can we calculate from the sagging of the weight how much the wire has stretched?

Explanation:
The heating of the wire by the burner flames made the wire expand, and this made the weight descend. The heating made the point of attachment of the weight move from P_1 to P_2. By measuring the lengths AP and PP_1, we are able to calculate the 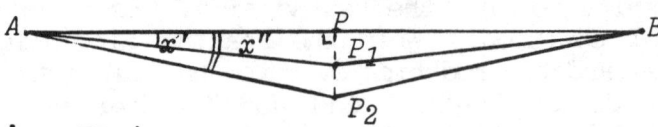 size of angle x' from tg x' = PP_1/AP, and the length AP_1 from sin x' = PP_1/AP_1. Similarly the angle x'' can be obtained from tg x'' = PP_2/AP_1, and the length AP_2 from sin x'' = PP_2/AP_2. The expansion of the wire is $AP_1 - AP_2$ times 2. **Another way is to use the Pythagorean theorem.**

The more burners are used for the heating, the more the weight will sag. Wires of different metals will expand differently and thus result in specific amounts of sagging distances.

HEAT

KINDLING POINT

7.19. BOIL WATER IN A PAPER CUP

Materials:
1. An unwaxed paper cup (cupcake baking cup suits).
2. Alcohol burner and stand, or candle.
3. Two cloth pins.

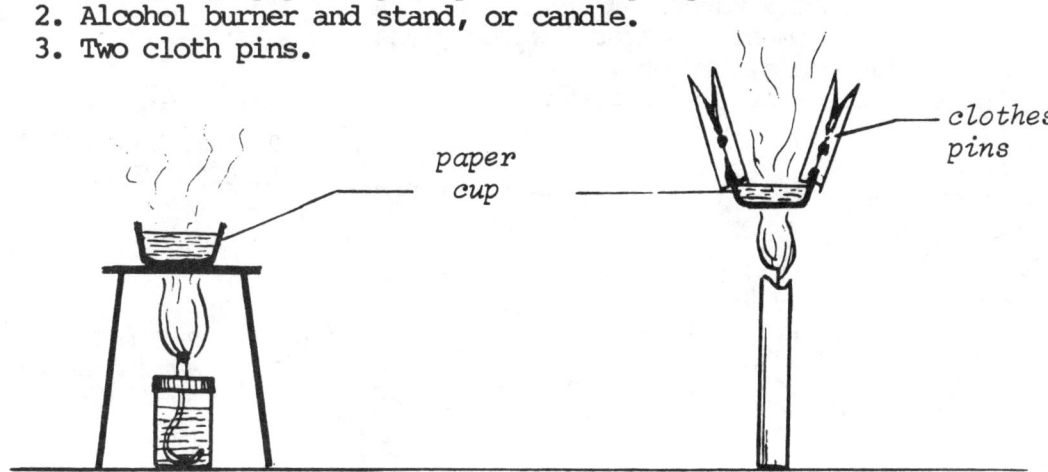

Sketch A   Sketch B

Procedure:
1. If paper cups are not available, a container can easily be made out of a paper card (3x5"), by folding the four corners up and clipping them with four paper clips.
2. Half fill the paper container with water and place it on the stand (see Sketch A) or use the cloth pins to hold the cup over the flame (Sketch B).
3. Light the alcohol burner or candle, and heat the water until boiling.

Questions:
1. Why doesn't the paper burn?
2. What would a waxed paper cup do if used?
3. Where does the heat of the flame go?
4. How high will the temperature of the water reach?
5. What temperature does it have to reach for the paper to catch fire?

Explanation:
The heat released by the flame is absorbed by the water in the paper cup. It is used to increase the temperature of the water from room temperature to the boiling point, which is 100° C. At this temperature the heat is then further needed to convert the water into water vapor. The temperature of the water will stay at the boiling point (100° C) and not increase that of the cup either, until all the water is evaporated. Because the **kindling point** of paper (the temperature at which paper catches fire) is much higher than 100° C (depending on the thickness of the paper), it will not catch fire before all the water is evaporated. In other words, wet paper will not burn until it is dry. This same principle is applied when "green" wood is burned in fireplaces: it is hard to catch fire.

HEAT — KINDLING POINT

7.20. THE COOL FLAME

Materials:
1. Carbon disulfide (CS_2) & Carbon tetrachloride (CCl_4) or ethyl alcohol & water.
2. A medium size beaker & watch glass.
3. Two glass stirrers.
4. A piece of window glass.

CAUTION

DO NOT USE EXCESSIVE AMOUNTS OF CS_2 OR ALCOHOL

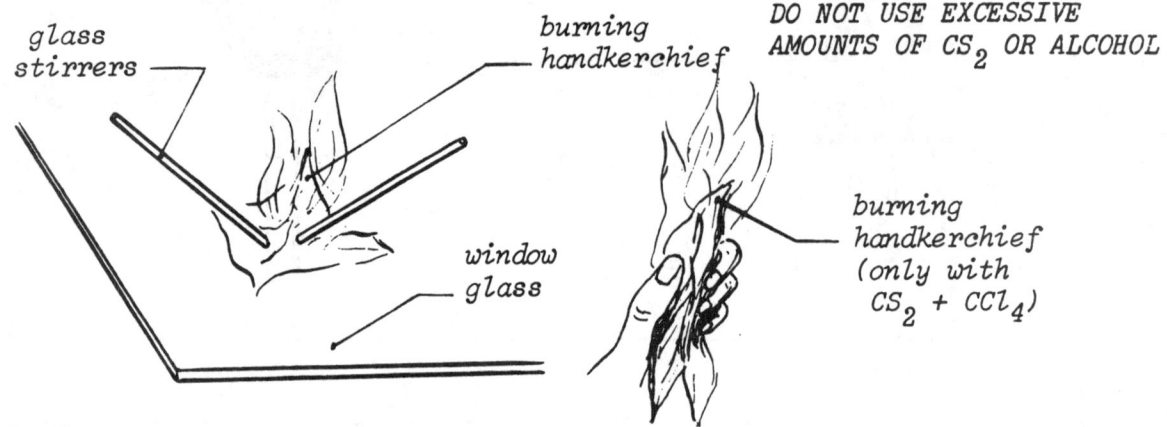

Procedure:
1. Mix the two liquids $CS + CCl_4$ or ethyl alcohol + water in the ratio of four to six in volume.
2. Pour a small sample of this mixture on the watch glass and test with a match for ignition. If it does not ignite, add a few ml at a time of CS_2 or alcohol to the main mixture, until a faint cool flame is sustained in the small sample.
3. You are now ready for the demonstration: Soak a handkerchief or dollar bill in the mixture and place it on the window glass. Light it with a match and keep it moving with the two glass stirrers until the flame is out (it is most effective to do the demonstration in the dark).

Questions:
1. What made it possible to hold the burning handkerchief when using the CS_2 + CCl_4 mixture?
2. What is the purpose of the CCl_4 or the water?
3. Why did the handkerchief have to be kept moving?
4. Will it work if the mixture of alcohol in water or CS_2 in CCl_4 were stronger or weaker?

Explanation:
The carbon disulfide, in the presence of carbon tetrachloride or the alcohol in the presence of water, gives off a flame with a rather low temperature. This low temperature is caused by the absorption of the heat by the evaporation of CCl_4 or water present. By constantly keeping the handkerchief moving, all parts of it are kept wet with the insulating CCl_4 or water. If it is not kept moving one part of it might dry out and a hot spot might develop. The more CS_2 or alcohol is used in the mixture the hotter the flame will be.

CHAPTER 8

HOW DOES MAGNETISM WORK?

OBJECTIVES

After dealing with and studying the concepts and sub-concepts in this chapter, the students should be able to:

a. recognize the correct explanation of an observed event based on each of the sub-concepts;
b. explain in their own words which of the sub-concepts is determining the course of an event;
c. distinguish true from false statements concerning each one of the sub-concepts;
d. identify the correct explanation of an event in daily life applying one of the sub-concepts;

all in relation to the following sub-concepts:

-- Materials that can be magnetized or attracted to a magnet are made up of or contain either iron, nickel, or cobalt.
-- Like poles repel and unlike poles attract each other.
-- Around a magnet there exists a magnetic field, the magnetic lines of which run from pole to opposite pole.
-- A magnet is made up of small dipole particles that are lined up in one direction.
-- The North geographic pole of the earth contains South magnetism and the South geographic pole North magnetism.
-- The North pole of a compass points North because man named this pole North.
-- Magnetic materials can be temporarily magnetized by induction.

MAGNETISM   MAGNETIC MATERIALS

8.1. THE FLOATING PAPER CLIP

Materials: 1. A rod magnet, stand and clamp.
            2. A paper clip and thread (transparent or black).

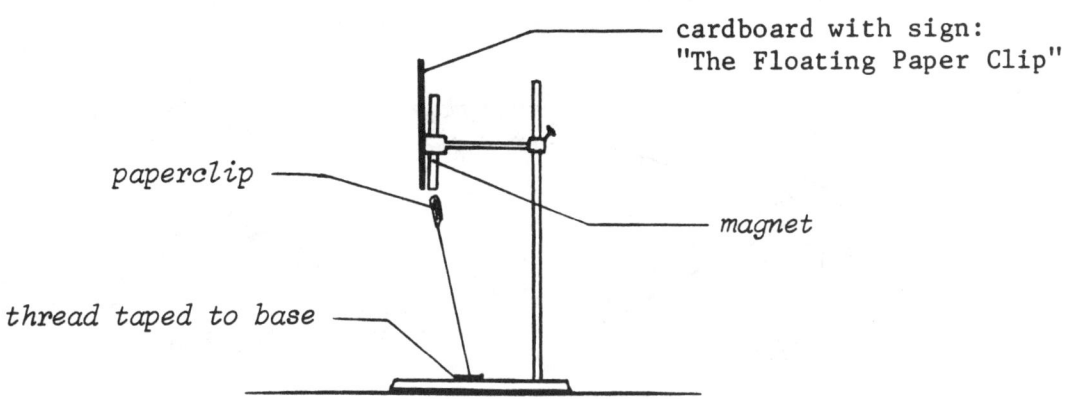

Procedure:
1. Clamp a rod magnet vertically and tape a cardboard sign, "THE FLOATING PAPER CLIP" in front of it. (The bottom edge should be flush with the magnet.)
2. Tie a paper clip to a thread and tape the other end of the thread to the base of the stand, such that the clip is still help up by the magnet, but leaving a gap between the two.
3. Show to students that no thread is holding the clip up by sliding thin objects through the gap (a ruler, a card, a comb, etc.).
4. Now take a pair of scissors and "cut" the magnetic lines that are holding the paper clip up.

Questions:
1. What materials could be slid through the gap without dropping the paper clip?
2. What materials will definitely "cut" the magnetic lines?
3. Which coins will go through the gap without dropping the clip?
4. Why do only some dimes (old ones) pass the gap without "cutting" the magnetic lines?
5. What were the scissors actually doing to the magnetic lines?

Explanation:
   The paper clip is held up by **magnetic lines** indeed. When the magnetic lines are prevented from going through the clip, it falls. Materials that can absorb the magnetic lines from the magnet are materials which are iron, nickel or cobalt or contain any amount of it (for instances in an alloy).
   Magnetism originates from within the atom. In the **magnetic materials**: iron, nickel and cobalt, the electrons around the nuclei although paired together, do not completely cancel out the magnetic fields. They could be considered as consisting of minute magnets that are randomly arranged. When these minute magnets are all lined up in one direction, the object can become a strong magnet.

MAGNETISM

RULE OF MAGNETS

8.2. WHICH POLE IS ATTRACTED?

Materials:
1. Two marked bar magnets.
2. One unmarked bar magnet.
3. A piece of string.

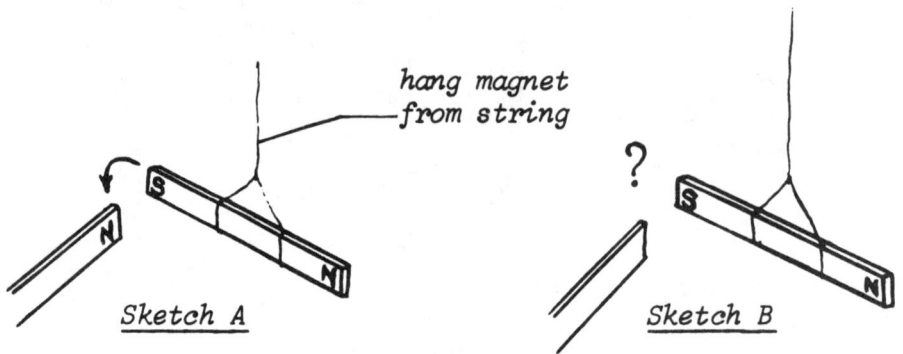

Procedure:
1. Tie a string to a marked bar magnet and let the magnet hang horizontally from the left hand (or from a stand).
2. Approach this free rotating magnet with another marked bar magnet in your right hand. Let students observe what the north and south end of the hanging magnet will do when another north or south end approaches it.
3. Now replace the marked bar magnet in your right hand with an unmarked bar magnet, and ask students: "Which pole will swing towards this approaching magnet?"
4. Wait for students' reactions (some anticipated answers: 'We don't know until you tell us what pole is coming near the hanging magnet').
5. Say: "If I tell you what pole I have in my right hand, can you tell me which one will be attracted to it?"

Questions:
1. Can you make a rule that all magnets will follow?
2. What would the hanging magnet do when approached with a regular steel bar?
3. What can you tell about the approaching bar if the N as well as the S pole swings towards it?

Explanation:
  The main purpose of this demonstration is to elicit the rule of magnetism -- **"Like poles repel and unlike poles attract"** -- from the students themselves. By showing them the phenomenon sketch several times, with different poles approaching the hanging magnet, they should be able to form and understand the concept themselves.
  By approaching the hanging magnet with an identical steer bar, which is **not** magnetized, both poles will be swinging towards it. This happens because the molecules in the non-magnetized bar are still randomly arranged and a **north** as well as a **south** pole of a magnet can attract the steel bar. A nail will have the same properties.

MAGNETISM   ATTRACTION &
                                                              REPULSION

8.3. THE MAGIC DANCER

Materials: 1. Two small disc magnets.
            2. A wooden stand, tape.

Procedure:
1. Construct a stand of wire in a wooden base and cut out a cardboard dancerine figure.
2. Hang the dancer on the stand with a thread.
3. Tape a disc magnet directly under the figurine and conceal another disc magnet at the end of the leg of the figurine.
4. Move the dancer a little and observe.

Questions:
1. Why does the figurine keep on moving?
2. How do the magnet poles have to be arranged?
3. Why does the base have to be made out of wood?

Explanation:
   The two concealed disc magnets are placed facing each other with like poles, so that they repel each other. This way the figurine keeps on bouncing away from the point vertically under it, causing it to move and twirl for quite an extended time. If the base of the stand were made out of iron or any other magnetic material, the magnet concealed in the base would lose much of its magnetism and the upper magnet would be attracted to any part of the base and would stop moving.

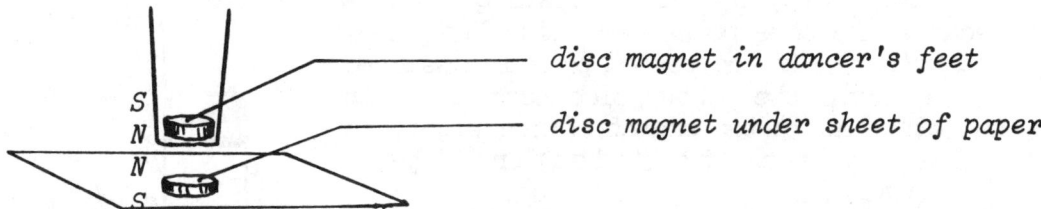

MAGNETISM

ATTRACTION & REPULSION

8.4. THE FLOATING DISCS

Materials:
1. Six or eight small disc magnets (or ring magnets).
2. A test tube (for the discs) or a glass tube in a one-hole stopper (for the rings).

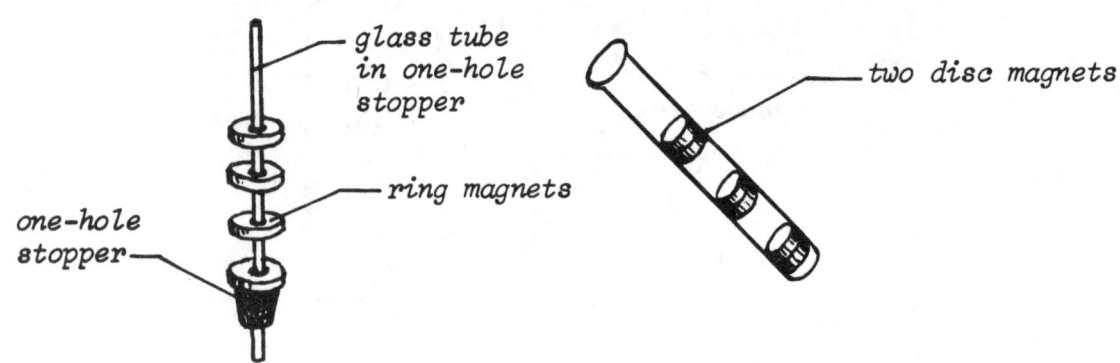

Procedure:
1. Stick the six or eight disc magnets together two by two.
2. Slide the first pair in the bottom of the test tube.
3. Slide the second pair in the tube, such that they repel the first pair. (If they are attracted to the first pair, take all four out, turn the two upper discs upside down and slide them back in the tube).
4. Do the same with the next two pairs of disc magnets.
5. If case ring magnets are used: slide them around the glass tube, such that they repel each other (see Sketch B).

Questions:
1. Why do the pairs of discs stick together?
2. Where would the poles of the disc magnets be located?
3. Assign a letter N or S for each of the poles.
4. How could we get the discs out of the test tube without turning the tube upside down?

Explanation:
The pairs of discs have unlike poles facing each other and therefore they attract each other. The poles of the disc magnets are located at its two circular flats, and the 'floating' occurs because the same poles face and thus repel each other. If the bottom pole is assigned to be North, the other poles have to be in the order as in the sketch on the right, in order to get the 'floating' pair arrangement.

The same can be done with the ring magnets.

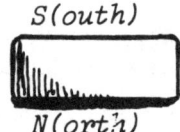

Side view of disc magnet

MAGNETISM   MAGNETIC LINES
                                                      OF FORCE

8.5. SEEING MAGNETIC LINES

Materials: 1. Two rod or bar magnets (and other shape magnets).
           2. A transparency or glass sheet.
           3. Iron filings.

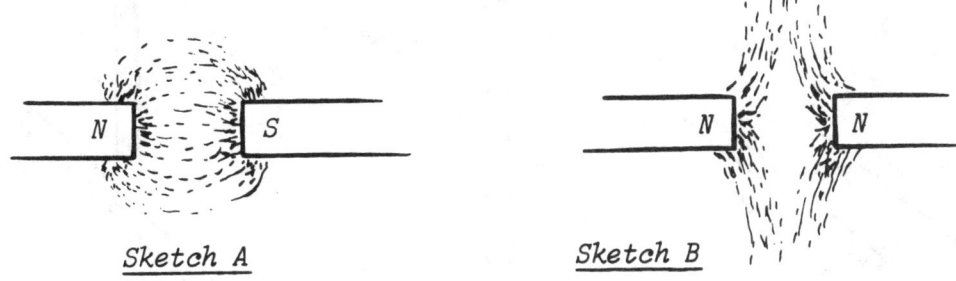

           Sketch A   Sketch B

Procedure:
1. Place two bar magnets with the unlike poles about 4-5 cm apart on an overhead projector (if not available: let students gather around).
2. Cover the magnets with a blank transparency and sprinkle some iron filings close to the poles especially. (When an overhead projector is not available, a sheet of window glass or white paper will do.)
3. After sprinkling the iron filings, tap lightly on the transparency (glass or paper).
4. Do the same (step 1 to 3) with the two like poles together or facing each other (Sketch B).

Questions:
1. What do the two different patterns of lines indicate?
2. How would the magnetic-line-pattern of a horseshoe magnet look?
3. How would the pattern of lines look, if the bar magnets were placed parallel to each other, with like poles next to each other? With unlike poles next to each other?
4. How would the magnetic line pattern of one bar magnet look?

Explanation:
Any thin sheet of material that is not magnetic (not iron, nickel or cobalt) can be used to show the magnetic lines with iron filings. The use of the transparency and overhead projector has the advantage, that it can be clearly shown to the whole class at one time. Students should gather around a little closer when it is not projected. These are the patterns that will show up for a horseshoe magnet and the parallel bar magnets:

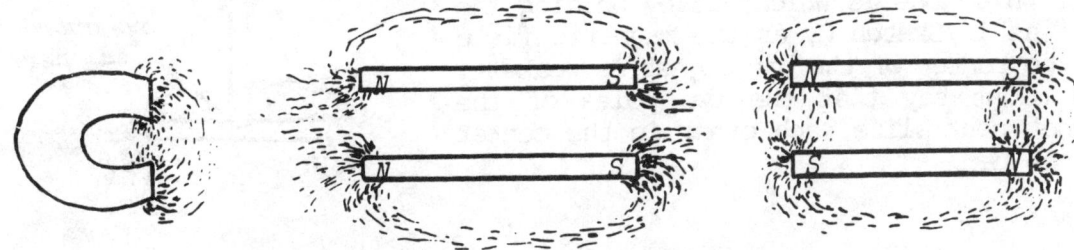

MAGNETISM

MAGNETIC LINES OF FORCE

8.6. WHICH ONE IS THE MAGNET?

Materials: 1. A strong bar magnet.
2. An identical steel bar (not magnetized).

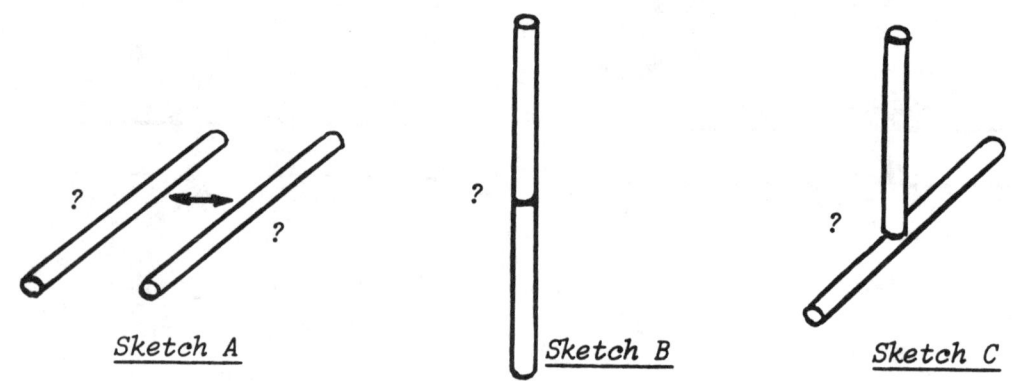

Sketch A   Sketch B   Sketch C

Procedure:
1. Show the students the two identical bars and say: "One of these bars is a magnet. How can we tell which one it is, without using any other materials?"
2. Carry out whatever the students want you to do (let one bar roll on the table attracted to the other bar (Sketch A) or put pole against pole (Sketch B), might be some suggestions.
3. If students do not suggest the set-up of Sketch C, give them a clue to think about the magnetic lines flowing around a bar magnet (Event 8.5.), and ask: "Where would the magnet be weakest in attraction force?"
4. Then show and demonstrate configuration C.

Questions:
1. Can we tell which one is the magnet with set-up A or B?
2. If the vertical bar attracts the horizontal bar (Sketch C), which one is the magnet?
3. Why would the end of the steel bar (the non-magnet) not attract the middle of the bar magnet?
4. How do the magnetic lines of force flow around the bar magnet?

Explanation:
Whether the bars are placed against each other as in Sketch A or B, there is no way to tell which of the two bars is the magnet, because there is no repulsion, as one of the bars is a regular non-magnetized steel bar. The only way to tell which one is which, is by holding the bars as in Sketch C, as the magnetic field in the center of the bar magnet is weakest. You might say that the two poles of the magnet neutralize each other in the center of the bar.

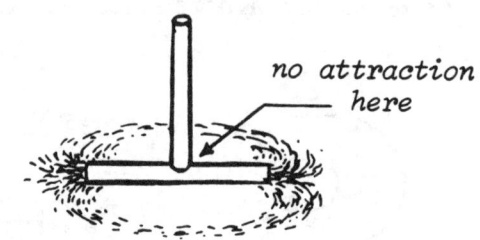

no attraction here

MAGNETISM — MAGNETIC LINES OF FORCE

8.7. THE MYSTERIOUSLY MOVING NEEDLE

Materials: 1. One strong bar magnet, a medium size sewing needle.
2. A wide and shallow glass or plastic tray (to hold water).
3. A small cork or small piece of styrofoam.

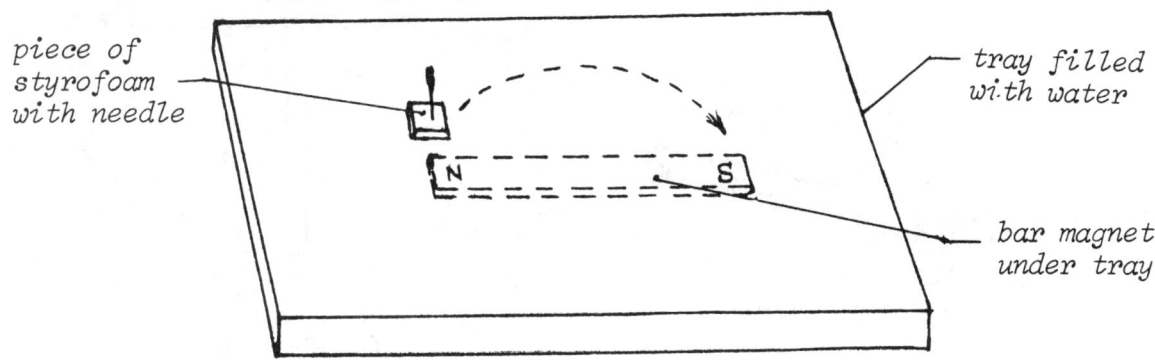

Procedure:
1. Magnetize the needle by rubbing the North pole of the rod magnet several times from the point towards the hole end (see step 1 under procedure for **Make a Needle Compass**).
2. Fill the tray with water (about 2 cm deep) and place it on top of the bar magnet (propping the sides so the water level stays horizontal).
3. Pierce the needle through half of a small cork or a small piece of styrofoam, such that the sharp end points vertically down.
4. Place this needle near the North pole of the magnet and observe! (make sure that the sharp point just floats 2-3 mm above the bottom of the tray. If it is not, just add some more water to the tray).

Questions:
1. What did you observe the needle doing?
2. What would it do if the needle is placed near the South pole?
3. Would rubbing the needle with the South pole of the magnet make any difference in movement?
4. What difference would it make if the needle was rubbed with the North pole but in the opposite direction (from hole to sharp end)?
5. Why doesn't the needle move in a straight line towards the poles?

Explanation:
By rubbing the needle with the North pole of the strong magnet from the sharp end towards the hole end, the needle itself will become a magnet, the sharp end being the North pole and the other end the South. After placing the needle vertically on the water surface, it is free to move, and it is thus repelled by the North pole and attracted by the South pole of the magnet (which is lying underneath the tray). It travels in a curved line following the magnetic lines of the strong magnet. The strongest field is closest to the poles of the magnet and thus the movement of the needle is fastest when approaching each of the poles.

A needle that is not magnetized would just move straight towards whatever pole is closest to the needle. What would happen with the needle if we had a horseshoe magnet under the tray of water?

MAGNETISM

**MAGNETIC MATERIALS
MAKE-UP OF A MAGNET**

8.8. THE TEST TUBE MAGNET

Materials:
1. A test tube and fitting cork.
2. A strong magnet and iron filings.
3. A small compass (or needle compass: Event 8.9).

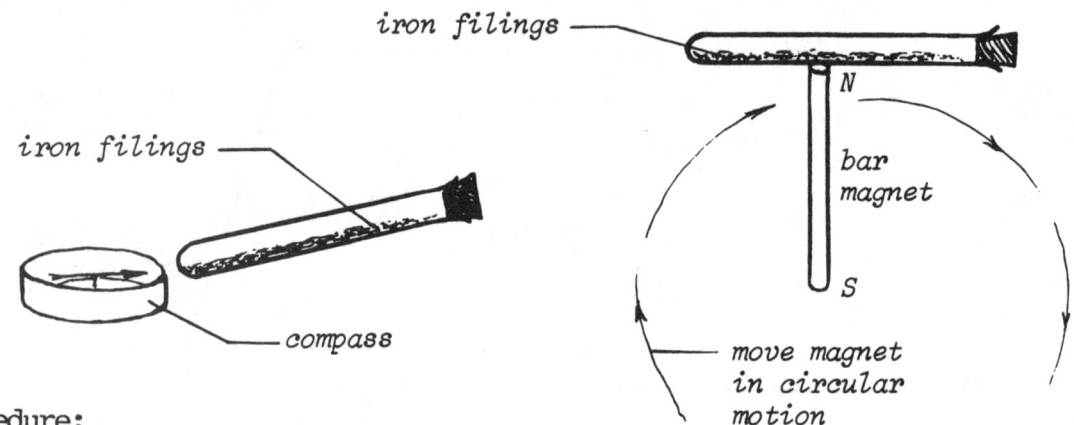

Procedure:
1. Place some iron filings in the test tube, cork it and shake the filings while holding the tube horizontally.
2. Bring the end of the tube close to the compass and show that the compass needle is not or very little attracted; also let students note that the same needle end is attracted to the test tube, no matter which end of it is approaching the needle (compass).
3. Now take the strong bar magnet and magnetize the iron filings in the test tube by stroking the bottom side of the horizontally held tube with one pole of the magnet in a circular motion (see Sketch).
4. After stroking the tube with the magnet for about 5-6 times, show that the test tube acts as a regular bar magnet by approaching the compass with the two ends of the tube.

Questions:
1. Why is the compass needle attracted to the iron filings before the tube was magnetized?
2. Why does the stroking of the test tube have to be done with a circular motion?
3. How can we know whether the iron filings in the test tube are magnetized or whether they are not?
4. How would the compass needle behave, if a regular bar magnet approached the compass?
5. What could we use if we did not have a compass?

Explanation:
A bar magnet is made up of very tiny particles that have very weak **dipoles** or North and South magnetism, just like the iron filings in the tube. They are scattered randomly in the bar. When these particles are lined up by stroking with a strong magnet, the bar becomes a magnet. This is demonstrated by the iron filings in the test tube. When the tube is shaken after it is magnetized, the particles are distributed randomly again, and no magnetism is left in the tube.

MAGNETISM   EARTH MAGNETISM
                                                    MAGNETIC RULE

8.9. MAGNETIC CONFUSION

Materials: 1. A small compass (one for each pair of students).
            2. A marked bar magnet.

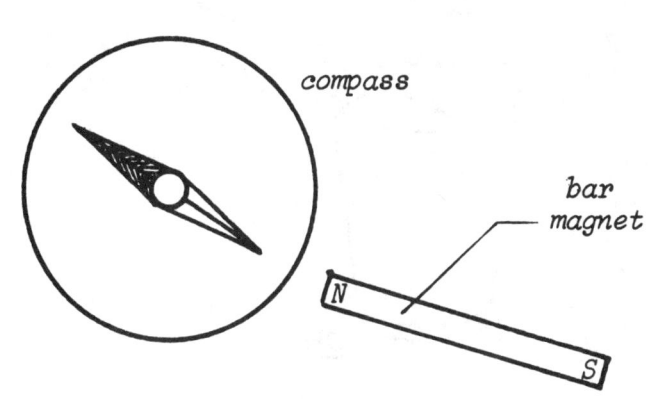
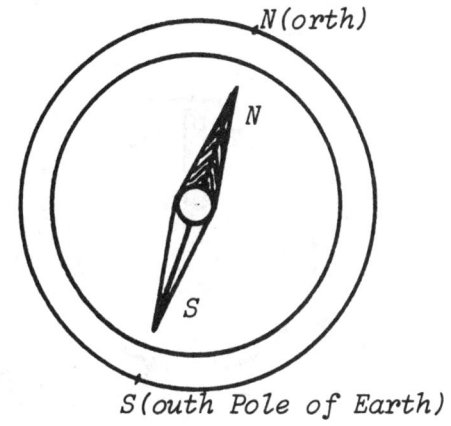

Procedure:
1. Distribute the compasses and ask: "Which end of the needle is the North end, and which one is the South end?"
2. "How can we find out?" Come around with the marked bar magnet and approach each compass with the North pole of the bar magnet.
3. Now that the students already know which end of the compass needle is North and which end is South, ask: "Which end of the compass points in the North direction?"
4. "Is North attracted by North? Is this not against our magnetic rule?"

Questions:
1. How did we find out which end of the compass was North?
2. If we did not have the marked bar magnet, how could we find this out?
3. If the marked bar magnet was hung on a string, which pole would point to the North pole of the earth?
4. If there were no marked bar magnets to find in the whole world (which was the case before magnetism was discovered), how would we label the first bar magnet?

Explanation:
   This activity poses the student with a real discrepancy, especially after learning the magnetic rule: **unlike poles attract and like poles repel.** The fact that the North end of the compass needle points to the North pole of the earth, is because this compass needle end is a north-seeking one and thus named **'North'**. The situation is then that, either the earth's North pole contains South magnetism, or that all North labeled bar magnets are actually the South pole. The most simple fact to remember for students is: that the North end of the compass needle is **named North**, because it is **North-seeking.**
   All bar magnets, when hung from a string, will line up in the North-South direction. The North pole of the bar will point to the North pole of the earth, because it is North-seeking.

8.10. MAKE A NEEDLE COMPASS

Materials: 1. A fine sewing needle, a piece of candle.
            2. A strong magnet, and a glass or cup.

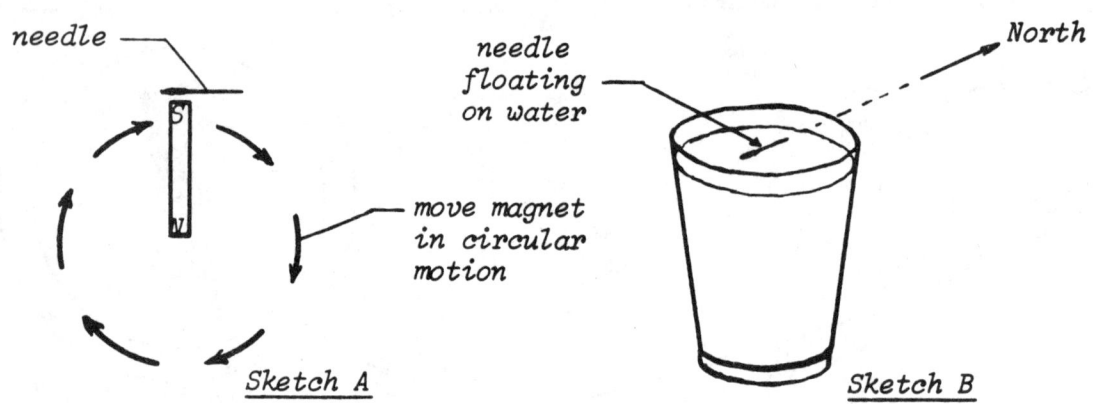

Sketch A   Sketch B

Procedure:
1. Take the needle between the fingers of one hand, and with the other hand rub the pole of the magnet against the needle in **only one direction**. After each stroke, move the magnet away from the needle and approach the needle from the 'eye'-side (Sketch A). Do this about 10 times.
2. Hold the needle between thumb and forefinger horizontally and approach the water surface as close as possible without touching it, then drop the needle. (When needle sinks, rub it against the candle before floating it, so that it would repel the water).
3. Once the needle is afloat, it will point only in one direction. Test this by approaching the cup with a magnet and removing it.

Questions:
1. When stroking the needle with the magnet, why can't we rub the needle back and forth against the magnet?
2. When rubbing the needle from eye to point with the south pole of the magnet, which end of the needle will point north?
3. Which end of a compass needle will point north?

Explanation:
   The rubbing of the needle with one pole of a magnet should be done only with one-direction strokes, because the purpose of the rubbing is to align the tiny particles of the needle in only one direction. When the rubbing is done with a back and forth motion, this alignment is not achieved as well. When the south end of the magnet is used to rub the needle from eye to point, the point will be the north pole of the needle, and thus this will point north. This is <u>not</u> because 'north' is attracted by 'north', but because the earth's north pole contains 'south' magnetism.
   Other ways to suspend the magnetic needle are: hanging it on a thread in a jar from a glass rod or wooden stick, which is resting on the opening of the jar; floating it on a wide flat cork disc on the water surface.

MAGNETISM   MAGNETIC INDUCTION

8.11. THE TEMPORARY MAGNET

Materials: 1. A strong bar magnet.
             2. A long iron nail.
             3. Paper clips or thumbtacks or small nails.

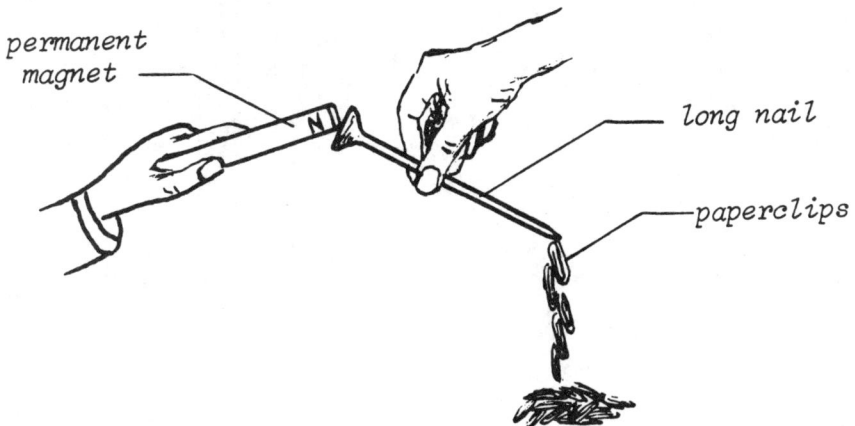

Procedure:
1. Place some paper clips, thumbtacks or small nails on a sheet of paper.
2. Touch the point of an iron nail to the heap of paper clips and see if it will pick up any of the objects (it should not pick up any).
3. Hold one pole of the strong magnet against the nail head and touch the other end to the small iron objects again (some will be picked up now: see Sketch above).
4. Remove the magnet now from the nail (the small objects will drop).
5. Repeat point 2, 3, and 4 to show the temporary nature of the nail as a magnet (induction).

Questions:
1. What made the nail become magnetic?
2. Could a copper or aluminum nail do the same thing?
3. If the north pole of the magnet were touching the nail, what pole would the nail point be?
4. What is the process of temporary magnetization called?
5. What other materials can be temporarily magnetized?

Explanation:
   The **magnetic field** surrounding the strong magnet can **temporarily magnetize** objects made of iron, nickel or cobalt. When this happens, it is said that the object is magnetized by **induction** and that the object is a **temporary magnet**. When the north pole of the magnet touches the nail head, this becomes the south pole and the nail point becomes the north pole.
   The particles (dipoles) of the nail are lined up in one direction and the whole nail becomes temporarily a magnet. As soon as the permanent magnet is taken away, the particles revert back to their random distribution and the nail's magnetism disappears.

CHAPTER 9

WHAT IS STATIC ELECTRICITY?

OBJECTIVES

After dealing with and studying the concepts and sub-concepts in this chapter, the students should be able to:

a. recognize the correct explanation of an observed event based on each of the sub-concepts;
b. explain in their own words which of the sub-concepts is determining the course of an event;
c. distinguish true from false statements concerning each one of the sub-concepts;
d. identify the correct explanation of an event in daily life applying one of the sub-concepts;

all in relation to the following sub-concepts:

-- Rubbing certain objects will result in the removal of electrons and a build up of positive or negative charges.
-- Certain uncharged objects are attracted by charged objects.
-- There are two kinds of static charges: positive and negative.
-- Certain objects may be charged by induction.
-- Capacitors are collectors of small static charges.

STATIC ELECTRICITY

ATTRACTION OF UNCHARGED OBJECTS

9.1. THE IMAGINARY SHELF

Materials: 1. Three or four balloons.
2. A clean blackboard or dry wall space.

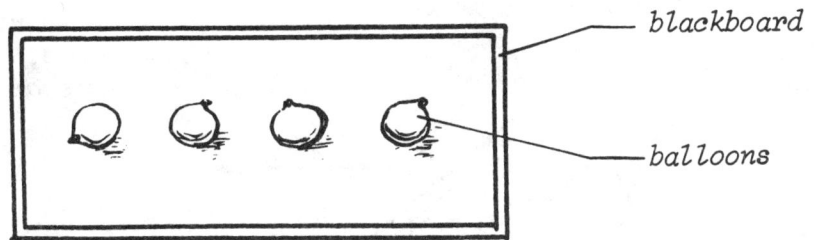

Procedure:
1. Blow three or four balloons up. Tell students that you constructed an imaginary shelf on the blackboard.
2. Rub the balloons carefully (do not press too hard) against your sleeve or shirt, as if cleaning the dust off the balloon, then stick it to the blackboard in a horizontal row.
3. When students catch on about the rubbing of the balloons, have them try to make a second lower 'shelf' with other balloons rubbed against their hair or shirt.

Questions:
1. How did the balloons get to stick to the board?
2. What did the rubbing do to the balloon?
3. Try rubbing the balloon against wool, silk, cotton, or hair. Do you find any difference in the static charge built up?
4. Why do you think the balloons will stick better to the wall on a very cold winter day than on a rainy day?
5. What are other materials that can be charged by rubbing?
6. Can the balloons stick to the walls indefinitely?

Explanation:
By rubbing the balloons, the material that was rubbed against it lost some electrons, and thus the balloon obtained an excess of electrons, which means that it has a negative charge. The easier the material used to rub the balloon is able to lose electrons, the easier the balloon gets charged. Wool, cotton or hair are the most common materials to charge the balloons. When the negatively charged balloon approaches the board, the negative charges are repelled and thus a positive charge is induced at the spot where the balloon touches the board. Because of this, the balloon sticks to the surface of the board. 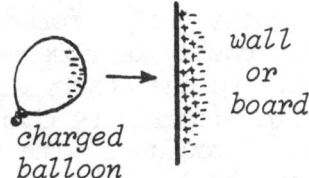 Over a longer period of time, electrons will transfer from the balloon to the board and so the balloon gets neutralized and falls to the floor. The losing of the excess charge occurs easier in damp weather, and this is the reason why it is easier to build up charges in winter.

STATIC ELECTRICITY

ATTRACTION OF UNCHARGED OBJECTS

9.2. THE MAGNETIC RULER

Materials:
1. A regular plastic ruler and handkerchief.
2. A short thin strip of paper.

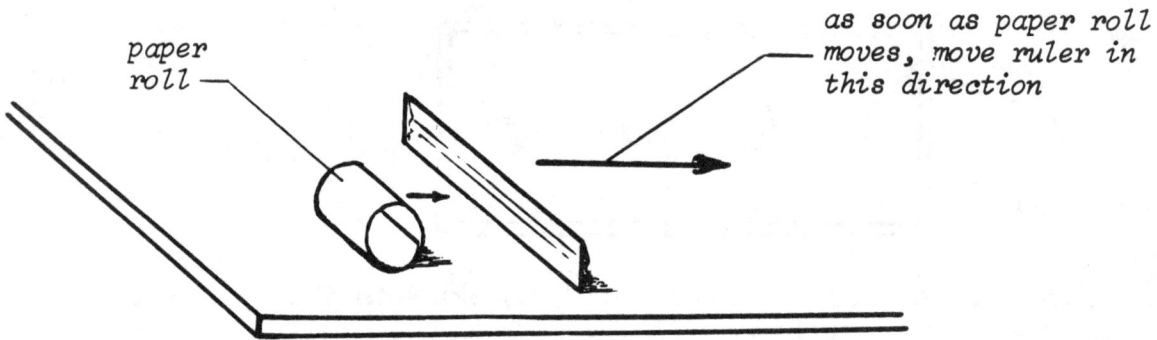

Procedure:
1. Make a small roll (about 2 cm diameter) from the strip of paper, so that it can roll over a smooth surface.
2. Take the ruler in one hand and the handkerchief in the other, and rub the ruler with it vigorously.
3. Approach the paper roll slowly until it moves, then move it away from the roll (try to keep pace with the rolling of the paper).

Questions:
1. What initiated the attraction of the paper to the ruler?
2. Is the ruler or the paper magnetic?
3. Can static electricity be considered a force?
4. Where did the initial energy originate?

Explanation:
  The rubbing of the ruler was muscle energy, or mechanical energy, turned into static electricity. The ruler is charged with an excess of negative charges and approaching the uncharged roll of paper, the latter gets attracted to the ruler. Almost anything can be rubbed and charged; some materials are easier to charge than others. The paper roll is attracted to the charged ruler, because the opposite charge is being induced in the paper, and an attraction occurs. Any attraction resulting in a movement of an object shows that a force is involved, thus **electrostatic energy is a force.** This can easily be observed, where the paper cylinder rolls from one end of the table to the other, as a result of the attraction of the charged ruler.
  Static electricity charges with their attraction and repulsion properties are often confused with magnetic properties. These are two separate and distinct entities. The first is obtained by the removal or addition of electrons by mechanical means (rubbing), and the latter by sorting or arranging the randomly distributed magnetic fields of the tiny particles (dipoles) of the material in one direction.

STATIC ELECTRICITY

ATTRACTION OF UNCHARGED OBJECTS

9.3. THE PAPER JUMPING JACKS

Materials: 1. A piece of window glass, a wool rag.
2. A plastic ruler and a handkerchief.

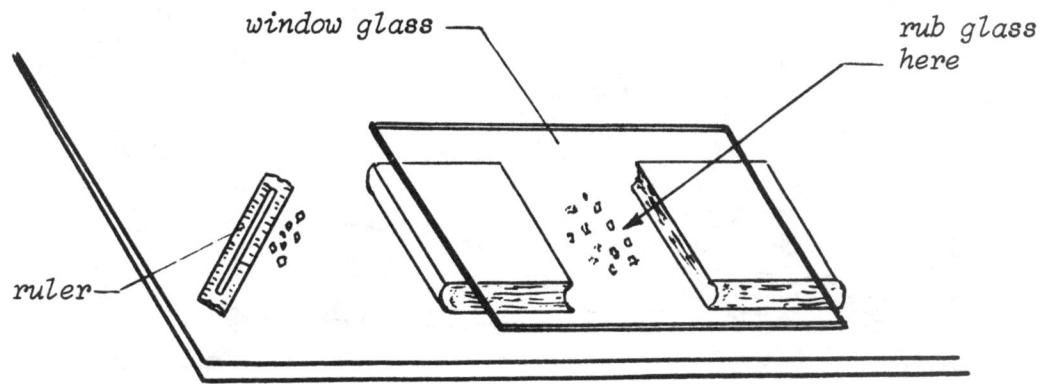

Procedure:
1. Place two books of the same thickness about 10 cm apart.
2. Put some small paper snippers between the books and cover them with the piece of window glass.
3. Rub the top of the glass with the wool rag (do not push too hard) and let students observe the paper snippers.
4. When window glass is not available, use the plastic ruler and rub it with a handkerchief. Approach the paper snippers and observe them jump up to the ruler and fall down again.

Questions:
1. What did the rubbing do to the glass or plastic?
2. Why did the paper snippers jump up and down?
3. What other materials can be used instead of the paper snippers?
4. What other materials can replace the glass or the ruler?

Explanation:

By rubbing the window glass or plastic ruler, negative charges are built up on the glass and ruler. The paper snippers are attracted to this charge, because the opposite charge is **induced** in them. At the moment that they touch the glass or ruler, the electrons can migrate to the paper; this gives it the same charge and thus they are suddenly repelled. As soon as they are back on the table, the paper snippers lose this excess charge to the table top, leaving them neutral, and the process can start all over again. This jumping back and forth from table to glass sheet can occur in very quick succession or it may take a longer time.

Other materials to replace the paper snippers would be: confetti, pepper, cork powder, fine dry sawdust, etc. Instead of the glass sheet or the ruler, we may also use: a blown up balloon, a transparency (for an overhead projector), a plastic comb, or any other plastic object, etc.

STATIC ELECTRICITY — ATTRACTION OF UNCHARGED OBJECTS

9.4. SEPARATE THE PEPPER FROM THE SALT

Materials:
1. Table salt and ground pepper.
2. A plastic petri dish, flannel or wool cloth.
3. A plastic ruler, flannel or wool cloth.

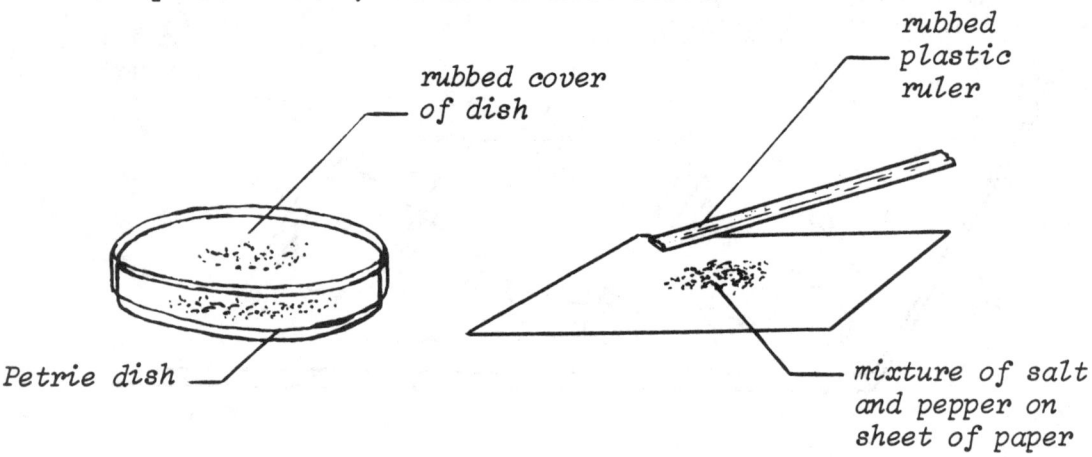

Procedure:
1. Place some salt and pepper in the plastic petri dish and put the cover on. Shake the two granulated substances and mix them. Ask the students: "How can I separate the pepper from the salt without using water?"
2. Rub the cover of the dish with the flannel or wool cloth (or your dry palm of the hand) and turn the dish upside down for a short moment (the pepper should stick to the cover).
3. The pepper can also be separated by placing the mixture on a sheet of paper and holding a plastic ruler that was rubbed with the wool cloth, over the mixture (the pepper should fly up and stick to the ruler).

Questions:
1. What made the pepper stick to the dish cover or ruler?
2. What did the rubbing do to the dish cover or ruler?
3. Why did only the pepper stick to the rubbed surfaces?
4. What would be another way to separate the two substances?

Explanation:
By rubbing the plastic surfaces with the wool cloth, these surfaces were charged with static electricity (from the excess electrons left on the plastic by the wool). The uncharged pepper particles got attracted to these surfaces, as opposite charges were induced in them, whereas no charges are induced in the salt particles.

Another way of separating the two substances is to sprinkle the grains of the mixture over the water surface in a container. The salt would sink and the pepper particles would float on the water surface, which could be scooped out. The salt will recrystallize after the water evaporates.

STATIC ELECTRICITY   ATTRACTION OF
                                                      UNCHARGED OBJECTS

9.5. THE PEPPER REPRODUCTION

Materials: 1. A plastic petri dish and lid (for each pair of students).
            2. A piece of wool cloth (for each pair of students).
            3. A pepper shaker.
            4. Paper and scissors.

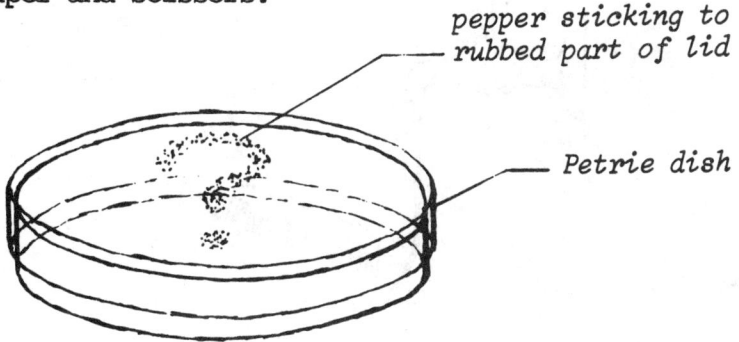

Procedure:
1. Distribute the petri dishes and the wool cloths to the students.
2. Have them place a pinch of pepper in each of the petri dishes, put the cover on them and scatter the pepper over the bottom.
3. Let the students take a piece of paper the size of the plastic dish and cut out a stencil from it (a letter or any shape).
4. Have them, while holding the paper stencil against the top of the dish, rub the open area with the wool cloth vigorously.
5. Now have them remove the stencil and invert the dish and then turn it right side up. Ask: "What do you see sticking to the cover?"

Questions:
1. Why did the pepper stick only to the rubbed parts of the lid?
2. What charge did the pepper have?
3. Why did the pepper stay attracted to the plastic?
4. What are the industrial applications of this principle?

Explanation:
   The paper stencil allowed rubbing of the plastic lid at certain confined areas. Only these areas were charged and thus the paper, which was uncharged, was attracted to the charged areas. The pepper particles stayed attracted, because they are electrical insulators and also because they are irregularly shaped. It is thus not so easy for the electrical charges to transfer themselves from the plastic to the pepper particles.
   Rubbing the plastic with the wool created negative static electricity, which induced positive charges on the upper side of the pepper particles. (see Sketch on the right).
   Industrial applications of this principle can be found in photocopying, removing of dust from exhaust fumes, industrial painting (paint particles negatively charged and target strongly positive charged).

STATIC ELECTRICITY   ATTRACTION OF
                                                      UNCHARGED OBJECTS

9.6. BEND THE WATER STREAM

Materials: 1. A large plastic comb.
            2. A wool, flannel, or felt cloth.
            3. A small water stream (from tap or small hole in a can).

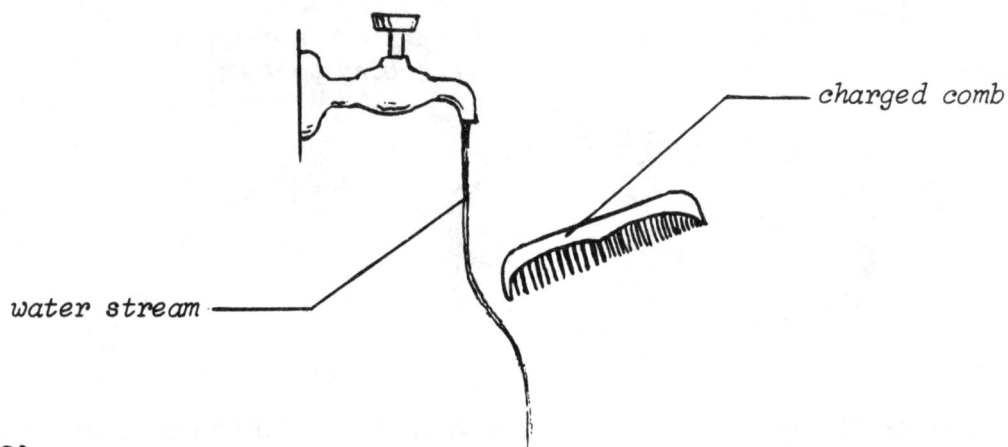

Procedure:
1. Open the tap and adjust it such that a small stream of water runs out of the tap (or hold a leaking can with water above a bucket).
2. Rub the comb with a wool, flannel or felt cloth and hold the back of the comb about 2 cm from the stream (move the comb slowly closer to the stream if it is not attracted to it).
3. Make sure that all students observe the stream being bent, as a side view. In order to do this, approach the stream from different sides.

Questions:
1. Why did the water stream bend towards the comb?
2. Was it a magnetic force that attracted the water?
3. What other materials can we use instead of the comb?
4. Would a larger stream of water also be bent?
5. What would happen if the charged comb got wet?

Explanation:
   The charged comb induces an opposite charge in the water stream and this latter therefore gets attracted by the comb. The mass has to be small enough for the electrostatic charge to move the mass. The static attraction force is relatively small and thus larger masses are not moved. This is why the water stream has to be small so that it can be attracted by the charged comb. A larger stream needs a much larger static charge to bend.
   It is suggested not to get the comb wet. The water would take away all charges and it would be very hard to get the comb recharged with static electricity. Other materials that can be used instead of the comb are: a balloon, hard rubber, a ruler, a glass rod rubbed by silk, etc.

STATIC ELECTRICITY

ATTRACTION OF UNCHARGED OBJECTS

9.7. THE ELECTRIC METER STICK

Materials: 1. A wooden meter stick.
2. A plastic comb, a piece of flannel or wool.

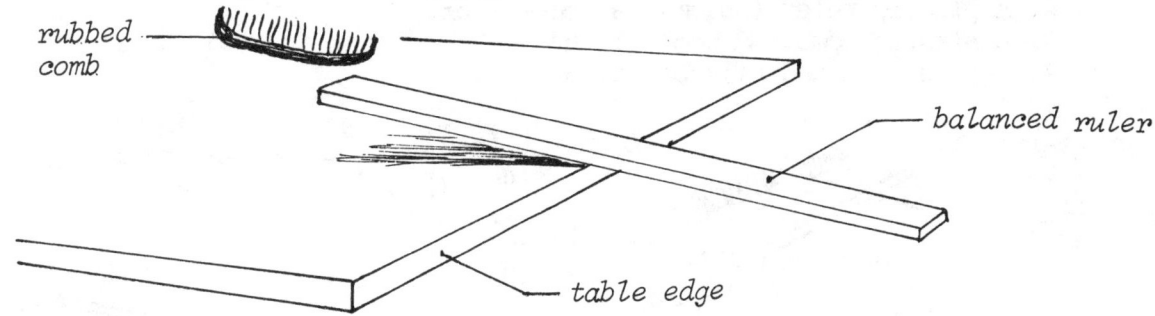

Procedure:
1. Find a table with a rather sharp edge and let half of the ruler protrude over the edge of the table.
2. Tip the ruler down over the edge, and by pulling it more or less over the edge, let the ruler balance in a slanted position. (see sketch above).
3. Rub the piece of flannel over the comb and approach the higher end of the ruler with the charged comb.
4. Now touch the upper end of the ruler with the charged comb and wait a bit till the ruler stops moving.
5. Rub the comb again and approach the same end of the ruler that was touched before. Which way does the ruler move now?

Questions:
1. Which way did the upper end of the ruler move when first approached with the charged comb?
2. Was the ruler end attracted or repelled by the comb?
3. After touching the same end of the ruler with the charged comb, was the ruler attracted or repelled by the comb?
4. If the comb was positively charged, can you explain in terms of moving electrons why the ruler behaved the way it did before it was charged? After it was charged (touched by the comb)?

Explanation:
The wooden ruler balances on the edge of the table on its centre of gravity and becomes a very sensitive balance. A minute force will tip the ruler up or down on one side.

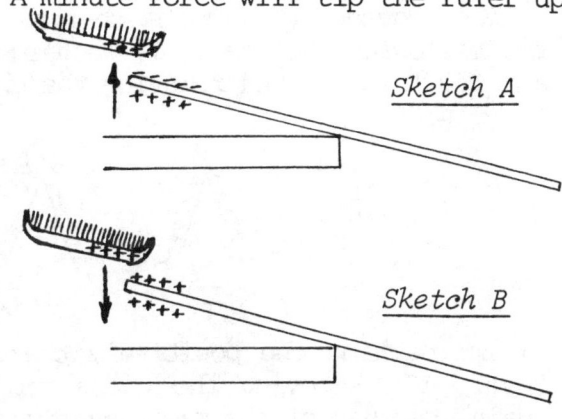

Sketch A

Sketch B

By rubbing the comb with a piece of flannel, it gets charged with static electricity (most likely positive). Approaching the uncharged ruler with the positive charge attracts the negative charges in the ruler towards the comb. As negative is attracted by positive, the ruler tips up (Sketch A). Touching the ruler with the comb withdraws the negative charges out of the ruler and leaves it positively charged. The second time we approach the ruler with the comb, the ruler moves down (see Sketch B).

STATIC ELECTRICITY — ATTRACTION & REPULSION

9.8. THE ALUMINUM FOIL ELECTROSCOPE

Materials:
1. A strip of aluminum foil, masking tape.
2. A plastic ruler (30 cm), an empty jar.
3. A plastic comb and wool cloth.
4. A glass rod and silk cloth.

Procedure:
1. Tape the ruler horizontally to the side of the jar like in the Sketch.
2. Fold a strip of aluminum foil double (about 2 cm wide and 20 cm long) and let it hang over the end of the ruler.
3. Make sure that the foil strips are hanging very close to each other (this can be done by folding or rubbing/smoothing the strips).
4. Rub the comb with a wool cloth and approach the top of the strips with it (strip ends open), and withdraw comb (strip ends close).
5. Rub the comb again with a wool cloth, approach the top of the strips and touch it (strip end will stay open).
6. Rub the glass rod with a silk cloth and approach the top of the charged electroscope with the glass rod (strip ends close), then withdraw the charged rod (strip ends open again).

Questions:
1. What happens to the charges in the Al-foil when approached with a positive charge?
2. Why did the Al-strips stay open when touched with the charged comb?
3. What happened to the charges of the foil when approached by the charged glass rod?

Explanation:
The sequence of flow of charges in the aluminum foil is best explained in the following sketches: I) Uncharged strips. II) Foil strips **induced**. III) and IV) Touching and removing the charged comb leaves foil strips charged.

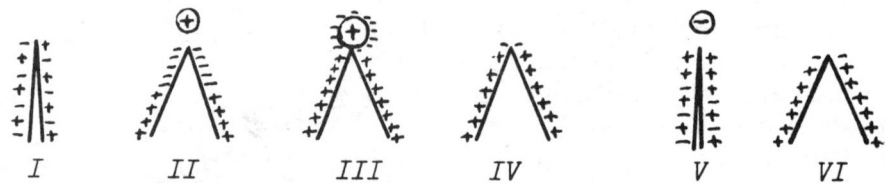

V) Approaching the positively charged foils with the glass rod neutralizes them. VI) Removing the glass rod makes the foil strips repel each other again, because of the same negative charge leaving the tips of the foil.

STATIC ELECTRICITY   ATTRACTION & REPULSION

9.9. THE CONFUSED PITHBALL

Materials:
1. A pithball and thread (the pithball can be made out of a styrofoam ball covered with pencil lead).
2. A plastic comb and a wool cloth.

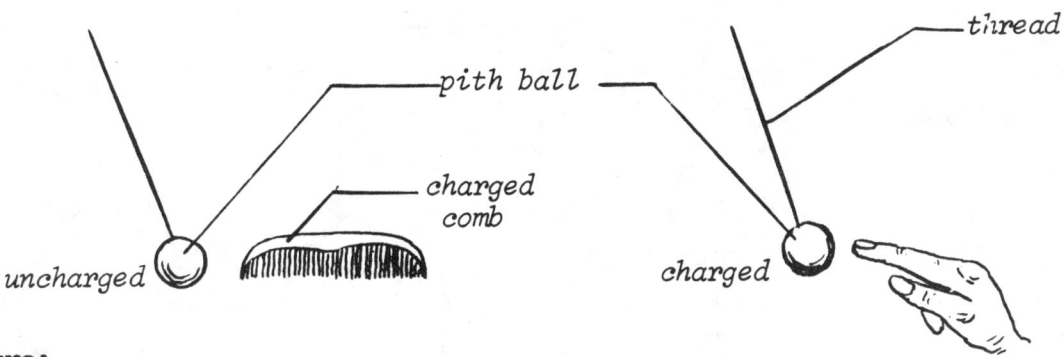

Procedure:
1. Suspend the pithball from a thread attached to a stand.
2. Rub the comb with the wool cloth and ask the students: "What will the pithball do when it is approached by the comb?"
3. Bring the comb near the suspended pithball and show that indeed the ball is attracted towards it.
4. Show this attraction several times without touching the ball. Now come a little closer to the ball with the comb and let them touch each other.
5. Now approach the charged ball with your finger.

Questions:
1. What caused the ball to be attracted by the comb?
2. If the comb was negatively charged, what charges were induced in the pithball?
3. What made the ball suddenly repel the charged comb?
4. What charge does the repelling ball have?
5. Why is the charged ball attracted by the finger?
6. What happens after the charged ball is touched by the finger?

Explanation:
1. When the comb is approaching the ball, there is an attraction, because a charged object attracts an uncharged object (the pithball) (Sketch 1).
2. As soon as the comb touches the ball, charges of the comb are transferred to the ball (Sketch 2).
3. The ball is now repelled because like charges repel (Sketch 3).
4. In bringing the finger close to the charged ball, it is attracted to the finger (uncharged object) (Sketch 4).
5. As soon as the ball touches the finger, the charges are neutralized and there is no attraction nor repulsion by the finger (Sketch 5).

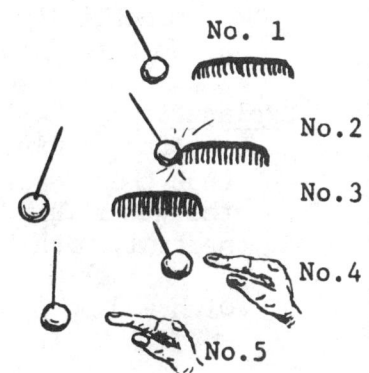

STATIC ELECTRICITY — ATTRACTION & REPULSION

9.10. THE BALLOON ELECTROSCOPE

Materials:
1. Two identically shaped balloons.
2. A wool cloth and thread.
3. A water sprayer.

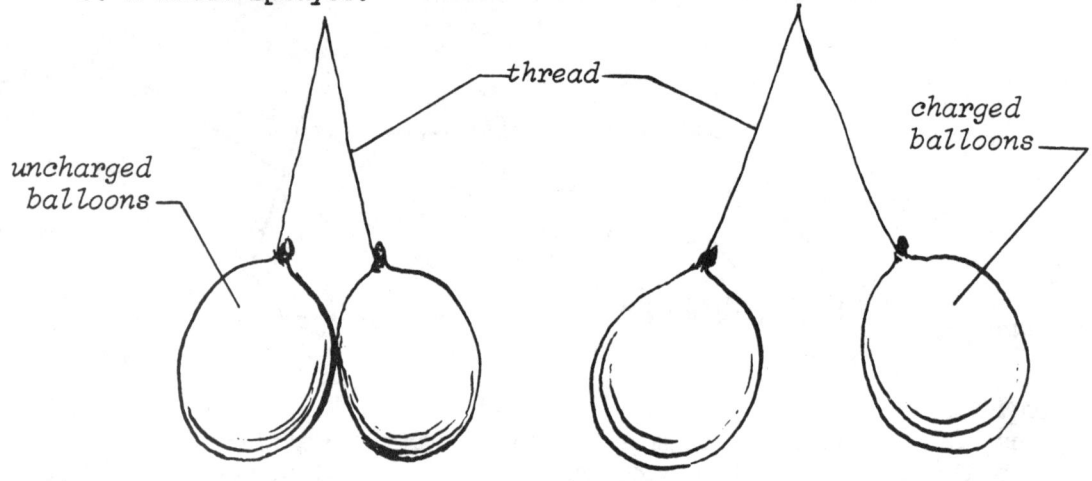

Sketch A   Sketch B

Procedure:
1. Blow two identical balloons up to about the same size and tie a thread to each of them.
2. Hold the two threads together and show that the balloons will hand against each other (without any rubbing beforehand) (Sketch A).
3. Now let a student hold the threads, rub the balloons with the wool cloth, and let them hang back (balloons will not touch: Sketch B).
4. Spray a mist of water against the balloons (they will fall back against each other like in Sketch A).

Questions:
1. Why did the balloons repel each other after the rubbing?
2. Why did the balloons fall back against each other after spraying?
3. How else could we let the balloons fall back against each other?
4. What needs to be done to have the balloon fall back?
5. During which time of year would it be best to do experiments on static electricity?
6. What is damp weather doing to the electric charges?

Explanation:
   After rubbing the balloons, they were charged with the same charge and thus they repelled each other. By spraying water droplets in their vicinity the water droplets carry the charges away from the balloons, rendering them neutral, with the result that they fall against each other.
   Another way to neutralize the charges on the balloons is to touch them with a damp or moist hand.

STATIC ELECTRICITY   INDUCTION

9.11. THE INDUCED CHARGE

Materials: 1. A pithball and thread.
            2. A plastic rod & fur, or glass rod & silk.

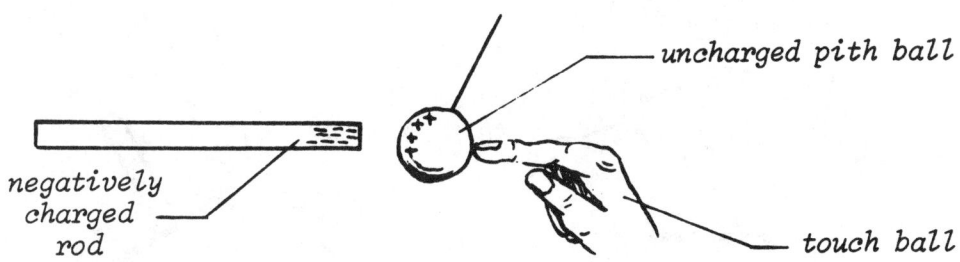

Procedure:
1. Suspend the pithball from a thread.
2. Rub the plastic rod with the fur (or the glass with the silk) to produce negative charges.
3. Approach the pithball with the charged rod (do not touch) with one hand, while touching the other side of the ball with the finger of the other hand (see Sketch above).
4. Remove the finger and then remove the charged rod.
5. Now approach the pithball with a finger: what happens?

Questions:
1. What was the pithball doing when approached by the charged rod?
2. What charges would be induced in the ball by the charged rod?
3. What did the touching with the finger do to the charges on the ball?
4. Why does the finger have to be removed first before the rod?
5. What final charge does the ball have?
6. What final induced charge would the ball have, if it was approached with a positively charged rod?

Explanation:
   Here follow three steps each for the induction of the pithball with a negative (first row) and a positive rod (second row):

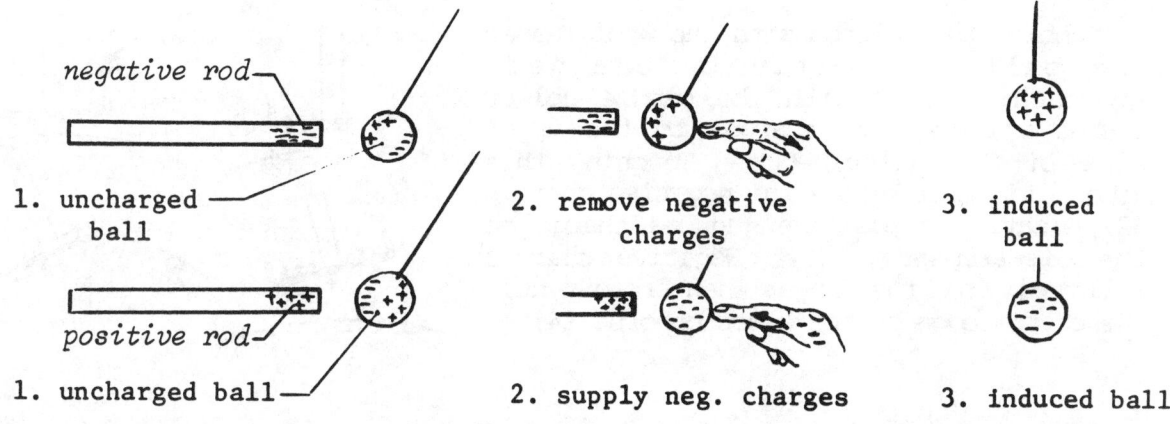

STATIC ELECTRICITY   INDUCTION

9.12. THE SPARK-PRODUCING FINGER

Materials: 1. A flat metal candle holder (or dish).
            2. A candle (medium size), a large round balloon.
            3. A wool or flannel cloth.

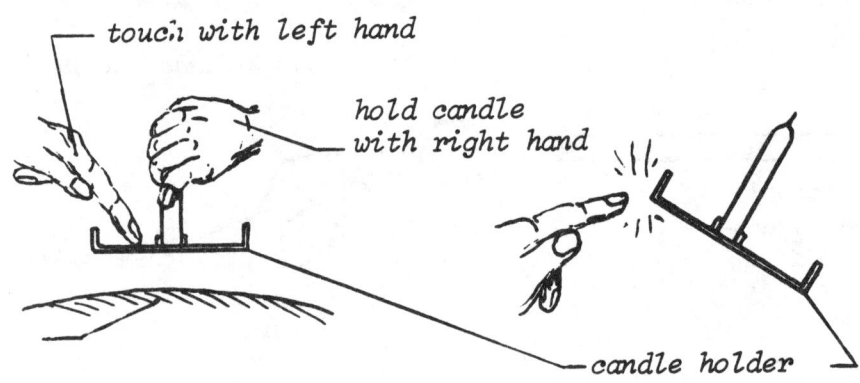

Procedure:
1. Inflate the balloon and rub it vigorously with the wool cloth.
2. Attach the candle to the candle holder, place the candle holder on the rubbed surface of the balloon, using the candle itself as a handle (hold with right hand).
3. Now touch the upper side of the candle holder with a fingertip of the left hand (a small spark may jump to your finger).
4. Lift the candle holder from the balloon, still holding on to the candle with the right hand.
5. Now approach the bottom edge of the candle holder with a fingertip of the left hand (a spark will jump).
6. Steps 2-5 may now be repeated several times.

Questions:
1. Was it actually the finger that produced the spark?
2. Where did the energy originally come from?
3. What type of energy was the original source?
4. What happened when the finger touched the upper side of the holder?
5. What charges jumped from the finger to the holder, when approaching the bottom of the holder?

Explanation:
Rubbing the balloon with the wool leaves the balloon negatively charged. Approaching this with the candle holder pushes all negative charges to the upper side of the holder (1). Touching this upper side will remove the negative charges (2), leaving a positive induced charge on the **electrophorus** (3). Negative charges will jump from the finger when brought near the bottom edge of the electrophorus (4).

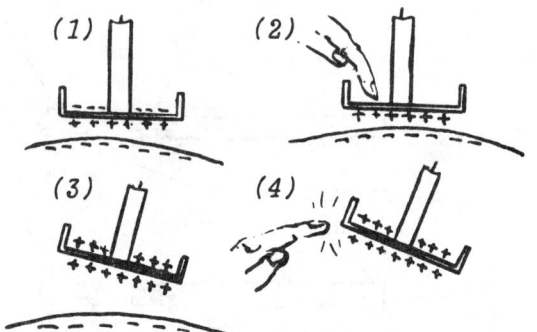

STATIC ELECTRICITY — CAPACITORS

9.13. THE ELECTROSTATIC STORAGE

Materials:
1. A gallon jar (wide mouth) & lid.
2. A spike (10 cm nail) & paper clips.
3. A one-hole stopper, aluminum foil.
4. A plastic rod & wool cloth (or Electrophorus: Event 9.11).

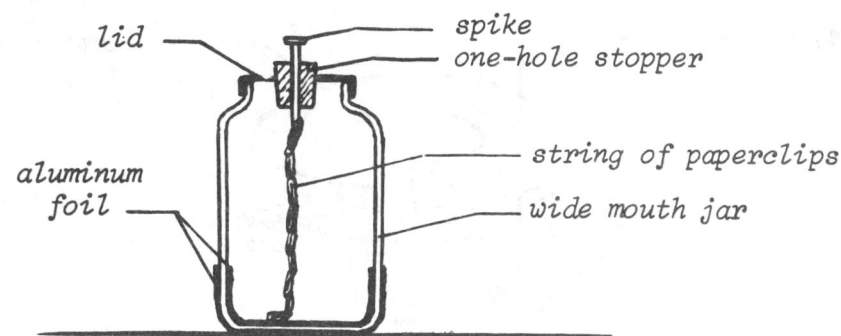

Procedure:
1. Make a hole in the jar lid the same size as the one-hole stopper.
2. Cover the bottom to about one-third of the sides up with aluminum foil, on the inside of the gallon jar.
3. Do the same thing with the outside of the jar (see Sketch).
4. Insert the spike in the one-hole stopper and attach a string of paper clips to the end of it, such that the end of the chain will drag on the bottom of the jar.
5. Now insert the stopper in the hole of the jar lid and close off the jar.
6. Charge up the jar with negative static, by rubbing the plastic rod with the wool and touching the spike (or with the electrophorus: Event 9.11) and do this about 10 times.

Questions:
1. What is the purpose of the chain of paper clips?
2. Would the storage jar work without the paper clips?
3. What type of static electricity would we store with the use of the electrophorus from 9.11?
4. What would be the danger if we used too small a stopper?

Explanation:
The two sheets of aluminum foil, separated by the bottom of the jar, act as a capacitor. The charges obtained by rubbing the wool over the plastic rod are small charges, which can be accumulated in this **Leyden Jar**. The purpose of the paper clips is to connect the inner layer of aluminum foil to the spike. It would be much better if the lid were made of insulating material (material that does not conduct electricity: plastic, hard rubber, etc.). With a metal lid, we must use a rather large rubber stopper in order to prevent charges from jumping from the spike to the lid. With the **electrophorus** (Event 9.11) the jar may be charged with 5 or 6 sparks, but with the rod and wool it will take 8-10 times of touching. When the jar is fully charged, we may short the outer foil with the spike with a wire connecting the two, and observe a large spark jump over.

STATIC ELECTRICITY   CAPACITORS

9.14. THE WATER CUP SPARK COLLECTOR

Materials: 1. A plastic cup, a plastic container (larger in diameter).
            2. Two pieces of bare copper wire.
            3. A glass rod & silk or a plastic rod & wool.
            4. A plastic or wooden cloth pin.

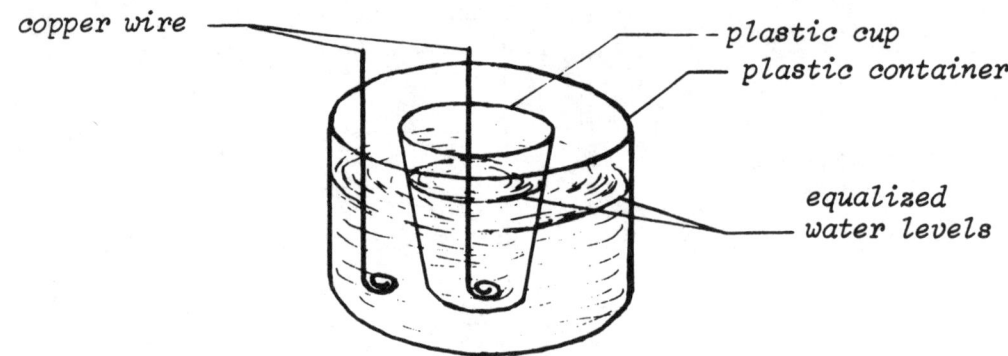

Procedure:
1. Place the plastic cup in the larger plastic container.
2. Fill the cup 3/4 full with tap water and fill the outer container with water to the same height.
3. Bend the two pieces of copper wire at one end, so that it can stand upright in the containers. Place one in the center and the other in the outer container, immersed in the water.
4. Charge up the capacitor now with the glass rod and silk, or with the plastic rod and wool, by touching the center wire after every rubbing. Repeat this about 10 times.
5. Hold the wire from the outer container with a cloth pin and move the tip closer to the center wire and observe the jumping of the spark!

Questions:
1. How similar is this apparatus to the Electrostatic Storage (9.12)?
2. In which part of the apparatus do the charges collect?
3. What other equipment may be used to charge up the collector?
4. What is the difference in using glass and silk or plastic and wool?
5. What would we need to increase, if we wanted to collect a much larger electrostatic charge?

Explanation:
   This apparatus is very similar to the Leyden Jar (Event 9.12). The only difference is, that the collecting surfaces are now, instead of aluminum layers, water layers that are separated by an insulating layer of plastic: the wall of the plastic beaker. Each of the layers is connected with the bare copper wire, and each one has the opposite charge of the other. The glass and silk will produce positive in the center, and so will the electrophorus (see Event 9.11), whereas the plastic rod and wool cloth will produce negative in the center wire.

CHAPTER 10

HOW IS CURRENT ELECTRICITY CREATED?

OBJECTIVES

After dealing with and studying the concepts and sub-concepts in this chapter, the students should be able to:

a. recognize the correct explanation of an observed event based on each of the sub-concepts;
b. explain in their own words which of the sub-concepts is determining the course of an event;
c. distinguish true from false statements concerning each one of the sub-concepts;
d. identify the correct explanation of an event in daily life applying one of the sub-concepts;

all in relation to the following sub-concepts:

-- The conductivity of metals and semi-conductivity of carbon.
-- Current electricity can only flow in a completed circuit.
-- A circuit may be controlled or broken by switches or fuses.
-- Current electricity may be transformed into heat.
-- Electricity can be produced from chemical sources.
-- A magnetic field exists around a wire which carries electricity.
-- A coil of wire can be turned into a strong magnet by electricity.
-- Electromagnets have many applications.

CURRENT ELECTRICITY CONDUCTIVITY

10.1. MAKE A CONDUCTIVITY TESTER

Materials:
1. A battery (dry cell) and masking tape.
2. A 1.2 volt bulb and bulb holder.
3. Three short wires (bent out paper clips will do).

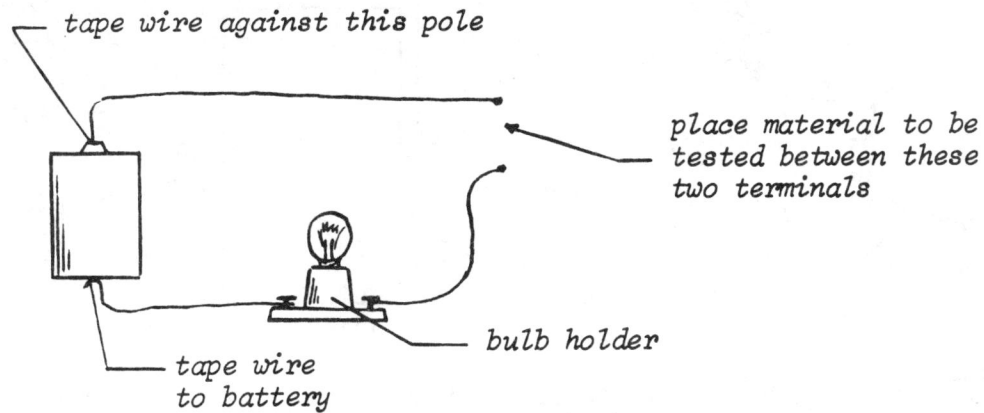

Procedure:
1. Connect one pole of the bulb holder to the negative pole of the battery (use masking tape).
2. Connect another wire to the other pole of the bulb holder.
3. Tape one end of the third wire to the positive pole of the battery.
4. The two loose ends are the test terminals. Test the light and when the two loose ends are touching each other, the light should go on.
5. Test any material for conductivity, by placing it between the two loose terminals.

Questions:
1. What should the light do when a conductor is placed between the two terminals?
2. If both terminals are touching one and the same object, and the light does not go on, what does it mean?
3. Which of the following materials are conductors: a penny, dime, nickel, quarter, aluminum foil, paper, wood, glass, the faucet, watch band, gold ring, ruler, pencil carbon, eraser, etc.?

Explanation:
The difference between static and current electricity is, that in the former case there is only a build-up of electric charges, but no flow of charges except in a spark (a short jump of charges). The latter needs a completed circuit for electrons to flow continuously. In the case of the **conductivity tester**, there is a break in the circuit between the two loose terminals. When a **conductor** is placed between these terminals, the **circuit is completed** and the light goes on. When a **non-conductor** is placed between the terminals, the circuit is still broken, because electrons cannot flow through non-conductors, and thus the light stays off. Objects that are made of metal are all electric conductors. Carbon of a pencil is a partial conductor, or semi-conductor.

CURRENT ELECTRICITY — CONDUCTIVITY & CIRCUITS

10.2. THE SECRET BURGLAR ALARM

Materials:
1. A battery (dry cell).
2. 1.2 volt bulb and a short solid bare copper wire.
3. Sewing thread and masking tape.

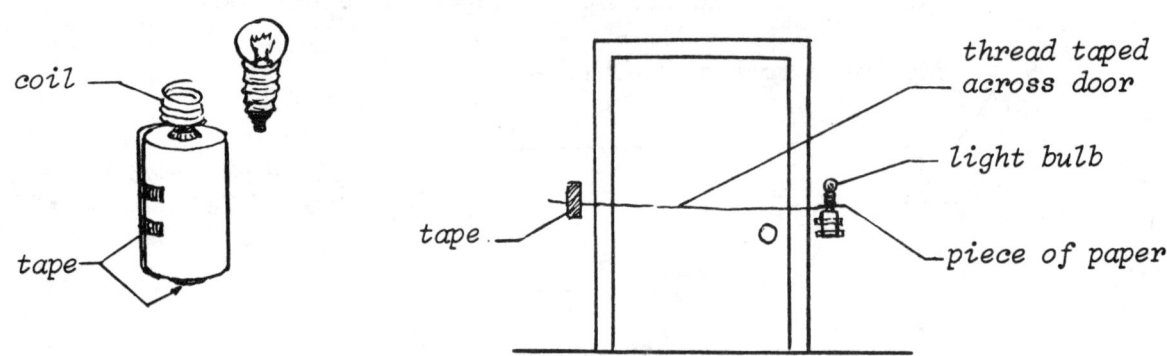

Procedure:
1. Take the copper wire and wind it along the groove of the bulb and bend the wire around the battery and tape the end of the wire against the negative pole of the battery (see Sketch A).
2. Turn the bulb to the right and make sure the bulb lights.
3. Place a piece of paper between the bulb and the battery. (Pull it out, and the light should go on.)
4. Tape a thread to the piece of paper (long enough to span across the doorway).
5. Tape the battery plus bulb to one sill of the door and tape the other end of the thread to the other side of the door sill. (Sketch B)

Questions:
1. What should the light do if someone comes in the door?
2. What is preventing the bulb from switching on?
3. What other materials may be used in place of the piece of paper?
4. How is a real burglar alarm triggered?

Explanation:
The function of the paper is to prevent completion of the circuit. When the alarm is set the light is not on and when the door is opened or when someone comes in the room the thread and the attached paper is pulled away. This leaves the circuit completed and the light is on. Other materials that may be used to break the circuit are non-conductors like: plastic, cardboard, cloth, etc.

A real burglar alarm is triggered by the breaking of a light or laser beam, which is invisible to the human eye, but needed to prevent the completion of the alarm circuit. Breaking of the light beam thus completes the circuit and the alarm sounds.

CURRENT ELECTRICITY   CONDUCTIVITY & CIRCUITS

10.3. LIGHT THE BULB

Materials: 1. A battery (dry cell) & a small bulb (1.2 V).
            2. Two short wires (paper clips or aluminum foil will do).
               (a set of materials for each pair of students).

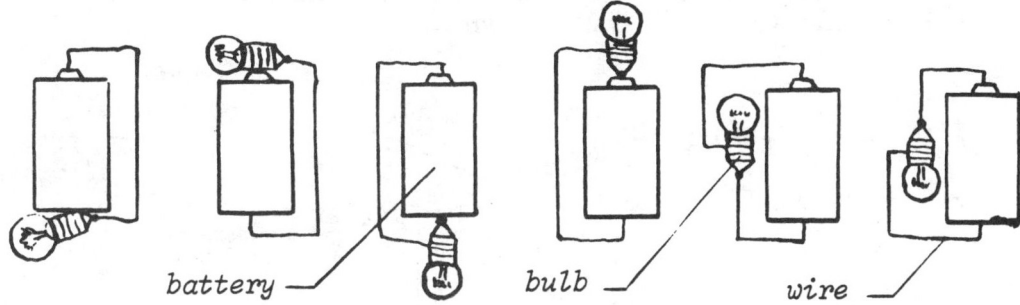

Procedure:
1. Provide each pair of students with a battery (dry cell), a bulb, and two short wires (or bent out paperclips or aluminum foil strips).
2. Ask the students to light the bulb, with two wires, then with one wire.
3. Ask: "In how many ways can you light the bulb?"
4. Place the above circuits intermixed with the following circuits on the board, and let students identify the correct ones.

Questions:
1. Will any set-up of the second set of circuits light the bulb?
2. Which of the schematics above are correct to light the bulb?
3. Which two poles of the battery and two poles of the bulb have to be connected to light the bulb?
4. Is it possible to light the bulb without wires or any other materials?
5. What other materials can we use instead of the wires?

Explanation:
  As long as one pole of the bulb is connected to either + or - of the battery and the other pole of the bulb to the free pole of the battery, the bulb will light. Without wires, the set-up on the right sketch might work, when the insulation of the rim of the battery is scratched off, and the battery is placed on a piece of aluminum foil (some of the bottom rim paint should be scratched off).

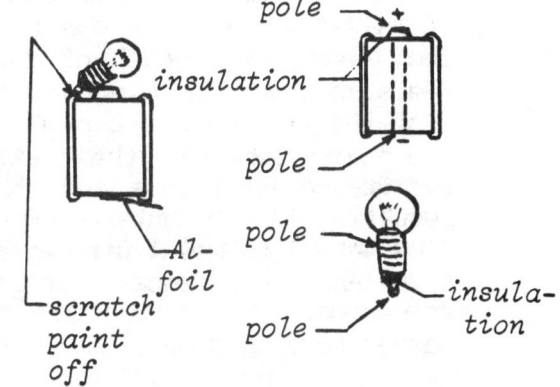

10.4. REGULATE THE CURRENT WITH A PENCIL

Materials:
1. A long pencil.
2. A 6 Volt battery & 3 alligator clip leads.
3. A bulb (6 V) and bulb holder.

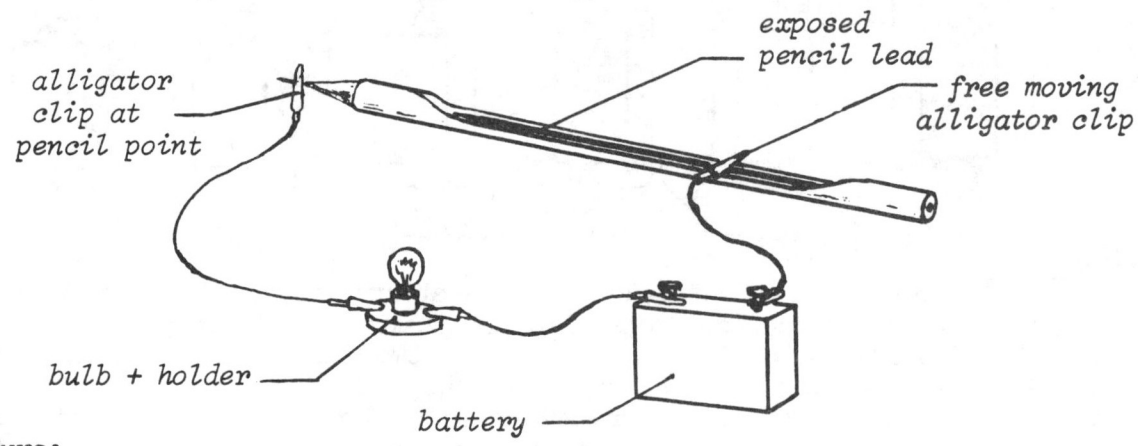

Procedure:
1. Cut the wood of the pencil away to expose the lead.
2. Connect the three wire leads with the alligator clips as follows:
   a. From the pencil point to one pole of the light bulb.
   b. From the other pole of the light to one terminal of the battery.
   c. From the other terminal of the battery to the open exposed pencil lead (free moving end of the wire).
3. Hold the free end of the third wire lead and slide the alligator clip over the exposed pencil lead. Observe the light bulb!

Questions:
1. Is pencil lead a conductor of current electricity?
2. Why did the light bulb glow dimmer and brighter?
3. What did we have to do in order to make the bulb glow dimmer?
4. Do you think there is more or less resistance for the current to flow through pencil lead compared to copper wire?

Explanation:
Pencil lead is a **semi-conductor of current electricity.** This means that not all the electrons can flow through, thus the voltage is hereby reduced. The longer the length of the pencil lead is in the circuit, the more the resistance for the electrons to flow through, and so the lower the voltage, with the result that the bulb is glowing dimmer. By sliding the free end of the wire closer to the stationary end of the pencil (pencil point), this resistance gets less and thus it gets easier for the electrons to flow, resulting in a brighter glow of the bulb.

A very long metal wire will have the same effect as the short pencil lead. A variable resistance--a resistance that can be changed at will--is called a **rheostat.** Rheostats are applied in dimmer switches, stove and oven temperature controls, and other appliances requiring current flow regulators.

CURRENT ELECTRICITY CIRCUITS

10.5. THE DIMMER AND BRIGHTER BULBS

Materials:
1. A battery (dry cell) and holder.
2. Two 1.2 Volt bulbs and bulb holders.
3. Four lengths of copper wire.

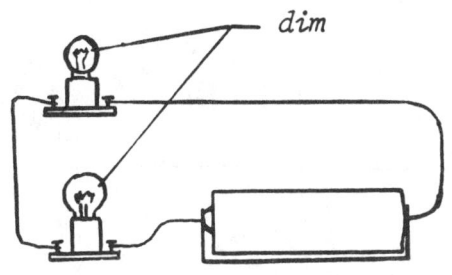

Sketch A : Series Circuit

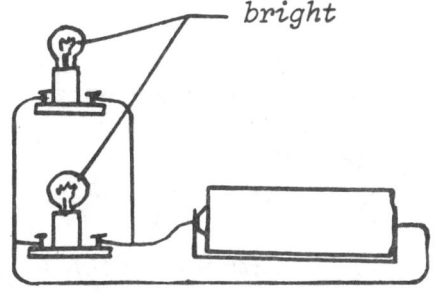

Sketch B : Parallel Circuit

Procedure:
1. Take a battery in its holder and two bulbs in their holders and connect them as pictured in Sketch A.
2. Turn one of the bulbs loose in the socket (both bulbs go off).
3. Change the circuit and connect the wires as pictured in Sketch B (the bulbs burn with a brighter light).
4. Now turn one of the bulbs loose in its socket (only this one bulb turns off and the other stays on).

Questions:
1. Why are the bulbs in the first circuit glowing dimly?
2. Why are the bulbs in the second circuit glowing brightly?
3. What is the reason that both bulbs are turned off when only one is turned loose in its socket in the first circuit?
4. Why does the second bulb stay on in the second circuit, when one bulb is turned off?
5. What type of circuit would we have in our home?

Explanation:
A glowing bulb in an electric circuit gives a certain resistance to the electron flow in the circuit. In the **series** circuit, the bulbs follow one after another in one wire, adding to the resistance in the current flow, and thus dimming the glow of the bulb. In the **parallel** circuit this is not the case. Each bulb is individually connected to the battery, independent from each other. This is the wiring system used in our homes. An example of a **series** and **parallel** circuit is found in the Christmas lights. Those with one wire in between the bulbs are wired in series, whereas those with two wires in between the bulbs have a parallel circuit.

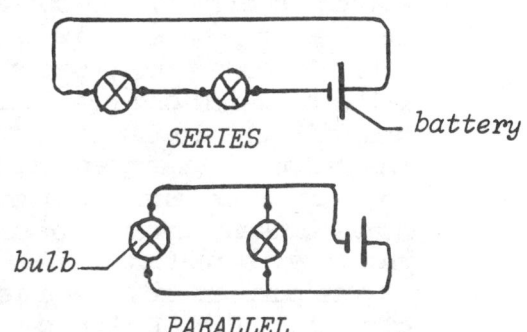

CURRENT ELECTRICITY

CIRCUITS

10.6. THE ALUMINUM FOIL FUSE

Materials:
1. A strong battery (6V) or five 1.2 V batteries.
2. Copper wire and aluminum foil.
3. Bulb (6V) and holder, and simple switch.

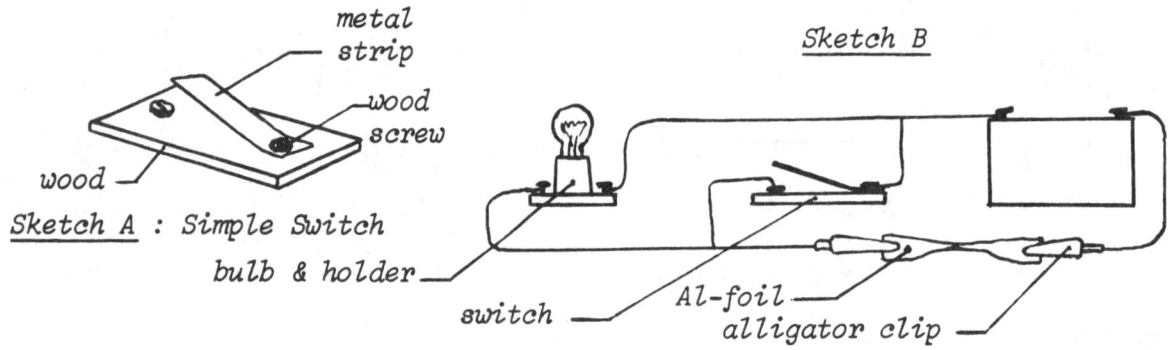

Procedure:
1. Cut a piece of aluminum foil (a gum wrapper will do) in the shape illustrated in the circuit of Sketch B.
2. Make a simple switch from a piece of wood, a metal strip (or paper clip) and two small wood screws (Sketch A).
3. Connect the aluminum foil fuse, the switch and the light bulb to the battery in a circuit pictured in Sketch B (when the switch is open, the bulb should be on).
4. Now draw students' attention to the light and the fuse: push the switch closed; the fuse blows and the light goes out (even when the switch is released, the light stays out).

Questions:
1. Why does the fuse not blow when the switch is left open?
2. Why does the fuse blow when the switch is closed?
3. What does the switch actually do to the circuit when it is closed?
4. What is the purpose of a fuse in an electrical circuit?

Explanation:
The switch in this circuit has the purpose to short the circuit. When the switch is closed, the current from the battery can flow much faster, because it does not have to pass the bulb. Electric current, just like a flow of water through pipes, seeks the path of lowest resistance. When there are two paths of which one has a higher resistance, the current will flow through the one with the lowest resistance: in this case the closed switch rather than the bulb. The sudden stronger current makes the thin aluminum foil too hot to stay whole and burns off. When the switch is now released, the circuit stays broken, because of the broken fuse, and thus the light stays off until the fuse is restored.

The purpose of a fuse is to prevent the start of a fire in case of a **short circuit**. This latter can heat the wires so much that the insulation might catch fire. A fuse will break the circuit immediately when shorted.

CURRENT ELECTRICITY

10.7. THE BALLOON FUSE

Materials:
1. A small balloon, aluminum foil, simple switch.
2. A strong battery (6V) or five 1.2 V D-batteries.
3. Bulb (6V), bulb holder, copper wire, steelwool.

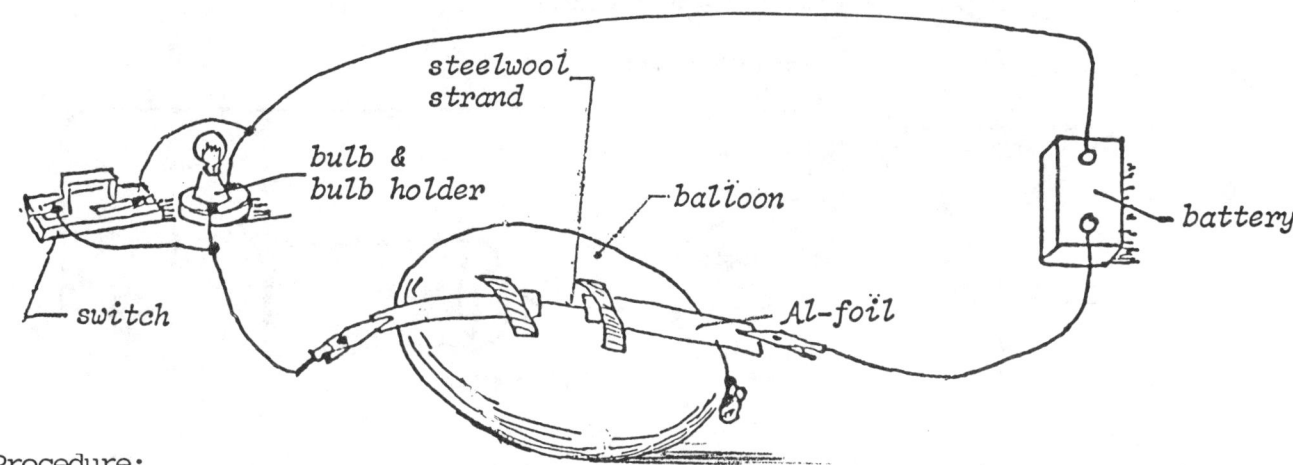

Procedure:
1. Cut two strips of aluminum foil and tape one strand of steelwool between the two strips against the balloon (see Sketch).
2. Use the copper wire to complete the circuit as laid out in the Sketch. (same circuit as in Event 10.6).
3. Show students that the light is burning because the circuit is complete.
4. Draw their attention now to the switch, which will make the short circuit when closed, and the single strand of steelwool on top of the balloon.
5. Insist on everybody's attention, then push the switch: BANG! The fuse broke and the light is off.

Questions:
1. What made the balloon pop when the switch was pushed?
2. What is a short circuit?
3. Which way would electricity rather flow, through the switch or the bulb?
4. Why did the light go out after the balloon popped?

Explanation:
The single strand of steelwool had the function of the fuse wire. When the circuit was shorted by pushing the switch, a sudden surge of electric current was flowing through the circuit causing the single strand of steelwool to heat up. As this strand of steelwool was taped against the balloon, it melted the rubber and bursted it. With the steelwool strand broken, the circuit in which the bulb is wired is thus incomplete, and so the light stays out.

In our daily life, in most electrical appliances, all automobiles and homes are protected with fuses. These fuses can let a certain amount of electricity pass through them (f.i. 10, 15, or 20 amperes). When a sudden surge of electrical flow occurs the fuse breaks, or in the new homes they are called **circuit breakers** which can be reset.

10.8. THE INFERENCE BOARDS

Materials:
1. Cardboard sheets (15 x 20cm), paper fasteners.
2. Metal clips (large), copper wire (bare), masking tape.
3. Light bulb & holder, battery & holder.
4. Wire leads with alligator clips.

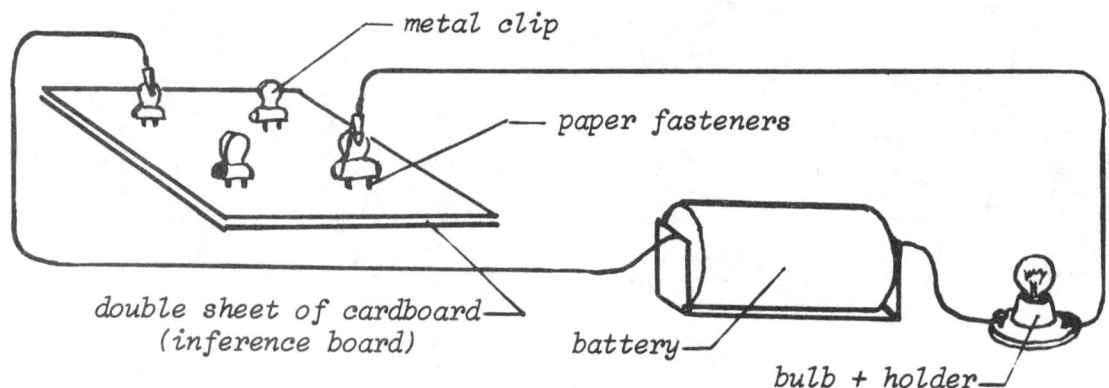

Procedure:
1. Make 4 to 8 different inference boards, by pushing paper fasteners through the cardboard sheets, bending the two prongs, connecting the fasteners with the bare copper wire (Sketch below), covering the back with another cardboard sheet (tape together) and placing the metal clips over each pair of paper fasteners. The 8 patterns are as follows:

2. Connect the bulb and battery with the wire leads as in the sketch above and clamp two terminals with the alligator clips (f.i. B & C), take clip A or D (one at a time) off the paper fasteners, and infer the wire pattern from the light bulb's off or on observations.

Questions:
1. When the alligator clips are connected to terminal B and C, and the light is on, what would be your inference?
2. With alligator clips on B and C, light on, taking off metal clip A, and the light goes off, what would be your inference now?

Explanation:
With the alligator clips of the testing instrument on B and C, the light will go on, because the current will pass through A. But as soon as the clip A is taken off, the light will go off. Taking off clip D will not affect the burning light bulb.

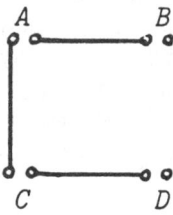

The connections behind the cardboard should be soldered.

10.9. CHANGE ELECTRICITY INTO HEAT

Materials:
1. Nickel-chrome wire and alligator clips stand.
2. A battery (6V), 3 wires, a simple switch.

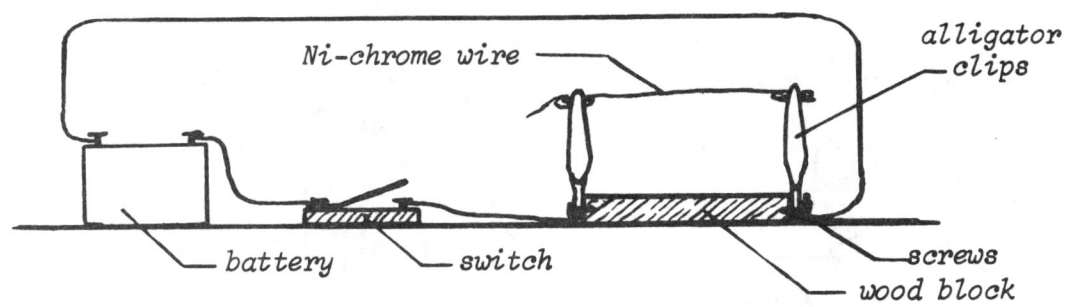

Procedure:
1. Connect the wires in the circuit as outlined in the sketch.
2. Clip the Ni-chrome wire between the alligator clips, making sure that they make connection (by doubling or tripling the wire where it is clipped).
3. Push the switch to complete the circuit and observe the Ni-chrome wire.
4. Show that the wire is actually red hot by touching a piece of paper to the Ni-chrome wire, while keeping the circuit closed.

Questions:
1. What is the original source of energy?
2. What is the electricity changed into?
3. What would a thin copper wire do in place of the Ni-chrome wire?
4. Where do we find an application of this principle?
5. What made the electricity change into heat?

Explanation:
   The 6 Volt battery is the original chemical source of energy, which is turned into electrical current. It passes through the Ni-chrome wire and is transferred into heat at that spot. The flow of electrons through the Ni-chrome wire may be compared to the flow of water through a narrow pipe. The current is flowing without resistance through the rest of the circuit, but in the Ni-chrome wire, the flow is hampered because of the small size of the wire (in diameter). The **flow of electrons** in this part of the circuit creates friction between the electrons and the wire molecules, resulting in the formation of heat.
   As the melting point of Ni-chrome is much higher than copper, the wire can glow without melting, but if a fine copper wire were used in its place, it would melt and break the circuit as a fuse would. This change of electric energy into heat is applied in electric stoves, ovens, toasters, heaters, dryers, kettles, etc.

CURRENT ELECTRICITY — SOURCES

10.10. THE LIQUID BATTERY

Materials:
1. A copper and a zinc (or magnesium) strip.
2. Dilute sulfuric acid, copper wire leads.
3. A beaker, a flash bulb & holder (or 0.2 V bulb).

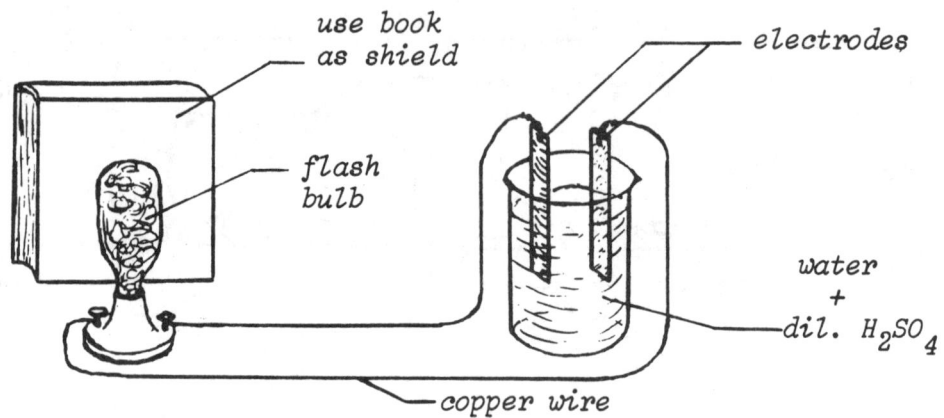

Procedure:
1. Fill the beaker half full with water and add dilute sulfuric acid till the beaker is 3/4 full.
2. Make small nail holes in one end of each of the electrodes and connect the copper wires from each of the electrodes to the two terminals of the bulb holder (**do not** dip the electrodes in the beaker yet!).
3. Stand a book in front of the flash bulb (to prevent students from getting blinded by the flash of light).
4. Hold the two electrodes, one in each hand, let students observe, and immerse the electrodes quickly in the liquid. (The electrodes should be gleaming clean: scraping with sand paper may help.)

Questions:
1. What made the flash bulb light?
2. What materials made up the source of the electric current?
3. Why was a shield necessary between the bulb and the observer?
4. What triggers the flash in a camera flash?

Explanation:
This battery is normally called a **wet cell**, because liquid is involved. Car batteries are usually wet cells: sulfuric acid and water acting as the **electrolyte** and lead (and lead oxide) as the **electrodes**. In this case the copper and zinc strips were the electrodes. Zinc (or magnesium) having more of a tendency to lose electrons compared to the copper, act together with the H_2SO_4 electrolyte as the source of electron flow.

Without the book as a shield, students might get temporarily blinded by the bright flash, especially when they look directly at the bulb. It is a good precaution to use a shield, as the flash is bright enough to be seen without directly looking at the bulb. When using a regular small bulb, this shielding is not necessary.

The flash of a camera flash is triggered by the built-in battery, either in the camera, the flash holder, or the flash cube itself.

CURRENT ELECTRICITY

10.11. ELECTRICITY FROM A LEMON

Materials:
1. A copper and a zinc strip (electrodes).
2. Two copper wire leads.
3. A galvanometer or a 0.2 Volt bulb.

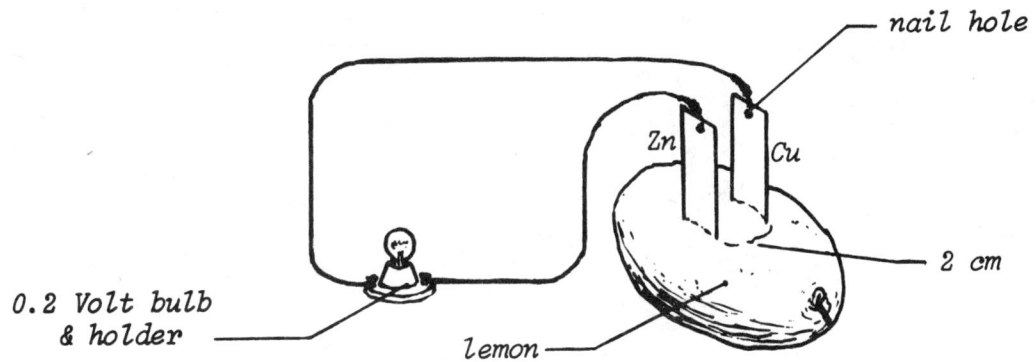

Procedure:
1. Make a small nail hole in the end of the metal strips.
2. Insert the copper and zinc strip into the lemon vertically, about 2 cm apart from each other.
3. Connect the wire leads to each of the electrodes and connect one of the other ends to one terminal of the bulb holder.
4. Get ready with the other loose end of the wire to touch the open terminal of the bulb holder. Before touching it, draw students' attention to the bulb (or galvanometer).
5. Touch the terminal with the loose end of the wire for short moments at a time (the current is so weak that you will not get electrocuted!)

Questions:
1. Where does the current come from?
2. What is current electricity actually?
3. Can we use other metals to make electricity with the lemon?
4. Would two strips of the same metal produce electricity?
5. What other fruit can be used instead of the lemon?

Explanation:
Current electricity is a **flow of electrons** through a completed circuit of conductors. The **source** of the electron flow in this case is the combination of copper and zinc strips in the citric acid of the lemon. Zinc has more of a **tendency to lose electrons** compared to copper, and thus connected through the wires and light bulb, the electrons flow from the zinc to the copper strip.

Other metal pairs that could be used to develop **potential differences** or **electromotor forces (EMF)**, are: copper and silver, copper and iron, copper and magnesium, etc. Other fruit that could replace the lemon are: orange, grapefruit; less potential difference comes from apple, pear and peach. It is the acidity (the citric acid) in the fruit, which is necessary to conduct the electrons from one electrode to the other.

CURRENT ELECTRICITY

SOURCES

10.12. THE COIN BATTERY

Materials:
1. Two pennies and two dimes.
2. Blotting paper or thick paper towel.
3. Copper wire and a galvanometer.
4. Common table salt.

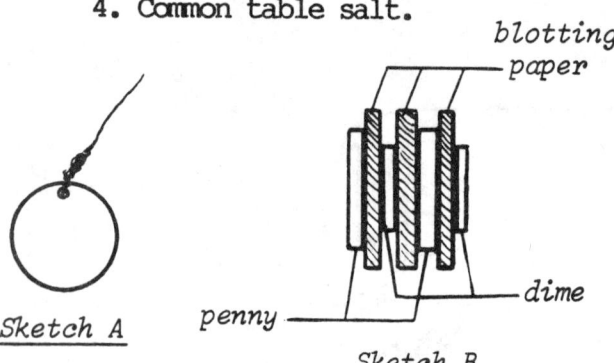
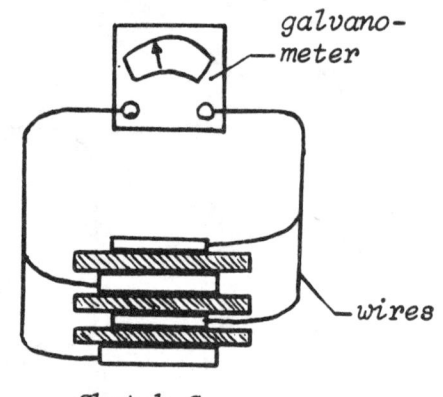

Procedure:
1. Clean the coins thoroughly with a brush, soap and water until gleaming.
2. Punch a nail hole close to the edge of the four coins, and attach a short and a longer copper wire to the two pennies, and do the same to the two dimes (see Sketch A).
3. Soak pieces of blotting paper in salt water and arrange the pennies and dimes between the blotting paper as in Sketch B.
4. Attach the wires to the galvanometer like in Sketch C. Before completing the circuit, draw students' attention to the galvanometer needle, then touch the terminal with the loose wire. (You will not feel any electric shock, as the currents are just very weak ones.)

Questions:
1. Why do the coins have to be cleaned?
2. Where does the electricity come from?
3. What is the source of the electron flow?
4. What is the function of the salt water?
5. Would using more pennies and dimes increase the current?

Explanation:
The dime and the penny act as silver and copper electrodes. The salt water in the blotting paper is the **electrolyte** needed to get a transfer of electrons from the silver to the copper. In this case, copper has a greater tendency to lose electrons than silver, and a flow of electrons is created from the copper to the silver electrode when the circuit is completed. The sodium (Na^+) and chloride (Cl^-) ions are the electron carriers in the electrolyte. Without the salt, the electrons cannot flow and thus no current is produced.

The larger the surface of the electrodes, the larger the **potential difference** between the electrodes. The increase in the number of pairs of coins would therefore increase the voltage as well as the current (amperage).

The common dry cell uses carbon and zinc as electrodes, with damp ammonium chloride (NH_4Cl) as electrolyte.

CURRENT ELECTRICITY

SOURCES

10.13. ELECTRICITY FROM A COIL?

Materials:
1. Bell wire, a large nail, a strong magnet.
2. A simple galvanometer (or compass and coil of wire).

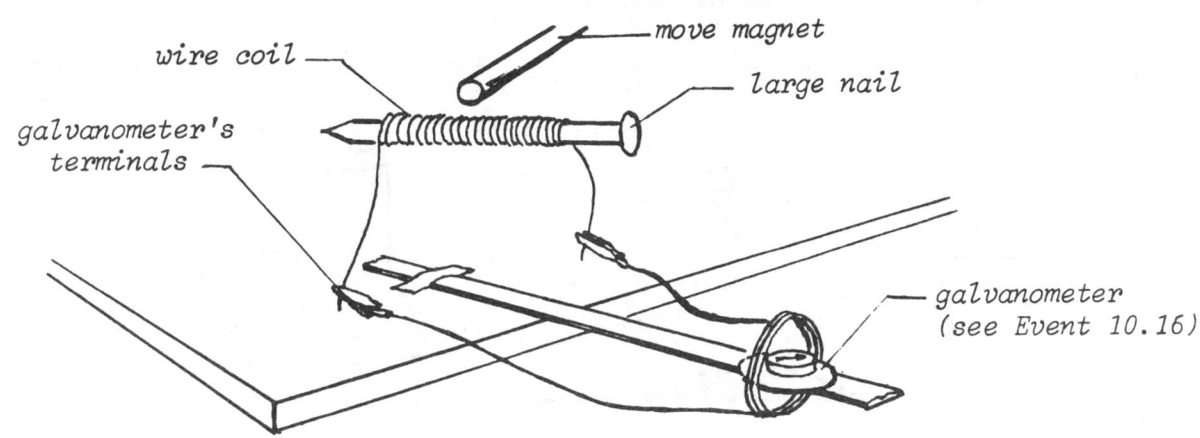

Procedure:
1. Wrap the bell wire around a large nail (or steel bolt) with at least 20 windings to make the coil.
2. Attach the ends of the coil to the terminals of the galvanometer (the simple galvanometer of Event 10.15 may be used).
3. Now take the strong magnet and move one pole of it back and forth near the coil, while observing the needle of the galvanometer.
4. Turn the rod magnet end for end and move the other pole back and forth near the coil. What does the galvanometer show?

Questions:
1. After reversing the magnet, what happens to the galvanometer?
2. What is the source of energy that is converted to electricity?
3. What factors are influencing the increase of the current output?
4. Where do we find this principle applied in daily life or industry?

Explanation:
This is an example of an **electromotive source of electricity**, where mechanical energy is converted into electrical energy. By moving the pole of the rod magnet back and forth near the coil, an electric current is produced in the coil, which is detected by the galvanometer (an instrument to detect weak currents). By moving the magnet in one direction along the coil and then back again, we can see that the deflection of the galvanometer needle changes direction, meaning that the current **alternates** from one direction into the other. The electrical generators in power plants are based on the same principle, where coils are moving in a strong magnetic field to produce **alternating current (AC)**. The bicycle dynamo is an example of a generator that produces **direct current (DC)**.

CURRENT ELECTRICITY — ELECTROMAGNETISM

10.14. THE NAIL MAGNET

Materials:
1. A large nail (10 cm long), bell wire (1 m long).
2. A 6 V battery (or 3 1.2 V D-size batteries).
3. A box of paper clips.

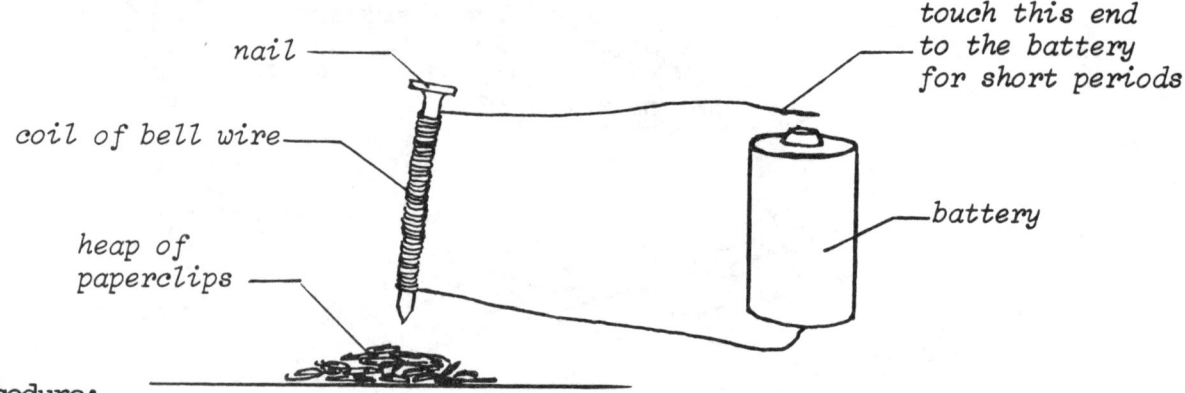

Procedure:
1. Wind the bell wire around the nail (about 20 times).
2. Scrape the insulation off from the ends of the wire and place the paper clips in a heap on the table.
3. Have someone hold the nail above the heap of paper clips, touch the two ends of the wire to the battery poles and let the student raise the nail. How many paper clips were picked up?
4. Wind another ten loops of the wire around the nail and complete the circuit again. Have the students count the number of clips picked up.
5. Increase the number of batteries (by placing them in series and connecting + pole to - pole of the next battery). Have students count the number of paper clips picked up.

Questions:
1. What made the coil become a magnet?
2. What effect did increasing the number of windings have on the magnetic attraction force of the nail?
3. What did two or three batteries in series do to the magnetic force?
4. Would removing the iron core of the coil weaken or strengthen the magnetic attraction?
5. How does the magnetic attraction of the coil change when the core is changed into a wooden, glass, or other kind of rod?

Explanation:
When an electric current flows through a coil, it acts as a magnet. By placing an iron rod inside the coil, the magnetic lines pass through the iron rod and the coil becomes a much stronger magnet, as long as the electric current flows through the coil. The more windings in the coil, the stronger the magnetic force. This force is also directly proportional to the magnitude of current.

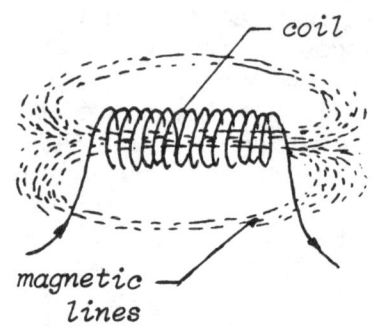

CURRENT ELECTRICITY　　　　　　　　　　　　　　　　　　　　ELECTROMAGNETISM

10.15. THE DEFLECTING COMPASS

Materials:
1. A D-size battery (1.2 V).
2. Copper bell wire (and a simple switch).
3. A magnetic compass.

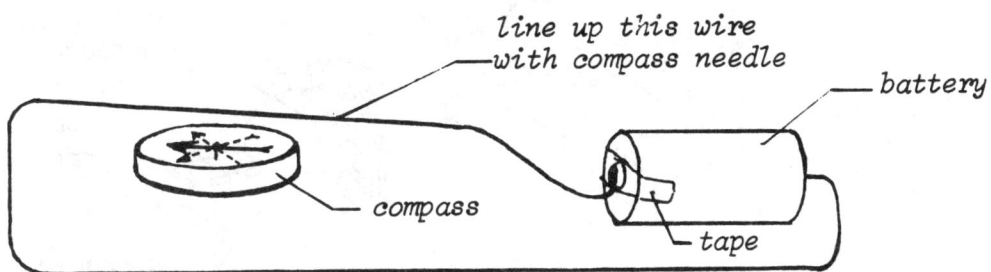

Procedure:
1. Distribute one compass, one battery and a length of bell wire to each pair of students.
2. Instruct students to do the following:
   a. Line up one wire coming from the positive pole of the battery with the compass needle by holding it above the compass (tape the wire end to the battery).
   b. Bring the other end of the wire to the negative pole of the battery but do not complete the circuit yet (use a switch if available).
   c. Pay attention to the compass needle, touch the battery with the loose end of the wire for short periods.
   d. Place the compass above the wire, complete the circuit, what happens?

Questions:
1. What did the compass needle do when the circuit was completed?
2. How did the compass needle deflect when it was above or under the wire?
3. What is the reason for the compass needle to move?
4. What would the compass needle do were a magnet to approach it?
5. What can we conclude about the surroundings of an electric current?

Explanation:
This demonstration shows that a **magnetic field** exists around an electric current. As soon as the circuit is completed, by touching the loose end of the wire to the battery, the North-pointing needle of the compass will deflect to the left (the way it is set up in the sketch). The left hand rule may be discovered by the students: **When holding the left hand over a compass, such that the flow of electrons is from the wrist to the fingertips, the North-pointing needle deflects along the opening thumb.** (Make sure that the palm is facing the compass.)

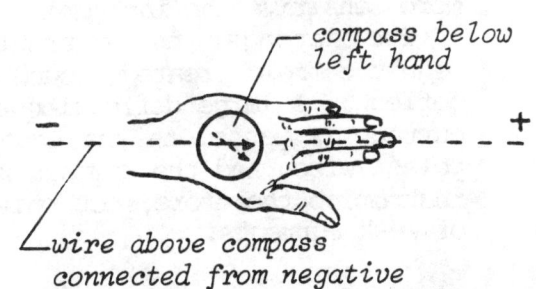

CURRENT ELECTRICITY — ELECTROMAGNETISM

10.16. THE SIMPLE GALVANOMETER

Materials: 1. A small compass (magnetic), bell wire, two alligator clips.
2. Cardboard, plastic ruler (30 cm) and masking tape.

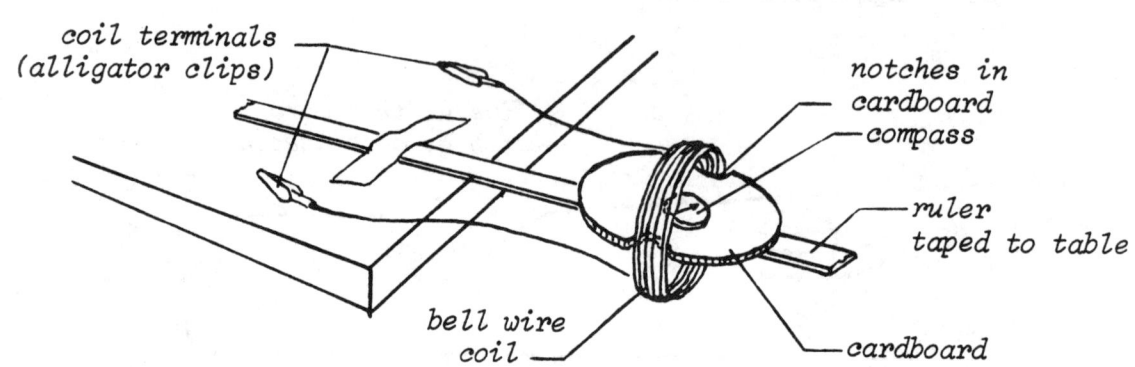

Procedure:
1. Cut out a circle (diameter about 10 cm) out of a sheet of cardboard, and make two notches opposite each other.
2. Wind the bell wire in a coil of at least 10 windings, with the same diameter (10 cm) and fit it in the two notches perpendicular to the cardboard circle.
3. Tape a ruler to the table top and tape the cardboard with the coil to the protruding end of the ruler (see Sketch).
4. Connect alligator clips to the two ends of the coil and leave this on the table top.
5. Tape the small compass in the center of the cardboard and line up the needle with the coil (by moving the ruler).
6. You are ready to test the galvanometer: any small current through the coil will move the compass needle.

Questions:
1. What makes a compass needle move in general?
2. What does the coil turn into when current passes through it?
3. What will make the galvanometer more sensitive?
4. What does a galvanometer measure?

Explanation:
When an electric current is led through a coil, the latter is turned into a magnet. A magnetic field is thus created around the compass, when a small current passes through the coil, and this field deflects the compass needle. The more windings in the coil, the stronger the magnetic field and thus the more sensitive the instrument.

The stronger the current passing through the coil, the stronger the magnetic field created around the coil and the more the compass needle will deflect. A large deflection therefore indicates a larger current. When the current is passed in the opposite direction through the coil, the magnetic poles switch and the compass needle deflects in the opposite direction. The instrument therefore, not only measures the strength but also the direction of weak currents.

CURRENT ELECTRICITY

ELECTROMAGNETISM

10.17. THE SIMPLE TELEGRAPH

Materials:
1. A simple switch, nail and bell wire.
2. A battery (6 V) and bulb (and holder).
3. A small block of wood, strip of tin, and small screw.

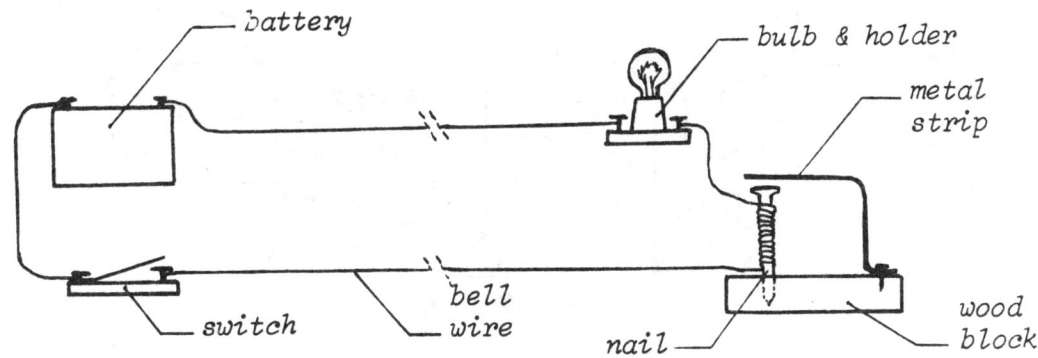

Procedure:
1. Hammer the nail 1/4 way into the block of wood, wind about 20-30 loops around the nail, and further connect the bell wire to the bulb, battery, and switch (see circuit in the sketch).
2. Bend the tin strip (cut from a tin can) in an S-form and screw it to the other end of the wood block, such that a small gap is left between the strip and the nail head.
3. Push on the simple switch for short periods of time to close the circuit and observe the light go on and the metal strip move down.

Questions:
1. What made the metal strip bend down towards the nail?
2. Would the metal strip still be dipping down, even when this 'nail switch' and light are several km away from the simple switch?
3. What does the light suggest about the circuit, when it is on?
4. Does the light bulb have to be part of the circuit in order for the metal strip to be attracted to the nail?

Explanation:
The closing of the simple switch completes the circuit, resulting in a flow of electrons through the coil of wire around the nail. This produces a magnetic field around the coil as the coil turns into a magnet, attracting the metal strip down towards the nail head. The completed circuit also makes the light go on. Without the light in the circuit, the metal strip still bends down towards the nail when the simple switch is pushed down. The completed circuit also makes the light go on. Whenever the simple switch is released, the circuit is broken, and the coil stops acting as a magnet, resulting in the release of the metal strip. As the strip has some spring action properties, it springs back up, as soon as the attraction forces disappear. This is the principle of the **telegraph**, where **Morse codes** are sent over long distances. Ships at sea communicate by Morse codes in terms of long and short flashes of light.

10.18. TURN ELECTRICITY INTO SOUND

Materials: 1. A simple doorbell and switch.
2. Two wire leads and a dry cell (6V).

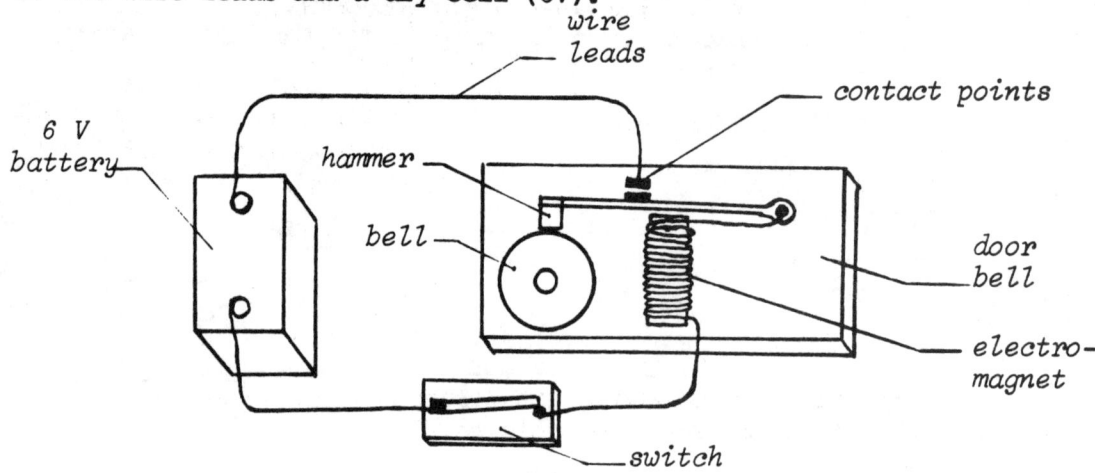

Procedure:
1. Take the cover off the doorbell and expose the electromagnet, bell, hammer and contact points.
2. Connect one wire to one terminal of the simple switch, and the other wire with a wire lead (with alligator clips) to the dry cell.
3. Attach the other wire lead between the simple switch and the free pole of the battery (circuit is pictured in Sketch).
4. Complete the circuit now by pushing the simple switch.

Questions:
1. What is accomplished in the circuit by holding down the simple switch?
2. As soon as electricity runs through the coil, what does it become?
3. What will the hammer do to the circuit when it hits the bell?
4. Why does the hammer move back towards the contact point?

Explanation:
When the simple switch is pushed down, the circuit is completed, and electricity flows through the coil. This turns the coil into a magnet, pulling the hammer towards it and in doing so it hits the bell. In the B position (see sketch on right) against the bell, the hammer actually breaks the circuit and the coil does not attract the hammer any more; thus it flips back towards the contact point, resulting in closing the circuit again, and the whole cycle starts over. This on and off switching of the circuit happens very rapidly and a vibration of the hammer results.

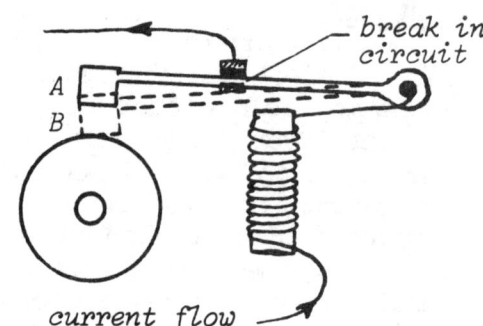

CHAPTER 11

HOW DOES LIGHT BEHAVE ?

OBJECTIVES

After dealing with and studying the concepts and sub-concepts of this chapter, the students should be able to:

a. recognize the correct explanation of an observed event based on each of the sub-concepts;
b. explain in their own words which of the sub-concepts is determining the course of an event;
c. distinguish true from false statements concerning each one of the sub-concepts;
d. identify the correct explanation of an event in daily life applying one of the sub-concepts;

all in relation to the following sub-concepts:

-- Light travels in straight lines.
-- The image of an object is seen exactly the same distance behind the mirror as the object is in front of it.
-- The angle of incidence is the same as the angle of reflection.
-- Light is refracted towards the normal when coming from a less dense medium into a denser medium.
-- When the critical angle is exceeded, the border between two different media will act as a mirror.
-- Real images may be formed by holding a lens between a light source and a paper card.
-- White light is scattered into a spectrum of rainbow colors when passed through a prism of denser material.

LIGHT   TRAVELS IN STRAIGHT LINES

11.1. MAKE A PINHOLE CAMERA

Materials: 1. A shoebox and onion skin paper (thin paper).
            2. A candle.

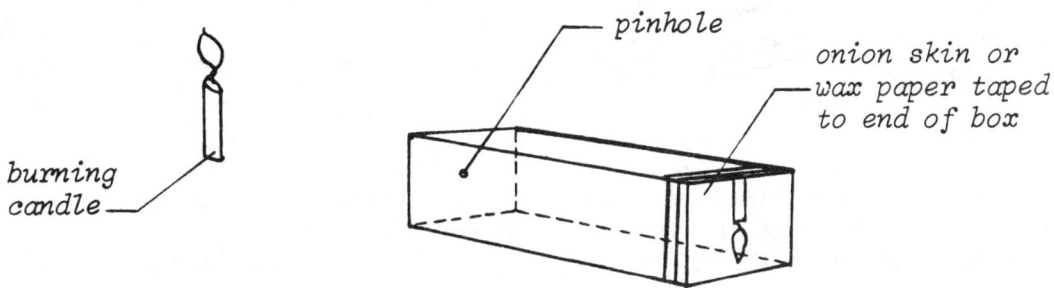

Procedure:
1. Take the shoe box and cut out the side of one end of the box and tape the thin onion skin paper to it.
2. Make a small round hole in the other end of the box and tape the lid to the box. (Make the hole a little larger than a pin hole).
3. Point the hole-end of the box towards a window or a light source: lamp or candle, and observe the image on the thin paper (prevent light from falling on the outside of the onion skin paper).

Questions:
1. What do you observe on the onion skin paper?
2. Why is the image on the paper always upside down?
3. What would happen to the image if the pinhole were enlarged?
4. How do the light rays travel from the object to the image?

Explanation:
   Light travels in straight lines. This is why the image is always inverted. The light from point A of the object travels in a straight line through the pinhole to point A' on the onion skin of the box, and point B to Point B' thus inverting the image. When the pin hole is enlarged the image will get blurred. The smaller the hole (aperture in a camera), the sharper the image.
   The onion skin paper at the end of the box may be replaced with a photographic plate and exposed for a short time. (The box has to be held steady.)

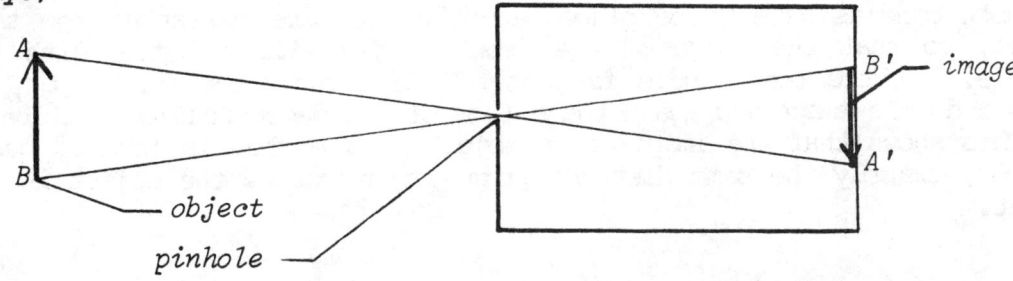

LIGHT — REFLECTION

11.2. THE COOL CANDLE FLAME

<u>Materials</u>: 1. A rectangular piece of window glass.
2. Two identical candles (same height, diameter).

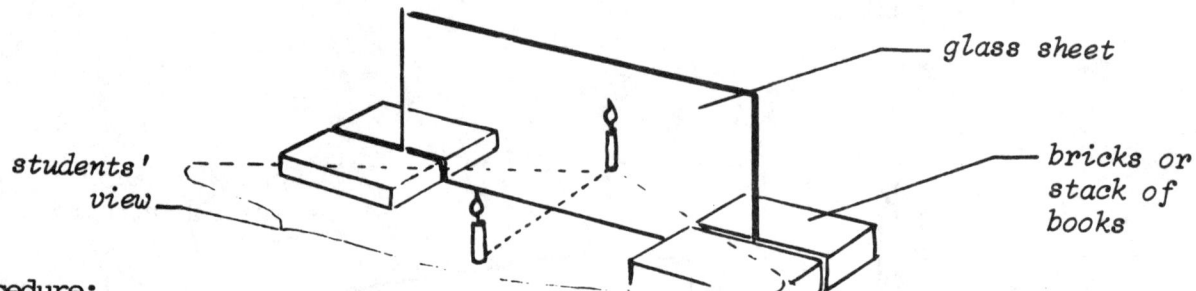

<u>Procedure</u>:
1. Mount the glass sheet vertically between four bricks or four stacks of books. (Or hold it in two mounts of molding clay).
2. Fix one candle in front of the glass sheet on the table (10 cm measured on the perpendicular to the glass).
3. Fix the second candle in the back of the glass sheet on the table (exactly the same distance from the glass). Block the student's view of the first candle with a large book or folder, and light **only** this candle (do as if you light both candles), then remove the folder. (Student's should sit in places, such that they cannot see the second candle without looking through the glass sheet--students' view, in sketch).
4. Place your (or a student's hand) above the second (unlit) candle for a very short while and withdraw. Increase this time 2 or 3 seconds (accompanied with an 'ouch' for dramatic effect).

<u>Questions</u>:
1. Why can you hold your hand above the second candle?
2. Where does the second candle have to be placed?
3. Why is it necessary to cover the student's view when lighting the first (front) candle?
4. Why can't the students see the second candle without looking through the glass sheet?
5. What happens if the second candle is not exactly the same distance from the glass sheet as the first?

<u>Explanation</u>:
   It is essential that the students think that both candles are lit; this is why the front candle has to be blocked from the student's view. This is the same reason why students may not see the second candle in plain view. Once the student knows that the candle behind the glass sheet is not lit, the whole effect of the demonstration is lost.
   Both candles have to be placed exactly the same distance from the glass sheet, so that the image of the front candle will coincide with the rear candle. If the rear candle is just a trifle off this exact spot, it will show a double image and again the effect of the demonstration will be lost.
   This shows that the image of an object in a mirror is located behind the mirror, exactly the same distance from the mirror as the object is in front of it.

LIGHT

REFLECTION

11.3. FUNNY REFLECTIONS

Materials: 1. Unframed square or rectangular mirror (for each pair of students).

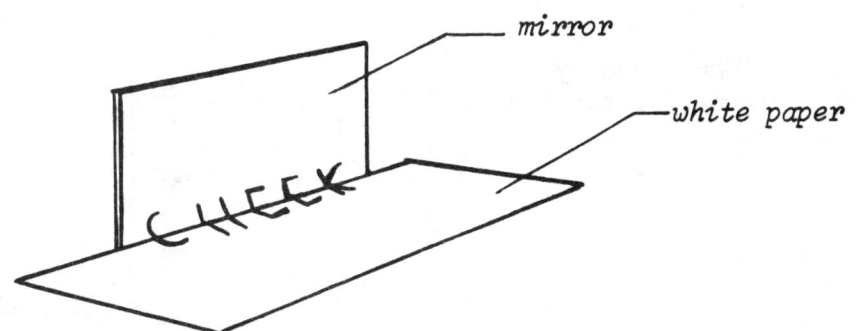

Procedure:
1. Print the following words each on a separate piece of blank paper in large even capital letters. CHEEK, BIKE, DECIDE, BOX, CHICK, CHOKE, BOOK, HIDE, CODE, DOCK, COOKIE.
2. Cut each piece of paper in half, right through the middle of each word printed on, leaving the lower half of the word on the paper.
3. Now place the mirror vertically against the paper (perpendicular on the table) and look in the corner of the paper and the mirror.
4. Do this for each one of the words.

Questions:
1. Will this work with any printed word?
2. Which of the letters, when cut in half, will appear whole with the mirror above it?
3. Which letters, when cut in half vertically, will give the other half in the mirror, when the half letter is held against it?

Explanation:
This activity will provide the student with the ability to recognize symmetrical objects and printed letters. It gives the student an opportunity to manipulate spatial relations. The image in the mirror shows that all images are actually formed behind the mirror.
It is only with the letters: B, C, D, E, H, I, K, O and X, that the above activity can be carried out. All other letters will not give the upper half in the mirror. The following can be cut vertically in half and the other half will show up in the mirror, when held against it: A, H, I, M, O, T, U, V, W and X, Y. This activity gives the students an idea of **symmetry** and **symmetrical** shapes.

LIGHT   REFLECTION

11.4. HOW HIGH TO PLACE THE MIRROR?

Materials: 1. A meter long narrow mirror or several smaller mirrors.
            2. A meter stick.

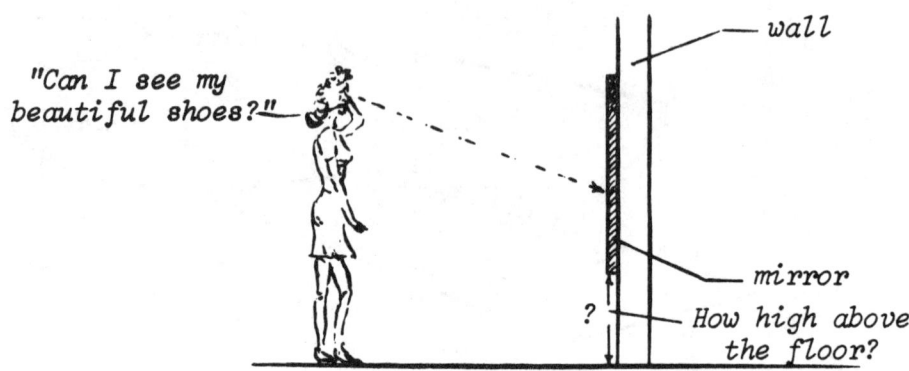

Procedure:
1. Pose the students with the question: "How high do I have to hang the mirror in order to see all of myself including my shoes?"
2. Record the student's predictions of the distance the mirror should be above the floor.
3. Let two students hold the mirror against the wall, a third student looks at her/himself in the mirror, and a fourth measures the maximum height of the mirror off the floor.
4. Attach the mirror at that height and let all the students see themselves in the mirror.

Questions:
1. Does the distance, from where you stand to the mirror, affect the image of yourself in the mirror?
2. How high off the floor does the mirror have to be hung to see your feet in the mirror when standing in front of it?
3. Would a shorter person have to move the mirror i order to see his/her feet?
4. Can you draw the paths of light rays that you see coming from your shoes?

Explanation:
Because the image of the person is exactly the same distance behind the mirror (AB=BC), the height BE is always half the height of the person (1/2CD). It does not matter where the person stands in front of the mirror. The height BE is not affected by the distance BC.

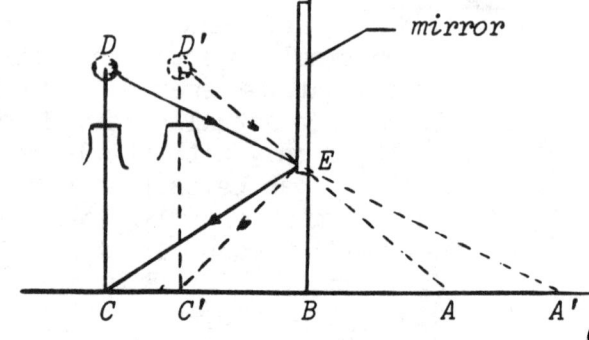

LIGHT   REFLECTION

11.5. LOOK AT YOURSELF AS OTHERS SEE YOU

Materials: 1. Two square or rectangular mirrors without frame (one set for each group of 6 students).

two mirrors held perpendicular to each other

Procedure:
1. Have one student stand the two rectangular mirrors perpendicular to each other on the table, facing away from him.
2. Let each student observe himself in the two mirrors by looking in the corner (joint) of them, and let them touch their hair with one of their hands or wink an eye.
3. Now let the student holding the mirrors place them in one plane (in each other's extension) and let the students observe themselves again.

Questions:
1. Why did the left hand appear on the right in the perpendicular mirrors?
2. How do light rays travel from object to mirror and back to the observer's eye?
3. What happened to your image when the mirrors were straightened?
4. How are the light rays travelling now?

Explanation:
   The light rays of the object (which is yourself) are reflected in 90° angles from one mirror to the other and then back to the eye. In a regular straight mirror, the rays bounce right back to the eye, without being reflected to another mirror. With the two mirrors perpendicular to each other, anything on the left of the object is reflected to the right hand side. All points of the object on the left side is seen on the right and vice versa (see sketch on right).

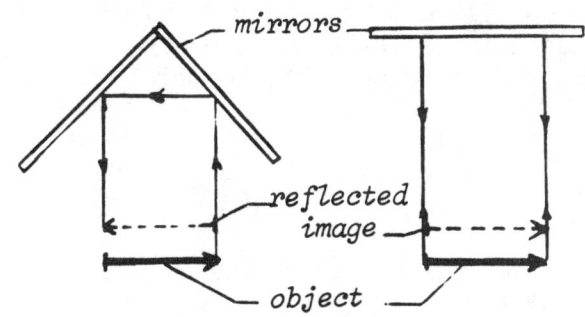

11.6. THE SWOLLEN FINGER

Materials: 1. A plastic vial (about 2 cm diameter)
(an olive jar is quite suitable)

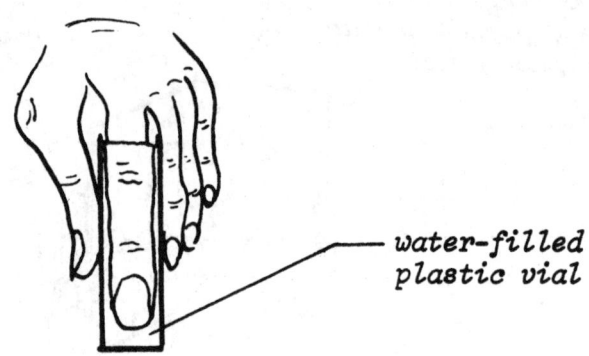

water-filled plastic vial

Procedure:
1. Fill the vial or jar about 3/4 full with water.
2. Immerse your forefinger in the water and show the difference in size with the other fingers.
3. Pour out the water and compare the size of the fingers again without the water in the vial.

Questions:
1. Why does the forefinger look so much larger than the other fingers?
2. Did the finger look larger without the water in the vial?
3. What is the basic cause for magnification?
4. Would we get magnification by looking through a rectangular jar?
5. What other liquids can we use in the vial?

Explanation:
   The basic reason why the finger looks so much larger is that the water in the vial acts like a magnifying lens. The shape of the vial resembles the shape of a convex lens. If the container were a flat rectangular container, this lens effect would be lost.
   The material inside the container may be compared with the glass, which the magnifying lens is made of. Other materials that have high **refraction indices** may be used to obtain this same effect.

LIGHT   REFRACTION

11.7. THE REAPPEARING COIN

Materials: 1. An opaque cup (foam or polystyrene).
              2. A coin (or any small object that sinks in water).

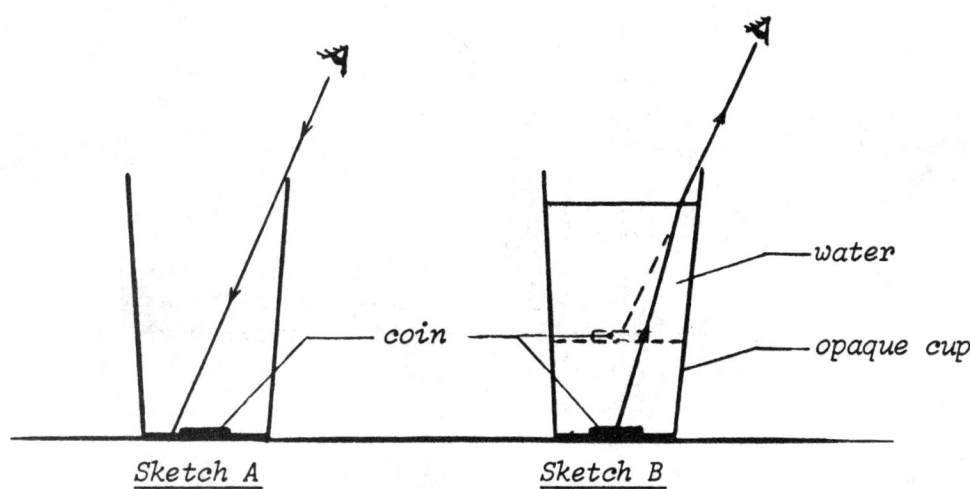

Procedure:
1. Let the students gather around the cup placed on a low stool.
2. Place a coin in the opaque cup and let the students move their heads down from where they can see the coin, so that they just cannot see it (because it disappears behind the rim of the cup), and let them hold their eyes steady at the spot (Sketch A--the coin may be taped to the bottom of the cup, so it won't slide when water is poured in).
3. Pour water into the cup until it is almost full: coin reappears!

Questions:
1. What made the coin disappear before the water was poured?
2. What made the coin reappear?
3. Did you have to move the position of your eyes to see the coin again?
4. What can you tell about the path of the light rays coming from the coin?
5. What other liquids can be used instead of water?

Explanation:
   The eyes were positioned, such that the light rays coming from the coin were blocked by the rim of the cup. By adding water to the cup, these same light rays from the coin now are **refracted away from the normal**, as it travels from a **medium with larger to a medium with smaller refractive index**.

   The **normal** is the line perpendicular to the surface separating the two media through which the light ray travels. Other liquids with high **refractive indices**, usually the denser ones, may replace water in this case.

   This phenomenon is the reason why some deep waters seem to look shallow.

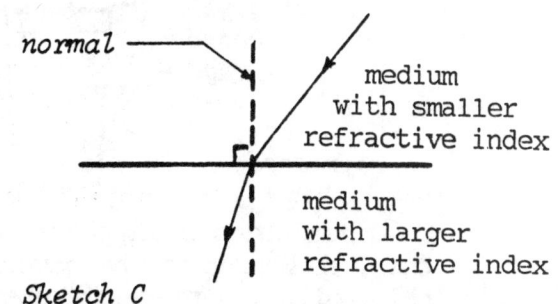

Sketch C

LIGHT — REFRACTION

11.8. THE INVISIBLE PENNY

Materials: 1. Two regular drinking glasses.
            2. A small saucer and a penny (or other coin).

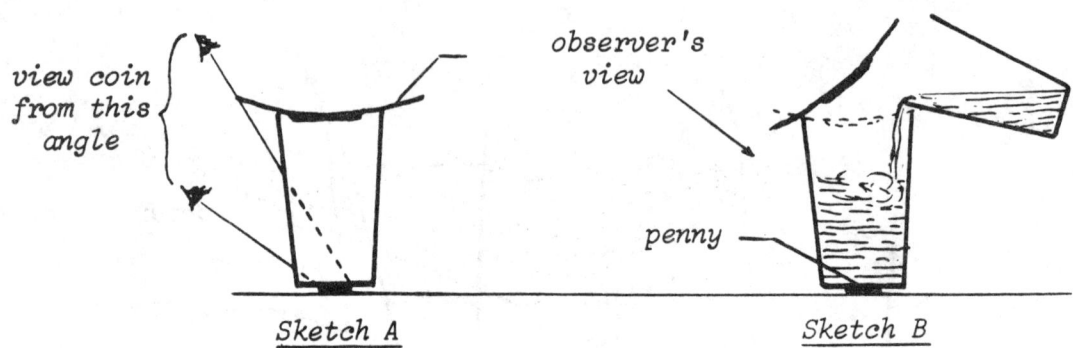

Sketch A   Sketch B

Procedure:
1. Fill one of the glasses almost full with water.
2. Cover the empty glass with the saucer and place the covered glass on the coin. Let the students observe the coin **only** from the side of the glass (Coin can be seen in Sketch A).
3. Fill this glass with water, tilting the saucer towards the observing students (see Sketch B), and place back the saucer.
4. Ask students: "Where did the penny go?"
   Move the glass to the side and reveal penny.

Questions:
1. What made the penny or coin invisible?
2. Where do the light rays of the coin travel?
3. Why do we need a saucer to cover the glass?
4. Why does the saucer have to slant towards the observer?
5. What other objects may be used in place of the coin?

Explanation:
   The coin is still visible with the empty glass and saucer on top of it, because the light rays are not refracted by the air filled glass, and the coin can be seen from the side of the glass. By filling the glass with water, the material through which the rays are travelling is denser and thus, when they hit the side of the glass they are being totally reflected or refracted very closely to the side of the glass. The saucer on top of the glass prevents seeing these rays and thus seeing the coin. The saucer has to be tilted towards the observer, while filling the glass with water, for the same reason.

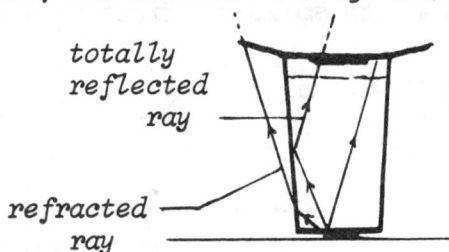

   Other objects in place of the coin would be, anything which is flat and thin, and that can be covered by the bottom of the glass; like a stamp, a thin button, a piece of paper, a washer, etc.

LIGHT — REFRACTION

11.9. THE WATERDROP MAGNIFIER

Materials: 1. A roll of wax paper.
            2. A small cup.

Procedure:
1. Tear off about 10 cm wide strips of wax paper from the roll and distribute this to each student or to each pair of students.
2. Let the students place the wax paper over small print or anything small that they want to see magnified.
3. Come around with water in a cup and put a drop on the wax paper--make a circle about 1/2 cm in diameter.
4. Let the students try out drops with different diameters.

Questions:
1. How can a water drop act as a magnifier?
2. Which drop magnifies better, the smaller or the larger?
3. What property of the drop makes the amplification larger or smaller?
4. What other material could we use to replace the wax paper?

Explanation:
   The water drop on the wax paper stays rounded on the top (**convex**) because of the water-repellent properties of the paper (**hydrophobe**). This is why the water drop resembles the shape of a magnifying glass. The smaller the water drop, the stronger the curvature, in other words the more curved the upper surface is, and thus the stronger the magnification. Any other material that is translucent and hydrophobe may be used to replace the wax paper. If wax paper is not available, overhead transparencies or glass sheets may be used. Regular writing paper might be used which is made translucent. (See "The Peek-a-boo paper," Event 11.16.)

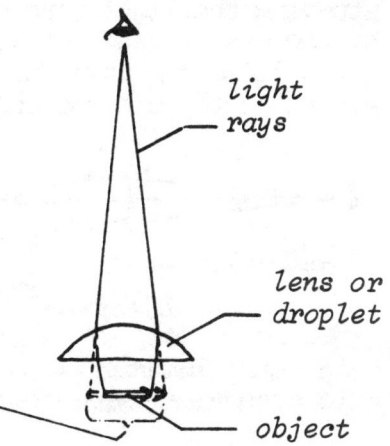

11.10. THE GLASS ROD MAGNIFIER

Materials: 1. Glass stirrers (one for each or a pair of students).
             2. Plastic vials or olive jars.

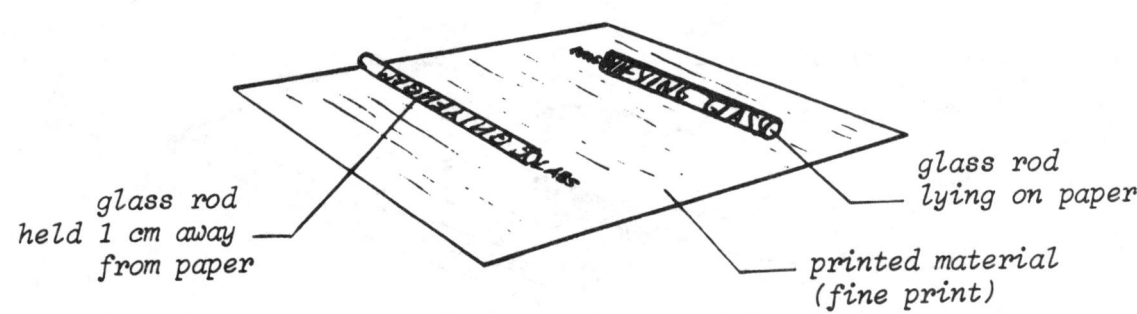

glass rod held 1 cm away from paper
glass rod lying on paper
printed material (fine print)

Procedure:
1. Distribute the glass stirrers to each or each pair of students. If glass rods are not available, fill plastic vials or empty olive jars with water, replace the top, and pass it around.
2. Let students place the glass rod or water-filled cylinder over printed material and look through the transparent cylinder.
3. Now let them hold the rod or cylinder about 1 cm away from the printed material (print is turned upside down).

Questions:
1. What made the rod or cylinder act like a magnifying glass?
2. How far from the printed letters could we hold the rod or cylinder before getting distortions in the letters?
3. What is necessary for the letters to be seen upside down?
4. How are the light rays travelling from the object to the eye?
5. What type of image is the magnified or inverted image?
6. Would a regular magnifying glass also invert the image?

Explanation:
   The best way to clarify the **magnification** and inverting of the image is by studying the light rays coming from the object and the ones that are detected by the eye. These latter are drawn in the sketches below.
   In both instances the image is a **virtual image,** one which is seen by the eye but not projected on the paper (which would be a **real image**).

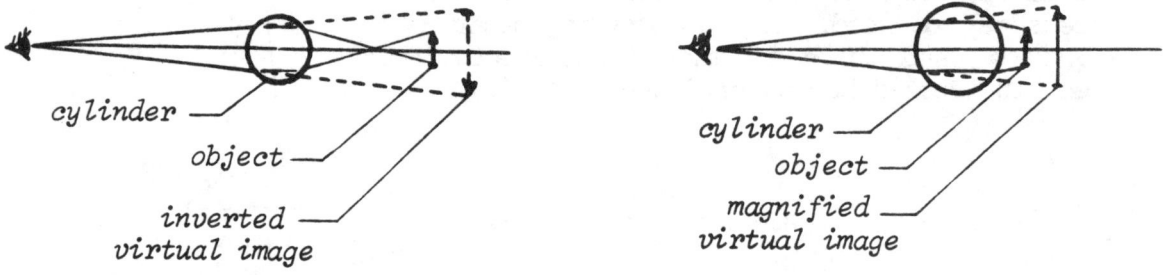

cylinder — object — inverted virtual image
cylinder — object — magnified virtual image

11.11. THE BROKEN PENCIL

Materials:
1. A medium size beaker or glass jar.
2. A pencil (or pen or straw).

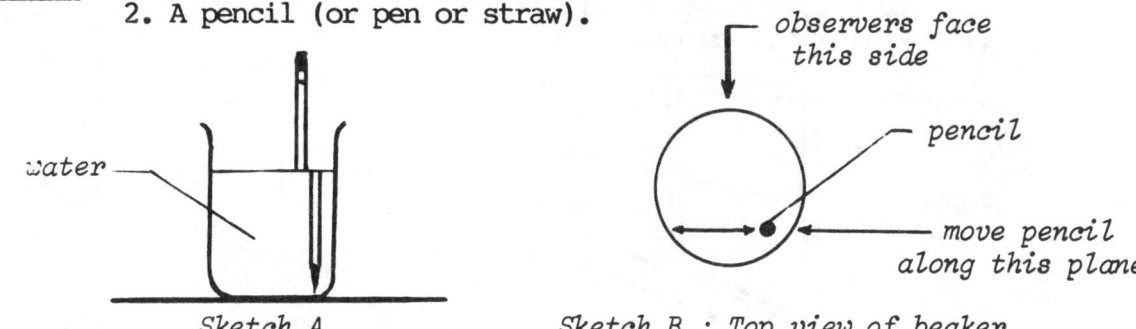

Sketch A

Sketch B : Top view of beaker

Procedure:
1. Fill the beaker or jar about 2/3 full with water.
2. Take a pencil, pen or straw and immerse it vertically about halfway into the water, and say to the students: "I have now broken the pencil (pen or straw)."
3. Move the pencil back and forth along a vertical plane somewhat closer to the demonstrator (see Sketch B) and parallel to the observer's eyes.

Questions:
1. Is the pencil actually broken?
2. Why does the pencil look like it is broken?
3. Would we get the same effect without the water?
4. What other liquids can be used for this demonstration?
5. What other objects can we use to immerse?

Explanation:

The light rays from the pencil travel from a denser medium (water) to a less dense medium (air). This makes the light ray refract **away from the normal** (see Sketch 1), and the pencil is seen at the extension of the refracted ray.

Sketch 1

Sketch 2 : Pencil in center

Sketch 3 : Pencil on other side

When the pencil is in the center of the beaker the ray is not refracted, because the light ray is travelling along the normal (Sketch 2). Sketch 3 shows the light ray from the pencil on the opposite side of the beaker.

LIGHT — REFRACTION

11.12. PRODUCE AN IMAGE OF THE WINDOW

Materials: 1. Two or three identical magnifying glasses.
2. A 3x5" paper card (7 x 12 cm).
3. A 15 cm or 30 cm ruler.

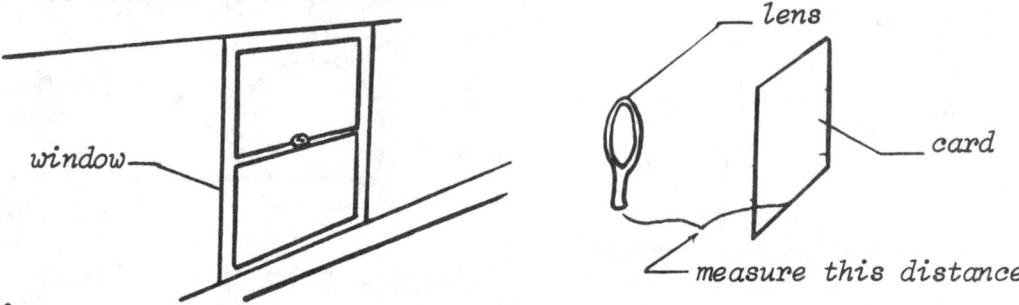

Procedure:
1. Divide the students in groups of three and distribute to each of the groups a set of the above materials.
2. Let one student hold the lens parallel to the window. (If there is no window, hold the lens horizontally parallel with the light in the classroom).
3. Let the second student hold the paper card behind the lens and move the card slowly farther away from the lens until a sharp image of the window (or light source) is formed.
4. As soon as a sharp image is formed on the card, let the third student in the group measure the distance between the lens and the card (image distance).
5. Place two lenses together and see how the image distance changes, and three lenses and see how the image distance changes with the number of lenses (measure from the middle of the combined lenses to the card for the image distance).

Questions:
1. What kind of lens is the magnifying glass?
2. How did the image distance change with the number of lenses?
3. What other objects can we use to produce an image of?
4. Would a near- or far-sighted person use glasses that can produce positive images like the magnifying glass?

Explanation:
Any magnifying lens is either a double convex or single convex lens. Far-sighted people wear convex lenses, and their glasses may be used to produce the same images like magnifying glasses do. As the light rays from the window are almost parallel, the image is formed at the focal point of the lens. Doubling the lens means cutting the image distance in half, as $\frac{1}{f_1} + \frac{1}{f_2} = \frac{1}{f_t}$

in which f_1 = focal distance of lens number 1
f_2 = focal distance of lens number 2
f_t = focal distance of total systems

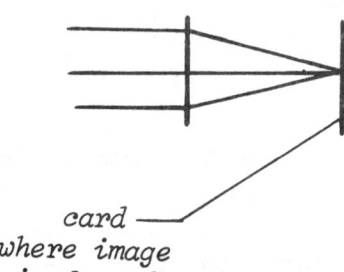

LIGHT

REFRACTION
TOTAL REFLECTION

11.13. WHY DO WE SEE TWO COINS?

Materials: 1. A penny or nickel (or any other coin).
2. A clear colorless drink glass, or beaker.

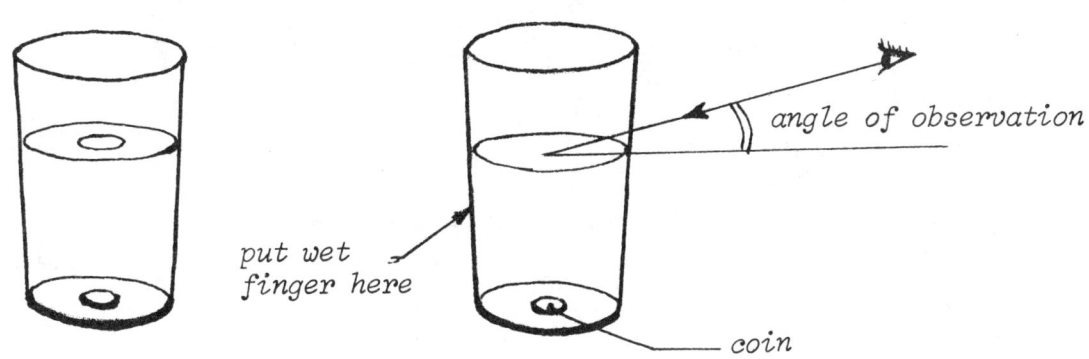

Procedure:
1. Fill the glass about three quarters full with water.
2. Drop the coin in the glass and maneuver the glass till the coin sits in the middle of the bottom.
3. Let the glass with the water and coin now stand still on a table and observe the surface of the water from about a 15-20 degree angle.
4. Touch the far side of the glass with a dry finger: does anything happen to the image of the coin?
5. Wet your finger, then touch the far side of the glass with the wet finger: what happens to the image of the coin?

Questions:
1. What makes us see another coin on the surface of the water?
2. Does the image of the coin get inverted?
3. What do we see when the angle of observation is larger?
4. Why is the image of the coin disappearing when the glass wall is touched with a wet finger?
5. Where is reflection occurring, and where refraction?

Explanation:
There is **total reflection** of the image of the coin occurring at the back wall of the glass, then a **refraction** occurs at the surface of the water.

If a wet finger is placed on the back wall where the total reflection is supposed to occur, a layer of water is actually placed on the outside of the glass. This equalizes the refraction indices on both sides of the glass, and total reflection is eliminated.

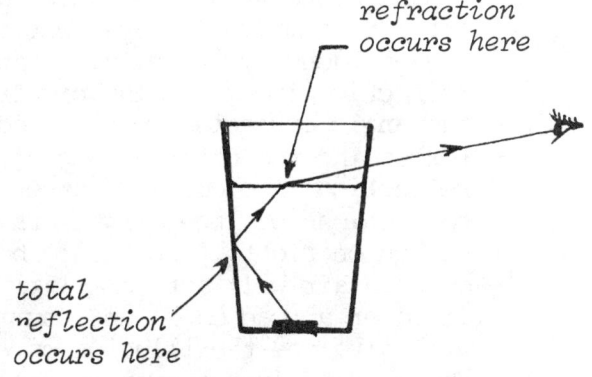

LIGHT

REFRACTION
TOTAL REFLECTION

11.14. USE WATER AS A MIRROR?

Materials: 1. A small candle & matches.
2. A 400 ml beaker or regular large drink glass.

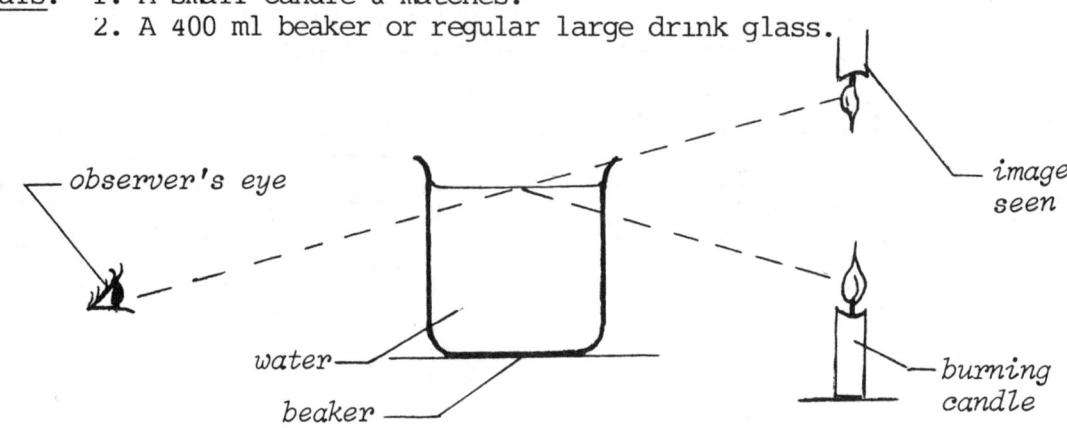

Procedure:
1. Light the candle and place it on a shelf or stand at about eye level in a dark corner of the room (when sitting it should be somewhat lower but still at eye level).
2. Fill the beaker or glass full with water.
3. Hold the beaker a little higher than the candle, and look up against the water surface, such that the candle flame will reflect in it - the height of the beaker may need adjusting in order to see the reflection of the candle flame (see Sketch above).

Questions:
1. Why do we see an inverted candle flame image?
2. How can water give a reflection like a mirror?
3. Why do we need a dark background to see the reflection?
4. What other substances will act similarly like the water?
5. What property was different between the water and the air?
6. Would different gases or gases of different densities be able to act like a mirror?

Explanation:
The boundary between water and air act like a mirror because water has a higher density than air and its **refraction index** is also higher. When the **angle of incidence** is too large (exceeding the **critical angle**), **total reflection** occurs rather than refraction of the light rays (see Sketch on right). Hot air bordering on cold air will act similarly. Hot air over a road on a hot summer day will give one the illusion of water on the road. Mirages over a desert are formed in the same way.

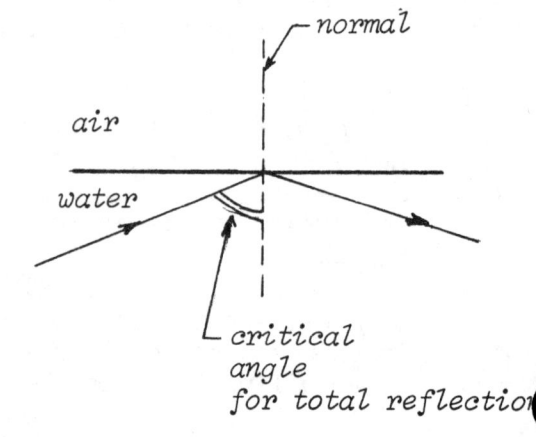

LIGHT

REFRACTION
TOTAL REFLECTION

11.15. THE REFLECTING BRICK WALL

Materials: 1. A long (minimum 10 m) wall on a hot sunny day.
2. A bright metallic object (like a shiny key).

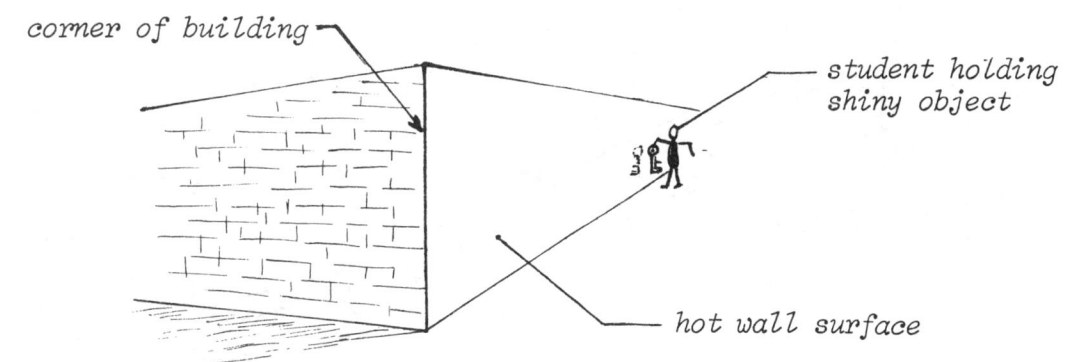

Procedure:
1. Take the class outdoors on a bright sunny day. Stand on one end in front of a brick or concrete wall against which the sun is shining.
2. Let one of the students hold a shiny metallic object, like a key or knife, about 5 to 10 cm from the wall, such that you can see it.
3. Observe an image of the object in the wall! (If you don't see the image, move in closer to the wall or let the student move the object closer to the wall).

Questions:
1. What made the wall act like a mirror?
2. Why does it have to be a sunny day, where the sun shines against the wall?
3. What other surfaces would act in the same manner?
4. In what other situations may we encounter the same phenomena?
5. Why would a shiny object work better than a dull object?
6. Why can we only see a reflection when we are so close to the wall?

Explanation:
When the sun shines against the wall, it heats it up and the hot wall heats the air surrounding it. This hot layer of air has a different **index of refraction** as compared to cooler air. It therefore acts as a different medium against which light can be totally reflected. This angle of total reflection occurs only at very large angles of incidence. This is why we have to be so close to the wall to see the reflection of the shiny object. A shiny object works better because it can easily be seen from such a distance.

This same phenomena occurs in the mirages that we see over deserts. The reflecting medium in this case is colder air, which hovers above the hot air over the desert sands. Another example is when we are driving on a hot day, we may see far off on the road a shiny spot as if it was wet or oily, but on approaching it, it suddenly disappears. Here the hot air above the road acted like a mirror, giving total reflection only at large angles of incidence.

LIGHT

REFRACTION
SCATTERING

11.16. MAKE RAINBOW COLORS

Materials: 1. A triangular prism (glass).
2. A blank piece of paper and sun rays.

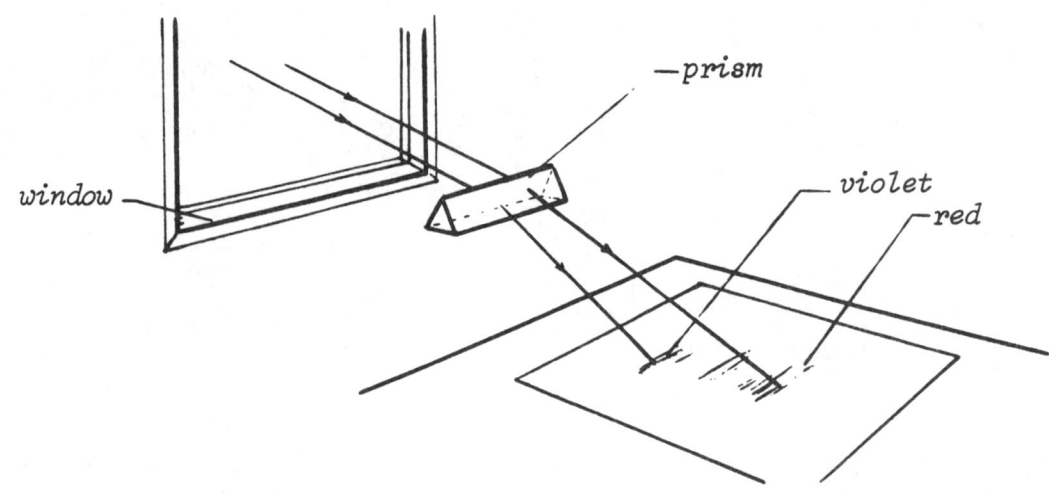

Procedure:
1. Place a blank sheet of paper on the table top where the sun rays shine through the window.
2. Hold a triangular glass prism in the sun rays and turn it around its lengthwise axis until rainbow spectrum falls on the blank paper.

Questions:
1. How can white light give the colors of the rainbow?
2. What are the colors of the rainbow?
3. Which color is bent most? (farthest away from the straight line?)
4. What source can we use instead of sun rays?

Explanation:
White light is made up of the rainbow colors: violet, blue, green, red, yellow and orange. The violet is **refracted** or bent most, and the red is bent the least (see sketch on right). This is why the sky stays blue when the sun sets in the west and becomes red. The moisture in the atmosphere would act as prisms and scatter or **separate** the blue light out. Viewing the sun rays from 90° angle, the color will appear blue (sky) whereas looking directly into the setting sun, the color is red.

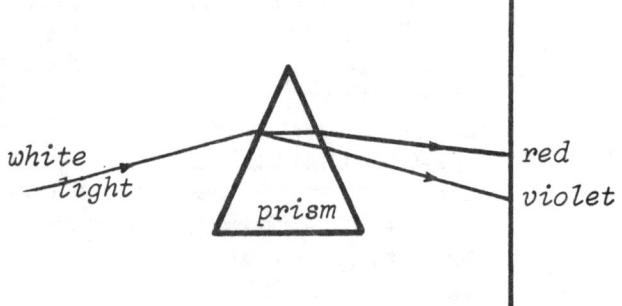

LIGHT

REFRACTION & SCATTERING
TOTAL REFLECTION

11.17. MAKE YOUR OWN RAINBOW

Materials: 1. A garden hose + sprayer nozzle (connected to water outlet).
2. Bright sunlight (when sun is low in the sky).

Procedure:
1. Take the class outdoors on a bright sunny day. Attach your garden hose to the outdoor tap, hold the spray nozzle in your hand and position yourself such that you face away from the sun (the sun behind you).
2. Squeeze the spray nozzle and produce a fine spray of water in front of you, and Voila! A beautiful rainbow!

Questions:
1. What are the conditions for seeing a rainbow?
2. When is a rainbow actually seen in nature?
3. What makes the different colours in the rainbow?
4. What colour is on top, what colour on the lower side?
5. Why is the rainbow always circular/ a segment of a circle?
6. Why is the rainbow never a square, triangle, or straight line?
7. Why does the sun always have to be behind us to see a rainbow?
8. Where do we see rainbows, besides in the sky?

Explanation:
The conditions for seeing a rainbow are: the sun has to be behind us, we have to look away from the sun towards fine droplets in the sky (clouds after a rain). These droplets can be a few meters, a few hundred meters or a few thousand meters or kilometers away from us.

The rainbow is always a segment of a circle, because it is a part of the circular base of a cone. These are the only droplets that we can see because of the **critical angle** that gives a **total reflection** of the sun ray falling into it. The blue colour is always on top since it is the most refracted one (with the largest refraction). Other places where we could see rainbows are: fountains, waterfalls in bright sunlight (provided we stand in the right position).

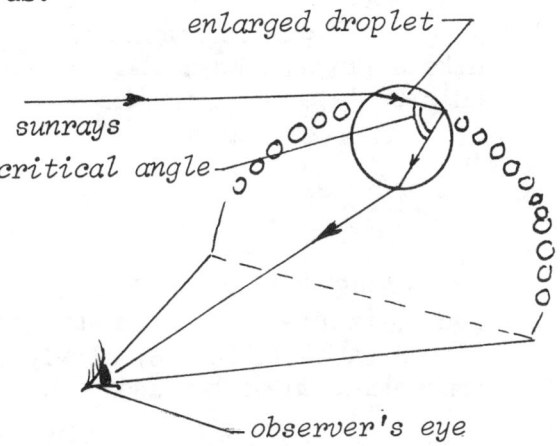

LIGHT

REFRACTION SCATTERING

11.18. THE SIMULATED SUNSET

Materials:
1. A beaker or glass jar.
2. Sodium thiosulfate ($Na_2S_2O_3$) and concentrated sulfuric acid (H_2SO_4).

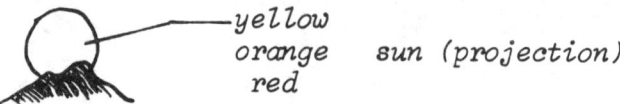

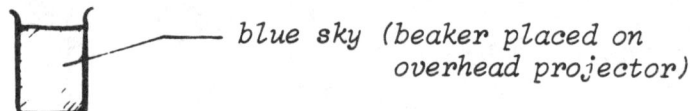

Procedure:
1. Cut a circular hole in a large piece of paper with the same diameter as the beaker or jar, place it on the overhead projector and a water-filled beaker over the hole.
2. Dissolve a scoop of sodium thiosulfate in the water until no crystals can be distinguished in the water.
3. Pour a few ml of concentrated sulfuric acid in the solution, stir, stand back and observe.
4. When the color of the beaker (seen from the side) starts to turn bluish draw students attention to the color of the sun (projection) and the color of the sky (side view of beaker). (When the sun's color gets orange-red, a torn piece of paper may be pushed under the beaker to simulate a mountain.)

Questions:
1. How do you think the bluish color in the beaker was created?
2. Why was the sideview of the beaker blue and the projection of it yellow or orange?
3. How is this demonstration similar to the actual sunset?
4. Why does the sun get red when it sets in the west?

Explanation:
White light is **dispersed or scattered** into the colors of the rainbow if it hits a prism. When the sun sets in the west we are looking through a much thicker layer of atmosphere compared to noon time. This means that when we look to the west, we are looking through more dust and moisture particles. Each of these particles acts like prisms, scattering white light into red at one end of the spectrum and blue at the other end. Blue has the shortest wavelength and refracts with the largest angle. The reaction of $H_2SO_4 + Na_2S_2O_3$ precipitated sulfur (S) slowly out of the solution, simulating the increasing concentration of the dust and water particles.

11.19. THE CONFUSED FLASHLIGHT

Materials:
1. A strong flashlight.
2. A large beaker & medicine dropper & stirrer.
3. A few ml of regular cow milk.

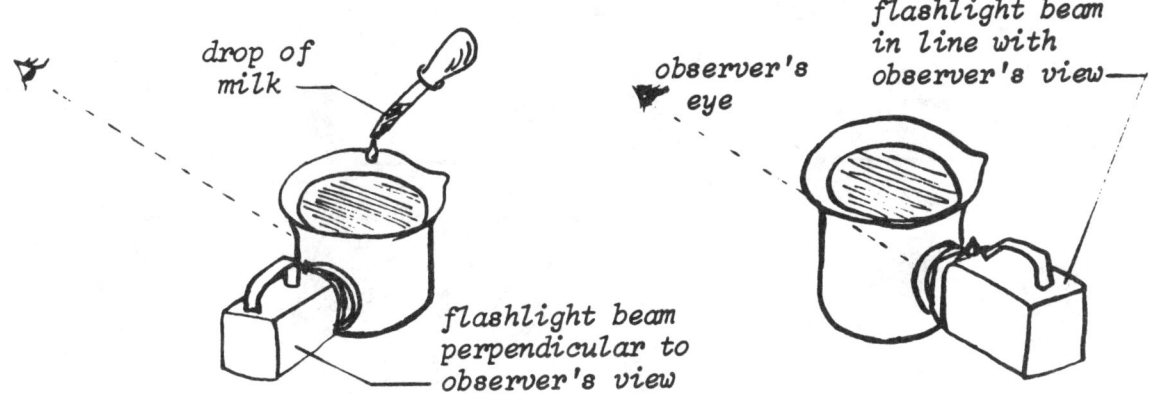

Procedure:
1. Fill the beaker nearly full with water.
2. With the medicine dropper, place one drop at a time of milk in the water and stir.
3. After each drop of milk added to the water, shine the flashlight through the water by holding it on the side of the beaker: let the light beams shine perpendicular to and into the student's view.
4. Keep adding the drops of milk to the water until the light shone into the student's view becomes red.

Questions:
1. What made the color of the light change?
2. What purpose does the milk have?
3. If the water were the earth's atmosphere and the flashlight were the sun what can the milk droplets be equated with?
4. Which of the colors do you think are refracted most? which the least?

Explanation:
Increasing the milk droplets in the water is like increasing the water and dust particles in the atmosphere. When the sun moves to the west approaching sunset, the atmosphere through which we are looking is much thicker. This means that we are actually looking through more dust and water particles. When we look to the west during sunset, we look directly into the sun rays, seeing the sun as a red ball. Looking up towards the sky when the sun is in the west, still leaves the sky blue. The sun rays are now perpendicular to the direction of our view. This difference in color is caused by the refraction of the blue light. The shorter the wavelength (blue), the more it is bent or refracted by milk droplets. The light with the longer wavelength (red) is not as much refracted and this is why the light from the flashlight is red when we look directly into it (through the beaker with milk). (See event 11.14.)

LIGHT

**TRANSLUCENCY vs OPAQUENESS
REFLECTION**

11.20. THE PEEK-A-BOO PAPER

Materials:
1. Any kind of writing or typing paper.
2. Regular frying oil (motor or machine oil will do).
3. Paper towel.

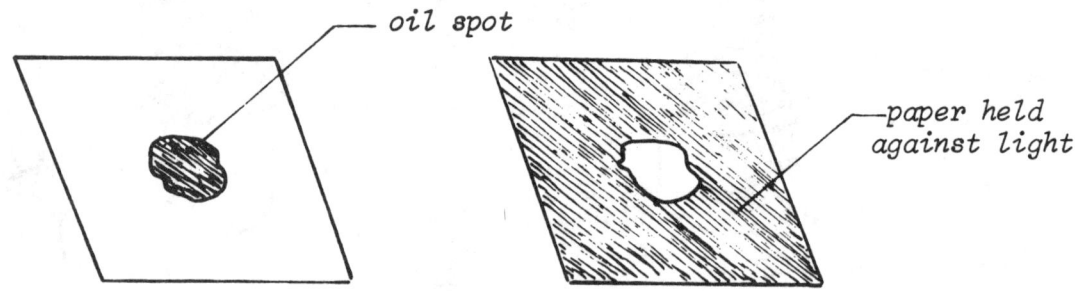

Procedure:
1. Distribute sheets of paper to each one of the students.
2. Dip a corner of paper towel in the oil and come around to each of the students.
3. Make an oil spot in the center of each of the paper.
4. Give the students some paper towel, and let them wipe the excess oil off the oil spot.
5. Let the students look through the spot by placing the paper over printed or written material.
6. Let the students hold the paper against the light and compare the lightness and the darkness of the spot with the rest of the paper.

Questions:
1. How can oil make opaque paper translucent?
2. What does it mean in terms of light rays when something is opaque or translucent?
3. Why is the oil spot dark compared to the rest of the paper when the light falls on the paper?

Explanation:
   The oil **penetrates** all the pores and fibers of the paper when it touches it. This makes the paper completely **impregnated** with oil and thus more uniform for the light to pass. Thus the light rays penetrate the oil spot and get **reflected** by the rest of the paper. This is why the oil spot looks darker, because the light rays are not reflected back. When the light source is behind the paper the opposite happens: light passes through the spot and it hits the observer's eye, whereas the rest of the paper blocks the light and appears darker.

CHAPTER 12

WHAT ARE THE PROPERTIES OF SOUND?

OBJECTIVES

After dealing with and studying the concepts and sub-concepts in this chapter, the students should be able to:

a. recognize the correct explanation of an observed event based on each of the sub-concepts;
b. explain in their own words which of the sub-concepts is determining the course of an event.
c. distinguish true from false statements concerning each one of the sub-concepts;
d. identify the correct explanation of an event in daily life applying one of the sub-concepts;

all in relation to the following sub-concepts:

-- Sound is produced by vibrating objects.
-- The loudness of sound is determined by the distance form the source, the intensity of the vibration, and the surface area of the vibrating object.
-- The pitch of note is determined by the mass of the vibrating object: the larger the mass, the lower the pitch.
-- In resonance, the resonating object has the same frequency of vibration as that of the sound source.
-- Sound does not travel through a vacuum.
-- Sound travels faster through liquids and even faster through solids compared to gases.
-- The speed of sound through air is around 340 m/s.

12.1. WHICH AMPLIFIER WORKS BEST?

Materials: 1. A cheap polystyrene comb for each student.

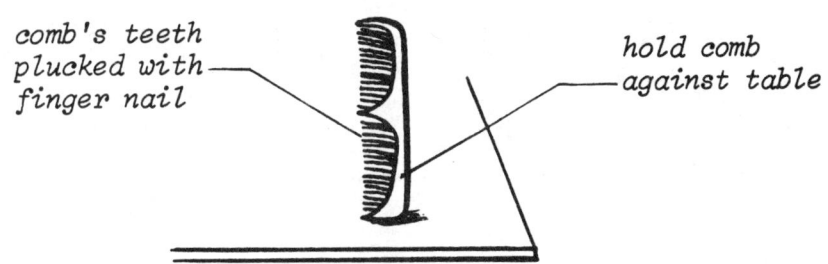

Procedure:
1. Hold a comb in the air and pluck the teeth with your fingernail.
2. Ask students: "How can I amplify the sound?"
3. Hold the end of the comb against the table top and pluck again.
4. Distribute different combs to each of the students and let them try to hold it against different material in the class and listen which of them gives the best amplification.
5. Point out to the students that the different lengths of the teeth give different pitches (simple tunes might be plucked by the teacher).

Questions:
1. How did the sound get amplified?
2. Why do we hear the sound louder when the comb is held against other materials?
3. Which factors are influencing the loudness of a sound?
4. What material is the best amplifier for the comb?
5. What made the different pitches in the plucking of the teeth?

Explanation:
In holding the comb against the table top while plucking the comb's teeth, it was not only the teeth and the comb that were vibrating but the whole table top vibrated with it. This means an increased surface of the vibrating object, which is the main reason for an increasing loudness or an amplification. The factors that are affecting the loudness of a sound are: 1. distance from sound source; 2. the amplitude of the wave, i.e., how hard the teeth are plucked; 3. the surface area of the vibrating object.

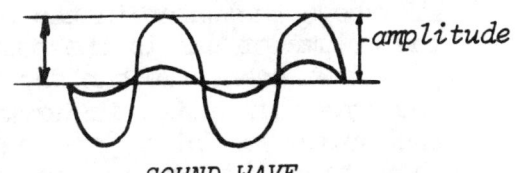

SOUND WAVE

In general wood sheets and wooden boxes are the best materials for quality in sound amplification. This is why sound boards (for the piano), guitar and violin bodies, and speaker boxes are all made out of wood.

12.2. THE REVERSING PITCH

Materials: 1. Three identical empty bottles.
2. A wooden stick or ruler.

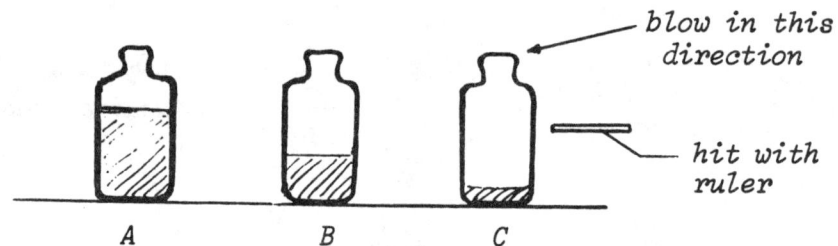

Procedure:
1. Fill the three identical bottles with different amounts of water, and ask the students: "Which will give the highest tone when I hit it on the side with the ruler?"
2. Lift the bottle with 2 fingers at the neck and hit the side of each bottle with the ruler and listen to the pitch of the tone.
3. Now ask the students: "Will bottle C always give the highest pitch?"
4. Blow over the bottle mouth, by placing your lower lip on the bottle rim and blowing a narrow stream of air almost horizontally over the mouth. Listen to the tone/pitch.

Questions:
1. What was vibrating when the bottles were hit on the sides?
2. What produced the tone when the bottles were blown into?
3. What is generally affecting the pitch of a tone?
4. Would different liquids of the same height in the bottles affect the pitch of the tone? When hitting the bottle? When blowing over the bottle opening?

Explanation:
When the bottles were hit on the side with the wooden ruler, the bottle and the water vibrated to produce the sound pitch. The more water there was in the bottle, the lower the pitch. When blowing over the bottle mouth, the air above the water in the bottle vibrated to produce the tone. The larger the column of air in the bottle, the lower the pitch. If we generalize over these two events we can say that **the more the mass of the material vibrates, the lower the pitch it produces.** This is why the lower and deeper tones on the guitar or piano are coming from the larger and longer strings. By shortening the strings (by placing the finger on a fret), the higher the pitch obtained.

By using different liquids in the bottles, but filling them to the same height, the pitch would probably be different when hit on the side (because of the different masses of the liquids), but stay the same when blown over the mouth (because the air column stays the same).

12.3. THE STRAW OBOE

Materials:
1. Two plastic drinking straws.
2. A pair of scissors.

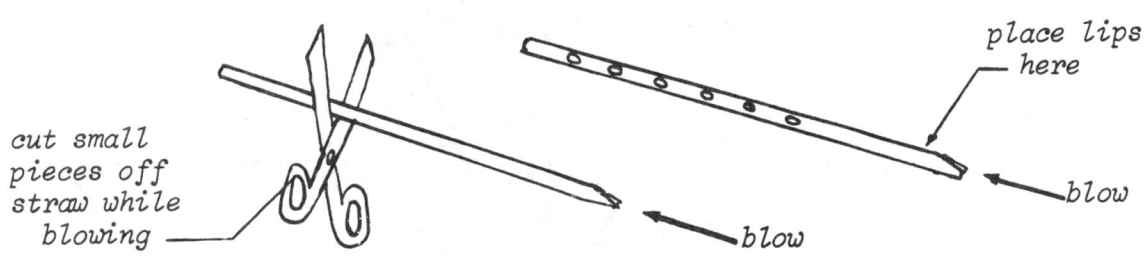

Procedure:
1. Cut small holes on one side of a straw, about 1 to 1.5 cm apart from each other.
2. Flatten one end of both straws and cut triangle pieces out.
3. Place this end of the straw with holes in the mouth, holding your lips just where the cut ends, and blow until an oboe sound is produced. (Shifting the straw somewhat in or out of the mouth might help obtaining the sound.)
4. Place three fingers of the left hand on the farthest three holes and three fingers of the right hand on the nearest three holes of the straw; and open or close holes for the different pitches.
5. Now blow the straw with no holes. Make the oboe sound, take the pair of scissors and cut small pieces off the straw while blowing.

Questions:
1. What was actually producing the sound?
2. What pitch did we get when all holes were closed, compared to when all holes were open?
3. What does opening or closing a hole in the straw really mean in terms of vibrating an air column?
4. Which musical instruments are based on this principle?

Explanation:
   This demonstration especially shows that the pitch of a note is determined by the length of vibrating air. By cutting the end of the straw we are actually making two reed-like protrusions, which when air blown through them, will vibrate and produce the oboe sound. The air column in the straw is vibrating with it and produces the pitch. With the holes in the straw we are able to make the vibrating air column longer or shorter, by either closing or opening a hole or several holes at a time.
   This principle is applied in the flute, clarinet, oboe, saxophone, and their varieties in soprano, alto, tenor and bass instructions.

SOUND   PITCH

12.4. PLUCK A RUBBER BAND

Materials: 1. A small medium thick rubber band.

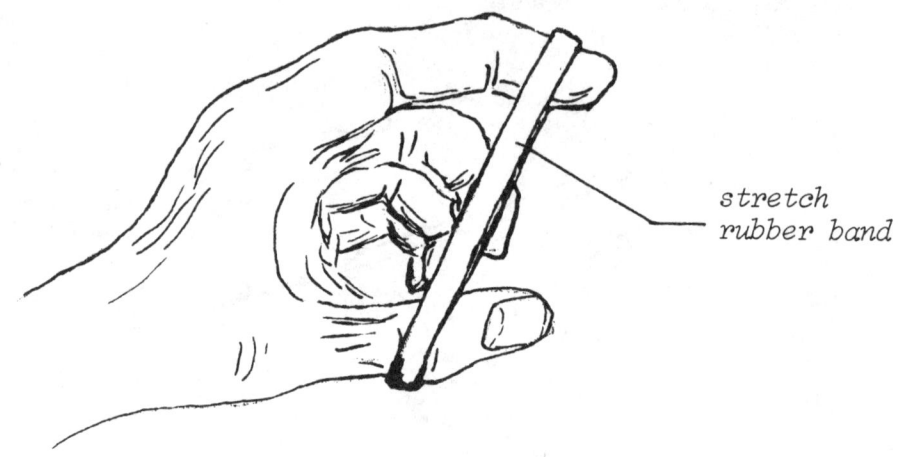
stretch rubber band

Procedure:
1. Place the rubber band between thumb and forefinger, and stretch it a little.
2. Hold it close to your ear and pluck it with the other hand.
3. Stretch the rubber band by widening the gap between thumb and forefinger and pluck again. What pitch do you hear now?

Questions:
1. Did the pitch of the sound change after widening the gap?
2. Which properties of the rubber band changed while being stretched?
3. What property changes when a guitar string is tightened?
4. Which properties do not change when a guitar string is tightened?
5. How do these properties compare to those of the rubber band?
6. In what ways can we change the pitch of a guitar string?
7. How could we change the pitch of the rubber band?

Explanation:
When the rubber band stretched and was plucked again, the pitch of the sound was staying about the same, if not getting a little lower. This is definitely contrary to what one would expect, which is a higher pitch for the stretched rubber band.

When a guitar string is tightened, the pitch becomes higher, because the tension is higher, thus the string vibrates with a higher frequency; the length and the density of this string, however, is staying constant.

With the stretching of the rubber band, all three properties: tension, length, and density change. The higher tension tend to increase the pitch, but this is compensated by the increase of the length, which tends to lower the pitch.

When the length of the rubber band is held constant, the pitch changes similarly like a regular guitar string. This can be done by stretching the rubber band over an empty open box.

SOUND

RESONANCE
& PITCH

12.5. THE PIPE AND STRAW TROMBONES

Materials:
1. Two rigid plastic tubes (one slightly larger in diameter than the other so that they can slide into each other, about 40-50 cm length each).
2. Two straws (plastic drinking), with the same specifications as above.
3. A propane torch or Bunsen burner.

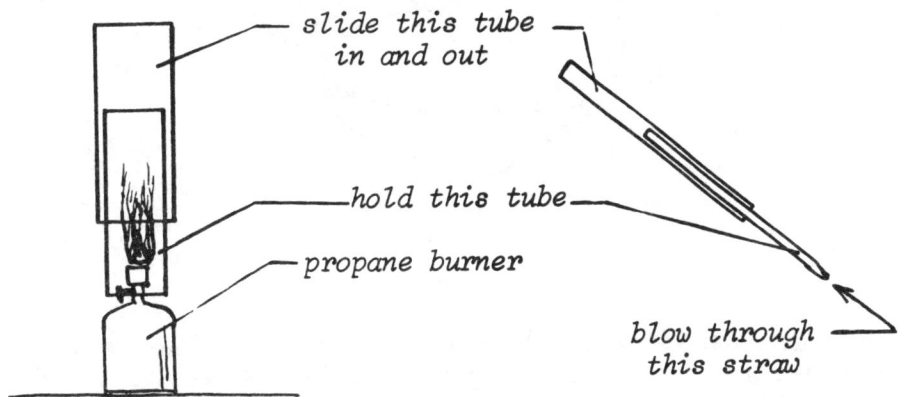

Procedure:
1. Light the propane torch or Bunsen burner and turn it to the largest setting with the air valve open.
2. Hold one of the rigid tubes vertically and lower it slowly over the flame until a resonating tone is produced.
3. Slide the larger tube up and down over the first one while holding them both over the flame (trombone effect).
4. Flatten one end of the straw and cut small triangle pieces out of each side to make a straw oboe (see Event 12.3).
5. While blowing through this oboe, producing the tone, slide the other straw over the first one in and out (trombone effect).

Questions:
1. How was the resonating tone in the rigid tube produced?
2. What changed the pitch in the sliding tubes?
3. What produced the tone in the straw oboe?

Explanation:
The heat of the flame in the tube suddenly expands the air in it and the air begins to **oscillate** up the tube, resulting in a **resonating** tone: a **standing wave** is created in a tube with open ends. The longer the tube, the longer wavelengths are produced in the standing wave and thus the lower the tone. The same goes for the sliding straws, except that the sound is produced in the standing wave and thus the lower the tone. The same goes for the sliding straws, except that the sound is produced by the **vibrating reeds**. These are made to vibrate by blowing the air stream between them.

12.6. THE SINGING GLASS

Materials:
1. A wine glass (or any other thin drinking glass with a smooth rim).
2. A little vinegar or dilute acid.

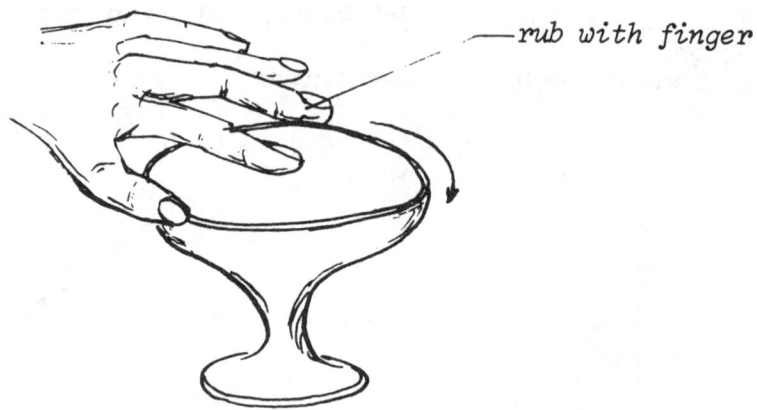
— rub with finger

Procedure:
1. Wash your hands with soap and place the wine glass on the table.
2. Dip the fore- or middle-finger in a little vinegar, hold the wine glass steady with the other hand (hold base down on table) and rub the rim of the glass with the dipped finger.
3. The rubbing should be done rather slowly (if the finger slips without giving off a sound, then either the glass rim is greasy or your finger is not quite clean).
4. Pour a little water in the wine glass and rub again. Notice the pitch change.

Questions:
1. What is the purpose of the vinegar?
2. Why do the hands and glass have to be clean?
3. What is the finger actually doing when it rubs the glass rim?
4. What is the pitch doing when water is added to the glass?

Explanation:
When the finger is rubbing against the rim of the glass, it is not supposed to just slide over it, but it catches and slides in very quick succession. It is just like hitting the side of the glass with a hard object, softly and in very quick succession. It almost works like a bell hammer. As soon as the first few vibrations are produced, they make the glass **resonate**, because the pitch produced is exactly the same as that of the glass itself.

By adding water to the glass, the vibrating mass increased and thus the pitch got lowered.

SOUND — RESONANCE

12.7. THE TUNED-IN WASHERS

Materials: 1. Two upright stands and 7 identical washers.
            2. Light string or sewing thread.

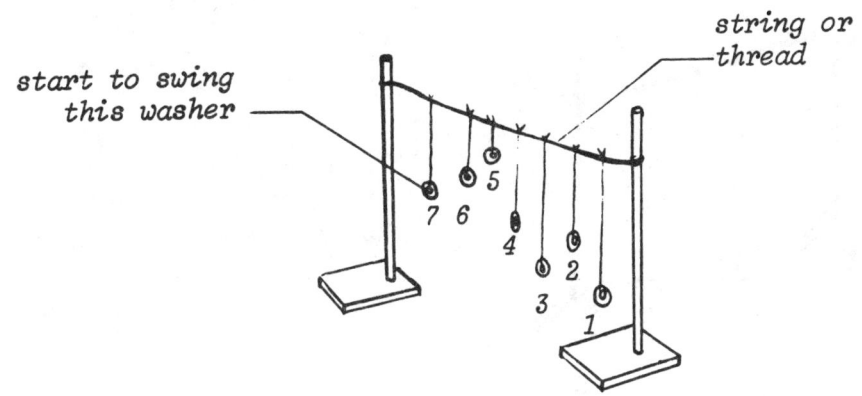

Procedure:
1. Place the two upright stands about 50 cm apart and attach a string between them at the same height.
2. Now tie the short lengths of string to the seven washers.
3. Hang the washers on the horizontal string at the following lengths:
   washer No. 1 - 20 cm.   No. 5 - 5 cm.
   washer No. 2 - 15 cm.   No. 6 - 10 cm.
   washer No. 3 - 20 cm.   No. 7 - 15 cm.
   washer No. 4 - 15 cm.   (somewhat farther apart).
4. Start to swing washer No. 7 and observe the other washers.

Questions:
1. Why did washer No. 2 and No. 4 swing with No. 7?
2. If washer No. 2 were swinging first, which one would swing with it?
3. If washer No. 1 were swinging first, which one would swing with it?
4. How can we compare this event with a sound source and resonating objects?

Explanation:
    This demonstration is another excellent illustration of the phenomenon of **resonance**. When washer No. 7 starts to swing, washers Nos. 2 and 4 will swing with it. Washer No. 7 can be compared with the source of vibration and washers Nos. 2 and 4 are **"tuned in"** with or have the same **frequency of vibration** as washer No. 7. The same happens when washer 2 or 4 is started with the swinging first. The washers that have the same string length will swing with the original "swinger." So when washer No. 1 is started to swing, then only washer No. 3 will swing with it.
    In resonance, the resonating object has to have the same frequency of vibration as the source. Sometimes the source of vibration can be the object itself and then it will increase its own vibrations. This was the case with the collapsing Tacoma bridge in Washington state, which started to vibrate because of a high wind that blew against the bridge.

SOUND   RESONANCE

12.8. THE SWINGING BOOK

Materials: 1. A paperback pocketbook.
2. A horizontal bar stand and string.

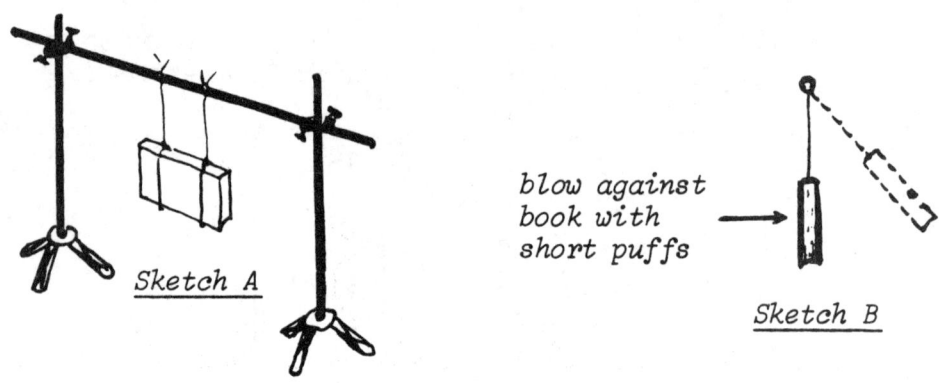

Sketch A

blow against
book with
short puffs

Sketch B

Procedure:
1. Tie two strings to the paperback pocketbook and hang it from a horizontal bar stand (see Sketch A).
2. Hold the book about 45° from the vertical line out and ask the students: "Would I be able to blow against the book until it is this far out?" (Anticipated answer: "NO").
3. Let the book hang back vertically and blow with huffs in phase with the swings of the book. (Just like pushing someone on a swing in the park.)

Questions:
1. Why is it not possible to blow the book to 45° from the vertical line with one huff?
2. What is most important to get the book swinging higher and higher with the huffs?
3. How can we stop the book from swinging without touching it?
4. How can we compare the swinging of the book and the blowing, with sound and resonance?

Explanation:
This is an excellent demonstration to illustrate what **resonance** actually is. When we deal with resonance there is always a source or **origin of vibration** (sound) and an object which is **resonating** with the source of vibration.

The huffing against the book can be compared with the source of vibration and the swinging of the book as the object which resonates with the source. When the huffing is in phase with the swinging book, it will swing higher and higher. In sound terms: when the sound source has the same **frequency** as the vibrating object, it will resonate with the sound source. In order to stop the swinging of the book, the huffing has to be done out of phase or in opposite phase with the swinging. Again, in terms of sound we would say: to stop the object from resonating, the source has to have a different frequency that the vibrating object, or the object should vibrate with a different frequency than the source of vibration.

SOUND   RESONANCE

12.9. THE SINGING BOTTLE

Materials: 1. Three or four identical bottles.
             2. A tuning fork.

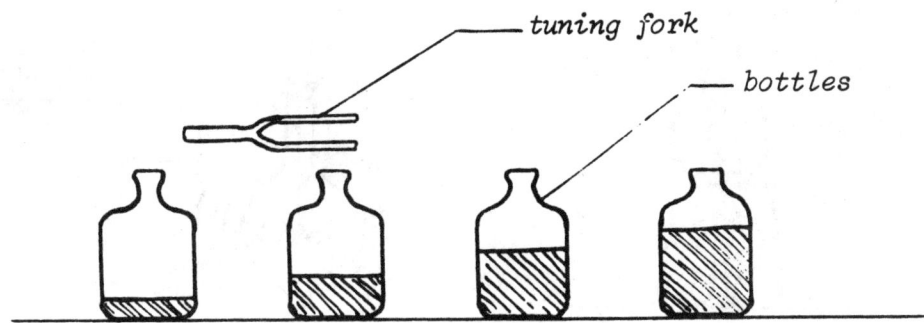

Procedure:
1. Place the four bottles on the table and fill them with different amounts of water.
2. Strike the tuning fork and listen to the tone.
3. Fill one of the bottles with just enough water, so that the air column above the water (when blown into) will vibrate with the same tone as the tuning fork.
4. Place this bottle back in row with the other bottles, strike the tuning fork and hold it over each of the four bottles (stopping a second above each one, then moving on to the next).

Questions:
1. Which bottle is "singing" with the tuning fork?
2. Why are the other bottles not "singing" with the tuning fork?
3. How can we make the other bottles "sing" too?
4. What is it that makes the bottle resonate with the tuning fork?
5. Where do we find this event applied in daily life?

Explanation:
Since only one bottle's water height was adjusted such that the air above it would vibrate with the tone of the tuning fork, this bottle is the only one that is **resonating** with the tuning fork. The columns of the other bottles were either too long or short to "sing" or **resonate** with the tuning fork. The **frequency** of the tone in these latter bottles is not the same as that of the tuning fork, and thus they do not vibrate with it. For a further explanation of resonance see Events 12.6. and 12.7.

An application of this event in daily life is encountered when a certain tone is sounded in front of a guitar or open grand piano; only certain strings that have the same pitch or octaves thereof, will resonate with the original tone or pitch.

SOUND MEDIUM OF TRAVEL

12.10. THE SOUNDLESS BELL

Materials:
1. A round-bottomed flask (or flat-bottomed).
2. A small bell (that fits through the neck of the flask).
3. A rubber stopper and short piece of iron wire.
4. A Bunsen or propane burner & stand, a cloth.

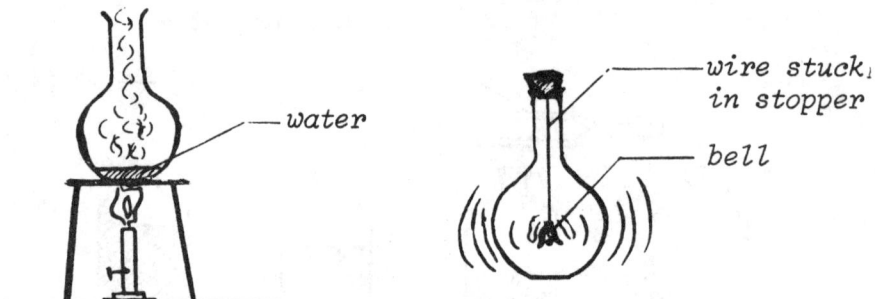

Procedure:
1. Attach the small bell to the wire and stick the other end in the rubber stopper. Adjust the length of the wire so that the bell will hang about in the center of the flask when closed off.
2. Close the flask off with the stopper and bell, shake the flask and let the students listen to the tinkle of the bell (a faint high note should still be detected).
3. Take the stopper off, pour about 20 ml of water in the flask, and heat it over the Bunsen or propane burner and stand until boiling.
4. Let the water boil for a minute or so and then shut the flame off, and immediately insert the stopper with the attached bell (hold the flask with the cloth to prevent burning of fingers).
5. Let the flask cool off for a minute, shake the flask again and try to detect the same faint high note of the bell.

Questions:
1. Why did the ring of the bell disappear after boiling water in the flask?
2. What is the purpose of boiling water in the flask?
3. If sound does not travel through a vacuum, could we hear rocket exhaust in outer space?
4. Why can sound not travel through a vacuum?

Explanation:
In order for human beings to hear a sound, sound waves or vibrations have to reach their ear drums, whether it is through a gas, a liquid or a solid. There has to be a **medium** in order for waves to travel from one place to the other. We hear sound and noises around us, because air is present everywhere. In the flask we created a partial vacuum. By boiling water, steam was produced and this pushed most of the air out of the flask. After the flask cooled, the steam condensed back into water and almost no air is left in the flask. This is why the bell could not be heard after the water has been boiled in the flask.

Any sound sources that are located in outer space (i.e., in a vacuum), are **not propagating any sound.** This is one of the reasons why we do not hear the tremendous sun explosions and the space travel sounds from the surface of the earth.

12.11. LISTEN TO THE POPPING BUBBLES

Materials:
1. A can of fresh soda pop.
2. A round-bottomed flask or drinking glass.
3. A tablet of Alka Seltzer (or other tablet which gives off gas bubbles when dissolved in water).

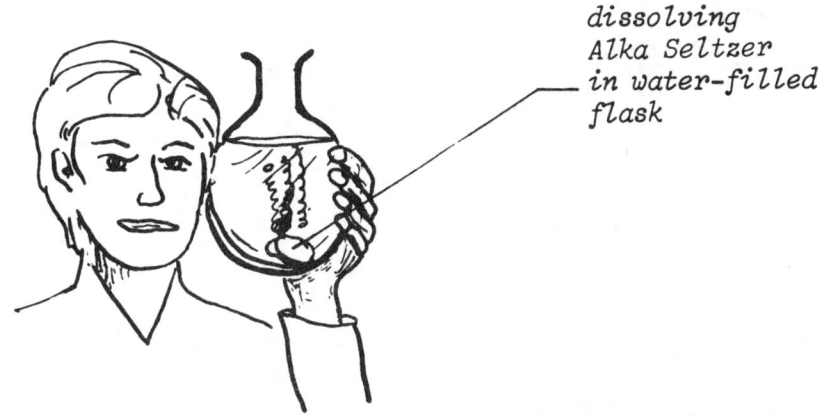

dissolving Alka Seltzer in water-filled flask

Procedure:
1. Open a can of soda pop and place it immediately against your ear (or pour the soda in the flask, and place the flask against your ear). If soda pop is not available, do point 2.
2. Fill the flask or drinking glass with water and plunge a tablet of Alka Seltzer in it. Place the flask or glass immediately against your ear.
3. Listen to the popping bubbles.

Questions:
1. What kind of sounds did you hear in the can or flask?
2. Why couldn't you hear the popping bubbles through the air?
3. What do you do when you want to hear something better?
4. What medium did the sound travel through?
5. Would you think that sound travels easier through liquid or gas?

Explanation:
   The source of sound in this case is formed by the bursting bubbles of carbon dioxide from the soda pop or the Alka Seltzer tablet. The bursting bubbles creates waves in the liquid, which reach the can or glass container. This vibrates in turn and the air close to the ear carries the vibrations to the ear drum.
   As soon as we bring the source of sound closer to the ear, we can hear it better. Like the ticking of a watch or someone's heartbeat, that we like to hear, we must bring the ear closer to the source. **Sound travels through liquids faster and easier compared to that in gases.** This is why the tapping of two rocks can be heard better under water than above water (try this in a swimming pool).

SOUND   MEDIUM OF TRAVEL

12.12. THE SODA CAN TELEPHONE

Materials: 1. Two empty soda pop cans.
             2. A string (same as length of the classroom).
             3. Nail & hammer, a can opener.

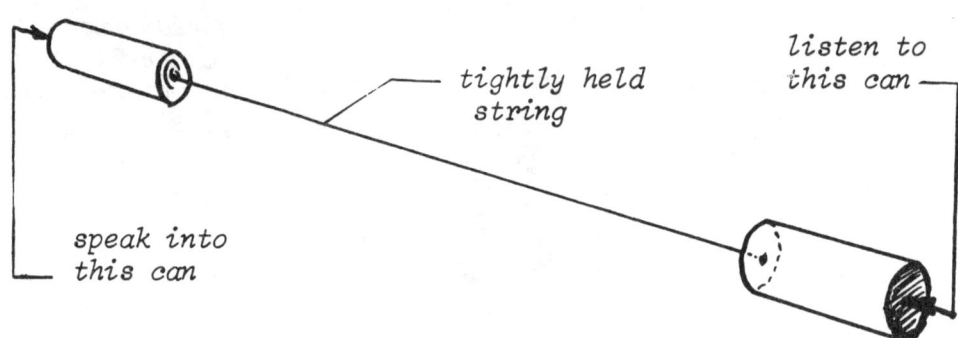

Procedure:
1. Cut one end of the two soda cans completely open with the can opener.
2. Punch a hole in the center of the other end with the nail and hammer.
3. Thread the string through the holes, such that the closed ends face each other and tie a large knot at the end of the string, so that it will not slip out of the hole.
4. Let two students stand at the far ends of the classroom, hold the cans keeping the string tight, and alternately speak softly and listen in the can (see sketch).

Questions:
1. How did the sound travel to the student's ear?
2. What did speaking in the can do to the bottom of the can?
3. Why did the string have to be held tightly?
4. Would this set-up work if we had a solid pipe or bar instead of the string between the two cans?
5. Could a whisper be heard from one end of the classroom to the other end, through the 'telephone'? Through the air?

Explanation:
   By talking into the can, the vibrations from the vocal chords make the air in the can vibrate. These vibrations are transferred to the bottom of the can, which in turn vibrate. The same vibrations are travelling along the string in **longitudinal waves**, making the bottom of the other can vibrate. The air in the receiving can is thus reproducing the exact vibrations of the first can, resulting in the same voice of the original sender. A whisper could not be heard through the air compared to a whisper through the 'telephone'. The waves travel through the solid string in the latter case, and it is much more facilitated. **Sound travels faster and easier through solids than through gases.**

SOUND   MEDIUM OF TRAVEL

12.13. THE COAT HANGER CHURCH BELL

Materials: 1. A pencil with eraser end.
             2. A short piece of string.
             3. A metal wire coat hanger.

hold eraser-end of pencil in ear

string

wire coat hanger

Procedure:
1. Tie the string tightly to the pencil (opposite eraser end) and to the hook of the coat hanger.
2. Place the eraser end of the pencil in the ear and while holding it there, swing the coat hanger and let it hit a solid object (or hit the hanging coat hanger with another pencil).

Questions:
1. What do you hear with the ear in which the pencil is placed?
2. Why didn't you hear the coat hanger vibrations through the air?
3. What was the source of the sound?
4. Describe how the vibrations reached the ear from the source.
5. What other objects could we hang from the string for a source of sound?
6. What would you do to hear someone's heartbeat?
7. What other applications of this principle do you encounter in daily life?

Explanation:
   The coat hanger hitting a solid object would **vibrate** and act as the **source of the sound**. The vibrations travel though the string and the pencil to the ear drum. As the string and pencil are solids, it is much easier for the sound waves to travel through them than through the air. It is the vibrations of the pencil that are immediately transferred to the ear drum, that make the sound so audible.
   Similarly, we place our ear against someone's chest, in order to hear his/her heartbeat. By placing our ear against the railroad tracks, we can hear a train approaching long before we can hear the train sounds through the air.

12.14. THE RESONATING BAR

Materials:
1. An aluminum solid bar or hollow pipe (3/4" or 2 cm diameter and about 3' or 1 m long).
2. A damp cloth.

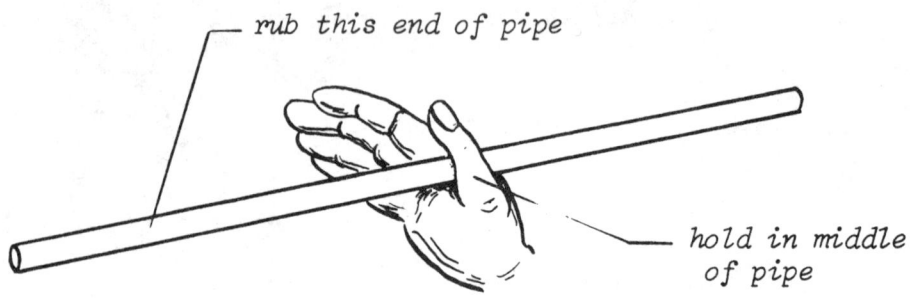

Procedure:
1. Let the aluminum bar balance on your right hand, by holding it between forefinger and thumb in the middle of the bar.
2. Rub the left hand side of the bar with the damp cloth, slowly and gripping the bar rather lightly (try this several times until a high pitched tone is produced).
3. If the rubbing does not succeed, tap the bar vertically on a hard surface or hit the end of the bar with a hard object.
4. Stop the vibration by holding the bar on either end of the center with the left hand.

Questions:
1. What did the rubbing of the bar do to it?
2. Why does the rubbing as well as the tapping produce the same tone in the bar?
3. Would hitting the bar on the side produce the same tone?
4. Why can the vibration be stopped on the side of the bar that was not rubbed or hit?

Explanation:
The rubbing with the damp cloth of the aluminum bar made the cloth slip and grab, slip and grab alternately, causing a **longitudinal** or **lengthwise wave** in the bar. The tapping at the end of the bar causes the same longitudinal wave, which will travel throughout the whole bar, not only on the side when the vibration originated. This is why the vibration can be stopped by holding even the end of the bar that was not rubbed. When the bar is hit on the side instead of at the end, a **transverse wave** is created, which has a much longer wavelength and thus a much lower tone or pitch compared to the longitudinal wave.

SOUND   STANDING WAVES

12.15. THE MULTIPLE TUNING PIPE

Materials: 1. An aluminum pipe (1" or 2.5 cm diameter and
               about 3' or 1 m long).
            2. A golf ball or other hard rubber object.

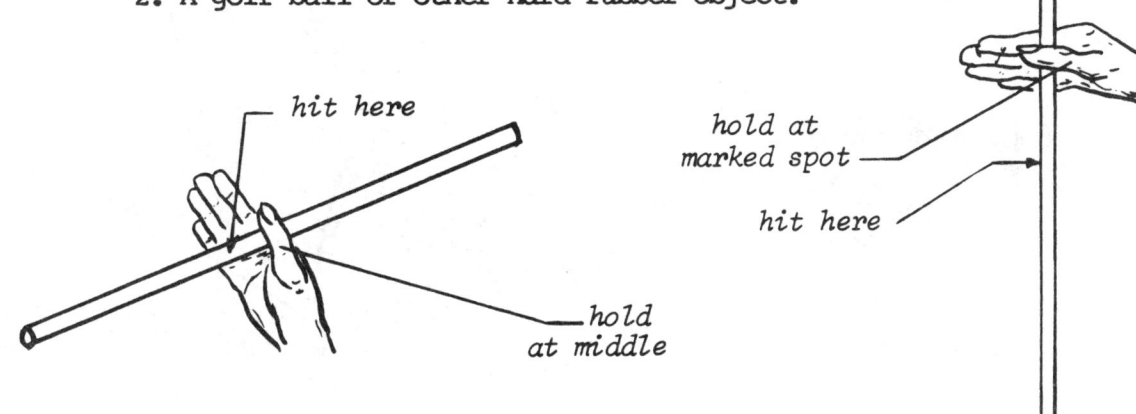

Procedure:
1. Let the pipe balance by holding it with the right hand between thumb and forefinger in the middle of the pipe.
2. Hit the pipe with the golf ball close to the middle of the pipe.
3. With a grease pencil or other marker, place marks on the pipe at places: 13, 25, 35, and 39 cm on each side away from the center (these are approximate distances: the correct spots may be found by holding the pipe at a spot and hitting it with the golf ball; the correct spot will be revealed by a pure resonating tone).
4. Hold the pipe between thumb and forefinger vertically at the marks and hit the pipe with the golf ball. What tones do you hear?

Questions:
1. Why did the pipe give different tones?
2. Did holding the pipe at different spots determine the tone?
3. Could the vibration continue by holding the pipe at more than one spot? More than two spots? Which of the spots?
4. Holding which of the spots would give the same tone?
5. Holding which of the spots on the pipe would give the highest pitch? The lowest pitch?

Explanation:
   By holding the pipe at one spot and hitting it with the golf ball, we are getting a vibration in the pipe, resulting in a **standing wave**, detectable by a certain tone/pitch depending on where it is held. The spot 25 cm away from the center gives the lowest pitch, whereas the spots 13 and 39 cm from the center give the highest pitch. This is because of the shorter wavelength produced.

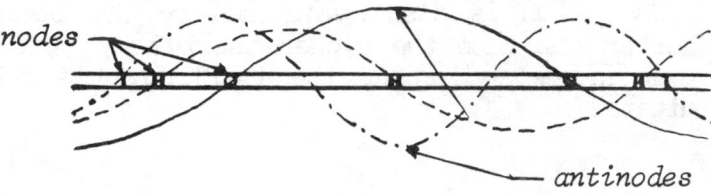

SOUND

RESONANCE & HARMONICS
BERNOULLI'S EFFECT

12.16. THE TWIRLING BUGLE

Materials: 1. A corrugated plastic tube (a swimming pool drain hose or vacuum hose; about 6' or 1.5 m long, 4 cm diameter).

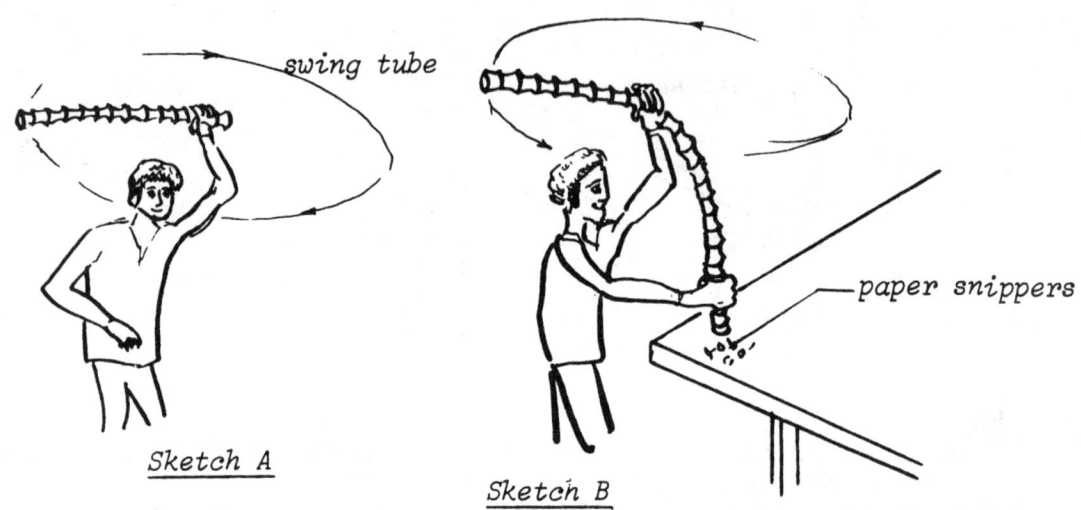

Sketch A

Sketch B

Procedure:
1. Hold the plastic tube in one hand at one end, and swing it above your head, first slowly, then faster and faster (different pitches will be heard--see Sketch A).
2. Tear a piece of paper up into a few small snippers and leave them in a heap on the edge of the table. Hold with one hand the stationary end of the tube above the snippers and with the other hand, swing the tube (the pieces of paper will be flying all over--see Sketch B).

Questions:
1. What made or produced the sound?
2. Why did we get different pitches by twirling the tube faster?
3. Why did the note skip in pitch; what is the next higher pitch called?
4. Which way is the air flowing in the tube?

Explanation:
   The air flows from the stationary end to the circling end, because the air flowing past the moving end of the tube reduces the air pressure (**Principle of Bernoulli**--see chapter on flowing air). As the air flows along the corrugations of the tube, it starts to **oscillate**. The corrugations determine the frequency of the oscillations and thus the tone. The faster the air flows through the tube, the easier the **higher frequency harmonics** are excited, thus producing the higher harmonic tones.
   The pitch is also determined by the length of the twirling tube. The shorter the tube the higher the pitch. With a little practice in slowing down and speeding up the twirling, the sound of a bugle can be easily imitated.

12.17. THE TWO-NOTE TUBE

Materials: 1. A length of rigid plastic or copper tubing (2-1/2" or 8 cm diameter, 1-1/2' or 50 cm long).

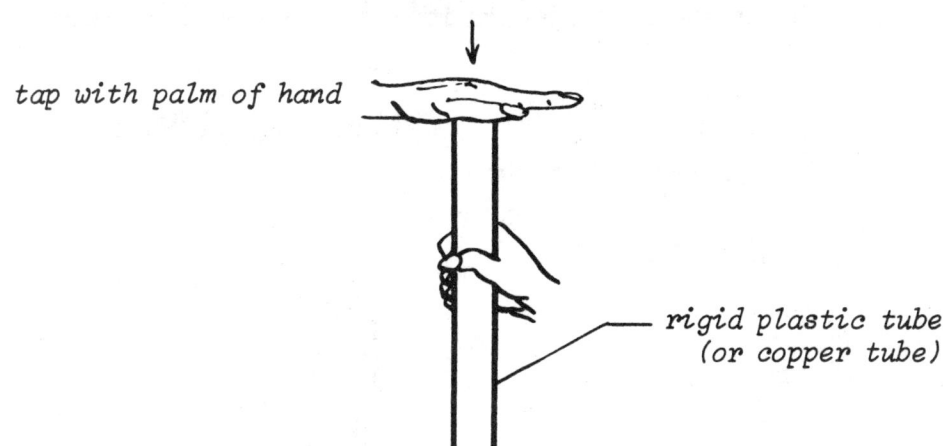

Procedure:
1. Hold the tube with your right hand vertically and tap the top end with your flat left palm (hold the palm against the tube): a certain tone is produced.
2. Now tap the top end of the tube with the flat of your left hand palm and release the contact with the tube immediately after the tap; in other words: move your hand immediately away from the tube (it is like tapping the tube very lightly): another note is produced.
3. Keep alternating the two kinds of taps and let the students try to distinguish the two notes.

Questions:
1. How was the tone in the tube produced?
2. Which of the two tones produced had a higher pitch?
3. How are the two tones or pitches related to each other?
4. What type of waves did we create with the two kinds of taps?

Explanation:
By tapping the tube at the top end, a vibration of the air within the tube was created, producing a tone. When the palm is kept against the end of the tube, the wave is occurring in a tube with a **closed** or **fixed end**, producing a low tone. When the palm is tapped against the tube only lightly, the wave occurs in a tube with **open ends**, producing a tone an octave higher.

This can be illustrated by holding a short slinky vertically at the top only (closed end wave), or in the middle and moving the top faster (open end wave - see sketch on the right).

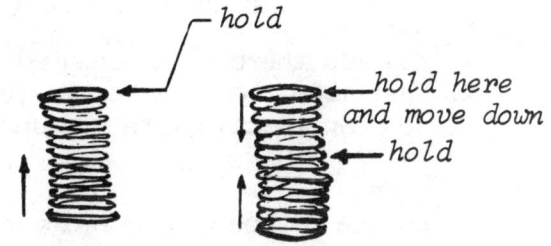

12.18. MEASURE THE SPEED OF SOUND

Materials:
1. A bright colored ball and string.
2. Two pieces of wood (to make a clapping sound).
3. A long measuring tape (or long rope and a short measuring tape).

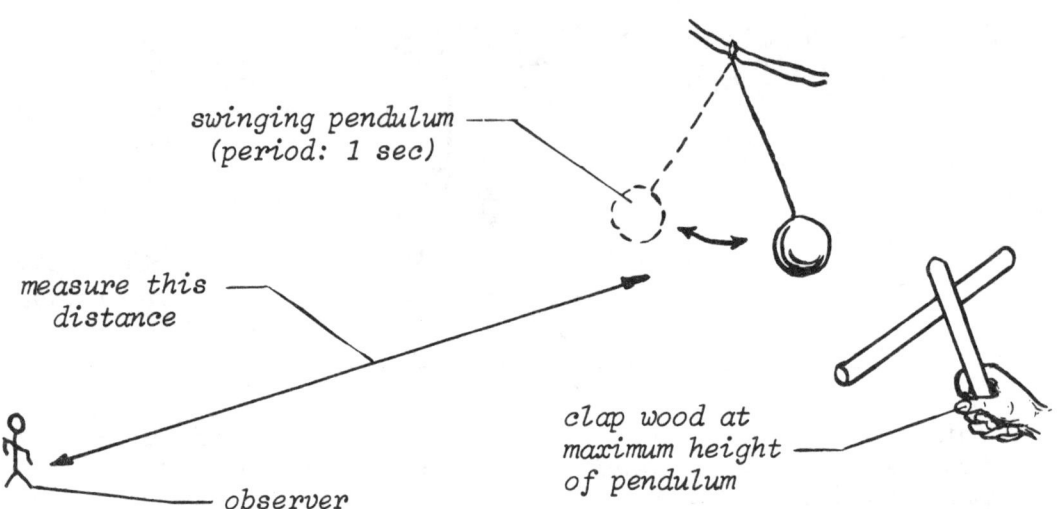

Procedure:
1. Make a pendulum from the bright colored ball and string, such that it swings a full period (back and forth) within one second (by adjusting the length of the string).
2. Take the class out to an open field (it should be 170-180 m long), attach the pendulum to a horizontal bar or tree branch and have one student clap the two pieces of wood together at the exact moment that the pendulum reaches the highest point of the swing.
3. Have another student stand next to the one clapping, to keep the pendulum bob swinging (by giving it a reinforcing push).
4. Once the clapping and swinging are synchronized, let the rest of the students walk back from the pendulum while observing the swing and listening to the clapping (soon the taps will get out of phase with the swings, and slowly get back in phase with the swings).
5. Stop moving away from the pendulum, as soon as the taps are synchronized again with the swings of the bob, and measure the distance from where you are now to the swinging pendulum.

Questions:
1. Why was there a lag created between the taps and the swings?
2. What made the taps become synchronized again with the swings?
3. How can we calculate the speed of sound from the data?

Explanation:
As the period of the swinging pendulum was one full second and the tapping of the wood was at each maximum height of the bob, the claps were 1/2 second from each other. The distance from the observer to the pendulum (170 m) divided by the time (1/2 second) would be the speed or velocity of the travelling sound (340 m/sec).

12.19. THE ECHO SPEED CALCULATOR

Materials:
1. Two pieces of wood (to make a clapping sound).
2. An open space and a large vertical flat surface outdoors (such as a high wall of the school building).

Procedure:
1. Take the class outside to the side or back of the school building.
2. Start from about 20 meters from the wall (facing the wall) and move very slowly backwards while clapping with the wood pieces.
3. As soon as an echo of the clapping sound can be detected by someone, stand still and double check whether indeed an echo can be heard.
4. If indeed an echo can be heard, measure the distance from the observer to the wall.

Questions:
1. How can we calculate the speed of sound in this event?
2. Could we start the clapping by approaching the wall from further out?
3. Why do you think we cannot hear an echo when we are too close to the wall or vertical surface?

Explanation:
The minimum time interval that the human ear can detect between two claps is 0.1 second. When this interval between the clap and the echo is shorter than 0.1 second, no echo is heard. This is why no echo is heard when we stand too close to the wall. By moving slowly backwards away from the wall, an echo will be heard at a certain moment. At this moment the echo came back within 0.1 second, the distance between observer and the wall is measured (about 17 m), and the speed can be calculated from: distance sound travelled (2 x 17 m = 34 m) divided by the time (0.1 sec) equals 340 m/sec.

When the speed of sound is already known, an approaching storm's distance may be estimated by counting the seconds between the lightning and the thunder, and multiplying this by the speed.

12.20. THE ELLIPTICAL WONDER

Materials:
1. Two thumbtacks, a pencil, and a marble.
2. A thick thread or thin rope.
3. A smooth table surface or large cardboard.
4. A long 3 cm wide cardboard strip and molding clay.

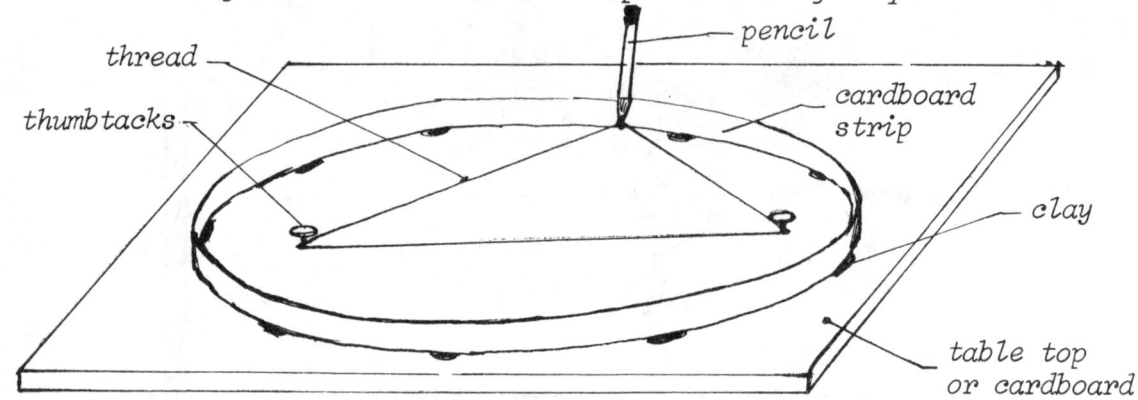

Procedure:
1. Press two thumbtacks about 30 cm apart in the cardboard or table top (this can be closer or wider apart depending on the size of the surface area of the table or cardboard).
2. Make a loop with the thread, where the total length of the thread is 5-10 cm longer than twice the distance between the thumbtacks.
3. Hold this loop of thread tight between the thumbtacks and the pencil tip (holding the pencil vertically) and describe a perfect elipse with the pencil on the cardboard.
4. Build a wall with the cardboard strip and the molding clay on this elipse by pressing blobs of clay between the strip and the table (or cardboard).
5. Roll a marble from one focal point (thumbtack) in any direction to the eliptical wall. Which way does it bounce back?

Questions:
1. Why does the marble always bounce back to the other thumbtack?
2. What points are the thumbtacks in the elipse?
3. How large an elipse can be constructed this way?
4. How would sound or light waves bounce off the eliptical wall?

Explanation:
The reason why a perfect elipse results by keeping the thread tight between the two focal points and the pencil is, because the sum of the two radii stays the same.

Another feature of the elipse is that the **angle of incidence** (i) of one radius is the same as the **angle of reflection** (r) of the other radius (see Sketch on right). This makes it possible that softly spoken words in one focal point can be heard at another focal point quite a long distance away from the first focal point.

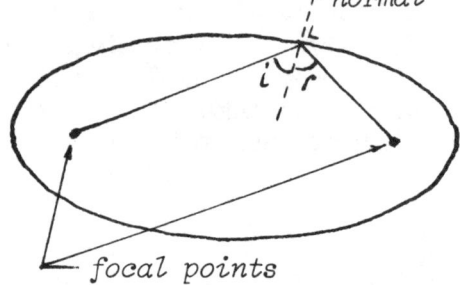

SECTION III

FORCES AND MOTION ON EARTH AND IN SPACE

This section consists of three chapters dealing with Newton's Laws of force and motion, applied on earth as well as in outer space. Some aspects of geology and astronomy are also discussed.

Chapter 13 deals specifically with the concept of gravity and the center of gravity of objects and systems, how torques can keep objects in balance, latent forces in rubber bands, the effect of sudden forces, simple machines, and the rate of falling objects.

Chapter 14 concentrates more on concepts and laws that can be applied to space science, like: inertia; Newton's First, Second, and Third Law; centripetal and centrifugal forces; conservation of momentum and plane of rotation; and weightlessness.

Chapter 15 deals with some aspects of geology and a few phenomena in astronomy. Erosion and weathering, volcanic action, and eclipses are some of the concepts dealt with.

In Science Learning .....

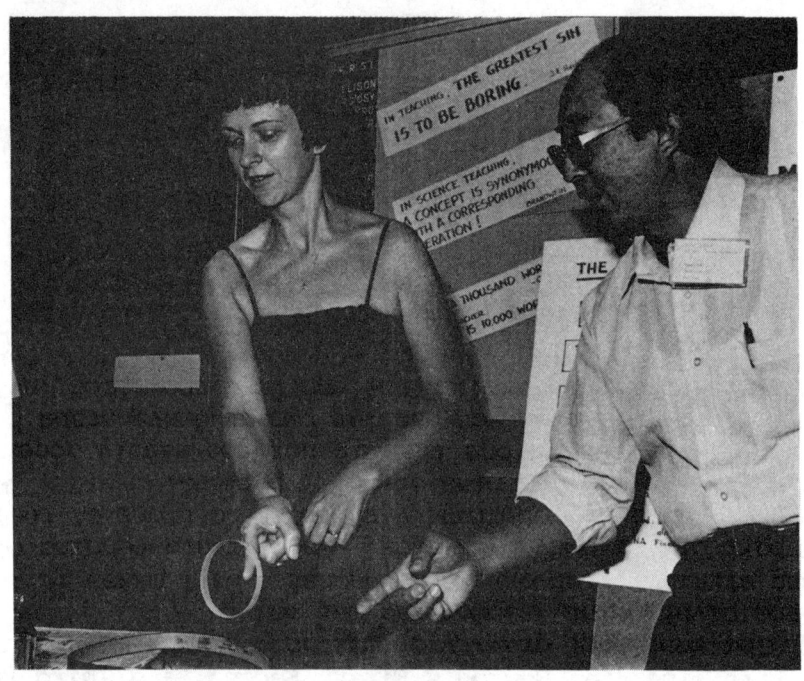

Involvement & Participation Give ....

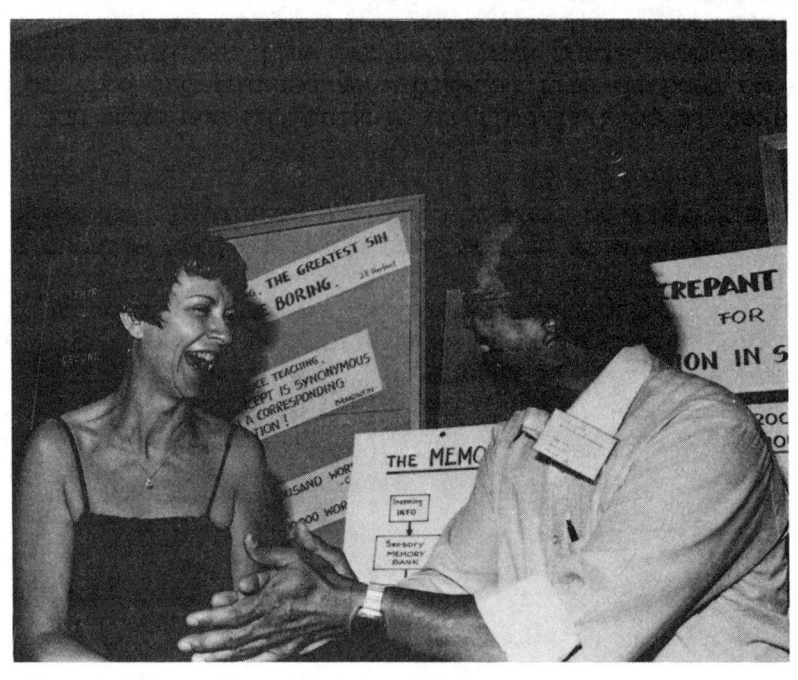

Reward and Satisfaction !

CHAPTER 13

HOW DO FORCES AFFECT THINGS ?

OBJECTIVES

After dealing with and studying the concepts and sub-concepts of this chapter, the students should be able to:

a. recognize the correct explanation of an observed event based on each of the sub-concepts;
b. explain in their own words which of the sub-concepts is determining the course of an event;
c. distinguish true from false statements concerning each one of the sub-concepts;
d. identify the correct explanation of an event in daily life applying one of the sub-concepts;

all in relation to the following sub-concepts:

-- Gravity pulls objects toward the center of the earth.
-- Falling objects fall at the same rate (in a vacuum).
-- If the center of gravity of a system is located below its point of support (pivot point), it is a stable system.
-- Spinning objects spin around their center of gravity.
-- A torque consists of two forces working in opposite directions, which cause an object to spin or rotate.
-- Water pressure increases with the depth of the water.
-- The frequency of a pendulum swing depends only on the length of the pendulum: the shorter the pendulum, the higher the frequency.
-- Corrugations, filling with compressed air, or weaving will generally strengthen the material.
-- In a first class lever the fulcrum is between effort and resistance.
-- In a second class lever, resistance is between fulcrum and effort.
-- In a third class lever, effort is between fulcrum and resistance.
-- Inclined planes and pulleys, like first and second class levers, are meant to reduce the effort the lift or move heavy objects.

FORCES   GRAVITY

13.1. ROLLING UPHILL?

Materials: 1. Strips of cardboard (about 3 cm wide).
            2. Paper cards and cellotape.

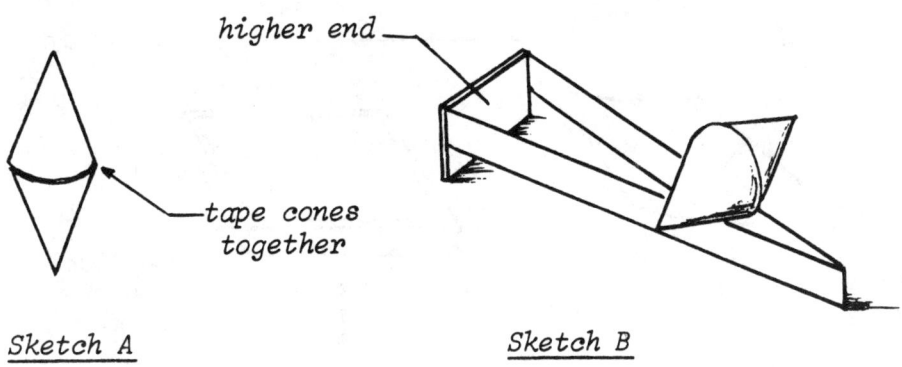

Sketch A   Sketch B

Procedure:
1. Make a cardboard ramp in the shape of a narrow V, and tape a wider cardboard strip to the open end of the V, so that this end will be about 2 cm higher (see Sketch B).
2. Construct two identical cones from the paper cards and tape them together (see Sketch A). Make sure that the surface of the double cone is smooth (especially where it is taped).
3. Place the double cone on the lower end of the V-shaped ramp and give it a starting push toward the higher end.

Questions:
1. How can an object roll uphill?
2. Does the double cone actually end up higher above the table?
3. Would a cylinder roll up the V-shaped ramp?
4. What force made the double cone roll toward the open end of the V?
5. What would a sphere do on the V-shaped ramp?

Explanation:
   This demonstration invites students to be very observant. It gives the illusion of the double cone rolling uphill. The fact is that the double cone actually rolls downhill. It started higher above the table at the lower end of the V-shaped ramp and ends up lower above the table at the higher end of the ramp. It can be observed by fixing the eye on the tip of the double cone or actually measuring the distance between this point and the table top before and after the rolling.
   It is because of the shape of the double cone and that of the ramp, that the object's **center of gravity** slightly gets lowered when rolling from the tip to the open end of the V-shaped ramp. A cylindrical shaped object would not be able to roll "uphill" on this ramp. A spherical shaped object, like a ball for instance, would be able to roll towards the higher end of the ramp, but drops down much sooner between the legs of the V-shaped ramp.

FORCES **GRAVITY**

13.2. WILL PAPER FALL LIKE A STONE?

Materials: 1. A hard cover book.
2. A sheet of paper (the size of the book).

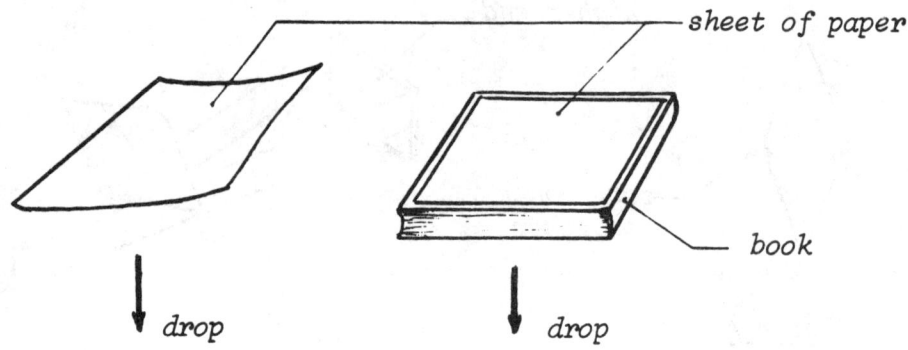

Procedure:
1. Drop a sheet of paper to the floor from arm's height (it will slowly fall, moving from side to side).
2. Drop a book separately (or beside the paper) to the floor from the same height (it will drop like a stone).
3. Now place the paper sheet on the book. Ask the students: "With the same gravity working on the paper and the book, will the sheet of paper fall at the same rate as the book?" (Anticipated answer: 'No').
4. Drop the book with the paper sheet on top of it (make sure that the paper's size is equal to or smaller than the book).

Questions:
1. Together why did the sheet of paper drop just as fast as the book?
2. Separately why did the sheet of paper drop slower than the book?
3. What would happen if the book were smaller than the sheet of paper and the two were dropped together?
4. What would happen if the sheet of paper were dropped in a vacuum tube?

Explanation:
Dropping the sheet of paper to the floor by itself will make it move from side to side and descend slowly, because of the **air resistance**. The book is heavy enough to overcome the air friction and it drops in that short distance as if there were no resistance. Together, our intuition says that the paper should stay behind the book in falling to the floor, but they fall at the same rate, because under the paper there is no air and thus no resistance. It is as if the paper fell in a vacuum.

FORCES

GRAVITY
NEWTON'S SECOND LAW

13.3. THE FALLING PENNIES

Materials:
1. Two pennies (or other coins).
2. A paper card (3x5" / 7x12 cm).

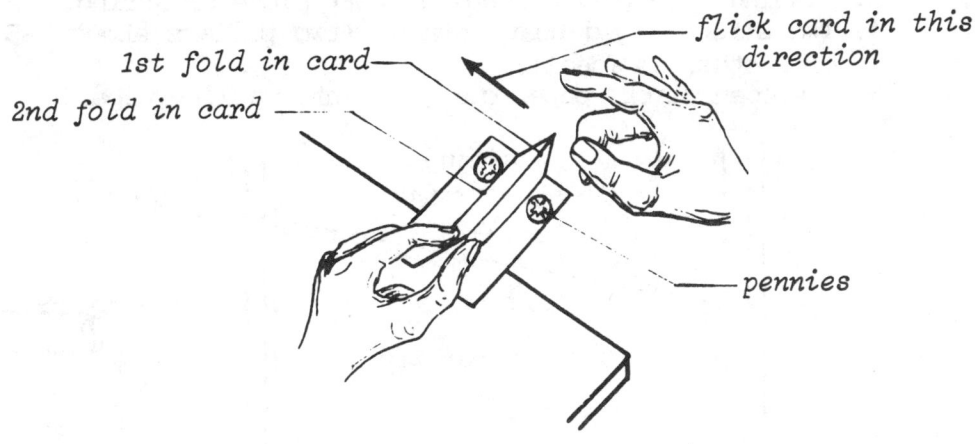

Procedure:
1. Fold the paper card in half, then fold each side one-third from the end outward.
2. Place the pennies on each side of the center ridge of the card and hold one end on the table edge (see Sketch).
3. Now flick the ridge of the card to the side with the middle finger of your right hand. This will fling one penny about 5 m away and at the exact same moment the other penny will drop straight to the floor.
4. Ask the students: "Which of the two falling pennies will hit the floor first?" (Anticipated answer: 'the straight falling one') Flick the card and listen to the click of the pennies hitting the floor.

Questions:
1. Did you hear one or two loud clicks of the falling pennies?
2. What would it mean if you heard only one click?
3. Does the direction of a falling object have any influence on the speed or rate of fall?
4. What was disregarded in this demonstration?
5. What can we conclude from this demonstration about the rate of fall of objects with different initial directions?

Explanation:
As both pennies were released at the same moment, the force of gravity started to work on the pennies at the same time. The downward component of the forces working on each of the pennies is therefore the same, and thus also the acceleration they obtain, resulting in reaching the floor in exactly the same time. This is why only one loud click (first click) is heard of the two pennies falling, regardless of the initial horizontal force imparted on one of them. When one of the pennies is replaced with a marble, the same one click will result, implying that all objects fall with the same rate (in a vacuum). In our case the air resistance was disregarded, because of the relatively short distance of fall.

13.4. HIT THE FALLING CAN

Materials:
1. An empty soda pop can.
2. A test tube and steel ball (which fits in the tube).
3. Fishing line (6-7 m long) & short piece of string.
4. Two strong rigid heavy stands (two pillars about 4-5 m away from each other is ideal).
5. A wooden stick, paper clip, splint, masking tape.

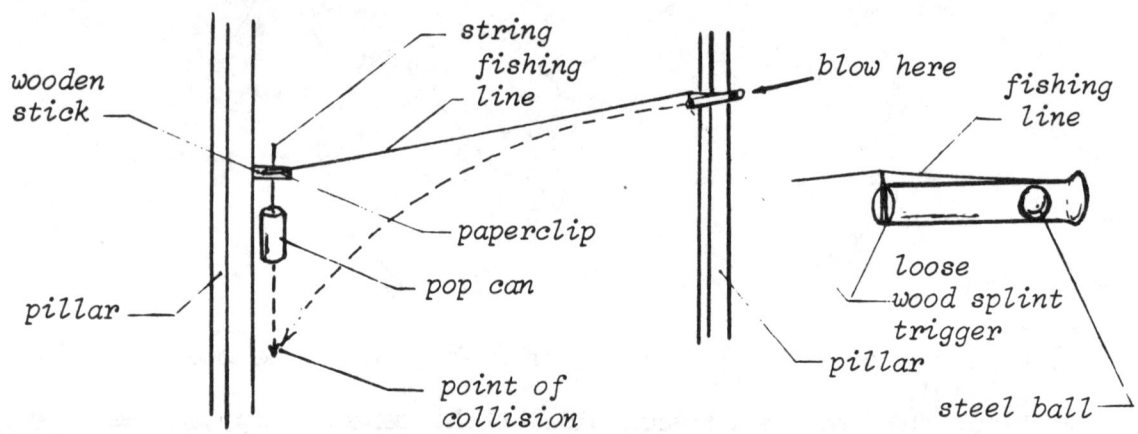

Procedure:
1. Break the bottom of the test tube by letting the steel ball roll down in it (a wide glass tubing could be used here instead).
2. Tape this tube horizontally about mouth-height to one of the pillars.
3. Punch a nail hole through the end of a 20 cm long wooden stick and get the fishing line though the hole.
4. Tie a paper clip to the end of the fishing line.
5. Tape the stick, slightly higher than the tube, to the other pillar.
6. Pull the fishing line rather tight from one pillar to the other (to where the tube is taped; the line should be slightly higher at the wooden stick end).
7. Tie a short string to the empty can and pinch the end of the string between the paper clip and the wooden stick. Set the trigger by placing a wood splint at the end of the tube (can should be held up when set, when wood splint falls the can should fall).
8. Place the steel ball in the tube and blow it hard in the direction of the hanging can (tube and can should be lined up).

Questions:
1. Why does the steel ball always hit the falling can?
2. Does it matter how hard we blow through the tube?
3. What force is pulling at falling objects?

Explanation:
Disregarding the **air friction** in this case (actually in vacuum), **gravity** is pulling at the can and at the steel ball with the same force, thus no matter whether the ball had a horizontal component or not, they fall at the same rate and should meet each other in their path of fall.

FORCES GRAVITY
 FREE FALL

13.5. THE MYSTERIOUSLY MOVING STEEL BALL

Materials:
1. A large steel ball (about 2 cm diameter) from a ball bearing.
2. A wooden lath (1 m long) or meter stick.
3. Two small cups (the size of large coke bottle caps).
4. Adhesive tape, and a little moulding clay.

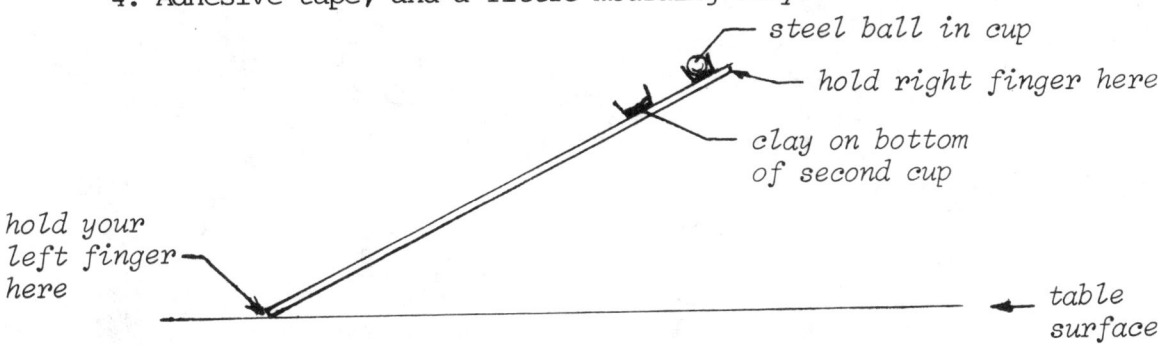

Procedure:
1. Attach one cup to one end of the lath and the other about 5 cm apart from the first (centre to centre), by using masking or scotch tape.
2. Place the steel ball in the first cup, and place some clay in the bottom of the second cup.
3. Hold the left end of the lath (the empty end) down to the table top with your left forefinger and lift the right end of the lath with your right forefinger (let the left end hinge around the left finger and the table surface, keeping the finger steady to keep the lath from sliding).
4. Hold the lath in about a 30 degree angle above the table surface, and suddenly release the right-hand end of the lath: the ball moved from one cup into the other! (If it did not work the first time, try again at larger angles or more sudden releases of your right finger).

Questions:
1. What made the ball move from one cup into the other?
2. Do falling objects fall with different rates?
3. Is the gravity acceleration not the same for all objects on earth?
4. If air friction was negligible, how would the acceleration of the lath compare to that of the steel ball?
5. Which point of the lath would fall at the same rate as the steel ball?
6. What does the right end of the lath have to do, in order for the steel ball to be able to move out of the first cup?

Explanation:
The centre of gravity of the lath with the cups attached is about 2/3rd away from the hinged end. This point falls at the same rate as the steel ball.

As point A (centre of gravity) is not as high above the table as point B (right end), point B has to fall faster than a regular free fall. Thus, the right-hand end of the lath indeed falls faster than the steel ball, and as the ball falls in a straight path, it ends up in the second cup.

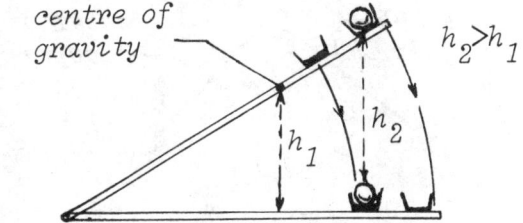

13.6. THE BALANCING PINS

Materials:
1. Three knitting needles, or iron wires (30 cm).
2. Four one-hole rubber stoppers.
3. A cork (medium size), a styrofoam ball or cube.

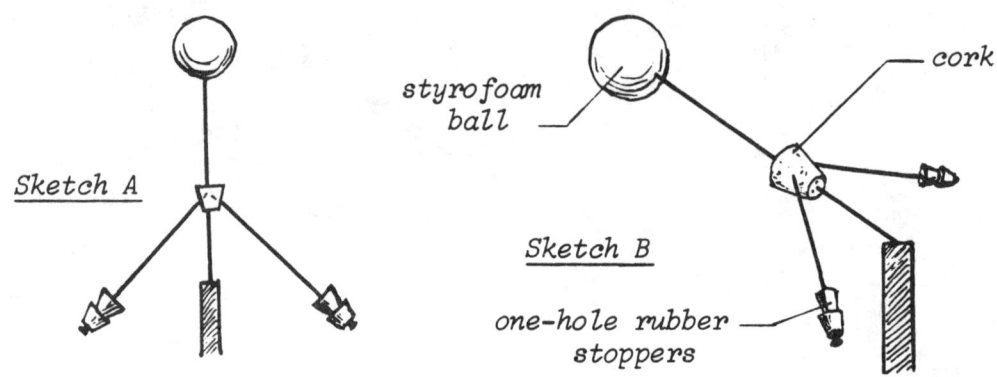

Procedure:
1. Stick one of the knitting needles, lengthwise, through the center of the cork, cut the head of the pin off, and stick the styrofoam ball to this end of the needle.
2. Let two of the stoppers fall in each of the remaining knitting needles through their holes.
3. Stick these pins rather slanted down into the side of the cork (Sketch A).
4. Let the whole system balance on your finger with the pin point as the pivot (use any protruding point as support).

Questions:
1. Why does the system of pins and stoppers stay up?
2. What is the purpose of the rubber stoppers?
3. What is the purpose of the styrofoam ball?
4. Where is the center of gravity of the system located? Without the pin with rubber stoppers? With the rubber stoppers?

Explanation:
The cork and styrofoam ball are very light in weight, compared to the rubber stoppers. By slanting the pins with the rubber stoppers down, the **center of gravity** is lowered below the pivot point. Another way of lowering this center of gravity of the system is to shift the cork down. The lower the center of gravity of the system, the more stable the system is.

It is because of this principle that race cars are built as low as possible, so that in sharp turns they will hug the road and not roll over.

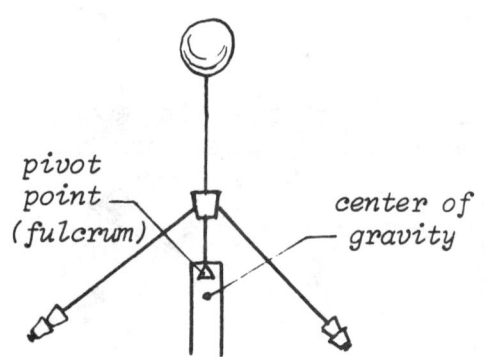

FORCES CENTER OF GRAVITY

13.7. THE BALANCING ACT

Materials:
1. A tall pop or wine bottle.
2. Two strainer scoops, 2 forks, 2 knives.
3. A barbecue fork and a metal skewer.

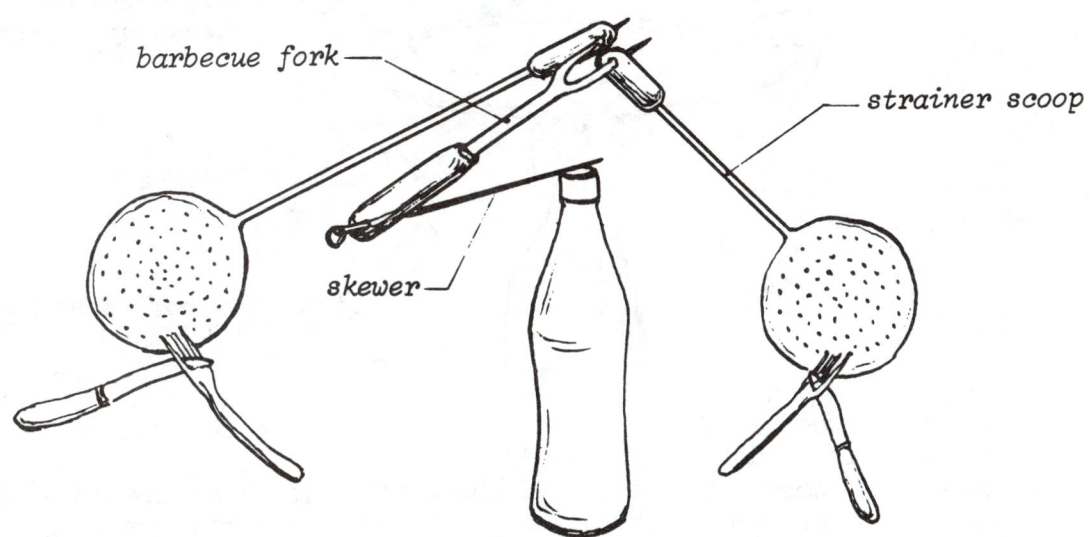

Procedure:
1. Attach the strainers through the handle holes to the barbecue fork, and push the skewer through the handle hole of the barbecue fork (see Sketch above).
2. While holding the system of utensils by the end of the barbecue fork, hang more weight on the strainers by pushing the fork teeth through the strainer holes and the knives through the fork teeth.
3. Balance the whole system of utensils now on top of the coke bottle by the end of the skewer.

Questions:
1. What makes the system stay balanced?
2. What is the purpose of the forks and knives?
3. Why doesn't the system stay balanced without the forks and knives?
4. What property is critical in deciding whether an object stays stable or unstable on a pivot point?
5. Where can we find applications of this principle?

Explanation:
The purpose of hanging the forks and knives on the strainers is to lower the center of gravity of the whole system, so that the system can stay stable on top of the pop bottle (pivot point). When the **center of mass** or **center of gravity** of a system is situated below the pivot point, the system is stable. **The lower the center of gravity of the system, the more stable the system is.**

An application of this principle is found with rope dancers. When they walk the ropes they carry long heavy bent poles (probably weighted at the ends) in order to stabilize themselves.

FORCES — CENTER OF GRAVITY

13.8. THE PLATE CAROUSEL

Materials:
1. A regular dining plate and 4 forks.
2. Two corks or a small raw potato.
3. A corked bottle and a needle.

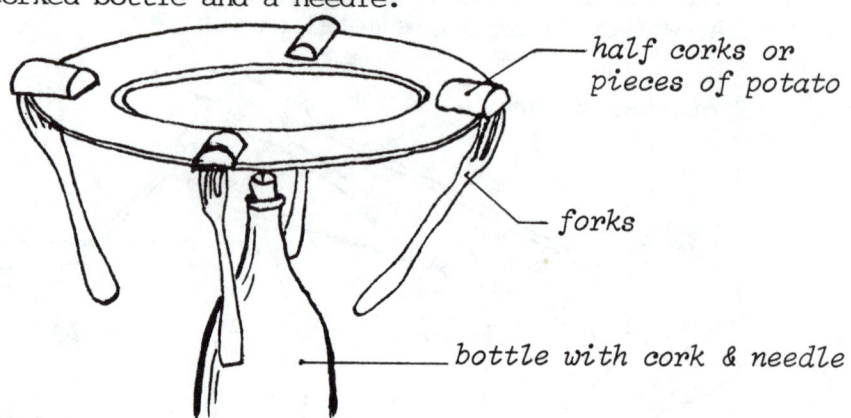

- half corks or pieces of potato
- forks
- bottle with cork & needle

Procedure:
1. Insert the needle vertically in the cork of the bottle and place the bottle in the middle of the table.
2. Cut the two corks in half with a sharp knife (so that the cut surfaces are smooth) or, when corks are not available, cut four equal pieces of potato out of a small raw potato.
3. Stick the four forks in the four half corks (or four chunks of potato) and hang them from the edge of the plate (see Sketch).
4. Balance the whole system of plate and forks on the blunt end of the needle (or the sharp end if you can push the blunt end in the cork).
5. Now gently blow against the forks (or tap against them) in one direction to make the plate rotate.

Questions:
1. Could the plate balance on the needle without the forks?
2. What is the function of the hanging forks?
3. What did the forks do to the centre of gravity of the system?
4. Where is the centre of gravity located of the plate without the forks?
5. Where is the centre of gravity of the system of plate and forks?
6. Would half corks or pieces of potato be better to make the system more stable, or does it matter at all?
7. Would smaller or larger pieces of potato be better to make a more stable system? Why?

Explanation:
  The plate on its own would not be able to balance on the needle, because its **center of gravity** would be located above the point of support (pivot point). By hanging the forks from the edge of the plate, the center of gravity of the system got lowered. It got lowered far enough below the pivot point that the system became stable. The more mass is placed below the pivot point, the stabler the system. Thus, when comparing the half corks with the pieces of potato, the first would be better, as the corks are much lighter in weight (less mass). The bigger the pieces of potato we use, the less stable the system, as more mass is placed above the pivot point - which is the top of the needle.

FORCES   CENTER OF GRAVITY

13.9. HANG A HAMMER ON A LOOSE RULER

Materials:
1. A hammer with a wooden handle.
2. A wooden or plastic ruler.
3. A short string or wire.

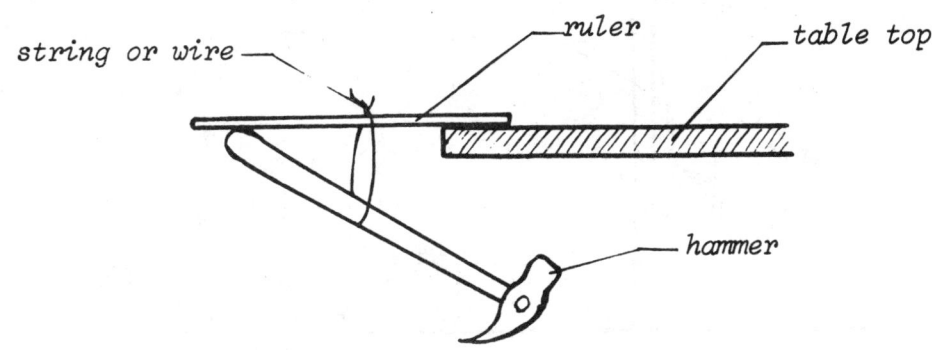

Procedure:
1. Take a flat wooden or plastic stick (the length of a ruler) and hang the hammer (with a wooden handle), with the iron part of it down, by means of a string.
2. Make a loop out of the string or wire (about 10 cm diameter), and slip it around the ruler and the handle of the hammer (let the end of the handle press against the ruler end).
3. Let the other end of the ruler hang from the edge of a table top.

Questions:
1. What made the heavy hammer stay up at the edge of the table?
2. Why do we need a hammer with a wooden handle?
3. Can we consider the ruler really to be loose?
4. Where is the center of gravity of the hammer alone?
5. Where is the center of gravity of the whole system of ruler, string, and hammer?
6. What is the difference between a stable and labile system?

Explanation:
This demonstration can only be carried out when the hammer has a wooden handle, as the **center of gravity** is located in the iron part of the hammer. The plastic or wooden ruler and string do not add much to the weight of the system on the handle side. They make the position of the center of gravity of the whole system shift just a little towards the right and upwards.

If this center of gravity is under the **pivot point (point of support)**, it is a **stable** system. If the center of gravity is above the pivot point, it is **labile** and it falls.

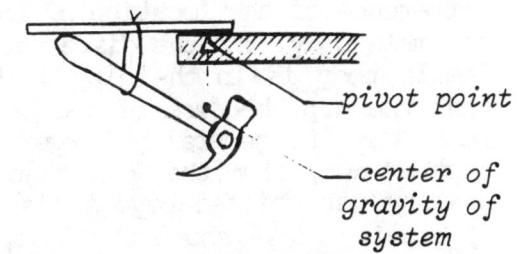

FORCES   CENTER OF GRAVITY
                                                          HUMAN BODY

13.10. ARE WOMEN STRONGER THAN MEN?

Materials: 1. A straight-back chair, some wall space.
             2. A few adult men and women (or teenage boys and girls).

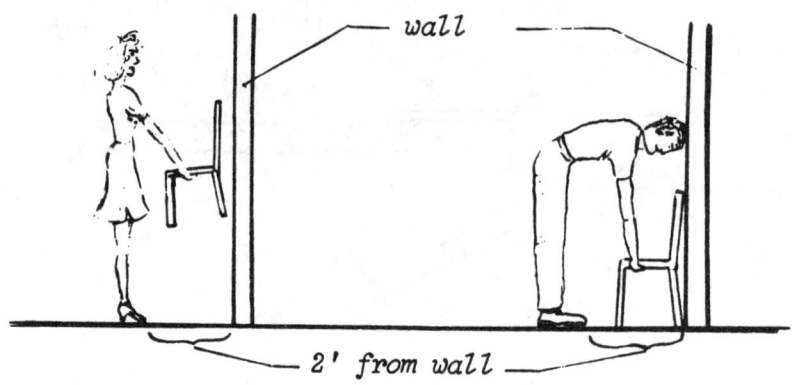

Procedure:
1. Show what the people have to do:
   a. Stand exactly two foot-lengths away from the wall (by measuring toe to heel off the wall with shoes on).
   b. Bend over with a straight back (turning in the hips) and let your head lean against the wall (see Sketch).
   c. Ask someone to hand you the straight-back chair (with the back against the wall).
   d. Lift the chair while your head is still leaning against the wall.
   e. With the chair in lifted position, try to straighten up (men cannot do this, but women can!)
2. Ask women and men to do the above alternately (hand them the chair after they have stepped back from the wall).

Questions:
1. Why can girls or women straighten up and boys or men not?
2. Is the 'weaker' sex really stronger?
3. What is the difference in body build of male versus female?
4. Where is the center of gravity of the body located?
5. What is the average shoe or foot size of females versus males?

Explanation:
   The result of this event has nothing to do with strength. It lies in the difference of the **location of the center of gravity** of the female compared to the male body, and the difference in foot size. The center of gravity of the female body is in the hip, while that of the male body is situation higher than the hip, because of the broader shoulders and narrower hips. The male foot size is generally larger than the female one, so he stands actually farther away from the wall than she does. When he bends over his center of gravity get shifted beyond his toes, but with the female person this center of gravity stays above the foot, and straightening up poses no problem. (See also Event 17.20.)

FORCES

CENTER OF GRAVITY
HUMAN BODY

13.11. THE UNREACHABLE CUP

Materials: 1. A cup or pop can (or any other small object).

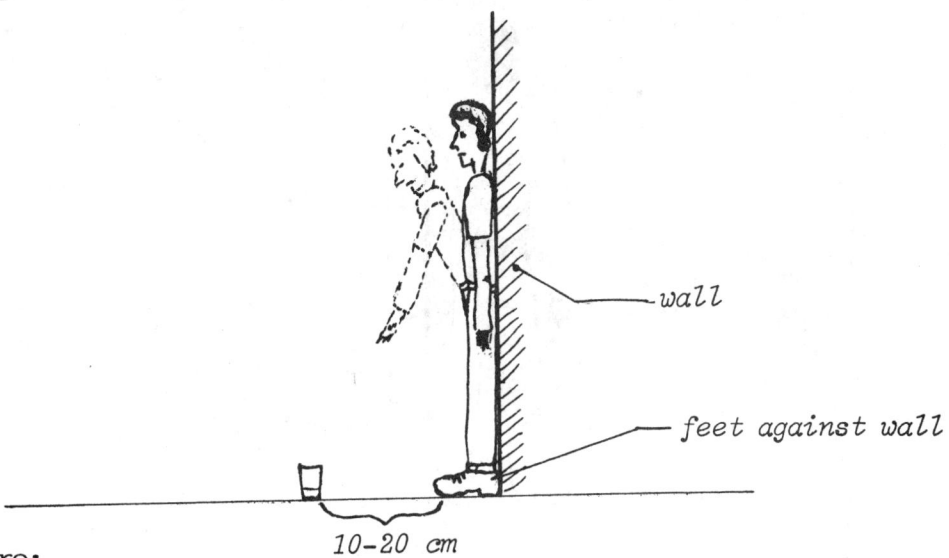

Procedure:
1. Place a cup or pop can about 20-30 cm in front of your feet on the floor and show how easy it is to pick it up without moving or bending your feet.
2. Let someone (student) stand straight with the back against the wall with his/her heels touching the wall.
3. Place a cup or pop can about 20-30 cm away in front of the student's feet on the floor.
4. Let the student pick up the cup from the floor without bending the knees and without falling forwards. Is it possible?

Questions:
1. What makes it so difficult to pick up the cup from the floor?
2. Why do we tend to fall forward when bending our body to the front?
3. Let the student stand away from the wall and let him/her pick up the cup from the floor without bending the knees. Observe the student from the side. What happens to the legs and lower part of the body?

Explanation:
When picking up an object from the floor without bending our knees, our legs and lower part of out body have to move backwards in order to stay balanced. The **center of gravity** of our body has to remain above our feet, which is the **pivot point**, supporting our body.

Standing straight with our back against a wall makes it impossible to move the lower part of our body backwards. The forward bending of the upper body shifts the body's center of gravity towards the front of the pivot point (our feet) and the whole body topples or falls forward.

If there is no wall or one that is not too easily accessible, like in an auditorium or the outdoors, you can let two students stand back to back and let them at the same time pick up the cup from the floor in front of each one of them.

FORCES CENTER OF GRAVITY
HUMAN BODY

13.12. STUCK TO THE WALL?

Materials: 1. Just yourself and the wall.

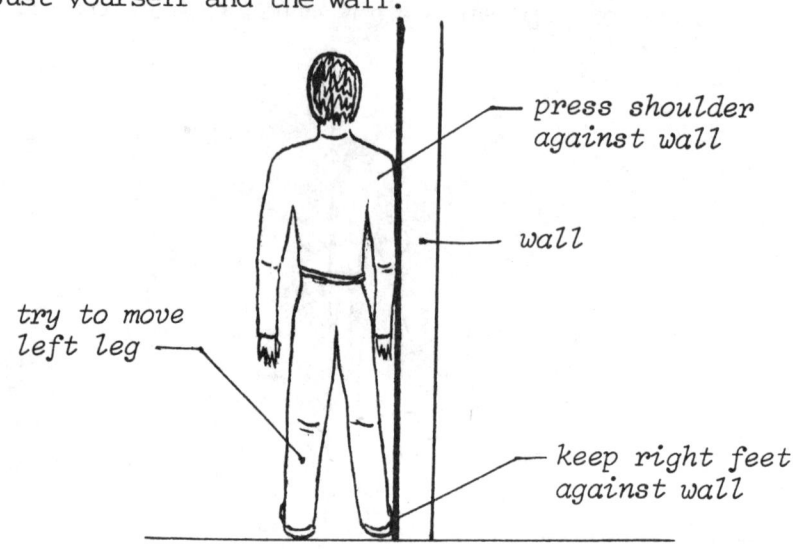

Procedure:
1. Stand up straight with your right foot and your right shoulder against the wall.
2. Try to move your left leg without falling or taking a step.
3. Now stand straight and touch the wall with your left foot and keep your left shoulder pressed against the wall.
4. Try to lift or move you right leg: IMPOSSIBLE!

Questions:
1. What makes it impossible to move your left leg when your right foot and right shoulder are touching the wall?
2. What did you have to do in order for your right shoulder to press against the wall?
3. What would happen if someone else forced your left leg to move while you are standing in that position?
4. Where is your center of gravity located when you are standing up straight?
5. Where does your center of gravity have to be located in order for you to be able to lift your left leg without falling?

Explanation:
In order to stand straight with your right foot and right shoulder touching the wall, your left leg has to push against the floor. This is necessary in order to move the upper body and your right shoulder against the wall. Because of the strain in your left leg it is impossible to move it away. If forced by someone else, you will fall towards the left, mainly because the center of gravity of your body is on the left side of your right foot, which would be the only support.

In order not to fall, your center of gravity should be located exactly above the pivot point or support – in this case: your feet. After your left leg is knocked away, your body needs to move towards the right in order not to fall, but the wall prevents this.

FORCES

CENTER OF GRAVITY
ADHESION

13.13. THE CENTER-SEEKING PAPER

Materials: 1. A small piece of writing paper (1/8 of letter size).
2. A wide shallow tray.

Procedure:
1. Draw a small parallelogram (about 2x3 cm) with its diagonals on the little piece of paper.
2. Wet the parallelogram (not the whole paper) with water by letting droplets fall from your wetted forefinger one at a time.
3. Fill the tray with water, lift the piece of paper carefully and let it float on the water surface.
4. Touch the tip of your forefinger to the water on the parallelogram at a spot other than the crossing of the diagonals. What is the paper doing?

Questions:
1. Why does the water not spread to the other parts of the paper?
2. What force keeps the water on the paper together?
3. When touching the wetted part of the paper near one edge, which way does the paper shape move to?
4. At what point does the finger end up above the paper shape?
5. What makes the wet paper shape move?
6. What makes the paper stop moving eventually?

Explanation:
The water drops on the paper are held together by **cohesive forces**. This is the reason why the water is not spreading to the other parts of the paper.

When the finger tip is touched to one side of the paper shape, it is pulled to the other side by a larger cohesive force (see Sketch A). When touching the opposite side, the movement of the paper goes opposite too (see Sketch B).

Whenever the paper stops moving, all **adhesive and cohesive forces are in balance**. It is also the center of gravity of the wetted shape. This can be used as a special way to find the center of gravity of an irregular shape.

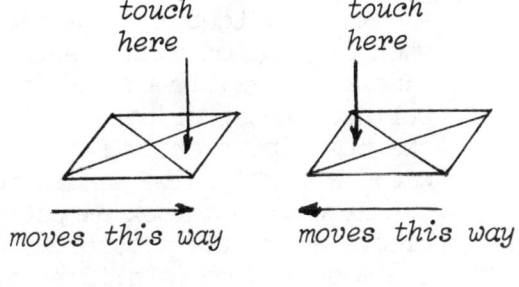

Sketch A   Sketch B

FORCES   CENTER OF GRAVITY
                                                        SHOCK ABSORBERS

13.14. THE STANDING MATCHBOX

Materials: 1. One full or almost full wooden matchbox.
               If used for class activity: 10-20 matchboxes needed.

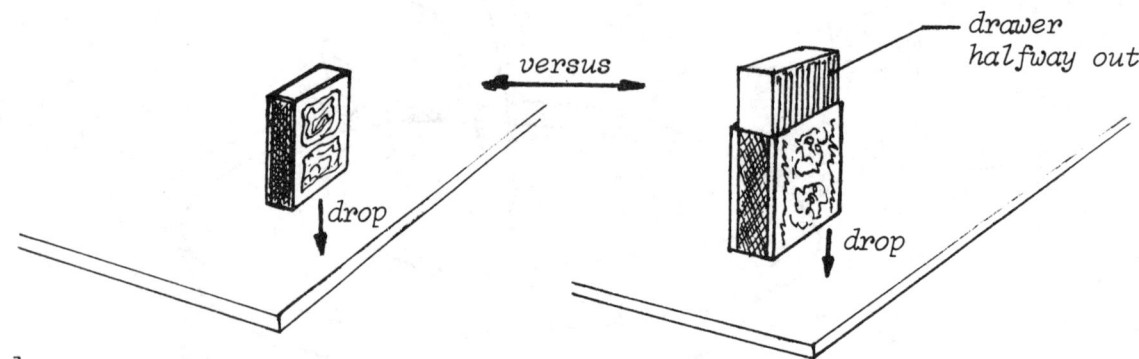

Procedure:
1. Distribute one matchbox per couple of students. Ask them to hold the box vertically above the table (about 15-20 cm above) and let go of the box. Then ask the question: "Can you drop it and leave it standing up?" or "Can you drop the box without it tipping over?" Give them time to try it out. Anticipated answer: "Impossible!"
2. Then you say: "Remember, nothing is impossible!" Take the box, open the drawer about half way out, hold the box vertically with the drawer up, and drop it from about the same height.
3. If you do not succeed in the first two or three tries, lower the height. Ask the students to try dropping the closed box ten times and counting the number of times it was left standing versus the half open box.

Questions:
1. Why does the closed matchbox almost invariably tip over when dropped?
2. What makes the half open box more stable?
3. What did we do to the center of gravity of the box by pulling the drawer half way out?
4. What happens to the center of gravity at the moment that the half open box hits the table surface?
5. How do people usually lessen the impact when jumping off a high ledge or loft?

Explanation:
   As the matchbox is made out of cardboard, it is somewhat springy. Thus when dropped on a hard surface it is bound to bounce sideways, as it is almost impossible to let it land exactly flat on the smallest rectangular surface. Most of the times it lands on an edge or a corner of the box, and the box tips over. If the box was made out of clay, it would not bounce, it would actually give at the point of impact and stand up straight.
   The half open box almost acts the same way as the clay cube. The outer sleeve acts like a cushion, as at the time of impact the drawer, where most of the mass is, keeps sliding in further, thus breaking the fall. Its **center of gravity**, which is located in the center of the drawer, keeps moving in further at the point of impact. This is exactly how people break their fall when jumping off high places, by bending their knees just after or at the point of impact.

FORCES   CENTER OF GRAVITY
                                                              AIR RESISTANCE

13.15. STAND A DOLLAR BILL ON YOUR FINGER

Materials: 1. A crisp dollar bill, or any other strip of paper of the same size
               or somewhat larger.

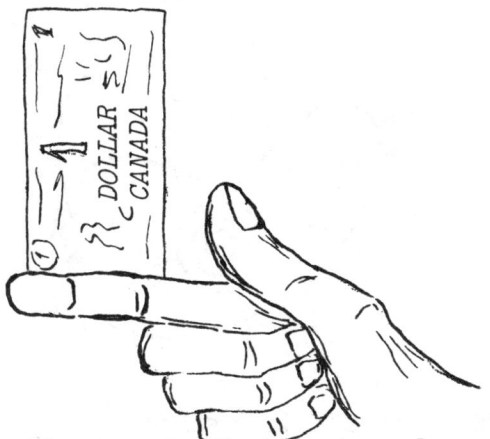

Procedure:
1. Take the crisp dollar bill and stand it vertically on the table top. Show the audience that it is pretty difficult to keep it standing.
2. Ask the students: "How long do you think can I keep it standing on my stretched-out finger?" Anticipated answer: "About 2-3 seconds".
3. Now bend/curve the bill somewhat length-wise and place it on your finger. Watch the tip of the bill and balance it (move your finger quickly to the left if the tip of the bill moves left, and to the right as soon as the tip of the bill moves right).
4. Let the students time how long you could keep the bill standing on your forefinger.

Questions:
1. How long can you balance the dollar bill on your finger?
2. Is there a limit to the length of time you can balance it?
3. What keeps the bill from falling over?
4. Would a longer or shorter strip of paper be easier to balance?
5. Would a wider or thinner (narrower) strip be easier to balance?

Explanation:
   This balancing act can also be performed with any sheet of writing paper, as long as there is a slight bend in the paper to make it stand rather stable on your hand (or finger). The larger the area of the paper, the higher the **air resistance** when moving it from left to right or the other way. There is a cushion of air surrounding it and holding it up, as it were. The taller the strip of paper, the higher the center of gravity of the piece, and thus the easier the balancing. The wider the strip of paper, the larger the surface area, and thus also the easier to balance it, as the air friction (or cushion) will keep it vertical.
   In juggling acts where the juggler balances a vertical pole on his head, the more mass is carried at the top of the pole, the easier it is to balance the whole system. This is because the top end will then be more inert, and it is easier to move the bottom end without moving the top end, and thus to move and place the support point under the center of gravity of the system.

FORCES   CENTER OF GRAVITY

13.16. THE WOBBLING CIRCLES

Materials: 1. A piece of cardboard, a pair of scissors.
             2. A drawing compass, a felt marker.

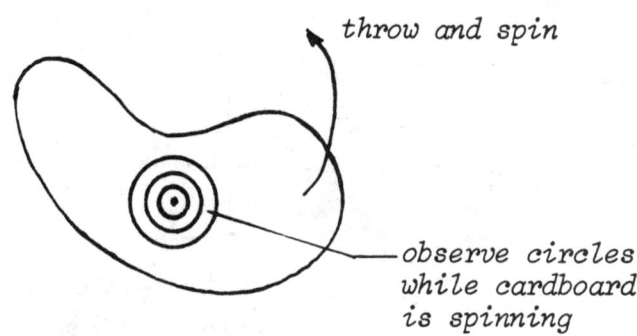

Procedure:
1. Cut the cardboard in a kidney shape.
2. Find the center of gravity of the cardboard shape by letting it balance on your finger and find the point where it balances: mark this spot.
3. Draw concentric circles around this spot with the compass and thicken the circles with the marker.
4. Draw exactly the same concentric circles on the reverse side of the cardboard, but around a spot 3 cm away from the center of gravity.
5. Have students now observe the following demonstration:
   a. Let the cardboard piece spin vertically in the air, by flipping it upward holding the piece by one of its ends, (use wrist action when throwing) and catching it when it comes down.
   b. Flip the board, so that the students would see the other side (be inconspicuous in doing this) and spin the cardboard again.
   c. Let students observe and follow the circles while the cardboard is spinning in the air.

Questions:
1. Why are the circles sometimes wobbling and sometimes steady?
2. How did I get the circles to wobble?
3. How can I find the center of gravity of the cardboard?
4. Around what point would an axe or hammer (with wooden handle) spin?

Explanation:
   Steps 1 through 4 of the procedure should be done before the demonstration without having the students observe it. **All objects spin around their center of gravity.** By drawing the concentric circles around the center of gravity on one side of the cardboard, the circles stay steady when the board is spun. When the other side of the board is shown to the students, the circles wobble, because it is off center and the circle actually spin around a point outside the circle's center making a wobbling motion.
   When an axe or hammer with a wooden handle is spun in the air, it will also spin around its center of gravity, which is very close to the iron part of the tool. This makes the handle seem to swing out much more than the metal part.

FORCES

CENTER OF GRAVITY
SPINNING OBJECTS

13.17. KICK A STRAIGHT LINE

Materials:
1. A wood block (2x4" about 1 ft long) or any other shape.
2. Four ballpoint pens.
3. A long piece of newsprint paper (or brown wrapping paper).

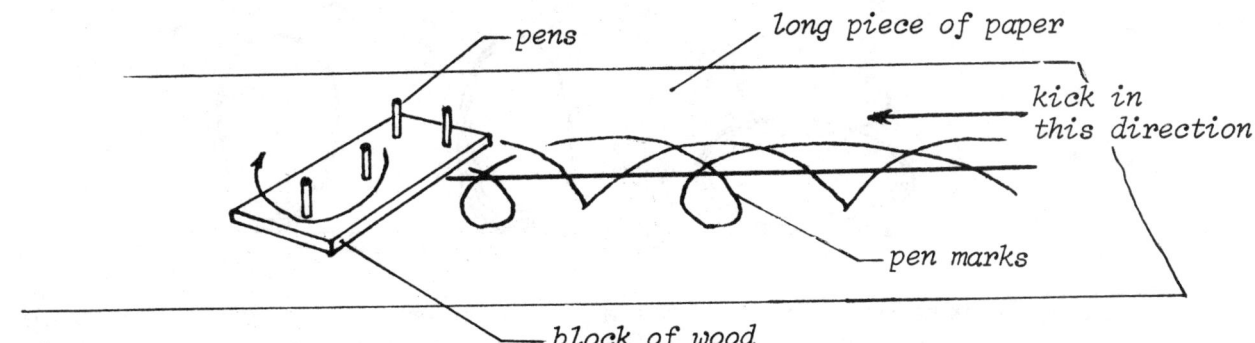

Procedure:
1. Drill three holes in the wood block: two at one end and one at the opposite end. The size of the holes should be such that the pens can be inserted rather tightly in the holes.
2. Insert three pens in the three holes, such that the point just sticks out on the other side of the board.
3. Now find the center of gravity of the total system by finding the balance point on the tip of your finger. Mark that point.
4. Drill the same size hole through this point and insert the final fourth pen the same way (make sure that all four pens are evenly touching the table top).
5. Place the wood block with the pen points on the long stretched out paper in the hall way.
6. Kick the block of wood, such that you send it spinning down the hall over the paper (kick towards the end of the block). Observe the pen marks!

Questions:
1. What kind of lines do you observe?
2. Why did only one line stay straight?
3. What kind of tracks would we get with only the first three pens inserted?
4. Would there be any difference in tracks if we eliminated one of the first three pens? (by retracting it such that the point does not protrude enough to make a track, but leave the pen in the hole).
5. What would happen to the straight line track if we took one of the first three pens completely out of the hole?

Explanation:
This demonstration is another proof that **spinning objects spin around its center of gravity.** The fourth hole in the block of wood was made exactly in its center of gravity. By kicking the block against one end in the direction of the long piece of paper, it was spinning down the hall way around its center of gravity. Thus the fourth pen did not move sideways, and a straight line was obtained. Without this pen no straight line could be obtained.

If one of the first three pens is completely removed, the fourth pen actually would not be located exactly at the system's center of gravity anymore, thus the straight line becomes also a wobbly line.

FORCES   CENTER OF GRAVITY
                                                          SPINNING OBJECTS

13.18. STAND A RAW EGG ON ITS HEAD

Materials: 1. One raw chicken egg.
            2. One hard-boiled chicken egg.
            3. A piece of paper towel or handkerchief.

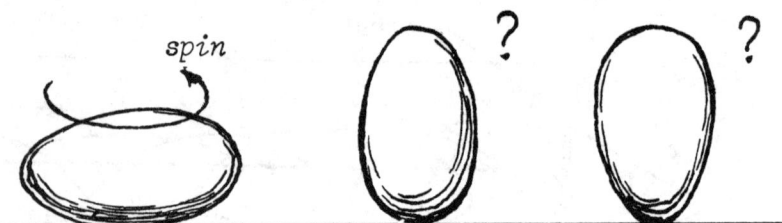

Procedure:
1. Place both eggs on the table and ask students: "How can I tell which one of the two eggs is hard boiled, without breaking the eggs?
2. Spin the eggs one at a time with the same force: Which egg spins much easier and faster?
3. Ask a student to stand the raw egg on its head on the table (with a table cloth). If the table has no table cloth, place the egg on a paper towel, napkin, or handkerchief. It is almost impossible to do.
4. Now hold the raw egg firmly in your hand and shake it vigorously for about 30 seconds, then place it immediately on its head on the table (on the paper or cloth). Voila! It stands!

Questions:
1. Why did the hard-boiled egg spin faster than the raw egg?
2. How else can we tell which of the two eggs was the raw one?
3. Where is the center of gravity of a raw egg located?
4. What was achieved by shaking the raw egg vigorously?
5. What made it possible for the raw egg to stand on its head?
6. Would it be possible to stand the hard-boiled egg on its head?
7. Would you expect it to be harder or easier to stand the hard-boiled egg on its head, as compared to the raw one?

Explanation:
A chicken egg consists of a yoke, egg white, and the shell. The yoke is where most mass is concentrated and thus the **center of gravity** of the egg is close to the yoke or in the yoke. When a raw egg is spun, the yoke swings out (see sketch) and swing back to the other side, thus slowing down the rotation.

When the egg is hard-boiled, the mass inside the egg is solid, thus the center of gravity is stationary and the egg can spin fast around this point.

Shaking a raw egg vigorously will make the yoke more mobile or movable inside the egg-white, thus after shaking the yoke can move down lower in the head of the egg. This lowers the center of gravity and the egg is therefore much more stable. Try to stand the egg on its smaller end: it's possible too!

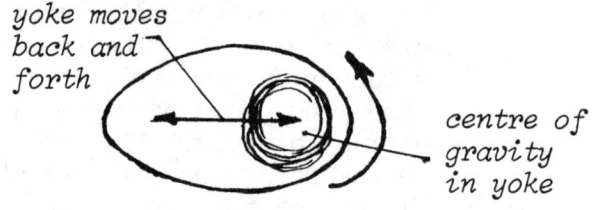

Top View

FORCES   TORQUES

13.19. THE FLOATING BELT HANGER

Materials:
1. A small piece of plywood cut in the shape pictured in sketch or a piece of coat hanger bent in an S shape.
2. A leather belt.

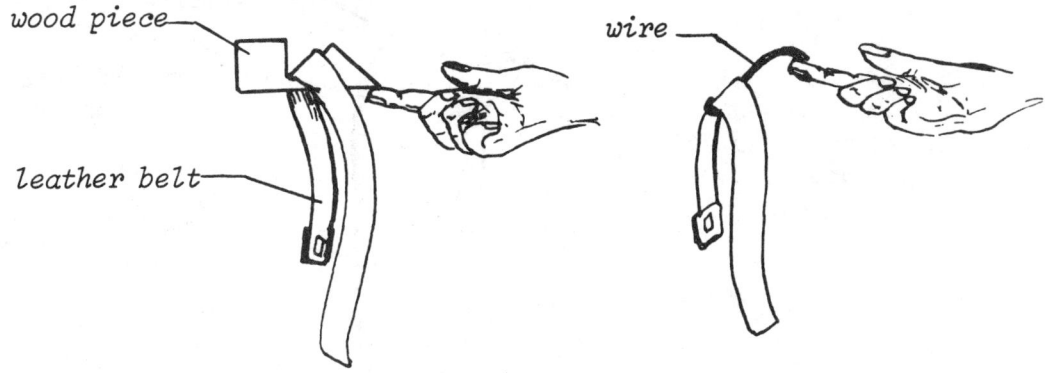

Procedure:
1. Ask students: "What will the wood piece do if I put it on my finger like this (see sketch but without belt) and let go of it?" (Anticipated answer 'it will fall').
2. Let go of the wood piece or wire and let it fall.
3. Now hang a leather belt on the incline of the plywood piece or at the end of the iron wire (make sure that it balances).
4. Hold both ends of the belt in your right hand and pull at it--the belt is held up!

Questions:
1. Why doesn't the belt fall?
2. Which forces are working on the piece of wood?
3. In what position is the belt only exerting a downward force?
4. Can belts made of other materials also be hung on the 'hanger'?
5. How steep does the incline have to be for the belt to stay up?

Explanation:
At this position (Sketch 1) the belt is exerting only a downward force and no other. The incline is now a horizontal ridge. As soon as the shape is turned around point P in a counterclockwise direction, the belt hangs on an incline. Now it exerts more than a downward force. It now exerts a **torque** in a clockwise direction. This means that **two equal but opposite forces that tend to rotate** are now working on the wood piece. The force which is exerting upward farther away from the pivot point P is the one that holds up the belt. The harder the pull downward on the belt, the larger this torque becomes, and thus the larger the force upward.

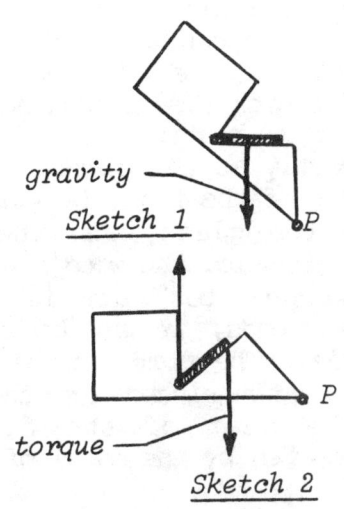

FORCES   TORQUES

13.20. THE FORK AND SPOON ACT

Materials: 1. A fork and spoon (regular dinner silver).
             2. A toothpick or wooden match, a drinking glass.

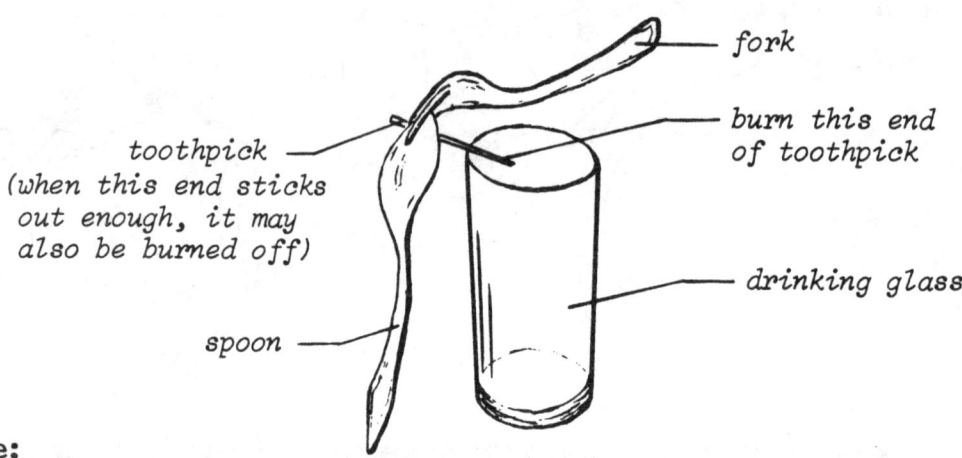

Procedure:
1. Attach the spoon to the fork by pushing it in between the teeth.
2. Place a toothpick or wooden match between two of the fork's teeth and let the system balance from the toothpick resting on the rim of a drinking glass (see sketch).
3. Once the spoon and fork are in balance on the glass rim, dramatize by burning off the end of the toothpick or match.

Questions:
1. What keeps the toothpick floating horizontally beyond the glass rim?
2. How far in or out can the toothpick protrude beyond the glass rim?
3. Why did the burning of the toothpick stop exactly at the fork or the glass rim?
4. What forces are working on the toothpick?
5. What other objects could be used in place of the toothpick?

Explanation:
   The spoon and fork are hanging almost on their sides from the toothpick and this is the reason why they are actually exerting a **torque** on the toothpick (see sketch on right).
   The burning of the pick stops exactly at the glass rim, because the heat of the flame is suddenly absorbed by the glass, and the temperature drops below the wood's **kindling point**. When the toothpick is protruding on both ends, it may be burned on both ends. The flame at the fork end will be extinguished for the same reason (the heat of the flame will be absorbed by the metal of the fork).

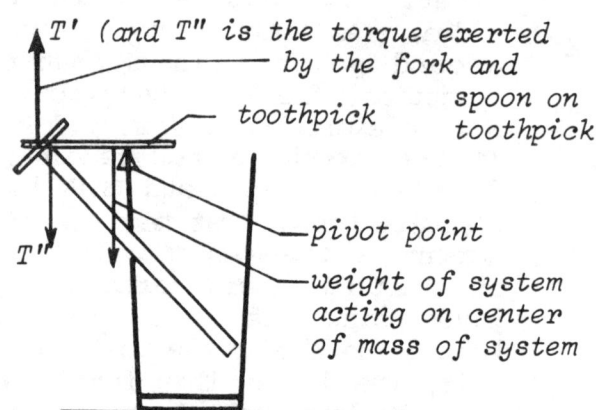

13.21. THE WEIGHTED PIPE

Materials: 1. A 1 m long water pipe (weighted at one end with lead).

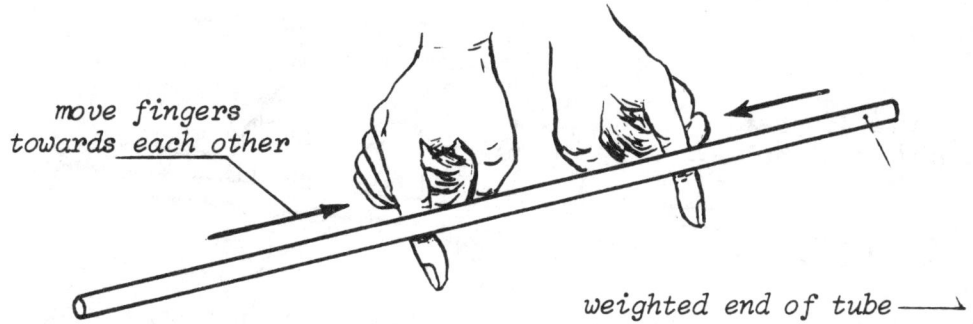

Procedure:
1. Take the water pipe and stuff one end with empty toothpaste tubes (do not show this to the students).
2. Place the weighted tube on your two stretched out forefingers at the far ends of the tube.
3. Move your hands slowly towards each other and ask: "Where will my fingers meet?" (they will not meet in the center of the pipe).

Questions:
1. Why did the fingers not meet at the center of the tube?
2. Holding the tube at the center of mass, which end of the tube is heavier? The short end or the long end?
3. Where are the actual forces working on the pipe in relation to the pivot point, which is the center of mass of the whole pipe?
4. Where would the fingers end up if the pipe were not weighted?

Explanation:
Under the weighted pipe, the sliding fingers will end up under the center of mass of the whole pipe, which is off center, more shifted towards the heavy end. The short end of the pipe is heavier than the long end, yet they are in balance, because the larger force (M_1) has a shorter arm (a) and the smaller force (M_2) has a longer arm (b) in relation to the pivot point (center of mass--see sketch on right). When the object is in equilibrium, **all clockwise couples ($M_1 \times a$) equal the sum of all counter clockwise couples ($M_2 \times b$).**

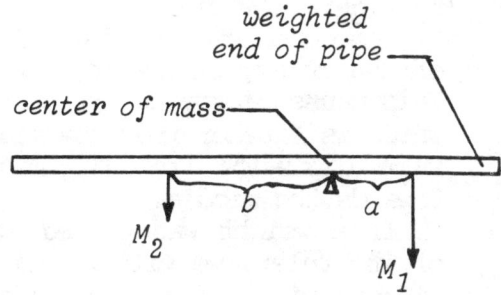

FORCES CENTER OF GRAVITY
FRICTION & TORQUES

13.22. THE GOLDEN MIDDLE

Materials: 1. A meter stick (one for each pair of students).
2. A weight (200 g) or rock.

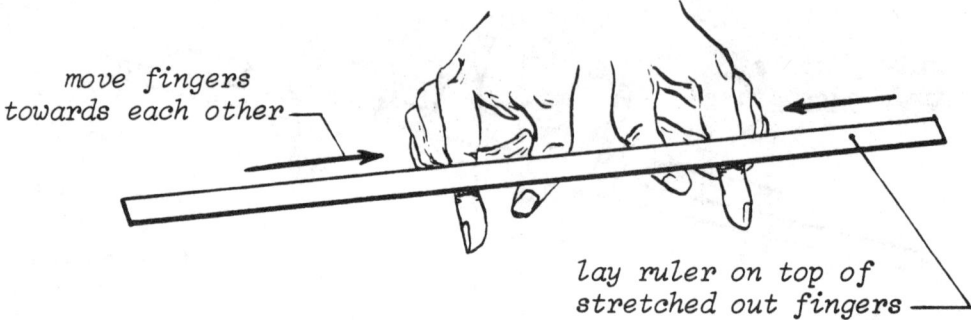

move fingers towards each other

lay ruler on top of stretched out fingers

Procedure:
1. Take the meter stick and place it in any position on your two stretched out forefingers.
2. Move the fingers towards each other: they always end up in the middle of the ruler.
3. Distribute the meter sticks to each pair of students and let them try to do the event themselves.
4. Place a weight or rock at 25 or 75 cm from the end, and let students bring their fingers together. Where do they end up?

Questions:
1. Why do the fingers always end up in the middle of the ruler?
2. Does it matter where the two fingers are placed initially?
3. Which finger moves more: the one closer or farther from the center?
4. What is keeping the finger closer to the middle from moving?
5. Where is the center of gravity of the ruler (without the weight)?
6. Between which finger and the ruler is there more friction: the one closer or farther from the middle?

Explanation:
Since the **center of gravity (center of mass)** of the ruler is in the middle of the ruler, it presses more on the finger which is closest to the middle. This causes a greater **friction** for this finger when moving it. So the finger which is farther from the middle would move first, until both fingers are the same distance from the middle, then both fingers move at the same rate towards the middle.

If a weight was placed at one end of the ruler, we will notice that the fingers will end up very close to the mass or weight, or rock placed on the ruler. This is because the center of gravity of the whole system (stick + weight) is shifted.

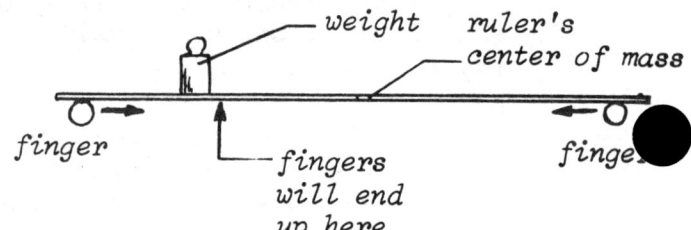

weight
ruler's center of mass
finger
fingers will end up here
finger

FORCES

ADHESIVE FORCES
FRICTION

13.23. THE MAGIC STRIP OF NEWSPAPER

Materials:
1. A 3-4 cm wide strip of newspaper (about 30 cm long doubled).
   (unfolded strip: 60 cm long).
2. Rubber cement, talcum powder (or any kind of flour).

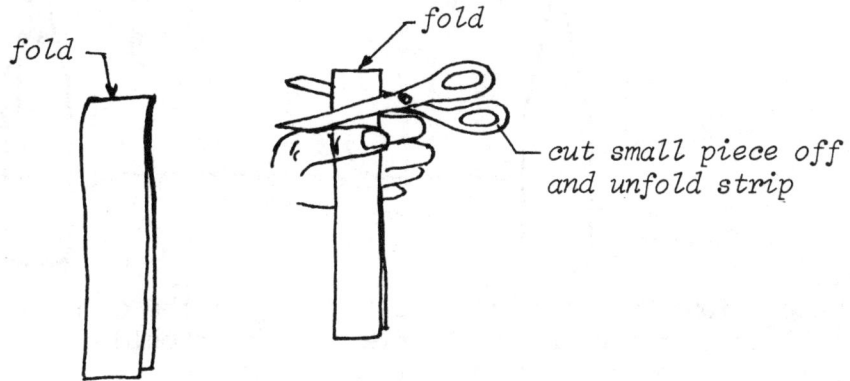

Procedure:
1. Unfold and lay the paper strip on a newspaper.
2. Cover one side of the strip with a layer of rubber cement and leave to dry.
3. Sprinkle some powder or flour over the dried rubber cement and wipe excess off the strip (make sure all of the paper is covered with the powder).
4. Now you are ready for the demonstration: show the audience the paper strip.
5. Hold the strip close to the fold. Take a pair of scissors and cut about 2 cm off from the top, hold one end of the strip and hold it up: it is one whole piece again!
6. This cutting procedure may be repeated several times.
   Let students come up with hypotheses of how this can happen.

Questions:
1. What makes the strip whole again?
2. After cutting, is the strip actually whole?
3. What is the function of the powder or flour?
4. What is the function of the rubber cement?
5. What other material can we use besides rubber cement?
6. What would happen if you cut the strip in an angle?
7. What hypotheses can you come up with to explain this event?

Explanation:
The function of the rubber cement is to make the two cut strips after cutting the long strip with the scissors, to **adhere** to each other. The powder or flour spread on top of the rubber cement makes the surface smooth and not sticky. The two rubber cement covered surfaces can be placed on top of each other without sticking to each other. At the edge where the paper has been cut with the scissors, however, the pressure was high enough that just the edge is sticking to each other. The two cut pieces then look like one whole piece again.

A variation to this demonstration is to cut both the top end and the bottom end, separate the top end and let it hang from one of the strip ends. It is an excellent demonstration to make the students think.

FORCES   FRICTION

13.24. THE INVISIBLE GLUE

Materials: 1. A plastic or glass bottle with a long tapered neck.
               2. A piece of thin rope, a small cork.

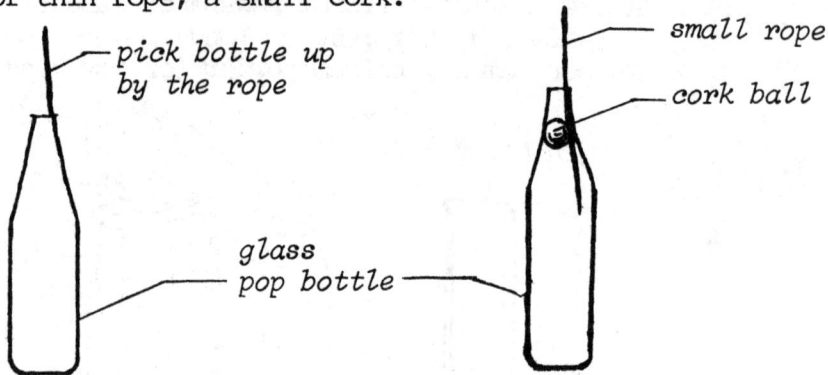

Procedure:
1. Preparation: Make a small ball (sphere) of the cork by cutting and filing until it just fits in the bottle neck (it should be a tiny bit larger so that you have to force the ball cork in).
2. Push the ball cork in the bottle and cover the bottle with paper or paint the whole bottle (so that it becomes opaque).
3. Now you are ready for the demonstration:
   Hold the rope in your hand and say to the audience: "I have some invisible glue in this small bottle" (an empty small bottle), "I´ll dip the end of this rope in it and let it stick to this large bottle".
4. Do the dipping and let the rope slide inside the opaque bottle, turn the bottle upside down, pull slowly at the rope until you feel some resistance, then turn the bottle rightside up, keep the tension in the rope, and let the bottle hang from the rope. Let students make inferences/hypotheses.

Questions:
1. What hypotheses/inferences can you make to explain the event?
   After students know about the cork ball:
2. What is the function of the cork ball?
3. Why does it have to be somewhat larger than the bottle neck?
4. What other material can we use instead of cork?
5. How can we get a regular cork that went in, out of a wine bottle?
6. What is the principle made use of in this demonstration?

Explanation:
    The reason why the bottle can keep hanging from the loose rope is **friction**. When the bottle is turned upside down, the cork ball rolls in the bottle neck and pinches the rope against the side of the bottle. By pulling the rope, it pulls the cork in even tighter as the neck is tapered. The **friction** between the rope and the cork is greater than that between the glass and the cork, so the cork gets pulled a little farther in the neck, thus pinching the rope. The rope can be pulled out of the bottle with a sharp yank. In place of the cork we can also use a rubber ball. A marble of a slightly smaller diameter than that of the bottle neck may also be used, but the rope has to be quite a bit thicker.
    Making use of the same principle, we can take a cork out of an empty wine bottle with a cloth napkin (serviette). Get one of the corners of the napkin inside the bottle neck, let the cork roll over the napkin, slowly pull the napkin until the cork tightens in the neck, then pull hard on the napkin; the cork will pop out!

FORCES   WATER PRESSURE

13.25. THE CARDBOARD BOTTOM

Materials: 1. A glass or plastic open tube (open at both ends).
               (about 2-3 cm diameter and 10-15 cm long).
            2. A glass or other transparent container (4-500 ml beaker).
            3. A stiff paper (or plastic) card, tape and thread, small beaker.

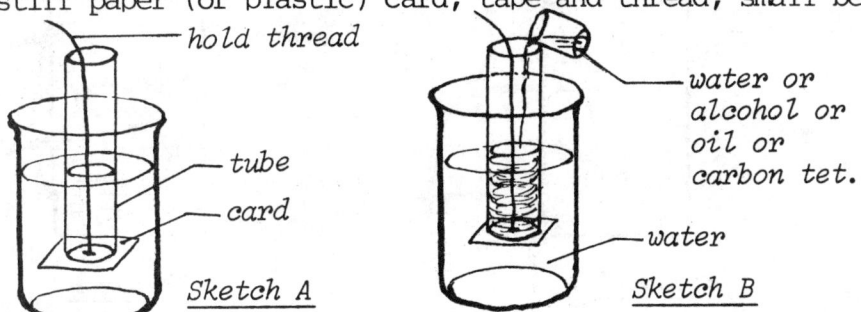

Procedure:
   1. Tape the end of the thread to the middle of the card, let the other end of the thread fall through the open tube, and hold the card by the thread against the bottom side of the tube (see Sketch A).
   2. Lower the tube vertically in the water filled glass container, while holding the card against the bottom of the tube, until about half the length of the tube is immersed.
   3. You can now let go of the thread (the card will be kept in place), and pour carefully and gradually some water in the tube from the small beaker (see Sketch B).
   4. Continue to pour water in the tube until the moment that the card falls away from the tube.

Questions:
   1. At what point does the card fall away from the tube?
   2. What kept the card in place even after releasing the thread?
   3. Why did the card not fall off even after pouring a little water inside?
   4. Would the card fall off if we poured alcohol or oil in the tube instead of water? How about pouring carbon tetrachloride?
   5. In order to pour the same amount of carbontet as the amount of water in the tube, what do we have to do with the tube (so the card would stay)?
   6. Is it possible to pour carbontet up to the level of the water (without the card to fall off)?
   7. What liquids can be poured in the tube up to the water level, without the card falling off?

Explanation:
   Water pressure is **exerted equally in all directions**. The card falls off from the bottom of the tube only when the water column above it is flush with the water level outside the tube.

   Lighter liquids (those of lower density) can stay up even higher above the water level, as it would take more of it to make the same weight of water. Heavier liquids will drop the card sooner. It is therefore impossible to fill the tube up to the water level with liquids of densities greater than 1. (see Sketch on right).

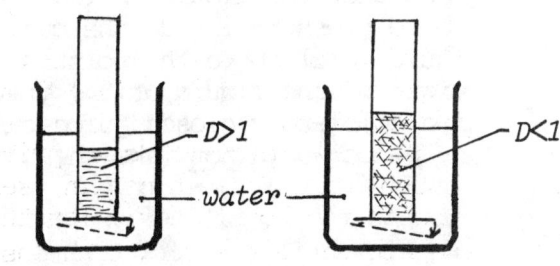

FORCES WATER PRESSURE

13.26. THE SQUIRTING WATER HOLES

Materials: 1. A large tin can (juice can) or milk carton.
            2. A medium sized nail, and bucket or sink.

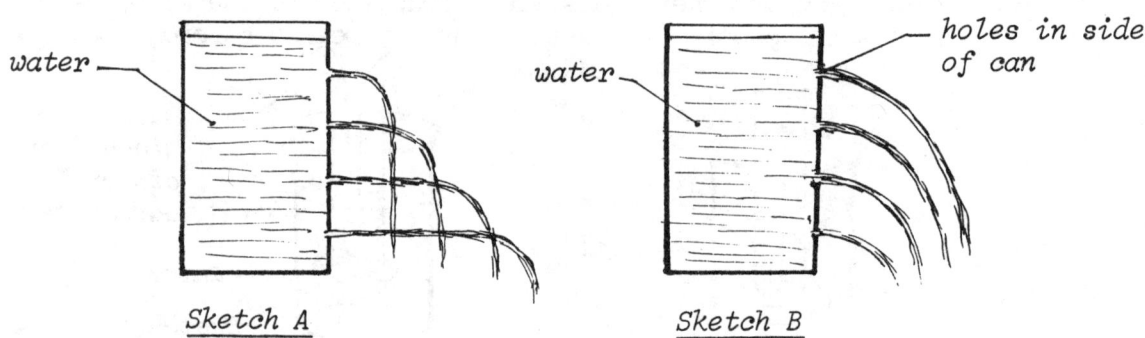

Sketch A   Sketch B

Procedure:
1. Punch 4 holes in the side of the tin can in a vertical line, about 3 cm apart from each other, the lowest one also about 3 cm from the bottom (make the holes the same size and shape).
2. Hold the can above the sink or bucket, have one of the student (X) cover the four holes with his/her fingers, and fill the can full with water.
3. Now ask the students: "Which hole will squirt water the farthest when student X takes away his/her fingers?", and let student X release the flow of water. Compare prediction and result.

Questions:
1. Which hole squirted the water out the farthest? In other words: which of the two sketches above is the correct one, A or B?
2. What factor determined the squirt distance from the can?
3. What are other variables that might influence the squirt distance?
4. How would different sizes of holes in the can compare?
5. Would the diameter of the can influence the squirt distance?
6. Would the height of the can or the height of the water level have more influence on the squirt distance?
7. How would water compare to alcohol or oil in the squirt distance?

Explanation:
   The correct sketch above is Sketch A, showing that the top hole in the tin can would give the shortest squirt, and the lowest hole the farthest. This immediately shows that the height of the water level is influencing the squirt distance. The more viscous the liquid, the less the squirt distance, this means that oil would not squirt as fast as compared to water. A smaller hole compared to a large hole, with all other variables held constant (the same), will give a larger squirt distance.
   Whether the liquid inside the can will squirt out at all, depends on the **liquid pressure** inside the can. This is caused only by the height of the liquid level above the point of puncture. This in turn is caused by the weight of the liquid or the gravity force. If there were no gravity working on a completely closed juice can, like in a satelite capsule, puncturing it at any place would <u>not</u> make the juice come out.

FORCES   WATER PRESSURE
                                                                          VORTEX IN A LIQUID

13.27. THE OUTPOUR RACE

Materials: 1. Two identical gallon jugs (with narrow neck).
             2. Sink or buckets.

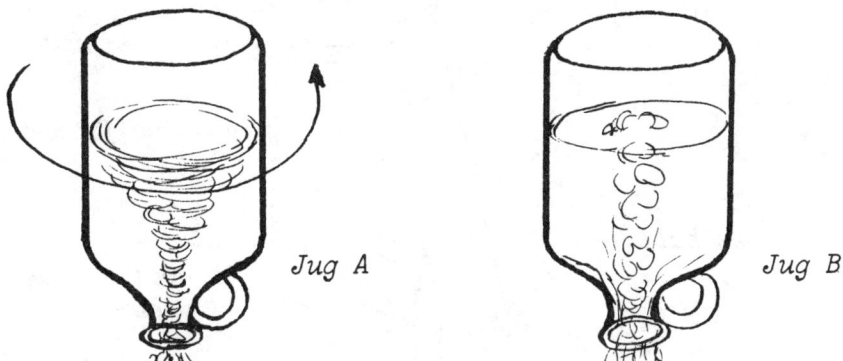

Procedure:
1. Fill both jugs three quarters full of water. Place them next to each other and show the students that they have the same amount of water.
2. Ask two students to come forward to pour the water out as fast as they can into the sink or the buckets (student A already got instructions beforehand to swirl the water before pouring it out: Hold your palm over the opening while holding the jug upside down, make a turning motion to swirl the water until it forms a vortex, then let go of your palm covering the opening).
3. The rules are: The pourers may do anything to the jug before and during the pouring of the water (except break it).
4. Have two other students time the outpour with a stopwatch or by putting their hands up as soon as all the water has poured out.

Questions:
1. Which jug invariably wins? Jug A or Jug B?
2. What does the swirling do to the water level?
3. Why does it take so long for water to pour out of the jug?
4. What is trying hard to get into the jug while pouring the water out?
5. What is usually the cause of the vortex formation in the water surface?
6. What would you expect the pressure to be in the center of a vortex, higher or lower than the periphery?
7. In what places in nature can you find vortex formation?

Explanation:
By moving the jug in a clockwise or counter-clockwise rotation before pouring the water out, a **vortex** is formed, indicated by a funnel-like shape of the water surface. At that moment the center of the vortex has a much lower pressure than the outer areas, making it easier for the air to come into the jug. Thus the water will pour out faster. Keeping the water swirling while the water pours out will make it pour out steadily as the vortex is maintained.

Vortex formation is found in nature in fast flowing rivers, where the water has to go around a corner or around a large rock. In pools below a waterfall we can also find vortex formation, caused by strong currents of water. As the pressure in the center of the vortex is lower than its surrounding outer area, swimming near a waterfall can be quite dangerous. When caught in a vortex, the best thing to do to get out of it would be to go with the flow, which is downward and come up to the surface at another spot.

FORCES — THE PENDULUM

13.28. HIT THE BOTTLE ON THE BACK SWING?

Materials:
1. A soft ball or base ball.
2. A thin string & hook or nail.
3. An empty coke bottle.

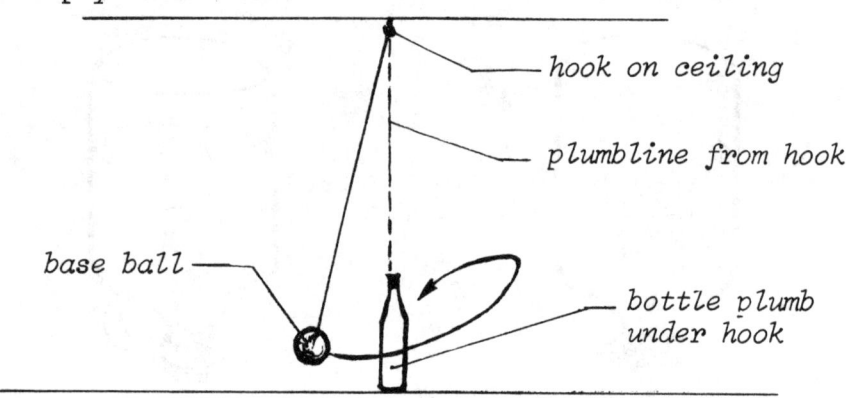

Procedure:
1. Tie the base ball to the string and hang it from the ceiling in such a way that it hangs about 2 cm higher than the bottle above the floor (do not make a knot in the string).
2. Let the ball hang very still and place the bottle plumb under the ball. Mark the position of the bottle on the floor with chalk.
3. Now lower the ball until it is about 10 cm above the floor, then put a permanent knot in the string (at the hook on the ceiling).
4. Bring the ball out towards you (about 2-3 m away from the bottle), swing the ball to the side of the bottle and try to hit the bottle with the ball on the back swing.

Questions:
1. Is it possible to hit the bottle on the back swing?
2. Why does the ball always swing back on the other side?
3. What is the energy of the swinging ball transformed into?
4. Will the ball eventually hit the bottle when it is allowed to swing more than once?
5. How should we swing the ball to obtain the least number of swings to hit the bottle?

Explanation:
The energy of the pendulum is transformed into heat at the point of attachment of the pendulum and in the air as friction. The amplitude or height of the swing is therefore almost the same for the first few swings.

When looking from the top of the pendulum and the bottle, we can decompose the movement of the pendulum along axes X and Y. As the swing stays almost constant along both axes, the ball can not hit the bottle on the back swing (for the first few swings). After a while the ball will fall back to the vertical position and eventually hit the bottle.

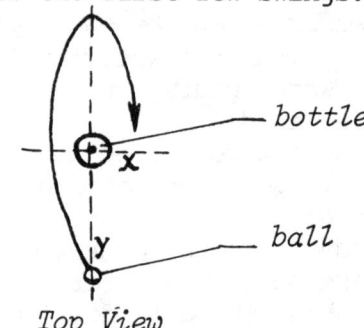

Top View

FORCES   THE PENDULUM

13.29. HOW MANY SWINGS CAN YOU GET?

Materials: 1. Fifteen washers (for a class of 30).
             2. Thin string or strong thread, and scissors.
             3. Wooden meter stick, a dozen small hooks (to place in ruler).

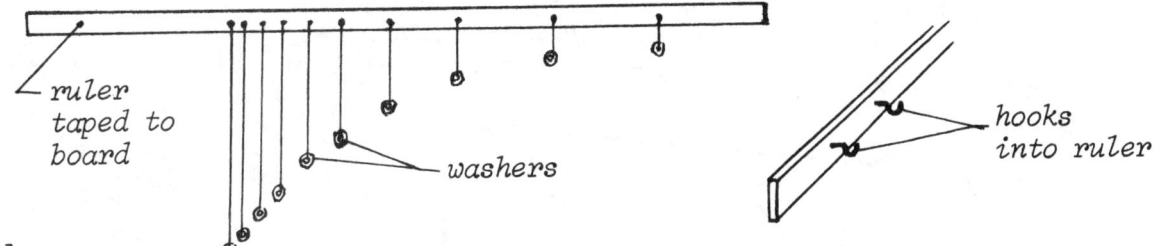

Procedure:
1. Place the small hooks into the wooden meter stick at the following spots: 39, 40, 42, 44, 47, 50, 54, 60, 66, 74, 94. To be safe, you may want to add a few more hooks about half or one cm away from each of the above points, then tape the ruler against the board (up high).
2. Tie a string to each of the 15 washers and make a knot at the washer and another knot the following lengths away: 60, 55, 50, 45, 40, 35, 30, 25, 20, 15, and 10 cm, and cut the string off 2 cm above this last knot, and make a small loop (to hang the pendulum on the hook).
3. Give each pair of students one pendulum. Let one student hold the pendulum with a steady hand, and let the other count the number of swings (whole periods) per minute (count the number of times it comes back to your hand in 30 seconds, then double it). Let them do this two or three times and take the average of the value.
4. Once they have the number, let them come to the board and hang their pendulum at the correct spot on the ruler (similar to plotting length of pendulum against frequency of swings).

Questions:
1. Which of the pendulums got the most number of swings?
2. What determines the number of swings per minute of a pendulum?
3. Would a different/heavier pendulum bob affect the frequency of swings?
4. Would a thinner or heavier string make a difference?
5. What are other variables that would affect the number of swings?
6. Which of the pendulums could be used as a timer?

Explanation:
   This activity is especially effective to demonstrate that the number of swings per minute (**frequency**) in a pendulum depends solely on the length of the pendulum. It is independent of the mass of the bob.
   Variables like air friction, friction of the suspension construction, movement of the suspension point (in our case whether the student has a steady hand or not), may have some influence on the slowing down of the pendulum and thus on the number of swings.

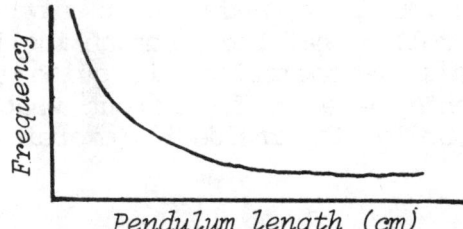

   When plotting the number of swings per minute or frequency against the length of the pendulum, we will get a very similar parabolical curve as the one on the board.

13.30. THE MAGIC COME-BACK CAN

Materials:
1. A tin can with an opaque lid.
2. A thick, long rubber band and a short piece of string.
3. A weight or stone.

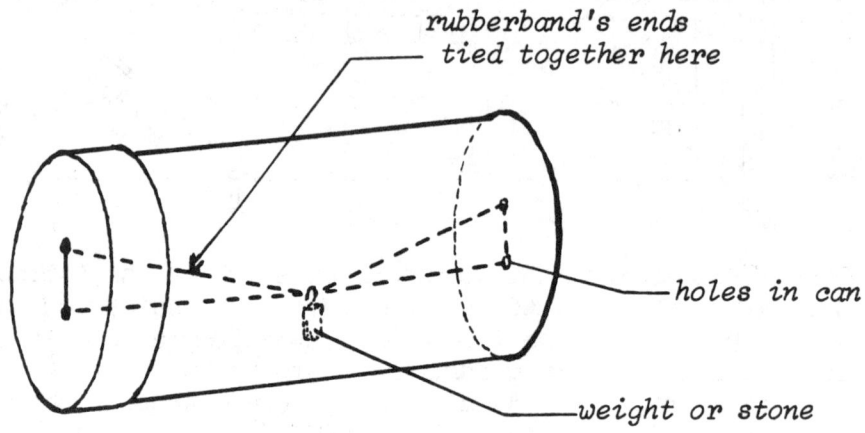

Procedure:
1. Make two small holes in the bottom and also in the lid of the can about 2 cm apart.
2. Cut the rubber band and push it through the holes, and tie the loose ends together inside the can (see sketch).
3. Tie a short piece of string around the weight or stone and tie the other end of the string to the center of the rubber band (do not let the weight or stone drag on the side of the can).
4. Place the lid back on the can, and you are now ready for the demonstration: Roll the can on a horizontal plane (the table top) and it comes rolling back!
5. Wind the rubber band up first, and let it roll uphill on an inclined plane.

Questions:
1. Why did the can roll back (or roll uphill)?
2. What is the purpose of the rubber band?
3. What else is inside the can besides the rubber band?
4. How is the construction inside the can?

Explanation:
   The purpose of the rubber band is to provide the rolling-back action. The weight or stone holds the rubber band stationary in the center of the can, so that when the can rolls, the rubber band twists itself, building up a **latent force** inside it, and when it untwists itself the cylinder rolls back. It can even roll up not too steep an incline.
   This demonstration is quite suitable for a **black box** activity, where students have to figure out what is inside the can that makes it move, without looking inside it (exercise on inferring).

FORCES

CENTER OF GRAVITY
SUDDEN STRESSES

13.31. THE INCREDIBLE STICK

Materials:
1. A thin and long pine stick (2 m x 1 cm x 1/2 cm).
2. Two wine glasses or small beakers.
3. Two pins, a thick meter stick (or other strong stick).
4. Two sharp kitchen knives & masking tape.

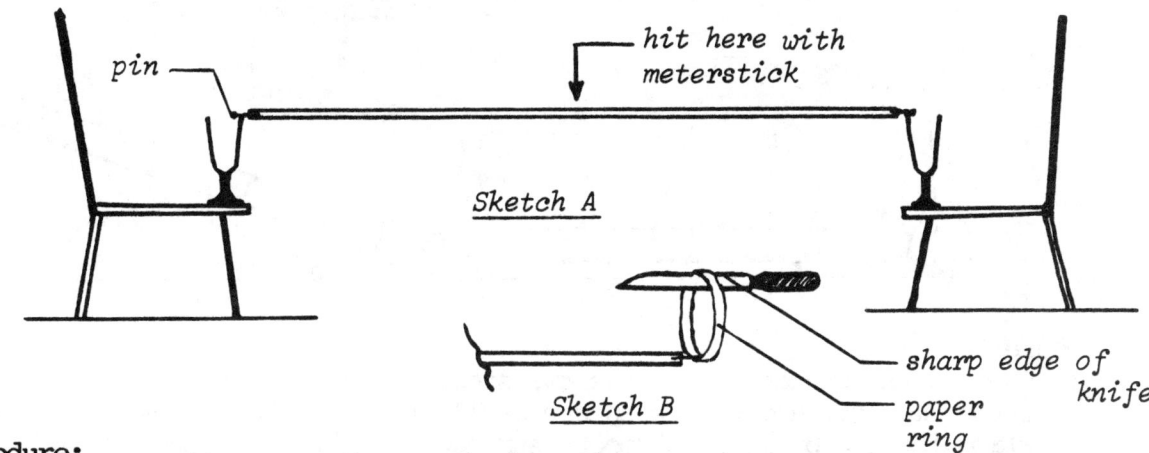

Procedure:
1. Push the pins into each end of the wooden stick.
2. Place the wine glasses or the beakers each on the edge of a chair or table, space them 2 m apart, and let the long wooden stick rest on the rim of the glasses suspended from the pins (see Sketch A).
3. An alternative would be to suspend the wooden stick on two thin paper rings hung on the sharp edge of the kitchen knives (taped against the back of two straight-back chairs--see Sketch B).
4. Hold the strong meter stick like an axe over your head and give the long horizontal stick a sharp blow in the center of its length.

Questions:
1. Why did the beakers not break or fall? (for alternative A)
2. Why did the paper rings not tear? (for alternative B)
3. Was the force exerted on the center of the stick large or small?
4. How did the two broken pieces of wood move when they were hit?
5. Around what point do rotating objects move?

Explanation:
   The force of the blow was so sudden in the center of the wooden stick, that it broke before the force could be exerted on the paper rings or on the glass rims. The descending cudgel hits the center of the stick with such a great and sudden force that the center part descends and breaks. The two broken pieces are actually being suddenly rotated, one in a clockwise and the other in a counterclockwise direction. As **rotating objects rotate around their center of gravity**, the far ends of the long stick actually move slightly upwards, as the two pieces start rotating. The result of the blow is spectacular: The center of the stick falls victim to the **sudden influx of energy**, whereas the ends remain pretty well stationary or even more slightly upwards.

FORCES

ELASTICITY
CONSERVATION OF MOMENTUM

13.32. IS THE BALL REPELLED?

Materials:
1. A strong rod magnet (or horseshoe magnet).
2. A small steel ball and a larger steel ball or two small identical steel balls.

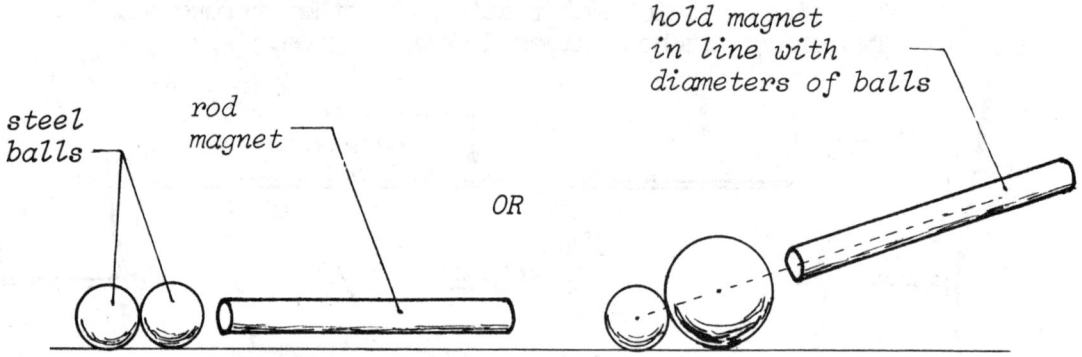

Procedure:
1. Show to the students that the two steel balls are both attracted by the strong magnet, and that both can easily hang on the magnet.
2. Place the two balls on a smooth surface, next to and touching each other.
3. Approach the balls slowly with the magnet from a direction which is in line with the two diameters of the steel balls (if one ball is larger than the other, approach the larger ball with the magnet).
4. Ask: "What made the second ball move away from the first one?"
   A clue: if the magnet is not held in line with the extension of the two diameters, the second ball will not be moving away!

Questions:
1. What would the balls do when they are approached by the magnet in an angle to the two diameters?
2. What would the balls do when they are approached from the smaller one (in case of two different sized balls)?
3. What would you expect the balls to do when using an overly strong magnet? Would they behave the same way?
4. How close does the magnet have to be before the balls are attracted?
5. Why do the two balls have to be touching each other? What would happen if they were not?

Explanation:
The first hypothesis that an observer will form when seeing the second ball moving away from the first is, that it does this because it is repelled by the magnet somehow (by a change of polarity, which is completely erroneous!).
Both balls are attracted by the magnet and at the time of impact the second ball is bounced back because of the momentum it has and the elasticity of steel metal. It moves away out of the magnetic field and is thus not attracted any more by the magnet. This bounce will not occur when the magnet is not held in line with the two diameters of the balls. An overly strong magnet will hold both balls in its magnetic field and the bounce will not occur.

FORCES   RIGIDITY

13.33. PIERCE A POTATO WITH A STRAW?

Materials: 1. A large raw potato.
            2. Two regular drinking straws.

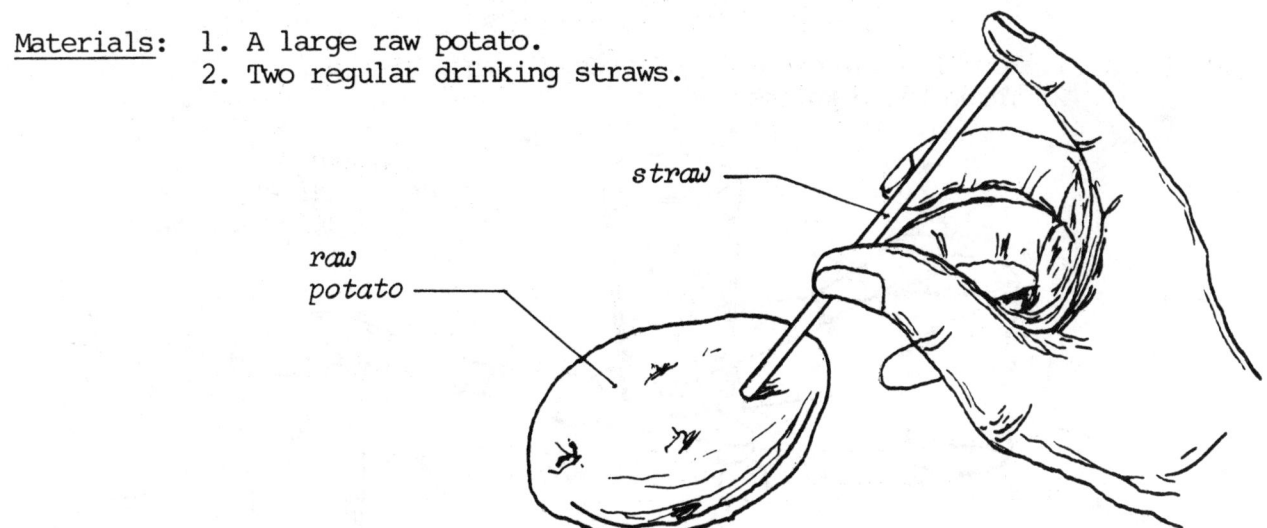

Procedure:
1. Hold one straw in your hand and push it against a hard surface, like f.i. the table surface and show the students that the straw is bending/folding.
2. Now hold the large potato in your left hand and the straw in your right (when you are right handed) with the forefinger tightly pressed against the end opening of the straw (see Sketch above).
3. Pierce the potato with one forceful stroke of the right hand. The straw goes right through the potato! This may be repeated.

Questions:
1. Why does the straw bend when pushed against a hard surface?
2. What is the purpose of holding the forefinger tightly pressed against the end of the straw?
3. What would happen if the forefinger was not held there?
4. What is inside the straw at the moment that it hit the potato?
5. Would the pressure inside the straw be higher or lower than the atmospheric pressure?
6. What are other examples where a higher pressure of air or water inside the object would make it more rigid and strong?

Explanation:
   The purpose of placing the forefinger tightly against the end of the straw is to prevent the air inside the straw from escaping. At the moment that the straw hits the potato, the air inside the straw is compressed and the higher pressure makes the straw rigid and strong. The flexible straw then acts like a solid spear and penetrates the potato very easily. If the forefinger was not placed at the end of the straw, it would just bend when hitting the potato with it, just like it did when hitting the table top with it.
   Other examples where a higher pressure of air makes the object rigid are: bicycle and car tires, inflated plastic or canvas beds and toys. Higher pressure of water inside the fibers and capillaries of plant tissue makes them more rigid and strong: fresh looking plants as compared to limp and drooping leaves because of water shortage.

FORCES

STRENGTH OF
WEAVED SUPPORTS

13.34. THE LOOSE KNIFE SUPPORTS

Materials: 1. Four regular drinking glasses.
             2. Three table knives.

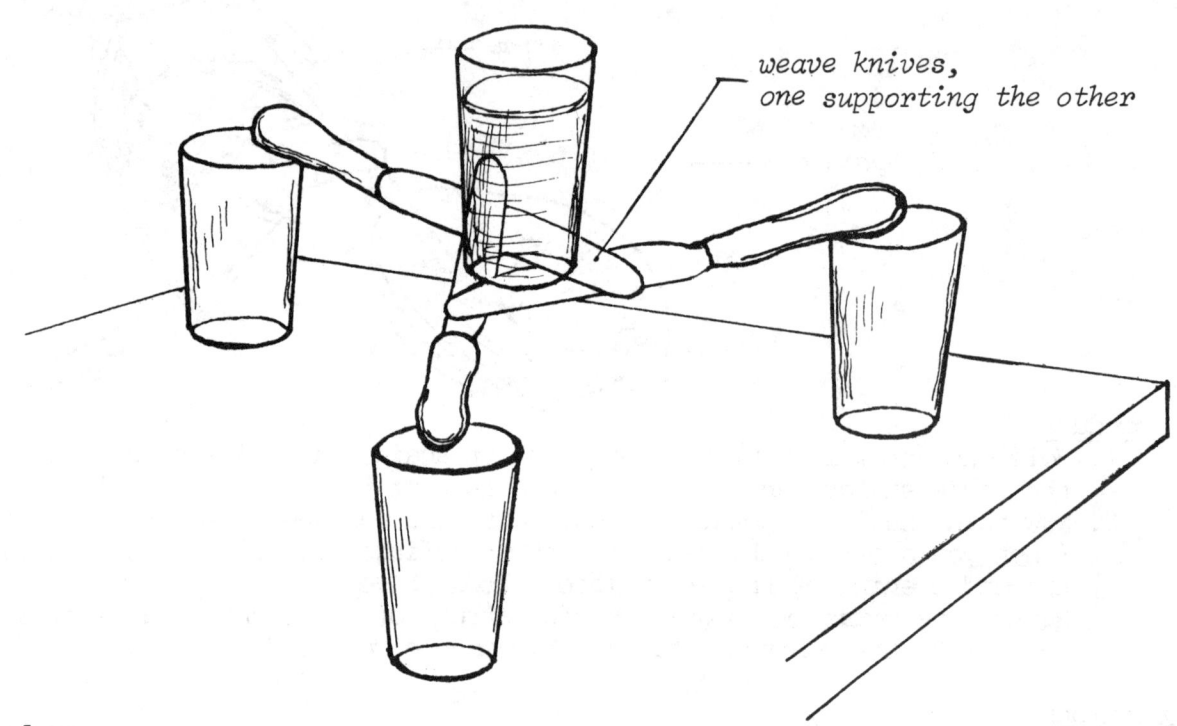

weave knives, one supporting the other

Procedure:
1. Place the three glasses in a triangle configuration on the table a little further than a knife long span away from each other.
2. Ask the audience: "How can I support the fourth glass in the middle of the other glasses at glass-height above the table with the three knives?"
3. Anticipated answer: "It's impossible!"
4. Weave the knives as shown in the sketch above, and a strong support is created, which can hold the fourth glass quite easily. You may even want to fill this fourth glass full with water.

Questions:
1. How can the three knives make such a strong support?
2. How are the knife ends held up?
3. What does the weaving of the knives actually accomplish?
4. What other materials could we use instead of the knives?

Explanation:
   The end point of each of the knives is resting on the middle of the next knife, which the weaving structure actually accomplishes. By weaving the knives they act like a solid metal plate, which is certainly strong enough to hold up a considerable amount of weight.
   Any other stiff material, like forks or spoons, or chop sticks may be used instead of the knives.

FORCES   STRENGTH OF
                                                          CORRUGATED MATERIALS

13.35. THE DOLLAR BILL BRIDGE

Materials: 1. A dollar bill (any denomination or other currency).
            2. Three drinking glasses (one empty, the others may be full).

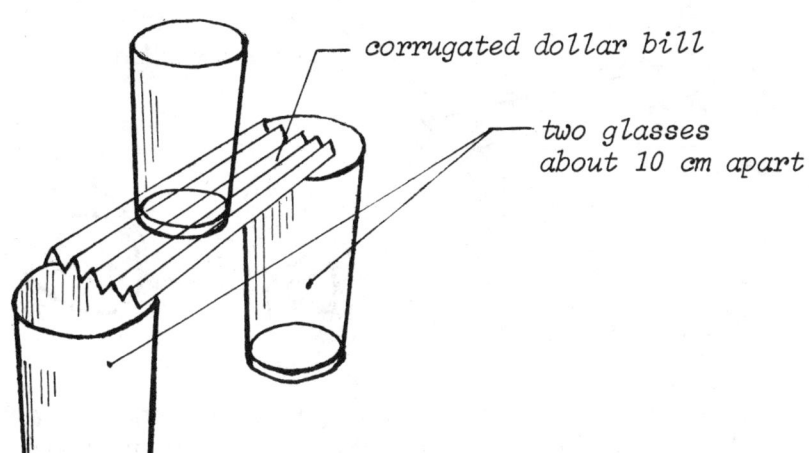

Procedure:
1. Place two drinking glasses (that may be full or empty) about 4 cm less than the length of the dollar bill apart from each other.
2. Show people the third empty glass and the dollar bill. Ask: "Can you support the third glass at glass-height in between the two glasses?" Anticipated answer: "It's impossible!"
3. Say: "Nothing is impossible!" Proceed to fold the bill in half lengthwise then fold this in half again, and once more in half. Unfold the bill and fold it now in and the next crease out as in pleads of a skirt (in zig-zag pattern when seen from the side). Make sure that the folds are sharply creased (use your nail to press).
4. Spread the pleated bar somewhat out and place it over the two glasses, bridging the two, and carefully place the third glass in the center on top of the dollar bill. Voila! The dollar bill bridge!

Questions:
1. What made the dollar bill so much stronger?
2. Where do we see this similar structure in nature?
3. Can you find similar zig-zag structures (corrugated) that strengthen different materials in our daily life?
4. Would more and smaller corrugations be stronger than fewer and larger corrugations in the dollar bill?

Explanation:
   Providing the dollar bill with the length-wise corrugations actually turned the bill into a strong beam across the two drinking glasses. There is an optimum number of corrugations which gives it an optimal strength. The larger the corrugations the less we can make from the width of the dollar bill, and the more corrugations the smaller they are and thus the weaker.
   In nature we find similar corrugations in celery stalks and other plant tissues. In our daily life we encounter the zig-zag or corrugated structures in bridges, plastic roofing, cardboard boxes - cardboard consists of two layers of heavy paper with a corrugated layer in between.

FORCES STRESSES IN PAPER

13.36. HOW LONG CAN YOU HOLD THE BURNING PAPER?

Materials: 1. A strip of paper (about 30 cm long, 4 cm wide).
2. A sharp edge of the table, and a match.

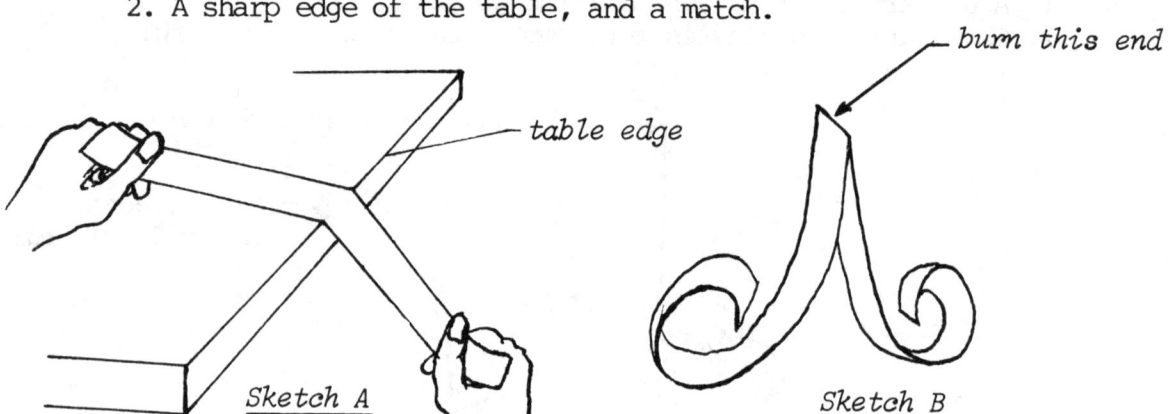

Procedure:
1. Take the ends of the paper strip, one end in each hand, and rub it over a sharp edge of furniture (edge of table or chair). Do this 2 or 3 times (see Sketch A).
2. Notice that the paper strip curls up. Now fold the paper in the center in the opposite direction of the curl (see Sketch B), and hold the two ends together. (You may want to tape the ends together if you like to leave it lying on the table beforehand).
3. Now you are ready to ask a student: "If you hold the strip of paper (at the taped end) vertically, and I burn the top, how long do you think it will take before you have to let go of the paper?"
Anticipated answer: "About half a minute".
4. Have a student hold the paper upright and burn the folded end. In about 3 seconds he/she will let go of the paper!

Questions:
1. What did the rubbing of the paper over the sharp edge do to it?
2. What made the student drop the paper so quickly?
3. How can we uncurl a curly strip of paper?
4. What did the burning of the folded end of the strip of paper do?
5. What would happen if we held the paper horizontally and burned it?
6. Would you hold the paper vertically downwards and start burning it from the bottom?

Explanation:
By rubbing the paper over the sharp edge, we caused the outside of the paper to stretch more, and thus a tension in the paper is created. When the paper is burned at the folded end, it separates and the two loose burning ends come curling downwards with lightning speed. Nobody would be holding on to a burning piece of paper if the flame suddenly moves towards the person. The immediate reaction of the person would be to let go of the burning paper.
Straightening out a curling strip of paper can be done by rubbing the outside of the curl over a sharp edge. Another way would be to apply heat to the piece of paper, by ironing for example.

FORCES   ROTATIONAL FORCES
                                                                  TORQUES

13.37. THE CONFUSED TWIRLING PAPER

Materials: 1. A strip of paper (about 15 cm long, 5 cm wide).
            2. A pair of small scissors.

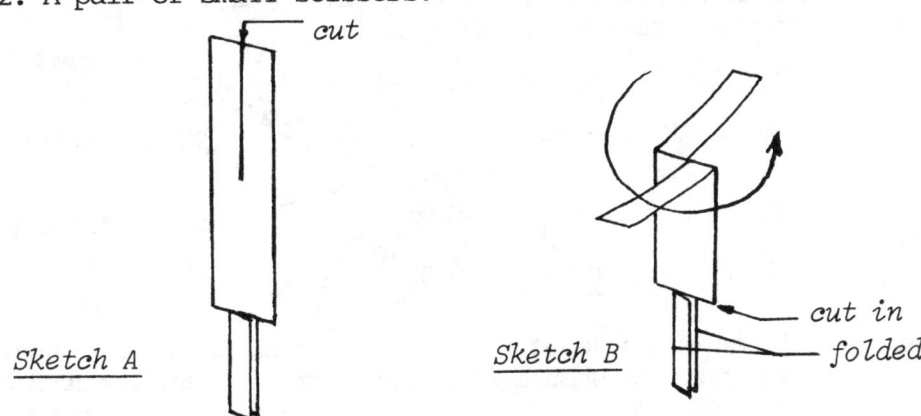

Procedure:
1. Take the paper strip and cut about 5 cm into the center of the strip and fold the two parts to opposite sides of the strip (see Sketch A).
2. At about 5 cm from the other end of the strip, cut crosswise one third into the width of the strip and fold the two parts on top of each other, making the bottom part thicker (see Sketch A).
3. Take the paper piece in the middle and drop it to the floor from standing height. Which way does the paper piece rotate? (See Sketch B)
4. Now fold the two "wings" to the opposite side. Drop the paper piece again from standing height to the floor. Which direction does it rotate?

Questions:
1. What makes the paper piece rotate when dropped to the floor?
2. What are the forces working on the paper piece?
3. What direction did it rotate the first time? And the second time?
4. What made the rotation change direction?
5. Can you find examples in nature making use of this principle?

Explanation:
   The forces working on the paper piece when dropped to the floor are caused by air friction. When looking at the piece from above (obliquely, like in Sketch B above), and the right wing is folded away from you and the left towards you, dropping it will make it rotate clockwise.
   When the "wings" are folded in the opposite way, the paper piece rotates counterclockwise. When we look at the forces working on the paper we note a torque working in both cases (see Sketch on right).
   Many seeds of different fruits in nature are equipped with "wings" which will promote the propagation of the seeds when released from the ripe dry fruit.

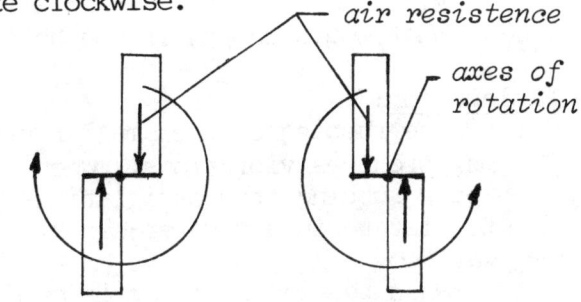

Top View

FORCES

ROTATIONAL VIBRATIONS
DIRECTIONAL OSCILLATION

13.38. THE YIP-YIP STICK

Materials: 1. Two wooden dowels (one somewhat thinner than the other).
   or two small pine sticks (about 1x1x15 cm).
2. A coffee stirrer and a small nail.

notches
small nail
small stick
coffee stirrer
notched stick

Procedure:
1. Make small notches on one edge of the pine stick about 1 cm apart from each other. The notches should be about 3 mm deep and about 2 mm wide.
2. Cut a 6 cm piece off from the coffee stirrer and make the same rounding edge on the cut end.
3. Nail the middle of this piece of stirrer to the top end of the pine stick, and wiggle it so that it will turn loosely around the nail (the nail should be hammered in only half way).
4. Hold the smaller stick in your right hand (if you are right handed) with your index finger pointing out (see Sketch) and hold the notched stick in your left hand.
5. Rub the notched stick with the smaller stick over the notches and let your thumb slide on the right side of the notched stick (stirrer will rotate counter-clockwise as seen from the performer's viewpoint).
6. Tell the audience that the stirrer (rotor) will turn the opposite way when you say "Yip-yip". Say "Yip-yip" and keep rubbing the notched stick but this time let your index finger slide against the notched stick. Observe the rotor rotate in a clockwise direction!

Questions:
1. What made the rotor turn?
2. What caused it to turn one way and what caused it to turn in reverse?
3. Did the saying of "Yip-yip" have anything to do with turning it in reverse?
4. Does the distance between the notches have an influence on the rotation?
5. Does the speed of the rubbing have anything to do with it?
6. Would rubbing the notched stick on one side, then turning it to the other side, have anything to do with it?
7. What would happen if you held the smaller stick another way?

Explanation:
The rubbing of the smaller stick against the notched stick with the rotor on, produces vibrations. When the thumb is held against the notched stick while rubbing, the oscillations take the form of an elipse causing the stick to vibrate in a counter-clockwise direction, and thus the rotor rotates that way too.

When the index finger is held against the notched stick, the oscillations are slanted in a clockwise direction. This changing of rotational direction can also be achieved by turning the notched stick slightly to the left or to the right, in other words by rubbing the notches on the left side as compared to rubbing it on the right side.

FORCES

SIMPLE MACHINES
FIRST CLASS LEVERS

13.39. TILT A HEAVY LOAD WITH ONE FINGER

Materials:
1. A long wooden stick (shovel handle) or an iron water pipe (3-4 m).
2. Two or three small wooden blocks.
3. A heavy piece of furniture.

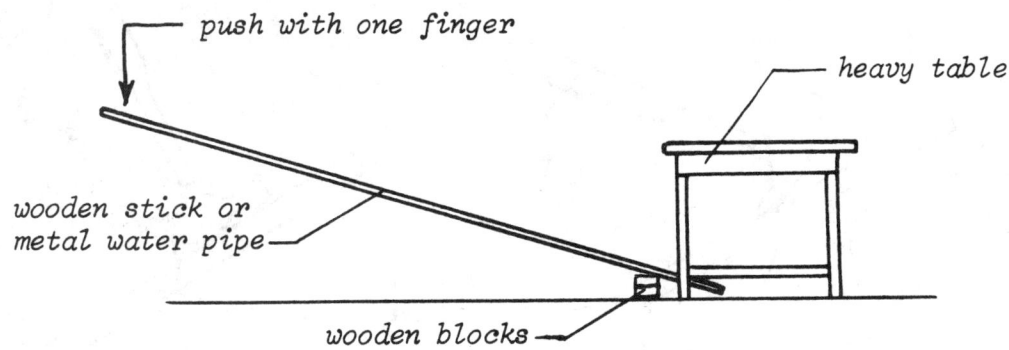

Procedure:
1. Have students try to lift one side of the heavy piece of furniture to give them an idea of its weight.
2. Place the two or three wooden blocks close to the side of the table to be tilted, place the end of the long stick under the table rung and use the blocks as a fulcrum for the lever (see Sketch).
3. Now push the long end of the stick with one finger down (as a heavy load, a desk or table, or a chair with a student sitting on it, may be used; make sure that a rung or horizontal bar is present, close to the bottom of the load to be lifted, to hook the lever on).

Questions:
1. What type or class of lever are we dealing with?
2. What functions as the resistance in this case?
3. What were the effort and the fulcrum in this lever?
4. What other examples can you name that are based on this class lever?

Explanation:
This **lever of the first class** has the **pivot or fulcrum** in between the **effort** and the **resistance**. The latter one being the heavy load (table, desk, or chair with someone seated on it) and the effort being the finger pushing on the long end of the lever, and the blocks of wood serve as the pivot or fulcrum.

Other examples applying the first class lever are: a pair of pliers, a pair of scissors, wire and chain cutters, most car jacks, teeter totters, crowbars, etc.

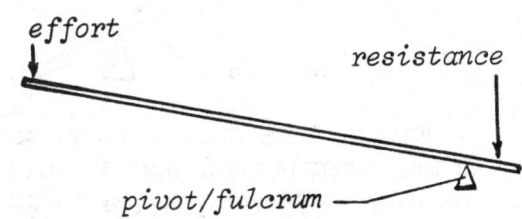

FORCES
SIMPLE MACHINES
SECOND CLASS LEVERS

13.40. IS THE HAMMER A LEVER?

Materials: 1. A hammer (which can be used to pull nails out).
            2. A wooden board and small common nails.

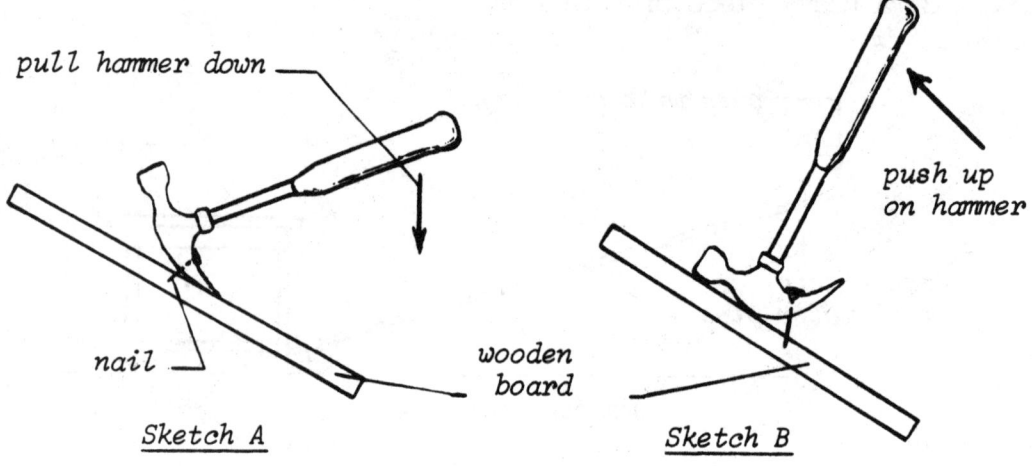

Sketch A   Sketch B

Procedure:
1. Hammer a couple of nails into the wooden board (half way) and pull one nail out by pulling down on the hammer (see Sketch A).
2. Pull the other nail out by pushing up on the hammer (see Sketch B).
3. Ask students the questions listed.

Questions:
1. Which of the two methods of taking out the nails was an application of the second class lever?
2. What class lever was the other method applying?
3. Where is the fulcrum, effort and resistance located in method A? Where are these in method B?
4. What other examples can you name that apply the second class lever?

Explanation:
    The best way to clarify where the forces are at work with the hammer pulling the nail, is to draw a sketch of both methods and the relationships and location of the forces and fulcrum.

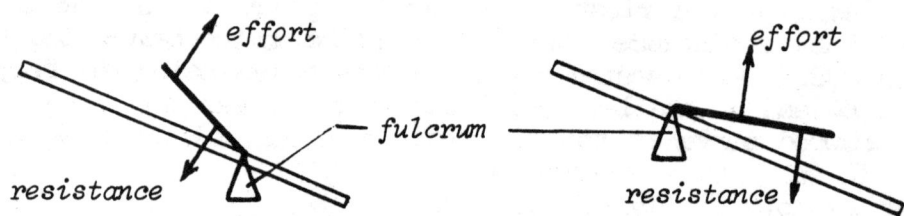

    Method A is applying the second class and method B the third class lever. Other examples of second class lever applications are: the wheelbarrow, crowbar, nutcrackers, paper cutters, two-hole punchers.

FORCES   SIMPLE MACHINES
                                                    THIRD CLASS LEVERS

13.41. THE THIRD CLASS BICEP

Materials: 1. Two pieces of wood (or two rulers).
           2. A small hinge (or masking tape), & thick rubber band.

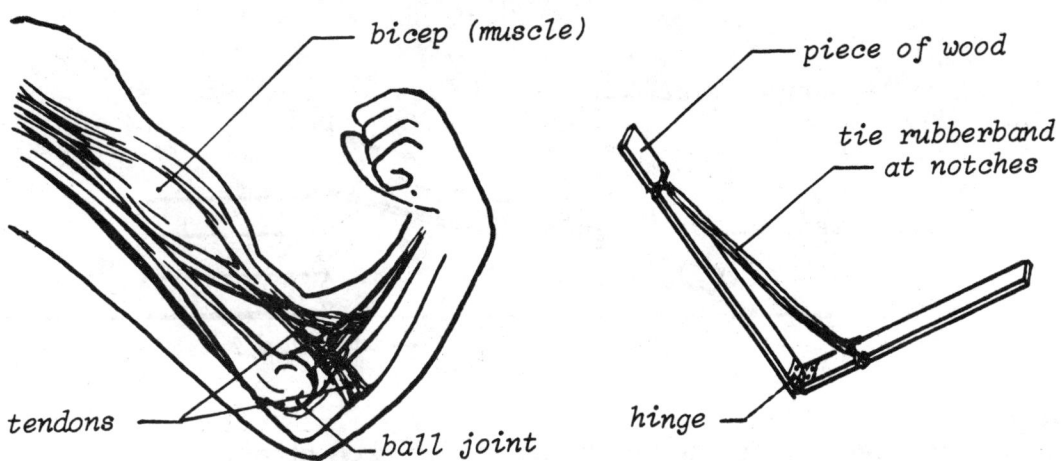

Procedure:
1. Attach the two pieces of wood together by means of the hinge (or tape the two rulers loosely together).
2. Cut grooves for the rubber band close to the free end of one arm and close to the hinged end of the other arm.
3. Tie the rubber band around each arm where the grooves are. You are now ready to show the bicep action.

Questions:
1. What class lever is the movement of the lower arm based on?
2. What does the rubber band simulate?
3. Locate the fulcrum, effort, and resistance in this arm lever.
4. Are other muscles in our body and the movement of the body parts based on the same class of levers?
5. What other examples can you name that are applying the third class lever in our daily life?

Explanation:
   The bicep is the **effort** in a third class lever when moving the lower arm upwards. This is illustrated by the rubber band when it is contracting. The hinge is the joint or **fulcrum** between the lower and upper arm, and the **resistance** is the weight of the arm alone or any additional weight that the arm is lifting (see sketch on the right).
   Other third class levers in our daily life are: a pair of tweezers, ice tongs, moving a hockey stick, using a shovel, broom, or rake, etc.

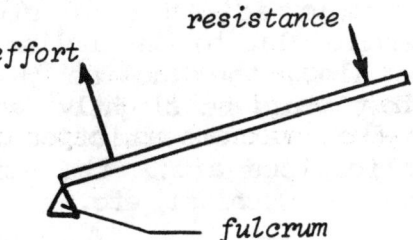

FORCES   SIMPLE MACHINES
                                                          INCLINED PLANE

13.42. WHICH ONE IS HEAVIER?

Materials: 1. A 200 g and a 300 g weight.
            2. A spring scale (500 g maximum).
            3. A 30 cm plastic ruler & a stack of books.

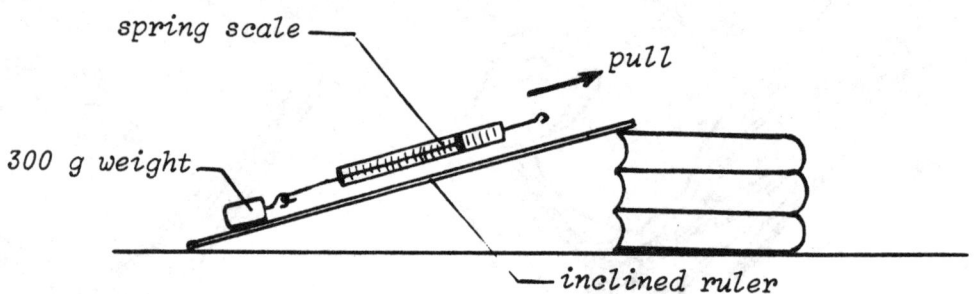

Procedure:
1. Place three or four textbooks on top of each other.
2. Place the two weights next to the stack of books and tell students that your task is to move the weights up on top of the stack.
3. Ask students: "Which one will take more effort to move?" Place the ruler on an incline (one end on top of the books and the other end on the table), and pull the 300 g weight over the ruler up on the stack of books with the spring scale (read the effort off this scale while moving the load).

Questions:
1. How much effort did it take to move the 200 g weight straight up?
2. How much effort did it take to move the 300 g weight over the incline on top of the stack?
3. Why was the effort much less than 300 g on the incline?
4. How are the force components working on the incline?
5. What are other applications on the inclined plane?

Explanation:
It takes much less effort than the full weight to move the 300 g mass over the inclined ruler, as the incline component force (W') is much smaller. The less steep the slope of the incline, the smaller the effort. The weight on the incline (W) may be decomposed into a component perpendicular to the incline (W") and one along the incline (W'). The effort would be slightly larger than W' (to overcome friction). Other applications are: the screw, the wedge, axe, chisel, etc.

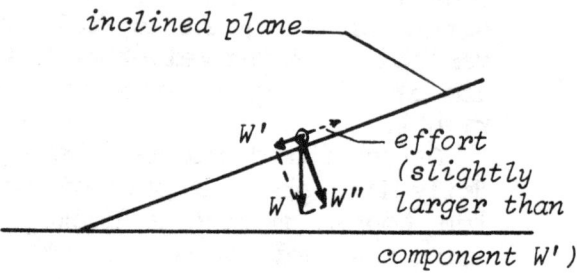

FORCES   SIMPLE MACHINES
                                                          PULLEYS

13.43. WHICH ONE WILL BE MOVED?

Materials: 1. Two clothesline pulleys.
            2. A sturdy rope and two strong sticks or pipes (50 cm).
            3. Two straight-back chairs.

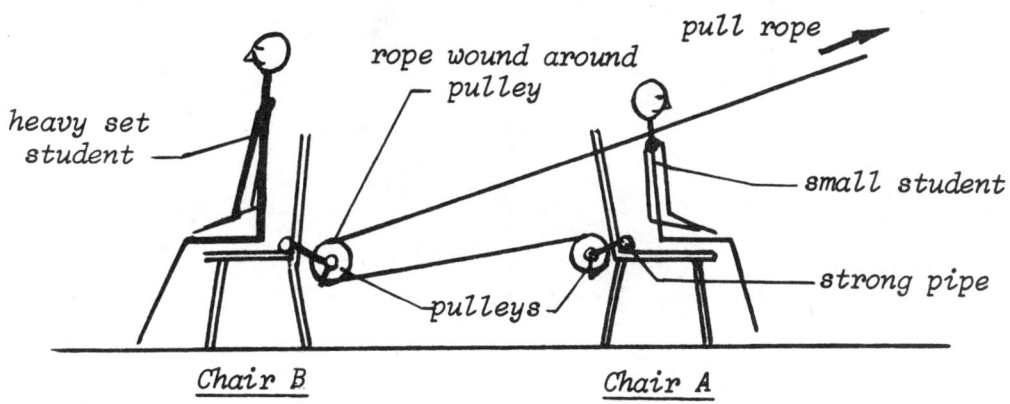

Procedure:
1. Attach each pulley to the center of each of the rods or pipes with a short piece of rope, and slip the rod between the seat and back of each chair, so that it catches on the frame when the pulley is pulled backwards.
2. Tie the end of a long rope to the other hanger of one pulley (see sketch Chair A), and thread the other end of the rope around the pulley of Chair B, and hold the end of the rope on the side of Chair A.
3. Have a tiny built student sit on Chair A and a heavy set student on Chair B, and ask: "Which of the two will be moved when I pull on the rope?" (anticipated answer: A). Pull and observe Chair B move!

Questions:
1. Which of the two students is heavier?
2. Why did Chair B, the heavier one, move and not Chair A?
3. Which of the chairs would move if the students were of equal weight?
4. Which one would move if the rope were also wound around pulley A, and the rope end attached to the floor?

Explanation:
There is only one rope pulling at Chair A and two ropes pulling at Chair B. **The force or tension in each of the ropes is equal** and thus the force pulling at Chair B is twice as large as that at Chair A. This is why even the heavier student sitting on Chair B is moving. If the rope were also wound around pulley A, Chair A would be moved (with the lighter student-- see Sketch B on the right).

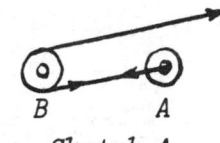
Sketch A

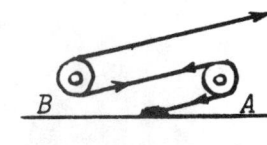
Sketch B

CHAPTER 14

WHAT DO WE LEARN WHEN DEALING WITH SPACE SCIENCE ?

OBJECTIVES

After dealing with and studying the concepts and sub-concepts of this chapter, the students should be able to:

a. recognize the correct explanation of an observed event based on each of the sub-concepts;
b. explain in their own words which of the sub-concepts is determining the course of an event;
c. distinguish true from false statements concerning each one of the sub-concepts;
d. identify the correct explanation of an event in daily life applying one of the sub-concepts;

all in relation to the following sub-concepts:

-- Inertia is the tendency of still objects to stay still or at rest.
-- The larger the mass of an object, the more inert it is.
-- Newton's First Law: a. An object at rest tends to stay at rest.
   b. An object in motion tends to stay in a linear motion.
-- Newton's Second Law: The acceleration of a moving object is directly proportional to the force exerted on it and inversely proportional to its mass ($F = m \times a$).
-- Newton's Third Law: When a body exerts a force upon a second body, an equal and opposite force is exerted upon the first body: for every action there is an equal and opposite reaction.
-- When an object is swung in a circular motion, there is an inward centripetal force working on the string.
-- The momentum of an object is equal to the product of its mass and the velocity it has.
-- When a moving object collides with another object, its momentum is conserved ($m_1 v_1 = m_2 v_2$).
-- The angular momentum of an object is unchanged unless an external torque acts on it (conservation of angular momentum).
-- In spinning objects, the more the object's mass is located towards the center of rotation, the faster the object spins.

SPACE SCIENCE   INERTIA

14.1. PUT THE COIN IN THE CUP

Materials: 1. A paper card and a coin.
            2. A cup or drinking glass.

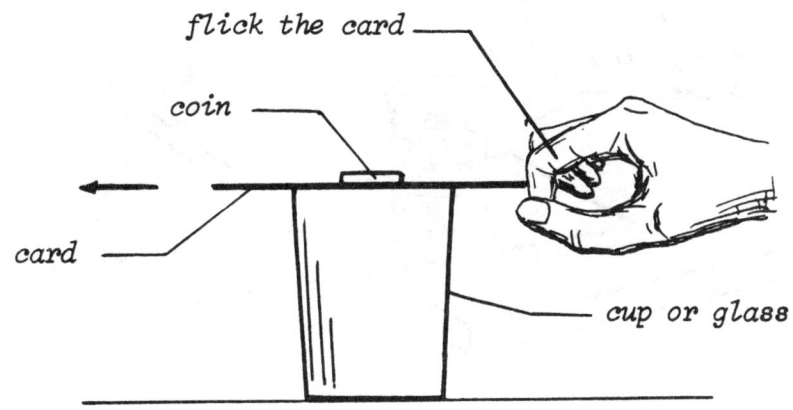

Procedure:
1. Cover the cup with the paper card and put the coin on top of the card.
2. Now ask the students: "How can I get the coin in the cup without lifting the card?"
3. After hearing all the different suggestions, flick the card with your forefinger in a horizontal direction.

Questions:
1. Why does the coin drop in the cup when the card is flicked away?
2. What held the coin back when the card was moved?
3. What happens to the coin if the card is pulled slowly?
4. Could we pull the card away rather than pushing it away?
5. Where do we see this event applied in daily life?

Explanation:
    This event is based on the common characteristic that all objects have and that is **inertia**. The coin lies **inert** on the card and by pushing the card suddenly away, the coin slides over the card and drops in the cup. The more sudden the movement of the card, the easier the coin will stay at rest. If the card is pulled slowly away, the coin will move with it. When pulling at the card instead of flicking it, the pulling has to be carried out with a sudden motion as well.
    We find this event in our daily life when we stand in a city or school bus, which suddenly starts to move. The bus moves forward with an abrupt motion and the standing person falls backward, because he/she had the inertia, which is also called **the tendency to stay at rest. The larger the mass of an object, the greater its inertia. The inertia of an object is directly proportional to its mass.**

SPACE SCIENCE   INERTIA

14.2. PULL THE TABLECLOTH

Materials: 1. A large beaker (or other glass container with a smooth base).
2. A long narrow cloth without seam (paper towel will do).

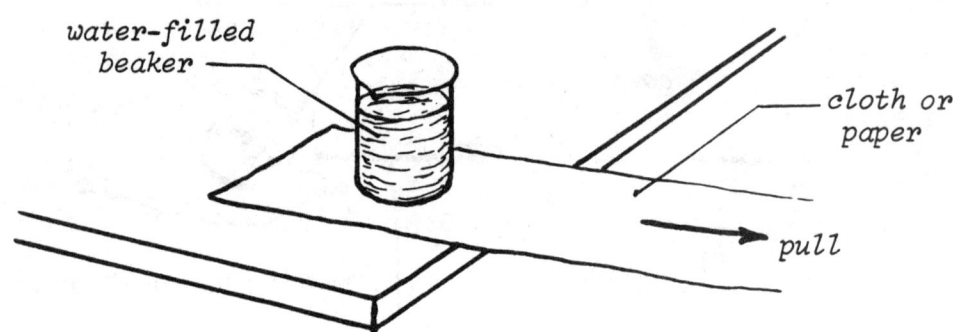

Procedure:
1. Fill the beaker with water and make sure that the outside of the beaker is bone dry.
2. Find a very smooth table top, wipe and dry it off, place the water-filled beaker close to the edge on top of the strip of paper.
3. Now pull at the paper strip, first slowly until the beaker comes to about 2 cm or so from the edge of the table, then pull with a sudden jerk (do not hesitate!).

Questions:
1. Why didn't the beaker fall off the table?
2. What made the beaker move closer toward the edge?
3. Why do we need a smooth table top for this demonstration?
4. What is the purpose of the water in the beaker?
5. What events in daily life make use of this principle?

Explanation:
This demonstration can be made most dramatic by placing the beaker about 50 cm from the edge of the table at the start and then pulling it to about 2 cm from the edge. This shows the students that pulling the paper will pull the beaker with it. The slow pull gives enough friction between the paper and the beaker, so that the beaker stays on the paper. The water in the beaker makes it heavier, and **the heavier the object, the greater its inertia or its tendency to stay at rest.** This is why it works easier with the water in it. The table top and the bottom of the beaker need to be smooth and dry, so that a minimum of friction exists when the paper is pulled from under the beaker.

An application of this first part of Newton's First Law in daily life is encountered when we move gravel or other heavy material with a shovel from one place to another: we pull the shovel with a sudden move out from under the gravel or in under the pile of gravel.

SPACE SCIENCE   INERTIA

14.3. THE IMMOVABLE PENNY

Materials: 1. A penny, paper card, and thin paper.
            2. A short glass tube, clamp and stand.
            3. A sharp long kitchen knife or hacksaw blade.

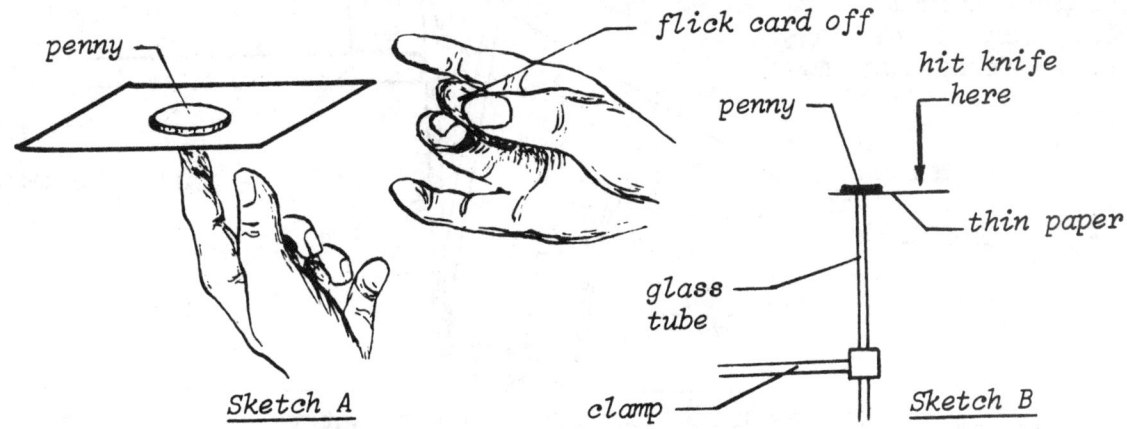

Sketch A   Sketch B

Procedure:
1. Cut a 3 x 5" card in half, place the penny in the center on the tip of the card, and balance the card and penny on your left index finger top (see Sketch A).
2. Flick the card off your finger with your right middle finger (do this horizontally in the plane of the card): penny stays on finger tip!
3. Clamp the glass tube vertically, place a short thin strip of thin writing paper on the top end of the glass tube, weighted by the penny (see Sketch B).
4. Hit the protruding end of the paper with the sharp edge of the kitchen knife or saw blade: penny stays on glass tube!

Questions:
1. What made the penny stay on the finger or on the glass tube?
2. Will the slow removal of the card or the paper have the same effect?
3. What other objects can we use instead of the penny?
4. What implications do you see when doing this on the moon?

Explanation:
   The removal of the card or paper has to be done with a quick action otherwise the friction factor will play a role. A slower movement of the card or paper will bring the coin with it. Also the flicking of the card has to be carried out, such that the card will slide off horizontally without tipping the coin.
   In hitting the protruding paper with the sharp edge of the knife, the paper is folded up exactly where the knife hits it, and the sharp edge grabs on to the paper and pulls it away.
   Other objects that can be used instead of the penny are: other coins, washers, etc. On the moon this event can be done much easier, as the mass stays the same but the weight is less, and thus the friction is less.

SPACE SCIENCE

INERTIA
NEWTON'S FIRST LAW

14.4. GET THE EGG IN THE GLASS

Materials:
1. One raw egg.
2. One egg holder.
3. One aluminum pie plate.
4. A large drink glass.
5. A household broom.

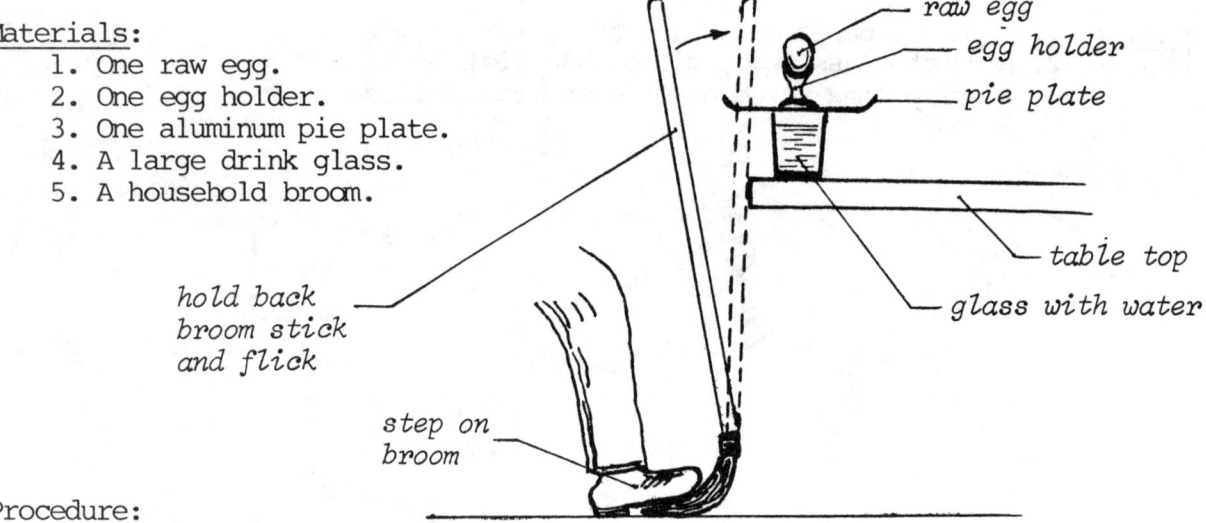

Procedure:
1. Fill the drink glass three quarters full with water.
2. Put the empty pie plate on the glass and center the egg holder and raw egg on the pie plate directly over the glass.
3. Place the glass with everything else on top, close to the edge of the table such that the rim of the pie plate hangs over the table edge (see Sketch).
4. Ask the students: "How can I get the egg in the glass with the broom without breaking it?" Anticipated answer: "Have no idea!"
5. Hold the broom directly in front of the set up, push down on it so the sweep part of the broom bends. Place one foot on the broom while holding back the wooden handle, then suddenly release the pole and flick it hard against the pie plate (see Sketch).

Questions:
1. What made the egg fall exactly in the glass?
2. What was the function of the water in the glass?
3. What would happen if we had a plate without a rim on the glass?
4. Why did the pie plate have to protrude over the table edge?
5. Would any other object in the egg holder end up in the glass?

Explanation:
The broom stick hit the edge of the pie plate without hitting the glass or the egg, because its motion was stopped by the table edge.
This sudden force moved the pie plate out from under the egg carrying with it the egg holder, as this was caught by the plate's rim (see Sketch). The egg was at rest and tended to stay at rest (first part of Newton's First Law). The water was needed to catch the raw egg and prevent it from breaking. Any other object on the egg holder would have fallen in the glass, provided it has a heavy enough mass (and thus have enough inertia).

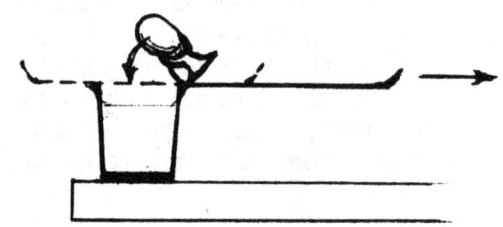

SPACE SCIENCE

DIRECTION OF FORCE
INERTIA

14.5. GET THE CHALK IN THE BOTTLE

Materials:
1. A flexible plastic hoop (cut from a polyethylene container).
2. A bottle with a narrow neck.
3. A small piece of chalk.

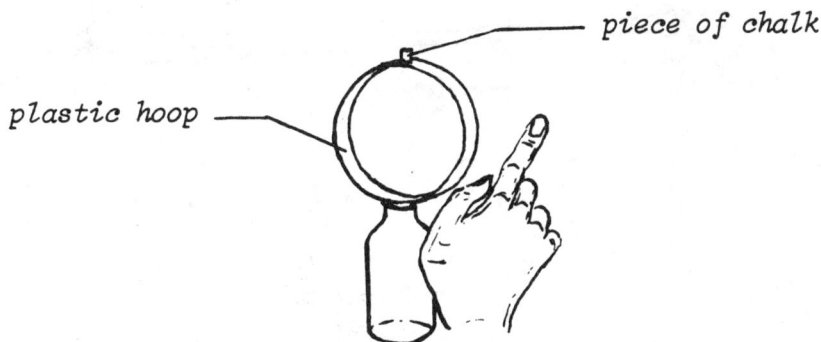

Procedure:
1. Cut a narrow (1 cm wide) hoop from a flexible polyethylene container.
2. Place it on the mouth of the bottle and center a small piece of chalk vertically over the opening of the bottle.
3. Flick the hoop with your finger from under the chalk by hitting the hoop on the inside (but holding the finger on the outside of the hoop before the flicking itself).
4. Let students observe carefully where you hit the hoop with telling them what you are doing.
5. Let students try to get the chalk in the bottle (those that know where to hit the hoop will be able to do it, others will not).

Questions:
1. Did you observe where the hoop was hit with the finger?
2. Why does the chalk fly off if the hoop is hit on the outside?
3. What force acts on the chalk when the hoop is hit on the outside?
4. What will the hoop do if hit on the outside or on the inside?

Explanation:
   The demonstration can be used to focus students' attention on careful and accurate observation skills. Without telling them where the hoop is hit, they should be able to find out for themselves.
   When the hoop is hit on the inside, it flattens and slips out under the chalk. When it is hit on the outside, the force is upwards, and the chalk gets pushed up (see sketches on the right).

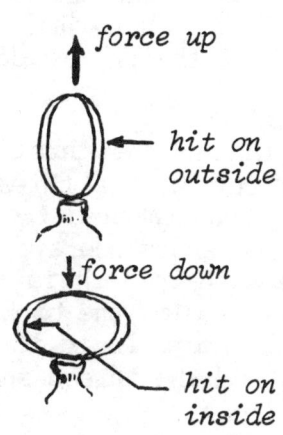

SPACE SCIENCE   NEWTON'S FIRST LAW
                                                                      INERTIA

14.6. BREAK THE STRING WHEREVER YOU WANT

Materials: 1. Two paperback pocket books (same thickness).
            2. A horizontal stand and sewing thread.

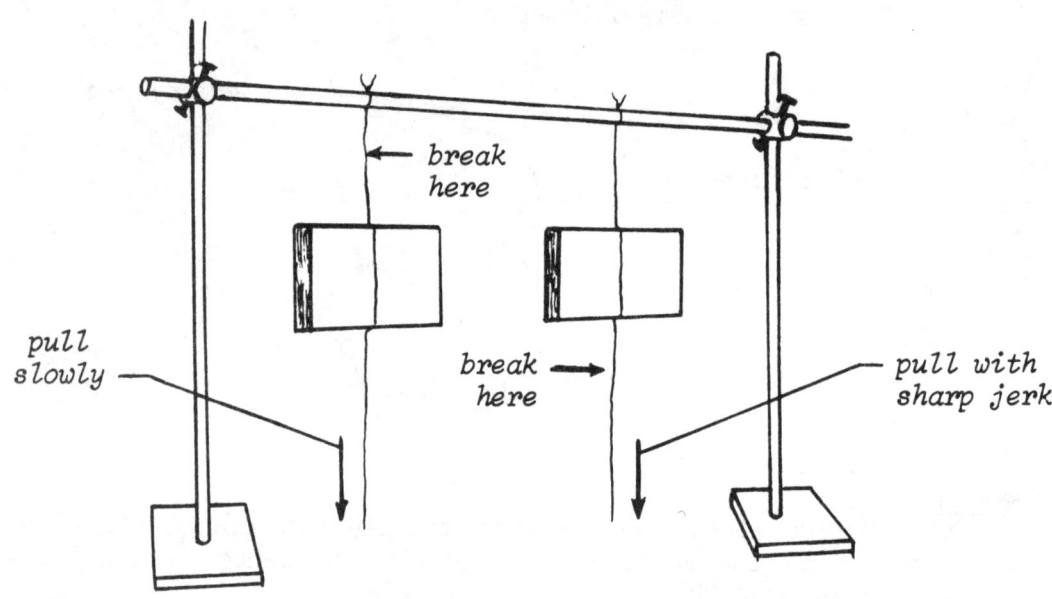

Procedure:
1. Tie threads around each of the books, such that a loose end comes out from each side of the book.
2. Tie one end of the thread around the horizontal stand and let both books hang side by side (see sketch).
3. Now ask the students: "Where do you want me to break the thread of book No. 1? Above or below the book?"
   If they answer: 'above the book,' pull slowly at the thread.
4. For book No. 2, pull the thread with a short jerk: thread will break below the book!

Questions:
1. Do you see any difference in the way the thread was pulled?
2. Why does a slow pull at the thread break it above the book?
3. Why does a sharp jerk break the thread below the book?
4. Which of the two breaks makes special use of the book's inertia?

Explanation:
   By pulling the thread slowly, we are not only putting a strain in the thread, but in the thread above the book, the book's weight adds to this pull. Thus compared to the strain below the book, this is much larger and the thread snaps wherever the strain is highest.
   When a sharp jerk is exerted on the thread, the inertia of the book keeps the strain below the book. Although there is some strain in the thread above the book, compared to that below the book, the strain in the latter is still higher, and the thread snaps below the book.

SPACE SCIENCE   NEWTON'S FIRST LAW

14.7. WHAT BREAKS THE THREAD?

Materials: 1. A 1 kg mass or a stone of about 1 kg weight.
            2. A medium size string and sewing thread.
            3. A stand and clamp.

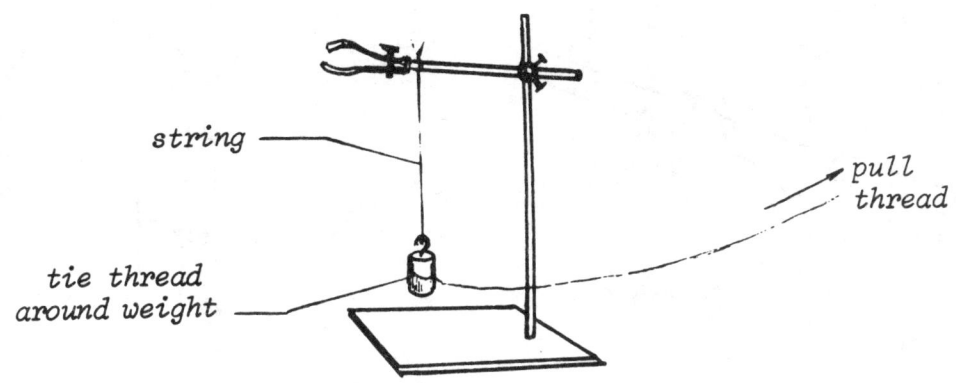

Procedure:
1. Hang the weight on a string tied to the clamp and stand.
2. Tie a length of sewing thread around the weight and pull slowly on the thread and show to students that it is possible to move the weight up high without breaking the thread.
3. Let the weight hang vertically and give the thread a sudden pull. Ask: "What property of the weight made the thread break?"
4. Tie another thread to the weight, pull the weight and let it swing back and forth, but let your hand swing in phase with it (holding the thread), then suddenly stop the hand. Ask: "What property of the weight made the thread snap?"

Questions:
1. Which part of Newton's First Law was the first break based on?
2. What events in daily life are based on this principle?
3. Which part of Newton's First Law was the second break based on?
4. Can you mention any events in our daily life that are based on this second principle?

Explanation:
When the thread was broken the first time around, it was done by pulling the thread with a sharp jerk. The weight was **at rest and it had the tendency to stay at rest.** (First part of Newton's First Law.) An application of this principle in daily life is pulling a table cloth out from under dishes, pushing a shovel under a heap of gravel, etc.

Breaking the thread the second time around was caused by a sudden stop of the moving hand, when the weight was still moving. **The weight was in motion and it tended to stay in motion** (second part of Newton's First Law). Events in daily life based on this principle are for example: the tightening of a loose axe on its handle, the sudden stop of a bus and the tendency of the passengers to keep moving forward, etc.

SPACE SCIENCE   NEWTON'S FIRST LAW

14.8. THE APPLE AND THE KNIFE

Materials: 1. Two fresh apples, a long kitchen knife.
              2. A wooden rod or stick (to hit the knife with).

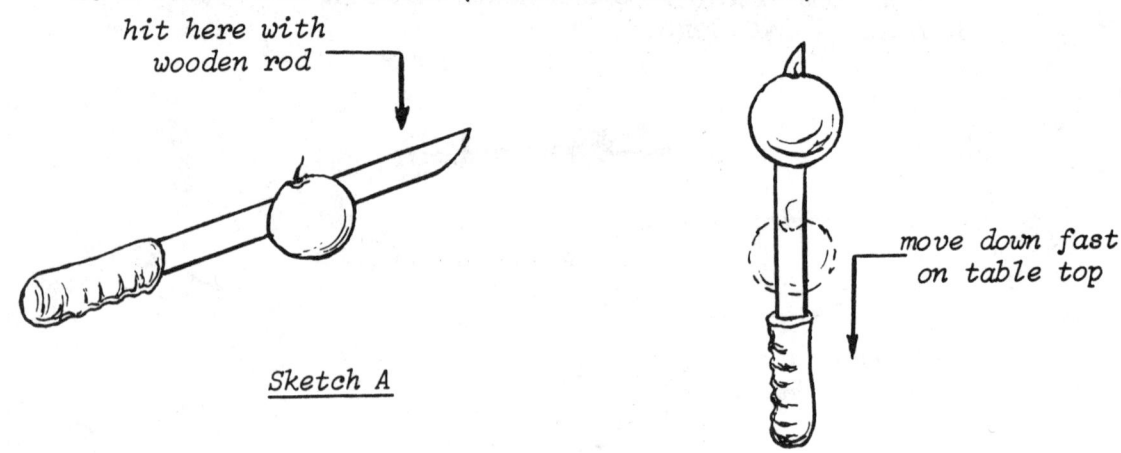

Sketch A

Sketch B

Procedure:
1. Press the kitchen knife about 1/4 way through the first apple, hold the knife horizontally and hit the end of it with the wooden rod (see Sketch A). The apple will split in half illustrating the first part of Newton's First Law.
2. Now push the point of the kitchen knife all the way though the second apple. Ask the students: "How can I get the apple to slide closer to the handle?"
3. Hold the knife vertically above the table with the point upwards. Move the knife fast to the table top, hitting it hard with the knife handle. The apple will slide closer to the handle, illustrating the second part of Newton's First Law (see Sketch B).

Questions:
1. Why did the first apple split in two halves?
2. Was it the apple or the knife that followed Newton's Law?
3. In sliding the apple, which of the two (apple or knife) followed the second part of Newton's First Law?
4. What other daily examples can you name that are based on the first part and second part of Newton's First Law?

Explanation:
    Newton's First Law states that: **an object at rest tends to stay at rest, and an object in motion tends to keep moving in a straight line.** The first apple split in two halves, because it tended to stay at rest when the knife was hit. The second apple slid closer to the knife handle because it tended to stay in motion when the knife was suddenly stopped by the table top.

SPACE SCIENCE

NEWTON'S SECOND LAW
ACCELERATED MOTION

14.9. THE FALLING WASHERS

Materials:
1. Six or seven metal washers)
2. A strong thin thread   ) for each demonstration.
3. An old film canister   )

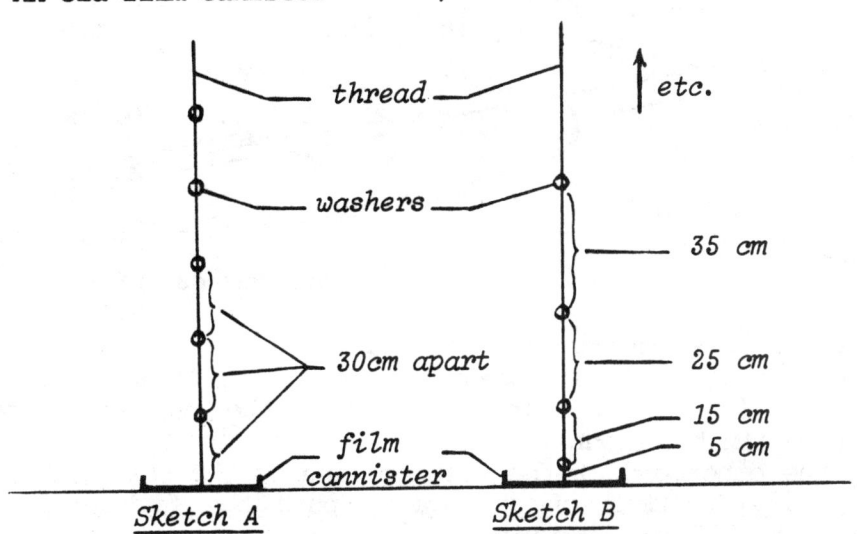

Sketch A   Sketch B

Procedure:
1. Tie the six or seven washers exactly 30 cm from each other on the thread and tape the end of the thread to the film canister, such that the first (lowest) washer is 30 cm above it.
2. Stand on a chair and hold the thread tight above the canister, while this latter is lying on the floor. Ask students: "What kind of tapping sound will the washers give us when I let them fall; even or faster and faster?" Now drop the washers (see Sketch A).
3. Tie seven washers to another thread the following distances from each other: canister to first washer 5 cm, the next six washers: 15, 25, 35, 45, 55, and 65 cm from each other respectively.
4. Repeat point 2 (see Sketch B).

Questions:
1. What kind of tapping sound did the washers give us in demonstration A, compared to that in demonstration B?
2. Which of the seven washers had the highest velocity when hitting the bottom?
3. What kind of motion is the free fall of the washers?
4. What gives a falling object its acceleration?

Explanation:
The falling washers are all subjected to gravity, which force imparts an accelerated motion to each of them. Newton's Second Law states that F = MA (F = force, M = mass, and A = acceleration), and as the force and mass of the washers are equal, the acceleration of each of the washers is the same. The higher, the larger the velocity of the washer hitting the canister. The distances between the washers in B were obtained from S = 1/2at (where S = distance, a = acceleration, and t = time).

SPACE SCIENCE   NEWTON'S THIRD LAW

14.10. THE BALLOON RACE

Materials: 1. Two long cylindrical balloons.
            2. A roll of fishing line (or smooth string).
            3. Two straws and masking tape.

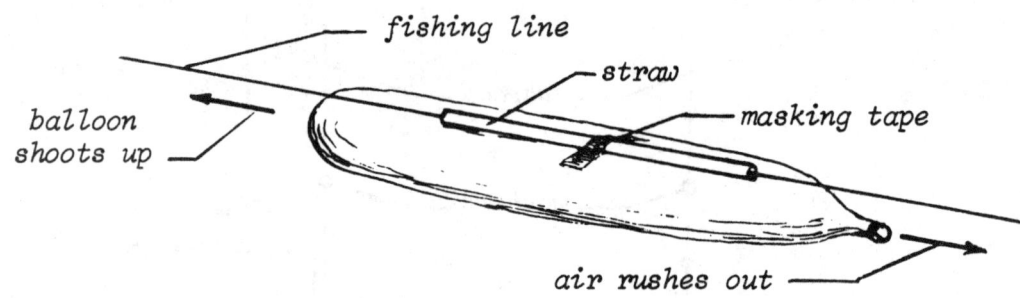

Procedure:
1. Divide the class in two groups: there is going to be a race!
2. Have one student of each group tie one end of 10 m length of fishing line to a pipe (or fire sprinkler) close to the ceiling at one end of the classroom (in the front).
3. Bring the other end of the fishing line to the rear of the classroom and have another student of each group push the fishing line through the straw, and hold the line tight.
4. Now have another student of each group blow up the balloon and hold the mouth closed (do **not** put a knot in it), and hold the balloon under the straw with its mouth facing the rear of the room.
5. Let another student tape the balloon to the straw (one strip of tape over the middle of the straw). You are now ready for the race.
6. Have the students release the balloon on the count of three.

Questions:
1. In which direction did the air of the balloon go?
2. What made the balloon shoot up?
3. Will a smaller or larger balloon shoot up faster?
4. If we want the balloon to shoot down, which way do we have to face the balloon's mouth?

Explanation:
   At the time that the balloon was held with its mouth closed at the rear of the room, all forces inside the balloon were balanced by an equal and opposite force. This will also be the cause when we put a knot in the balloon's mouth or in the case of a closed pressured chamber in space (see Sketch 1 on the right).

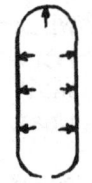

all forces   resultant
in balance   force: up

   When the balloon is released, the downward force is eliminated, and the resultant force is upwards. This is quite similar to the operation of a pressure chamber in a rocket or satellite. When a left turn has to be made, a valve on the right hand side of the vehicle is opened.

SPACE SCIENCE   NEWTON'S THIRD LAW

14.11. THE MATCH MISSILE

Materials: 1. A box of wooden matches, a box of paper clips.
            2. A roll of aluminum foil.

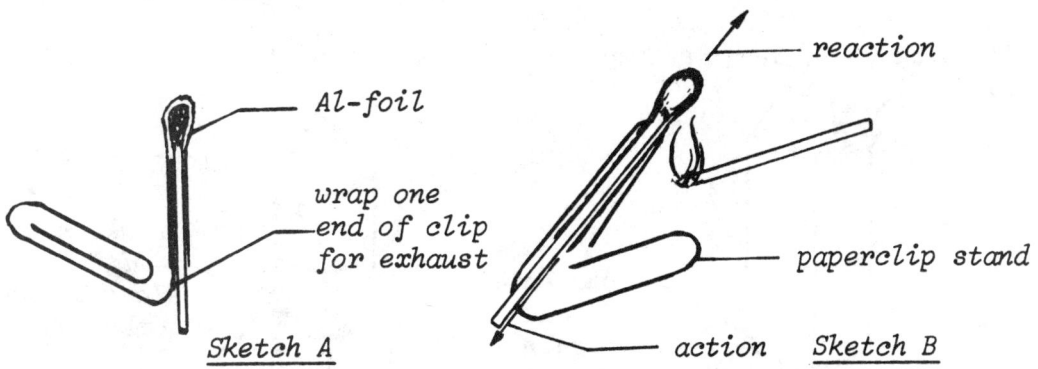

Procedure:
1. Distribute two matches and two paper clips and a strip of aluminum foil to each pair of students (let students work in pairs)
2. Demonstrate to students how to make the missile:
   a. Wrap the match airtight with one end of the paper clip against the match stick (see Sketch A) and slide the clip out (this provides the exhaust pipe).
   b. Bend out another clip to make a stand and lay the wrapped match against it (see Sketch B).
   c. Heat the wrapped match head with another lit match.
3. Let students try to make the farthest reaching missile.

Questions:
1. What makes the "missile" shoot up?
2. What would happen if we heated the wrapped match without providing for an exhaust?
3. Does the angle of the bent up clip make any difference in the distance that the missile can reach?
4. What are other factors that influence the reach of the missile?

Explanation:
   When wrapping the match (missile) with aluminum foil, we need to lay one leg of a paper clip (bent out) against the wooden stick, to provide an exhaust for the expanding gases when the match ignites. Sometimes when this exhaust is not working properly, a bursting of the aluminum foil near the match head will occur.
   Factors that influence the reach of the missile would be:
   a. the weight of the missile--in our case the Al-foil, the less used, the lighter, but the less strong to keep the gases in.
   b. the match itself--some have larger heads, which would provide more power (try wrapping in an additional match head).
   c. the angle of blasting--45° angle is optimal; larger or smaller angles with the horizontal plane will give shorter distances.
   This event is based on Newton's Third Law, which states that: **For every action, there is an equal and opposite reaction.**

SPACE SCIENCE   NEWTON'S THIRD LAW

14.12. THE MILK CARTON SPRINKLER

Materials: 1. An empty milk carton (2 l) or large tin can.
             2. A length of sturdy string.

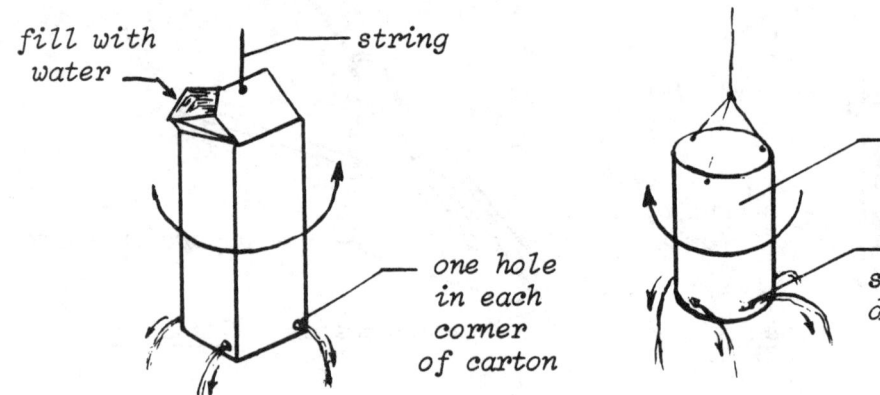

Procedure:
1. Make four small holes, one in each right hand lower corner, in the sides of the milk carton.
2. Tie the string through a hole in the top of the carton, to hang it over the sink from a support (or from your hand).
3. Fill the carton full with water over a sink and let it hang from the string, and observe!

Questions:
1. Looking from above, which way does the carton turn?
2. Is it possible to turn the carton in the opposite direction?
3. Where do the holes have to be placed in order for the carton to move in the opposite direction?
4. When would the carton stop turning?
5. What principle is this event based on?

Explanation:
   This event is based on the Third Law of Newton: **action is reaction.** The action here is the water spouting out from each lower right hand corner of the side of the carton (see Sketch 1).
   The reaction is an equal and opposite force on the carton (see Sketch 2), which makes the carton turn counterclockwise. In order to make the carton turn clockwise, the holes have to be placed in the left hand lower corner of each side.
   A tin can could replace the carton by making nail holes near the bottom and slanting the nails in one direction.

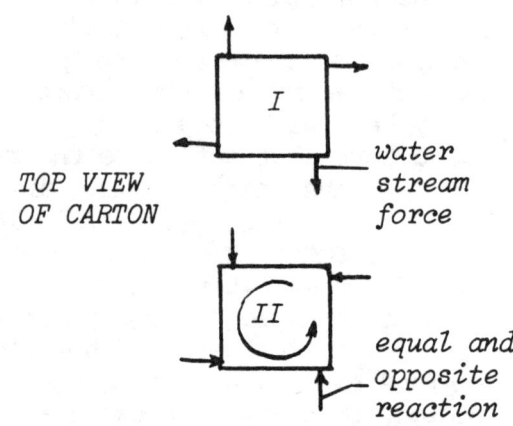

SPACE SCIENCE   NEWTON´S THIRD LAW

14.13. THE STRAW ROCKET

Materials: 1. A plastic flexible bottle.
            2. Two drinking straws of two different sizes.
            3. Some molding clay and construction paper.

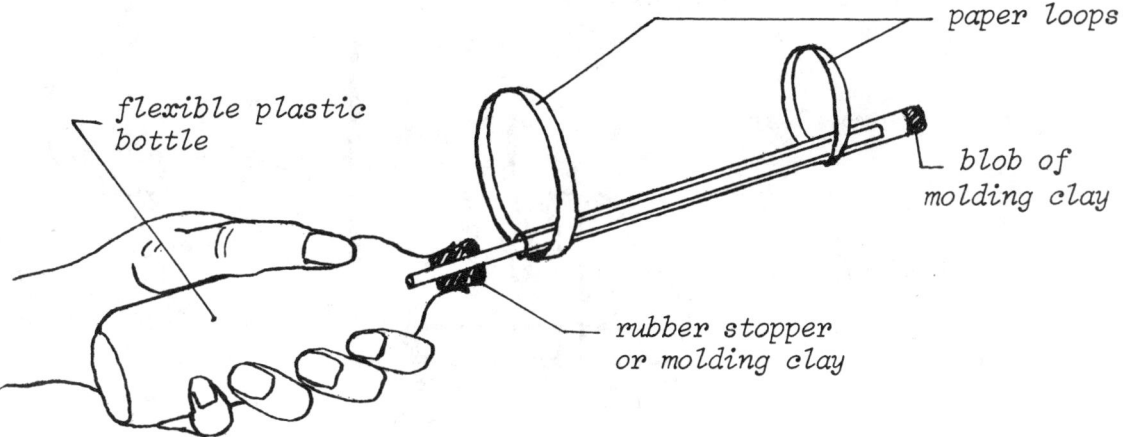

Procedure:
1. Prepare the 'bottle starter' by putting a roll of molding clay around one end of the smaller straw and fitting it into the mouth of the plastic flexible bottle - test for leaks by plugging the other end of the straw with a finger, then squeezing the bottle: this should feel hard to squeeze when there are no leaks.
2. Prepare the rocket by taping two paper loops to the larger straw. One smaller loop to the front end and a larger loop to the back end. Stuff the front end of the straw with a blob of clay.
3. You are now ready to launch the rocket. Place the larger straw over the smaller one, hold the paper loops on the top side of the rocket with another straw, and squeeze the bottle with a sudden motion.

Questions:
1. What made the rocket move forward?
2. Why does the smaller straw have to fit tightly in the bottle?
3. What is the function of the blob of clay in the larger straw?
4. What is the function of the paper loops?
5. What initial action gave the rocket its energy to shoot forward?

Explanation:
   By squeezing the plastic bottle, air is blown through the smaller straw into the larger straw. As this latter one is plugged at the front end, a higher pressure is built up in the larger straw and this makes the rocket shoot forward. As Newton´s Law states: **For every action there is an equal and opposite reaction.** The action here is air shooting out of the rear of the rocket (larger straw), and the reaction is that the rocket moves forward. The function of the paper loops is to keep the straw floating in a horizontal direction.
   Toys that shoot plastic balls by squeezing a plastic gun, and air guns (Beebee guns) are applications of this basic principle of **action is reaction.**

SPACE SCIENCE — NEWTON'S THIRD LAW

14.14. BLOW YOUR OWN SAIL?

Materials:
1. A roller skate, a small motor and propeller.
2. Six 1.5V C-size batteries & 2 short lengths of bell wire.
3. A piece of cardboard.

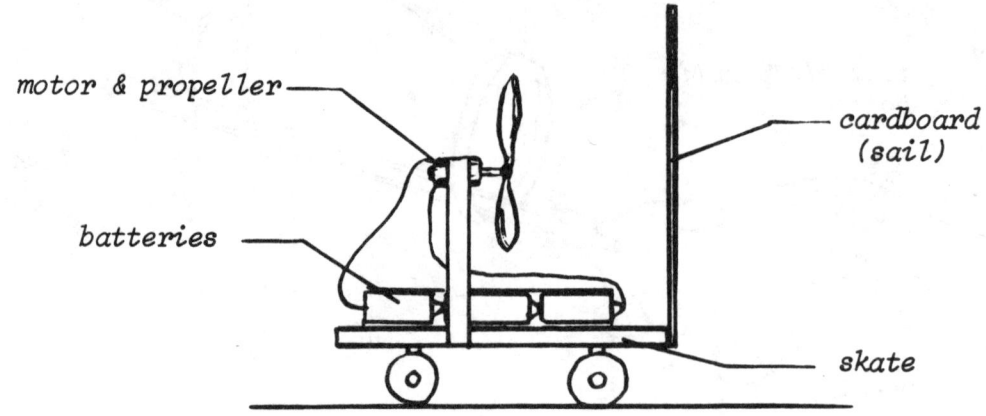

Procedure:
1. Mount the motor and propeller by means of wooden or metal strips on one end of the skate.
2. Tape the six batteries three by three in series together, and place them on the skate. Connect the wires from the batteries to the motor (leave the connection of one lead off for the moment).
3. Tape a piece of cardboard to the other end of the skate, such that the propeller blows air against the cardboard.
4. Ask the students: "What will the skate do when the fan is turned on?" Now complete the circuit and turn on the propeller.
5. Ask the students: "What will the skate do when the sail is taken off?" Take the cardboard away while the propeller is still turning.

Questions:
1. What forces are acting on the skate with fan on and sail on?
2. What forces are acting on the skate with fan on and sail off?
3. Why does blowing your own sail in a sailboat on a windless day not make you move forward?
4. What is the propeller blowing against when the sail is taken off?

Explanation:
With the propeller blowing against the cardboard, two forces of the same size but in opposite direction are working on the skate. When the cardboard or sail is taken away, that force is also eliminated, leaving one force in the opposite direction of the blowing as resultant force. For the same reason, blowing your own sail on a windless day will not work, as two equal and opposite forces are working on the boat. The same two equal and opposite forces are present when we push down on the table top. We feel the table pushing back against our hand.

SPACE SCIENCE

CONSERVATION OF MOMENTUM
NEWTON'S THIRD LAW

14.15. THE RECOILING SKATE

Materials:
1. A rollerskate and a stone.
2. A heavy rubber band and a piece of string.

Procedure:
1. Cut the rubber band and attach the two ends to the front sides of the skate.
2. Tie a string to the center of the rubber band and pull it back like a catapult (see sketch) and attach the other end of the string to the back of the skate.
3. Place a medium size stone in front of the rubber band and cut the string with a pair of scissors (make sure that you do not touch the skate when cutting the string).

Questions:
1. What did the skate do when the stone was flung to the front?
2. How far did the skate recoil, and how far did the stone fly?
3. Would the skate recoil more or less with a larger stone?
4. What events in daily life would be similar to this one?
5. What other objects or set-ups could we use instead of the skate?

Explanation:
The action in this demonstration is the flinging of the stone by the rubber band to the front of the skate. The reaction is that the skate rolls backwards. The Third Law of Newton states that the force flinging the stone is equal to the force pushing the skate back. Because the masses of the two objects are not equal, the accelerations or velocities of the objects are also different. The momentum imparted on the stone is equal to the momentum of the skate: $M_1V_1 = M_2V_2$, where M_1 = mass of stone, V_1 = velocity of stone, M_2 = mass of skate, V_2 = velocity of skate.

This **conservation of momentum** principle shows clearly in this event, where the stone has a much higher speed than the skate.

Other equipment that can replace the skate would be: a wooden board on marbles or straws. The rubber band and string could be attached to the board with thumbtacks (see sketch on right).

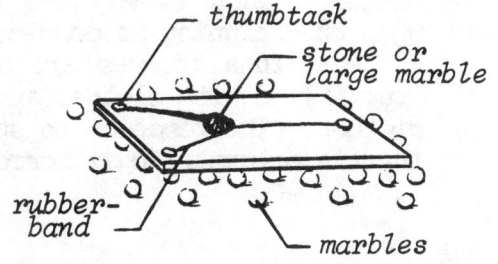

SPACE SCIENCE   CENTRIPETAL FORCE

14.16. THE STICKY PENNY

Materials: 1. A wire coat hanger, a penny (or any other coin).

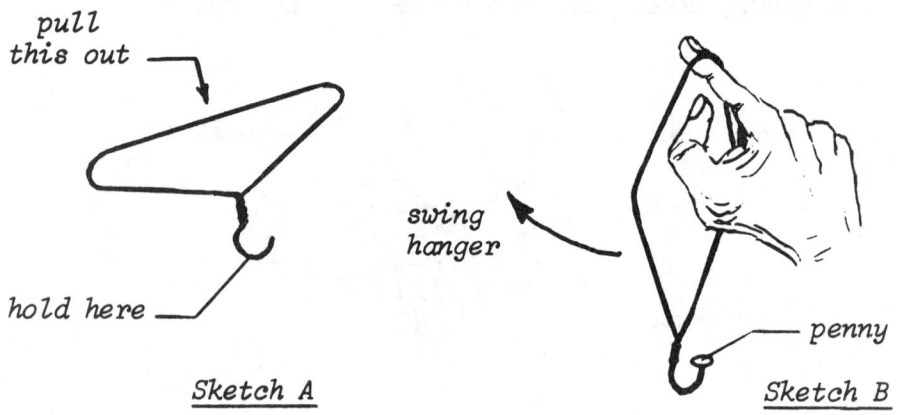

Sketch A   Sketch B

Procedure:
1. Pull the middle of the longest straight part of the hanger out (see Sketch A) and shape it into a long narrow shape (Sketch B).
2. Let the hanger hang on your right index finger (left, if you are left handed) and place a penny on the end of the wire (sometimes this end needs to be filed, especially if it is not flat).
3. Start to swing the hanger--first slowly back and forth, then do full loops (this needs a little practice, without dropping the penny).
4. Swing it slower (when ending the swing) and try to catch the penny.

Questions:
1. How could the penny stay stuck against the end of the hanger?
2. What kept the penny against the coat hanger?
3. Would the coin still stick to the hanger if the swinging was done much slower?
4. What events in our daily life are similar to the moving penny?

Explanation:
   The spinning of the penny gives it a centrifugal force holding the coin against the end of the hanger. The faster the spinning, the larger this centrifugal force; the slower the spinning, the smaller this force is, that keeps the coin up against the hanger. This is why the coin falls when the spinning is done too slow.

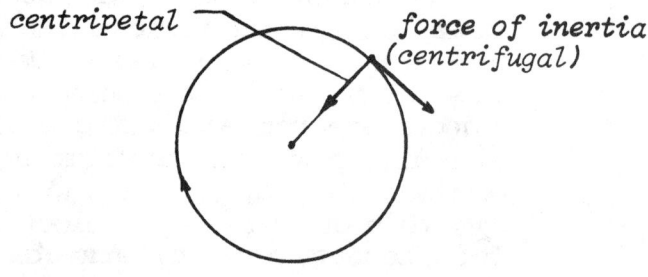

   Centrifugal forces are applied in the automatic cloth washer, where the drum filled with clothes and water spins fast to separate the water from the clothes. **(The tendency to stay moving in a straight line causes this force of inertia or centrifugal force).**

14.17. THE FLYING WINE GLASSES

Materials:
1. Three or four plastic wine glasses.
2. A sturdy plastic try (or cardboard sheet).
3. A strong flexible rope.

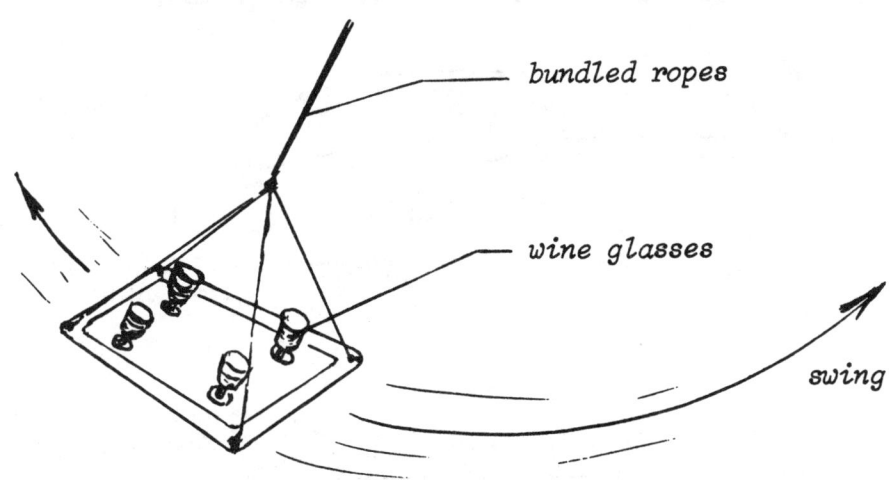

Procedure:
1. Drill four small holes in the four corners of the plastic tray.
2. Tie four pieces of rope to each corner of the tray (through the hole), bring them together to the center above the tray while it lies horizontally on the table, and tie a knot in all four strings together.
3. Tie another knot at the end (about 1 m away) of the bundled rope.
4. Place the four wine glasses on the tray and fill them 3/4 full with red colored water (or real wine for that matter).
5. Start swinging the tray with the glasses and swing full loops, making sure that the tension in the rope is maintained.

Questions:
1. What made the wine stay in the glasses even when inverted?
2. What kept the glasses pressed against the tray?
3. What is the tension in the rope, while swinging, an indication of?
4. What would happen if we lost this tension in the rope, while the tray was in an upside down position?
5. How can we keep a rather even tension in the rope when swinging?

Explanation:
While in motion, the wine as well as the wine glasses tend to move in a straight line (second part of Newton's First Law). As they are kept in a circular motion, they are pressed against the tray (centrifugal force) and an equal center seeking force, **centripetal force**, is felt as a tension in the rope. The faster the swing, the stronger the tension in the rope. A rather even tension in this rope may be obtained by swinging the tray back and forth in the beginning and making the full loops when the swing almost reaches the horizontal position. (Practice by swinging a stone tied to a string.)

SPACE SCIENCE — CENTRIPETAL FORCE

14.18. A MEASURE OF CENTRIPETAL FORCE

Materials:
1. A one-hole rubber stopper.
2. A piece of glass tubing (about 15 cm long and nicely rounded off at the ends).
3. Metal washers, a strong string, paper clips.

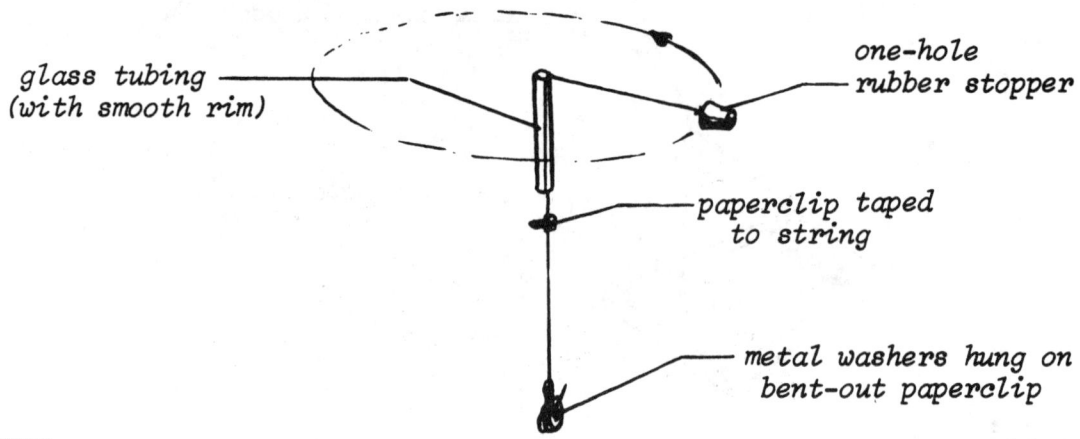

Procedure:
1. Thread the string through the glass tubing and let about 50 cm comes out on each end of the tubing.
2. Tie the one-hole rubber stopper to one end and a paper clip to the other end of the string. Bend the paper clip leg somewhat out.
3. Hang a few washers on the bent-out paper clip and tape another paper clip about 30 cm from the paper clip end to the string (as an indicator to see whether the string is sliding up or down the tube).
4. Twirl the rubber stopper above your head by holding the glass tubing and swinging the stopper. Try to keep the paper clip indicator at the same height by adjusting the twirling speed.
5. Add more washers to the bottom paper clip: "What do you have to do with the twirling speed?"

Questions:
1. What is the relationship between rotational speed and number of washers?
2. What relationship is there between speed of rotation and the diameter?
3. What relationship is there between diameter of swinging object and the number of washers?
4. How can we measure the centripetal force of the swinging stopper?

Explanation:
Holding the diameter of the swinging object (length of string above the tubing) constant, the faster the speed of rotation (number of revolutions per second), the more washers have to be hung on the bottom paper clip. The same goes for increasing the diameter when the speed is held constant. With the same number of washers, the smaller the diameter, the higher the speed of rotation, in order to keep the forces in balance. The centripetal force is the total weight of the washers hung on the paper clip.

SPACE SCIENCE   CENTRIPETAL FORCE
                                                                 GRAVITATION

14.19. THE GRAVITY MACHINE (I)

Materials: 1. An old variable speed turntable.
            2. A wooden stick (double the width and length of a 30 cm ruler).
            3. Two identical empty jam jars (with lid).
            4. Two medium size candles.
            5. Epoxy glue or some other strong cement.

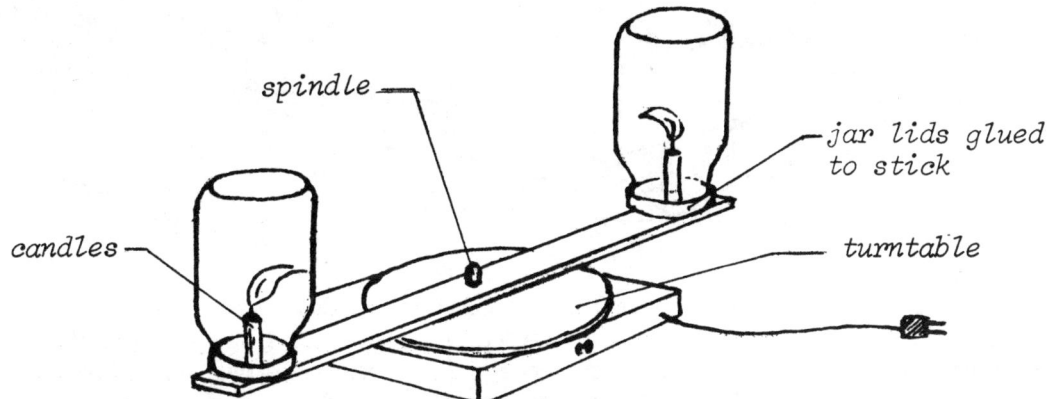

Procedure:
1. Make a small hole in the center of the wooden stick, such that it will fit tightly over the spindle of the turntable.
2. Cement the two jar lids upside down on each end of the wooden stick with epoxy glue or other strong cement.
3. Attach each candle to the center of each of the jar lids (with candle wax).
4. Light the candles and show what the flames are doing when the turntable is switched on (use the low speed for this).
5. Ask students to predict what the candle flames will do when the glass jars are screwed on just before the table is turned on.
6. Screw the glass jars over the burning candles into the lids and immediately switch on the turntable (using higher speeds). Observe the flames!

Questions:
1. What made the flames point towards the center of the turntable?
2. Which way does a candle flame point in still air? Why?
3. What does the spinning of the jars do to the air molecules in the jar?
4. What part of the jars would contain denser air?
5. In which direction did the spinning create a gravitational force?

Explanation:
The spinning of the jars on the turntable created a centripetal force (center seeking). The air molecules in the jars are put in motion and tend to stay moving in a straight line. As the jars move in a circular motion, the air molecules tend to move toward the outside (farthest from the spindle), thus making the outside part of the jars more dense. Since candle flames always point towards less dense areas of air - in still air they point upwards as the heat makes the air above the flame less dense - in the jars they point towards the inside part or towards the center of the turntable.

In space, satellite stations' gravity will most likely be created in the same fashion. Other applications are found in amusement parks where we find rotating drums and motorcycles riding against the walls of a drumlike pit.

SPACE SCIENCE

CENTRIPETAL FORCE
GRAVITATION

14.20. THE GRAVITY MACHINE (II)

Materials:
1. Two identical empty jam jars (with lid).
2. Two ping-pong balls, thread, and tape.
3. A sturdy wooden board, a turntable.

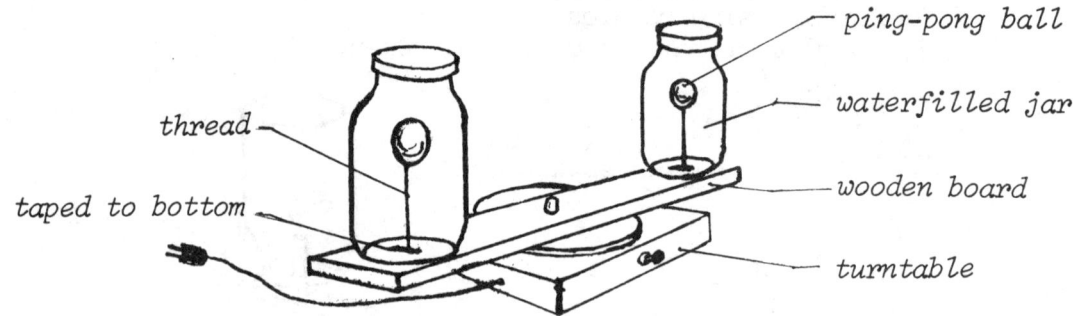

Procedure:
1. Drill a small hole in the center of the wooden board, such that it will fit tightly over the spindle of the turntable; and fit the board on it.
2. Tape the two jars right side up against the ends of the board (see Sketch).
3. Attach a short (about 10 cm) piece of thread to each of the ping-pong balls, and attach the other end of the thread against the bottom of each of the jars (make sure that when the thread is held up tight, the ball will only come up about three quarters of the jar's height).
4. Fill the jars full with water (the ping-pong balls should be totally submerged under water and hanging up from the thread), and close tightly.
5. Before rotating, ask: "Which way will the ping-pong balls move to, when the table rotates?" Anticipated answer: "Outward, or against the direction of the rotation". Now switch the turntable on, and observe!

Questions:
1. Which way did the ping-pong balls move towards during rotation?
2. Why did the ping-pong balls hang upwards in the water?
3. What would you expect the ping-pong balls to do if they were hung from the lids in air-filled jars?
4. What would golf balls do, when hung by threads from the lids in water-filled jars? In air-filled jars?
5. Which way would a helium-filled balloon (tied to a thread) in an accelerating bus move to? Forwards or backwards or stay stationary?

Explanation:
Just like the candle flames in Event 14.19. THE GRAVITY MACHINE I, the ping-pong balls are lighter than water, and tend to float towards areas of lesser density. As soon as the water-filled jars are being rotated, the water molecules are flung outward, making the outside of the jar denser and the inside (toward the center of the turntable) less dense. This is the cause of the balls to be flung inwards.

If the ping-pong balls were suspended from the lid in air-filled jars, the balls would fling outwards, as they are heavier than air. So will golf balls in water or air behave. But helium-filled balloons in a suddenly moving bus would move forward when the bus accelerates, and move backwards when the bus suddenly stops.

SPACE SCIENCE   CONSERVATION
                                                                 OF MOMENTUM

14.21. THE FUNNY MARBLES

Materials: 1. A plastic ruler (30 cm long, with center groove).
            2. Seven identical marbles.

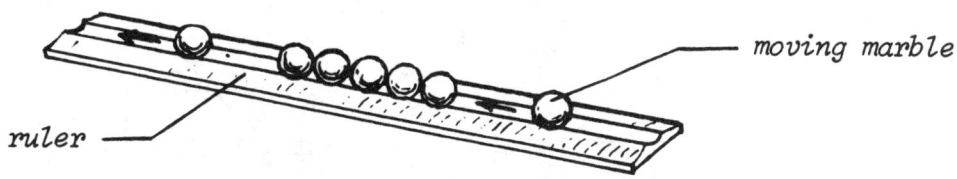

Procedure:
1. Place the seven marbles in the groove of the ruler all next to and touching each other.
2. Now take one marble and let it roll from about 10 cm away, with some speed against the other six (only one will move away).
3. Place the marbles back on their original position and do the same with two marbles bumping against the remaining five.
4. Now separate four and before letting them roll against the remaining three, ask the students: "How many marbles will move away?"

Questions:
1. When one marble bumps against six, why does only one marble move away?
2. Are the other five marbles moving much after the collision?
3. How many would move away if five marbles were pushed against two?
4. Would a marble twice as heavy also move only one away?
5. Would the end marble move also faster when one is hitting the row with a faster speed.

Explanation:
This event demonstrates the **conservation of momentum**. If all the marbles are identical in mass and size, whatever number rolls against a row of stationary ones, will move the same number away from the stationary row. These last moving marbles are just taking over the momentum that the first marbles were imparting to the row of stationary marbles. **The momentum of a moving object is the product of its mass and its velocity (mv).** When this is imparted to another stationary object, this second object will have the same initial momentum ($m_1v_1 = m_2v_2$). When both objects have the same mass, the velocity of the second object (v_2) will be the same as the first object's velocity (v_1). When the moving ball is twice as heavy ($m_1 = 2m_2$), then two balls with masses m_2 will move away. A faster moving marble will impart the same speed to a marble of the same mass. A marble with half the mass of a stationary one will only impart half the speed to the heavy marble.

Applications of this principle are encountered in head-on collisions of trucks and cars, where the truck driver almost always survives the accident but not the car passengers.

SPACE SCIENCE   CONSERVATION
                                                                 OF MOMENTUM

14.22. THE CRASHING SKATES

Materials: 1. A pair of skates (boards on marbles or straws will also do).
             2. A brick or other heavy weight.

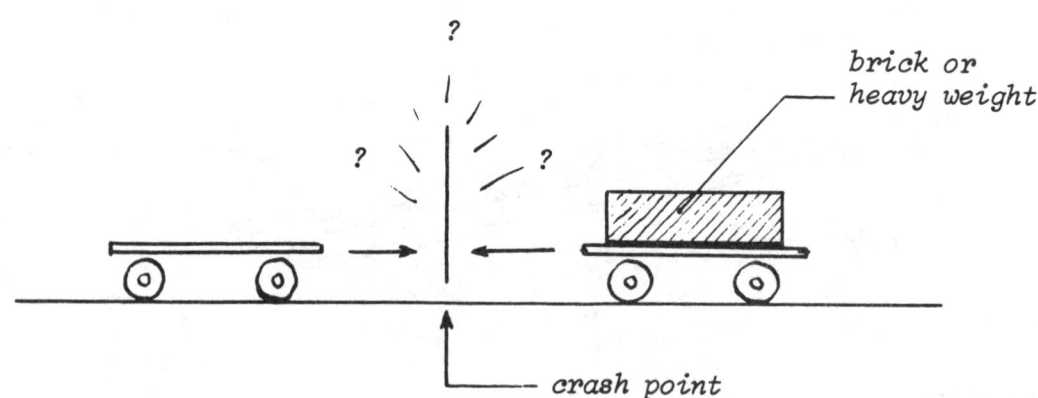

Procedure:
1. Place a brick or other heavy weight on one of the skates.
2. Hold the two skates about a meter apart on a smooth surface (table or floor surface will do. Each skate can be held by one student).
3. Ask the students to push their skates toward each other with about the same speed on the count of three.

Questions:
1. Which of the skates keeps moving in the same initial direction?
2. Which of the skates has a larger momentum?
3. If an egg were tied with a rubber band to each of the skates, which one do you think would most likely break?
4. What vehicles on the road can the skates be compared with?
5. In a head-on collision between a car and a truck, why does the truck driver usually survive the crash and not the car driver?

Explanation:
   The skate with the brick on top has a much larger mass and thus a much larger momentum. At the point of impact, the object with the larger momentum will keep on moving in the same direction, whereas the one with the smaller momentum suddenly has to reverse its direction of movement. This is the main reason why in a head-on collision of a truck and a car, the truck driver usually survives the crash. At the point of impact, the truck with a much larger mass than the car, keeps on moving forward, although with a much slower speed. The car, however, is suddenly pushed in a reverse direction and the people in this car are flung forward. The velocity of the people upon impact is the sum of the speeds of both vehicles. If for example the truck was driving 100 km/h and the car 90 km/h, at the point of impact, the people in the car would have a speed of 90 plus a little less than 100--it is like having a speed of 190 km/h hitting a brick wall, whereas the truck driver's speed would probably only be drastically reduced.

SPACE SCIENCE

CONSERVATION OF MOMENTUM

14.23. THE COLLIDING STEEL BALLS

Materials:
1. Two medium size steel balls (about 1 cm diameter).
2. A plastic ruler (30 cm long), metal washer.
3. Rather stiff wire (coat hanger wire), masking tape.
4. Four sheets of carbon paper.

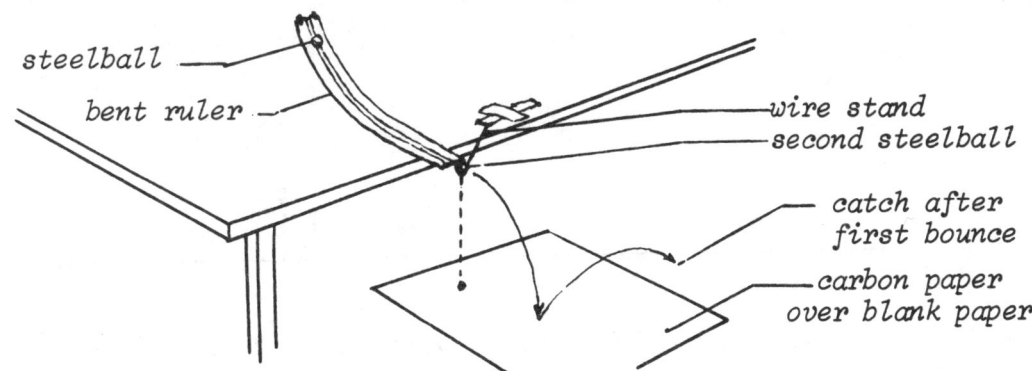

Procedure:
1. Make two 90° hooks at both ends of the stiff wire, such that the length from hook to hook is exactly the length of the ruler.
2. Bend the plastic ruler and keep it bent by hooking the wire behind the ruler and bending the wire with it (keep the center groove free).
3. Tape the end of the ruler to the table edge (by taping only the back and sides of the ruler and keeping the center groove free).
4. Make a small steel ball holder (stand) from a paper clip and tape this to the edge of the table, such that a steel ball can sit just off center in front of the groove (see sketch).
5. Hang a metal washer from a string attached to the paper clip stand and find the point plumb under the point of impact.
6. Tape four carbon sheets on top of four white sheets to the floor, such that the rolling steel balls will hit the paper.
7. Let a ball roll off the ruler 4-5 times and make sure that you catch it after the first bounce. Place a second steel ball on the stand in front of the groove and let the first ball collide into it, from the same height as when it was rolling by itself (catch both balls after their first bounce with the help of a student). Do this 4-5 times.
8. Take the sheets of carbon paper off and draw vectors from the point of impact (plumb under the end of the ruler) to where the single ball fell and to where the two balls together bounced on the floor.

Questions:
1. What is the relationship between the vectors?
2. What do the vectors indicate about the force or momentum imparted from the first ball onto the second steel ball?

Explanation:
The vector of the ball rolling alone, is like the resultant of the two smaller vectors of the 2 balls falling together (see sketch).

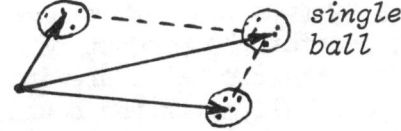

SPACE SCIENCE   CONSERVATION OF MOMENTUM

14.24. HOW HIGH WILL THE BALL BOUNCE?

Materials: 1. One golf ball.
            2. One ping-pong ball.

Procedure:
1. Hold the golf ball about waist high above a smooth hard floor or about half a meter above the table, and let it fall. Let students observe the height of the first bounce.
2. Do the same with the ping-pong ball and let the students again notice the height of the first bounce.
3. Now place the ping-pong ball on top of the golf ball and ask the students to predict how high the ping-pong ball will bounce up.
4. Drop the two balls simultaneously on the hard surface and observe! (sometimes the ping-pong ball will shoot off in an angle; repeat until both balls fall vertically on top of each other).
5. Invert the order of the balls and ask: "What do you expect the balls to do now?" Drop the balls and observe!

Questions:
1. What made the ping-pong ball bounce up so high?
2. Was the height of the bounce the sum of the heights of the separate balls bouncing? Was it higher than the sum?
3. Where did the ping-pong ball get its energy from, to bounce so high?
4. Why did the balls hardly bounce up when the golf ball was on top?
5. What would happen if we used a tennis ball instead of the golf ball?
6. How would the bounce of two ping-pong balls compare to our first bounce?

Explanation:
   The height of the bounce of each individual ball is about three quarters of the original height. When the ping-pong ball falls together with the golf ball (vertically underneath it), one would expect the bounce to be the sum of each ball individually making it twice as high, but it bounces up much higher. This is caused by the much larger mass of the golf ball as compared to the ping-pong ball. As the **momentum** (of the golf ball **is conserved**, it is actually imparted on the ping-pong ball ($M_1 V_1 = M_2 V_2$). Because the mass of the ping-pong ball is much smaller, it has to have a much larger velocity (V_2) in order for the product of mass and velocity to stay the same.
   When the two balls are inverted, there is only a small momentum imparted on the golf ball after the bounce, which is not large enough to bounce the golf ball back, thus both balls fall dead on the surface.

SPACE SCIENCE

CONSERVATION OF MOMENTUM
RECOIL

14.25. THE TEST TUBE CANNON

Materials:
1. A thick walled test tube + cork or rubber stopper.
2. Tall stand, thin wire, Bunsen burner.
3. Protective goggles.

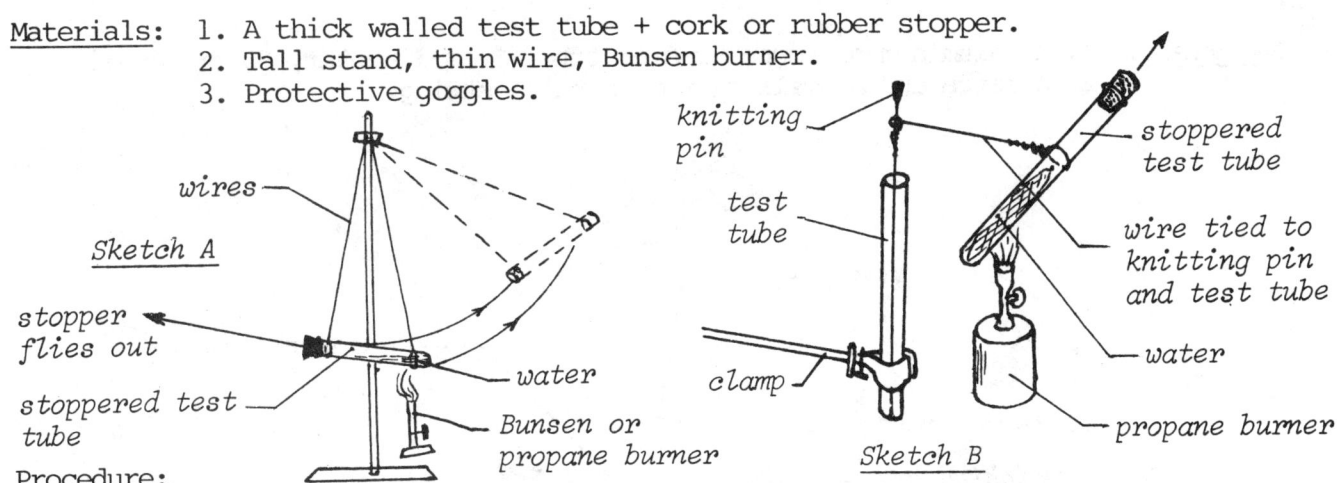

Procedure:
1. Tie two wires to the test tube and hang it from the tall stand as shown in Sketch A (somewhat slanted with opening up).
2. Pour a few ml of water in the tube and stopper with the cork or solid rubber stopper. **CAUTION: DO NOT INSERT STOPPER TOO TIGHTLY!!**
3. Put your safety goggles on and start heating the tube. Make sure that the mouth of the tube is facing away from people and breakable material.
4. Keep heating to boil the water - hold the burner with an outstretched arm and keep your face as far as possible from the tube, and wait for the POP! Observe the movement of the test tube!
5. An alternative way for suspending the test tube is shown in Sketch B.

Questions:
1. What made the test tube move at the time of the "POP"?
2. How does the use of a rubber stopper compare to a cork?
3. Will the use of more or less water influence the amount of recoil?
4. How would a longer or wider test tube influence the recoil?
5. What is the danger in inserting the stopper too tightly?
6. Will the tightness of the stopper influence the amount of recoil?

Explanation:
By boiling the water in the test tube, vapour is being created, and thus pressure is built up, as the test tube is stoppered tightly. The vapour molecules move faster and faster as energy is supplied to the steam by the heat, and at the moment that the pressure is high enough to overcome the friction of the stopper against the glass, the stopper POPS out.
The heavier the mass of the stopper (rubber as compared to cork), the larger the recoil distance, as the momentum ($M_1 V_1$) of the stopper is equal to the momentum of the test tube ($M_2 V_2$) in which M = mass and V = velocity. When only M_1 is increased, and all other variables are held constant, it will result in an increase of V_2.
Using a wider test tube means increasing the stopper's mass, but also that of the test tube, thus the recoil would most likely stay the same. Using a longer test tube would change M_2 and thus the amount of recoil is not as large. This can be compared to using a revolver and a rifle (less recoil in a rifle).

SPACE SCIENCE

CENTER OF GRAVITY
ANGULAR MOMENTUM

14.26. THE SPINNING PLANETS

Materials:
1. An aluminum or copper rod or tube (1 cm diameter, 50 cm long).
2. A large and a small styrofoam ball, string.

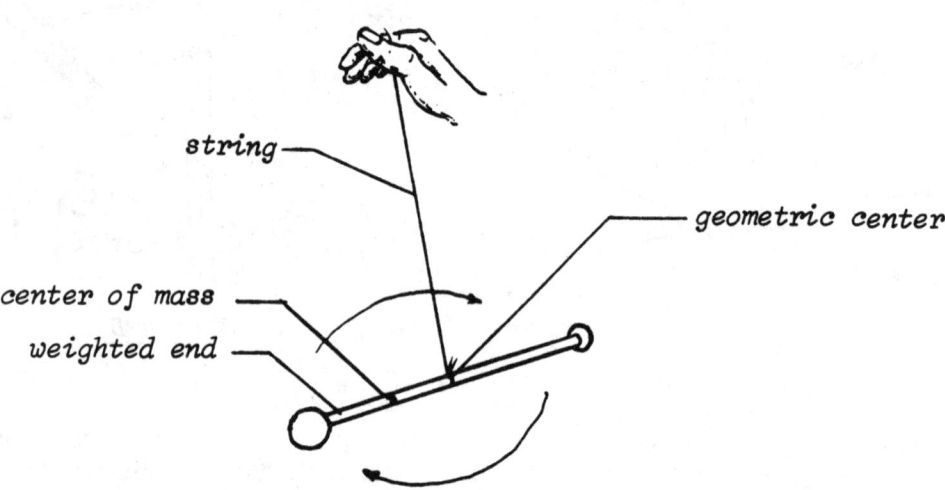

Procedure:
1. Stuff one end of the tube with the lead (solder wire would do fine) and attach the larger ball to the weighted end and the smaller one to the other end of the tube.
2. Find the balance point of the tube and mark the center of gravity by drawing a line around the tube with a marker.
3. Tie the string to the geometric center of the tube and show that the tube cannot balance horizontally at that point. Ask: "How can I balance the tube horizontally from the string?"
4. Now spin the tube while holding the string with the other hand, and keep spinning the tube around (the tube is now hanging horizontally from the string). Observe the marked center of mass/gravity!

Questions:
1. Did the geometric center stay in the middle of the rotating system?
2. Around what point did the system spin?
3. What kept the pipe hanging horizontally from the string?

Explanation:
All spinning objects spin around their center of mass, and this is why the marked spot stayed at the same place while the tube was spun. The geometric center, not being the center of mass, because the tube was weighted at one end, had to spin around the center of mass. This is the reason why we could keep the tube spinning. As the tube was spinning in a horizontal plane, it was actually kept horizontal and hanging horizontally from the string, even though it was not hanging from its center of balance (center of mass/gravity). The **plane of rotation** of the tube happened to be in a horizontal plane. If the tube could be spun at a higher speed, this plane of rotation could take up any position.

SPACE SCIENCE   ANGULAR MOMENTUM

14.27. THE PAPER CARD BOOMERANG

Materials: 1. A stiff paper card (5x8" / 12x20 cm).
            2. A pair of scissors.

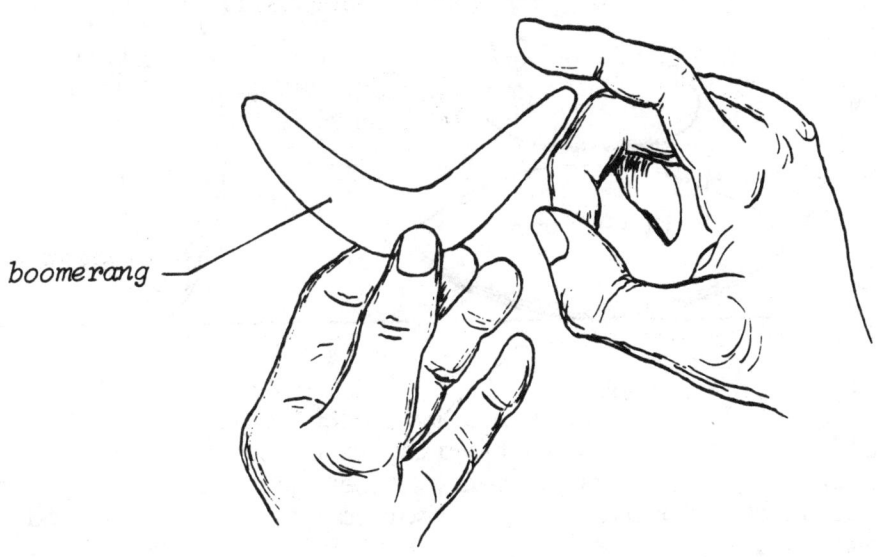

Procedure:
1. Cut the paper card in the shape of a boomerang with a pair of scissors (see sketch).
2. Hold the paper boomerang with two fingertips at the bend, and slightly turn the legs upwards (by stroking them).
3. Hold the boomerang at an angle of 45° very lightly between index finger and thumb of your left hand, and flick it away with the middle finger of the right hand (see sketch).

Questions:
1. What made the boomerang return by itself?
2. What is most essential for a boomerang to be able to return?
3. If the paper boomerang were not spun, would it return?
4. Why does the boomerang have to be thrown slightly upwards?
5. Would other shapes work to make a boomerang?

Explanation:
   When we flick the boomerang, a double movement is imparted to it: a spinning or fast rotation and a general upward motion. The spinning forces it to rise obliquely and **conserve its angular momentum in the plane of rotation** until the general upward momentum is exhausted. At this point it is still spinning, but instead of rising further, its weight causes it to descend. However, it does not drop straight down, but as **it tends to conserve its angular momentum** and also **its plane of rotation**, it slides back to the thrower in almost the same plane. The air resistance helps the boomerang to stay in its plane of rotation and it actually rides on the air, returning to the man who threw it in the first place.
   "Frisbees" may be thrown like a boomerang based on this same principle, when a partner for a catcher is not available.

SPACE SCIENCE

ANGULAR MOMENTUM
MOMENT OF INERTIA

14.28. THE SPINNING FOOTBALL

Materials: 1. An American football (actual or toy size).
2. A smooth table top.

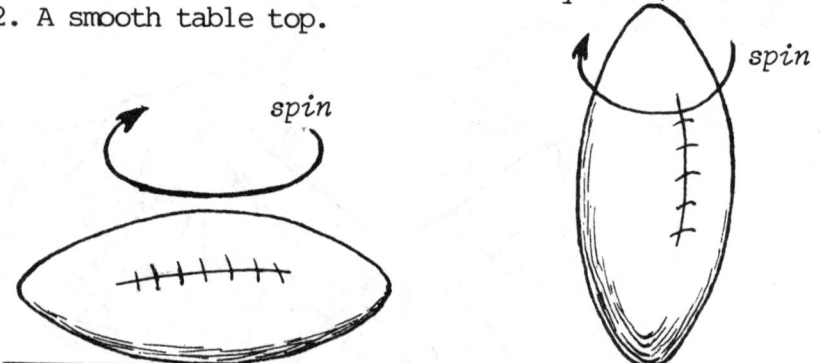

Procedure:
1. Place the football horizontally on the table top.
2. Hold it down firmly with the fingers of one hand and give it a quick spin. Observe! (If nothing happens, it was not spinning fast enough: try again).
3. Instead of a football, a hard-boiled egg may be used and it will behave the same way.

Questions:
1. What did you observe the ball doing?
2. What made the football spin vertically?
3. What would happen if you spin the football slowly (horizontally)?
4. What would happen if you spin the ball vertically?
5. Why doesn´t the football stay spinning vertically?
6. What would happen if you spin the football fast vertically?
7. Why can´t you spin the football fast horizontally and keep the ball spinning horizontally?

Explanation:
It takes more energy to rotate or spin the football horizontally as compared to vertically. This is because the mass of the football is spread out further from the center of gravity which is also the center of rotation, when it is in the horizontal position. There is a tendency for a system to move toward a **state of lowest internal energy** or **enthalpy**, which is the case when the football moves from the horizontal to the vertical position. But in order to spin at this position a certain energy level is required. This is why a slow spin at the horizontal position will just keep it spinning horizontally.

We find this principle applied in our daily life in figure skating, where the skater will spin faster when he/she pulls in his/her arms closer to the body while spinning. (**Conservation of Angular Momentum**).

During the high dive in the swimming sport, we see that divers keep their body closely tucked in when doing forward or backward saltos or flips. When they want to slow down the rotation, they stretch their body and this usually happens just before they enter the water.

SPACE SCIENCE

ANGULAR MOMENTUM
PRECESSION

14.29. THE HUMAN GYROSCOPE

Materials:
1. An old bicycle wheel.
2. A wooden board (about 50 x50 cm), an old broom stick.
3. A hand full of steel ball bearings.

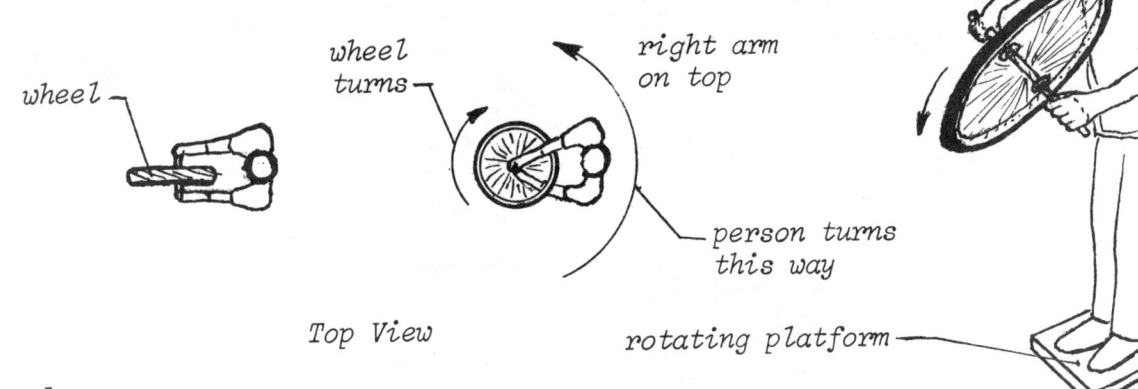

Top View

Procedure:
1. Cut a couple of pieces of broom handle and attach them to the axle on each side of the bicycle wheel for handles (drill holes in one end of the wood and screw the wood in the axle end).
2. Place the wooden board on the ball bearings on a smooth floor surface and have someone stand on the board.
3. Hand the wheel to the person standing on the board and let him/her hold the wheel vertically while you spin it with a downward motion.
4. Now ask the person holding the spinning wheel to rotate the plane of rotation from the vertical position to a horizontal position by turning it to the left. (What happens to the person?)
5. Now ask him/her to bring it back to the vertical position and keep turning it to the right to a horizontal position (What happens now?).

Questions:
1. Which way did the person holding the wheel turn, clockwise or counter-clockwise (top view) when turning the spinning wheel to the left?
2. After turning the wheel to the left, in what direction does the wheel spin? Clockwise or counter-clockwise? (looking from the top).
3. After bringing the wheel back to its original vertical position, what happened to the person holding the spinning wheel?
4. In which directions did the wheel and person rotate by turning the spinning wheel to the right?

Explanation:
By turning the spinning wheel to the left, the wheel was turning in a clockwise direction and the person on the platform rotated counter-clockwise : **conservation of angular momentum.** The faster the wheel spins and also the more mass the wheel has (attaching lead weights to the wheel rim would be effective), the greater the angular momentum; thus the more difficult to turn the **plane of rotation** (flywheels in ships). The turning to the left of the person when the wheel is tipped to the left is called **precession**. This occurs f.i. in motorcycle riding: in order to turn left the motorcyclist tips the vehicle to the left and almost doesn't have to move his steering. A right turn is made by tipping the vehicle to the right.

SPACE SCIENCE

ANGULAR MOMENTUM
MOMENT OF INERTIA

14.30. THE TIN CAN RACE

Materials:
1. A variety of round tin cans of soup, vegetables, fruit cocktail, tomato juice, including solid dog food.
2. A one meter long inclined plane (a wooden board propped up at one end with a stack of books).
3. A stop watch (optional).

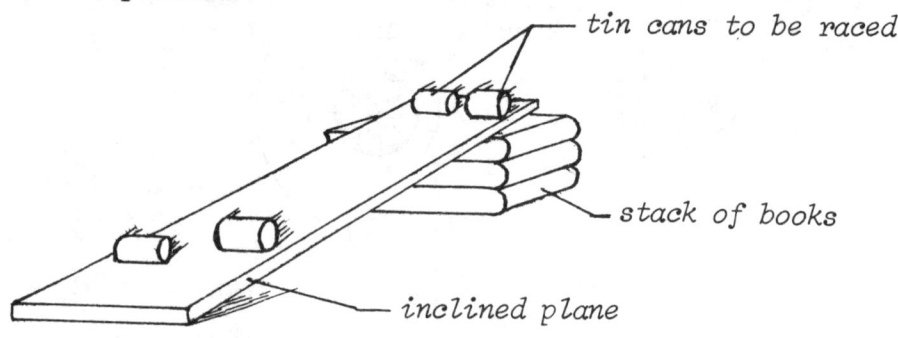

Procedure:
1. Set up the board as in the Sketch to form the inclined plane.
2. Divide the class in groups of at least three students in each group: one student to release the cans at the top of the plane, one to catch the cans at the bottom end of the plane, and one to record the time or to observe which of the two cans won the race.
3. Let each group of students choose a tin can that they think should roll the inclined plane the fastest and be the winning can (take average of three).
4. Let two groups at a time race their tin can against each other. The winner of this race will compete against the next group, etc. until all groups have participated. Declare the overall winner!

Questions:
1. Which of the tin cans is the overall winner?
2. What is the consistency of the contents of the winning can?
3. How did the tomato soup compare to the chunky soup?
4. How did the larger tin cans compare to the smaller ones?
5. Were the heavier cans always winning from the lighter ones?
6. What are the variables involved in determining the rolling speed?
7. Which are the manipulated and responding variables?
8. How can we control the other variables?

Explanation:
   The variables involved in the tin can race are: rolling speed (responding), size, contents, weight of can, incline of the plane, etc. Any of the latter variables can be the manipulated one. Say we wanted to manipulate the size of the can, then we take two different sizes of cans containing the same thing. If we take, f.i. a can of soup and the solid dog food (with the same size), then we are manipulating the contents (in this case the consistency). Invariably, the can with the solid contents will win the race, as the **moment of inertia** is the smallest to overcome. Cans containing loose chunks of mass, like chunky soup or fruit cocktail are most likely the slowest, as they have the tendency to get the mass on the periphery of the can and thus the highest moment of inertia. (Hollow and solid cylinders can also be used to compare rolling speed).

SPACE SCIENCE

14.31. THE CUP OF COFFEE DROP

Materials:
1. A styrofoam cup, a large wide and deep bucket.
2. Coffee or intensely coloured water.

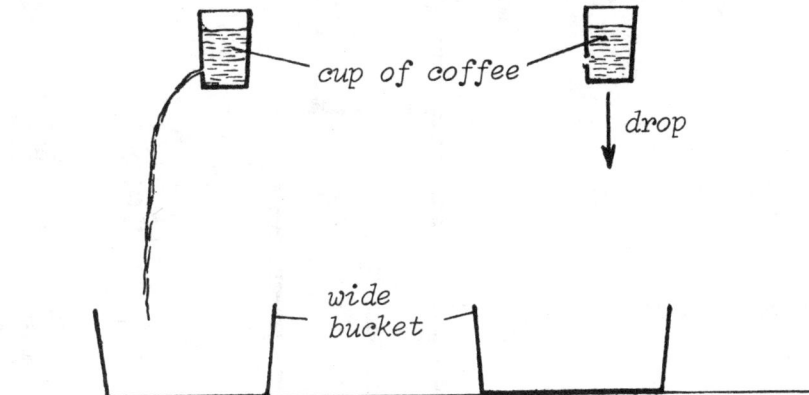

Procedure:
1. With a pencil point poke a hole near the bottom in the side of the foam cup. Place a finger over it and fill the cup with coffee or coloured water (if coloured water is used, make sure it is intensely coloured).
2. Stand and hold the liquid-filled cup over the bucket, which is placed on the floor in front of you. Let go of your finger over the hole and show that the liquid squirts out of the cup.
3. Place your finger back over the hole and tell students that you are going to drop the whole cup. Let them follow the falling cup and observe closely whether any liquid is squirting out of the hole while the cup is falling. Now drop the cup and simultaneously remove your finger which was covering the hole.

Questions:
1. What makes the liquid squirt out of the hole in the beginning?
2. Would the strength of the squirt be the same on the surface of the moon, all other variables being equal?
3. Why did the liquid stop squirting during the fall?
4. What would happen if the finger covering the hole is removed a little sooner than the release of the cup?
5. If this same liquid-filled cup is held in an orbiting satellite, would any liquid squirt out of the hole? Would it be possible to hold the liquid in an open cup?

Explanation:
When the liquid-filled cup is held stationary, the liquid is pulled down by the earth's gravity causing a liquid pressure at the point of the opening in the side of the cup, resulting in the squirt. When the cup is falling, the liquid pressure is suddenly eliminated, and no liquid is squirting out.

All variables being equal, when this liquid-filled cup is held on the surface of the moon, the liquid will only squirt about one sixth of the distance from what it was on the earth, as the gravity pull is about 1/6th of the earth's.

In a space satellite, no liquid will come out of the hole at all, just like in the falling cup, all materials actually are weightless. It would be very difficult to keep liquids in an open container in a satellite.

14.32. THE FALLING ELEVATOR

Materials:
1. A bathroom scale.
2. A fast moving elevator.

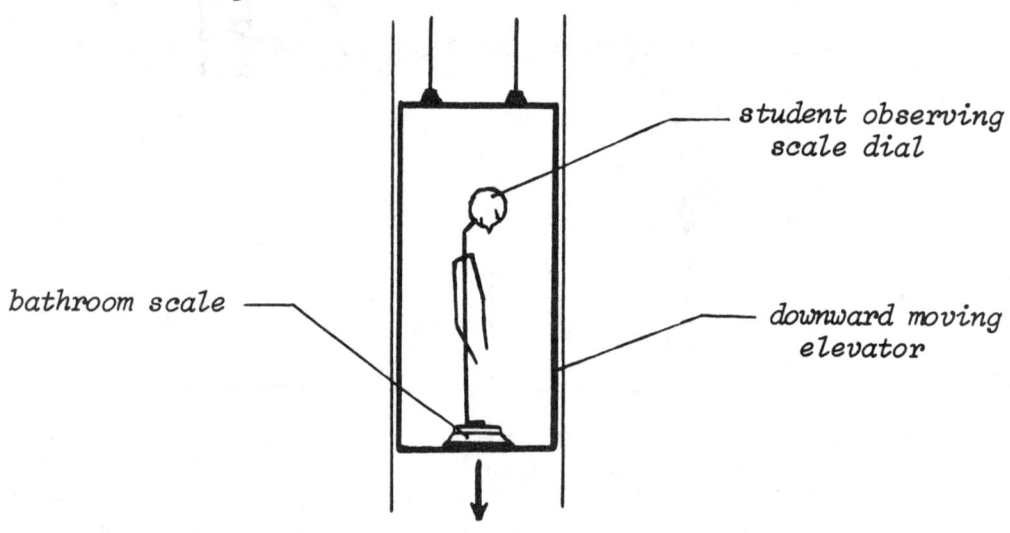

Procedure:
1. Take the students in small groups into a fast moving elevator.
2. Let one student stand on the scale in the center of the elevator and have the other students observe the indicator of the scale.
3. Send the elevator up and down, and draw attention to the weight change at the start and end of the upward or downward motion.

Questions:
1. What did the scale indicate at the start of the upward motion?
2. What was your weight at the beginning of the downward motion?
3. How do you usually tell, without looking at the lights (indicators), which direction the elevator takes you?
4. What would happen to your weight if the elevator were to fall freely?
5. What is it that causes things to have weight?
6. Does weightlessness have any effect on the mass of objects?

Explanation:
It is the attraction of objects by the earth that gives them their weight. This weight is measured by the deflecting needle of a scale, usually based on the degree of stretch in a spring. In the motionless elevator, the weight of the student is the same as that on the solid floor. But, as soon as the elevator starts its upward motion, an added force upward is exerted on the scale, which increases the student's weight. This force, however, is only exerted during the accelerated part of the upward motion.

When the elevator moves down, a downward force is exerted, resulting in a smaller upward force on the scale, and thus showing a lower weight. If the elevator were to fall freely, a **constant acceleration** and thus a force downward is exerted on the scale, eliminating the upward reactive force of the student's weight, resulting in weightlessness of the person.

CHAPTER 15

WHAT ARE SOME PHENOMENA ABOUT THE EARTH AND THE MOON?

OBJECTIVES

After dealing with and studying the concepts and sub-concepts in this chapter, the students should be able to:

a. recognize the correct explanation of an observed event based on each of the sub-concepts;
b. explain in their own words which of the sub-concepts is determining the course of an event;
c. distinguish true from false statements concerning each one of the sub-concepts;
d. idnetify the correct explanation of an event in daily life applying one of the sub-concepts;

all in relation to the following sub-concepts:

-- Dissolved salts in ground water are deposited when the water evaporates, as in stalactite and stalagmite formation.
-- Some rocks are softer and more soluble in dilute acids than others.
-- Erosion and weathering may occur by plant growth or the freezing of water.
-- Continents and ocean floors stay at the same level in spite of erosion and deposits of material, because of the liquid mass in the inner core of the earth (isostatics).
-- Volcanic and geyser eruptions are caused by pressures in deeper lying layers of the earth.
-- The phases of the moon are caused by the different positions of the moon in relation to the earth and to the sun.
-- A solar eclipse is caused by the moon's shadow on the earth.

EARTH SCIENCE

EROSION & WEATHERING

15.1. MAKE STALACTITES & STALAGMITES

Materials:
1. Magnesium sulfate (Epsom salt).
2. A large beaker and stirrer.
3. Two small beakers, a thick water-absorbent string or cloth.

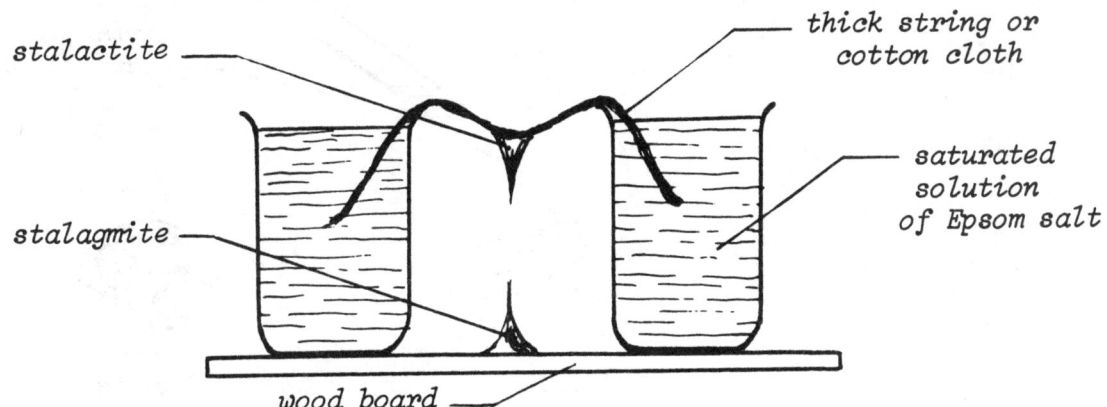

Procedure:
1. Make a saturated solution of magnesium sulfate in water in the large beaker, by dissolving as much of the powder as you can in about 200 ml of water (until some solid stays undissolved).
2. Fill the two small beakers with the saturated solution and place a thick water-absorbent string or cloth between the two beakers, such that the center part hangs somewhat lower than the beaker's rim (see sketch above).
3. Let stand for a few days and observe.

Questions:
1. Which cone is called stalactite and which stalagmite?
2. How were the cones formed?
3. What material are the stalactite and stalagmite made up of?
4. What other materials can we use instead of the string or cloth?
5. Where in nature are stalactites and stalagmites formed?

Explanation:
Stalactites and stalagmites in nature may be found in **underground caves**. Groundwater which contains dissolved salts and minerals, drips from the ceiling of the cave. While the drop hangs down, the water evaporates and leaves some of the salts deposited on the ceiling, forming the stalactites. Similarly, the stalagmites are formed on the bottom of the cave. The water drops containing the dissolved salts evaporate and keep depositing the salts on the same spot, leaving a cone of salts that were dissolved in the water.

This demonstration is a simulation of the stalactite and stalagmite formation in caves, which may have taken centuries to build up. Water-absorbent paper towel, blotting paper, etc. may be used instead of the string or cloth.

EARTH SCIENCE   EROSION &
                                                                 WEATHERING

15.2. CAN STONES DISSOLVE?

Materials: 1. Pieces of limestone, marble, granite.
             2. Vinegar or dilute hydrochloric acid.

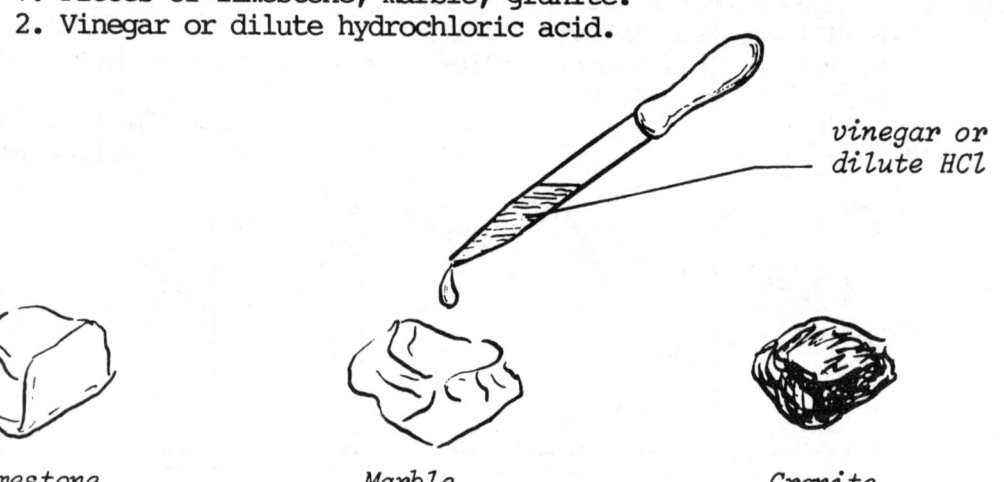

Procedure:
1. Observe the pieces of limestone, marble, and granite. Ask: "How do they differ? How do we know which is which?"
2. Place a few drops of vinegar or dilute HCl on a flat surface of each of the rocks and observe for chemical reaction.
3. Let stand for a few minutes, then place again a few drops of acid on each of the rocks. Ask: "Which one reacted most?"
4. Scratch the surface of each of the stones with the fingernail or a pen knife. Ask: "Which one is the softest? Which the hardest?"

Questions:
1. How can we distinguish which stone is which?
2. What are the characteristics of each of the stones?
3. Which of the stones would be eroded most when subjected to the same erosive forces in nature?
4. How would you tell the difference between common salt and marble?
5. What other methods are geologists using to identify minerals?

Explanation:
   The limestone is the softest of the three and granite the hardest. limestone can most likely be scratched by the human fingernail (number 4 in the hardness scale), whereas the other two cannot. Limestone reacts most with the vinegar or dilute hydrochloric acid. Marble does react with the acid, but not as readily, because of its more compact structure. Chemically they are the same (calcium carbonate). Other methods that geologists use to identify minerals are: breaking the rock and observing the crystal structure with magnifying glasses or microscopes.

EARTH SCIENCE

EROSION & WEATHERING

15.3. CAN PLANTS BREAK ROCKS?

Materials:
1. Dried lima beans, red beans, or corn seeds.
2. Two small flowerpots & soil.
3. Plaster of Paris (or 'Polyfilla').
4. A piece of window glass (the size of the pot).

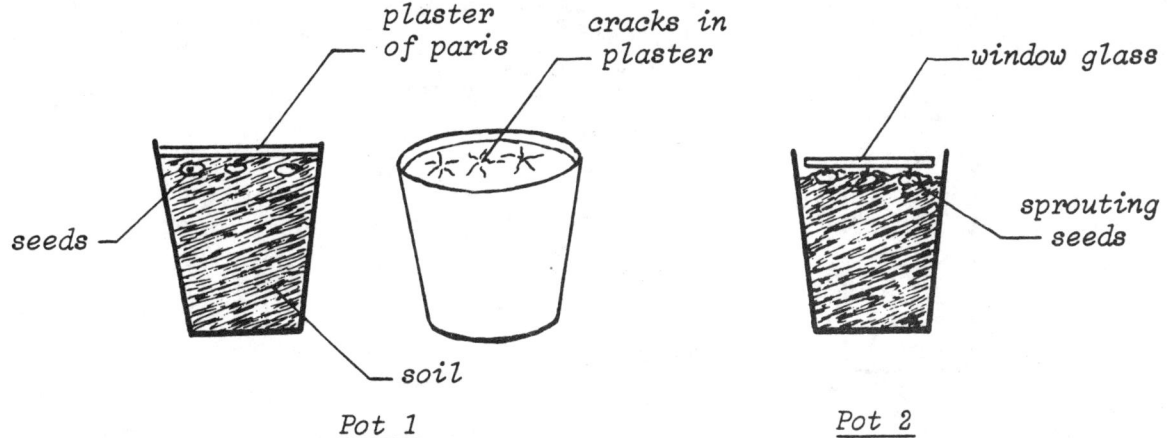

Pot 1   Pot 2

Procedure:
1. Soak six to ten beans or seeds in water and let stand overnight.
2. Plant the seeds just under the soil surface in the pot, and water.
3. Cover the soil with 1 cm thick layer of plaster of Paris (or Polyfilla); cover the other pot (with planted seeds) with window glass.
4. Observe and examine the two pots daily (in pot 1, cracks will appear in the plaster; in pot 2, the glass will be lifted).

Questions:
1. Why did the seeds have to be soaked before planting?
2. What happened to the plaster of Paris after a few days?
3. What did you observe the window glass was doing after a few days?
4. Can we find places where plants have broken through asphalt, brick or cement?
5. If sprouting seeds have such strength, can you imagine how strong the roots of tall trees are?

Explanation:
This demonstration may be used as an activity for students to discover the strength of sprouting seeds, and how they can cause rocks to move out of their path of growth or break them up into smaller pieces.

Take the students out on the school yard to find plants that have grown through cracks in the sidewalk, lower edges of the wall, or other such places.

Roots of tall trees can easily break foundations of concrete when the trees are planted too close to the buildings. Large rocks can similarly be broken up by the growth of roots, causing **pulverization or erosion** of the rocks.

EARTH SCIENCE

EROSION & WEATHERING

15.4. HOW CAN WATER BREAK ROCKS?

Materials:
1. Two empty jars (one with a screw lid).
2. A piece of sandstone, limestone, and pumice.
3. Four small polyethylene bags.

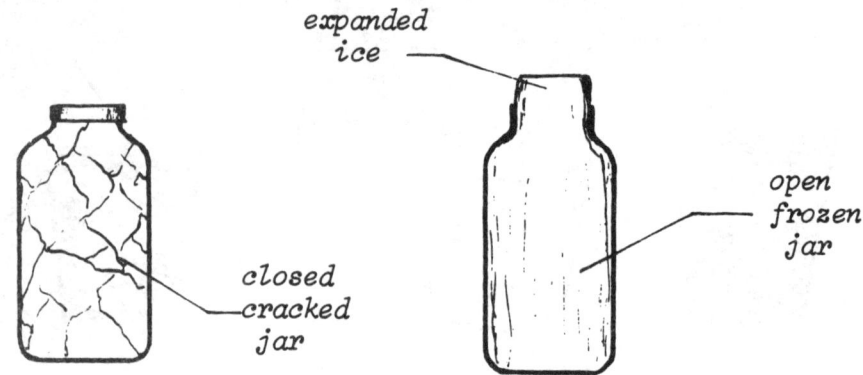

Procedure:
1. Fill the two jars brimful with water.
2. Close one off with the screw lid, leave the other open, and place both jars in the freezer, after wrapping them in a plastic bag.
3. Observe what happened to the jars after three or four hours, or at the end of the day.
4. Soak the pieces of rock in water and place them in the freezer, after each one is wrapped in a plastic bag.
5. Observe what happened to the rocks at the end of the day.

Questions:
1. What happened to the two jars with water after the water froze?
2. Would the closed jar also break if it was not completely filled?
3. How much air space do we need above the water in order for the jar not to break in the freezer?
4. How similar is this breaking of the filled and closed jar in the freezer to the breaking up of rocks in nature?
5. In what regions of the world does this process of erosion occur?

Explanation:
   The jar that was closed off broke into pieces after the water froze. This happens because water expands when it changes into ice. This freezing action is so strong, that it can break basement walls when the drainage outside the basement is not adequate.
   Similarly, when water seeps into cracks or pores of rock, especially of pumice, and the temperature drops below freezing, the water freezes, expands and breaks up the rock. This type of erosion only occurs in regions of the world where temperatures drop below 0° C in winter.

EARTH SCIENCE

ISOSTATICS
EROSION

15.5. WHY DO ERODED MOUNTAINS KEEP RISING?

Materials:
1. A thin and a thick block of wood.
2. Sand and a scoop, a short piece of string.
3. A transparent plastic container (or aquarium).

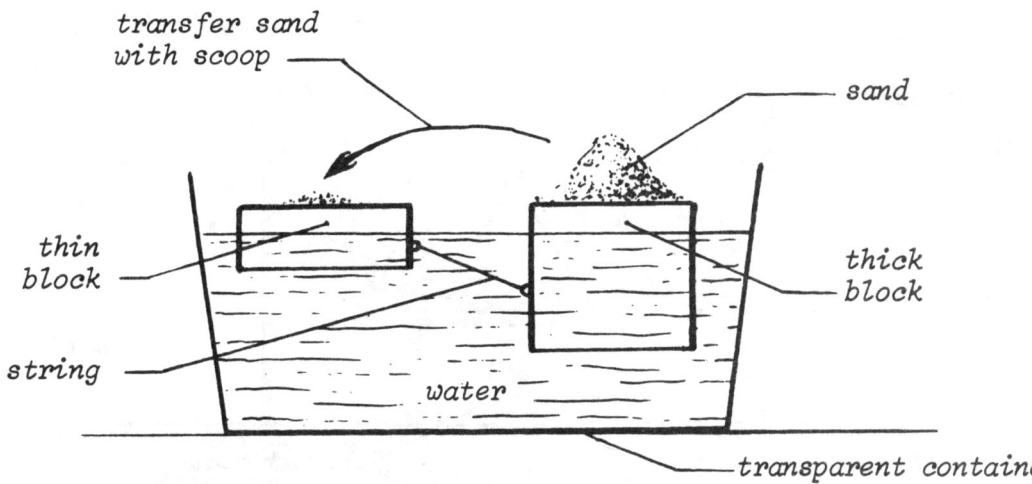

Procedure:
1. Tie two short pieces of string to the wooden blocks in such a way that about 10 cm is separating the pieces of wood.
2. Fill the large container with water and let the two wooden blocks float on the water surface.
3. Place a heap of sand on the thick block, such that it just floats above the surface of the water.
4. Transfer the sand little by little to the thin block with the scoop. Observe the rising and sinking of each block.

Questions:
1. What do the two blocks represent?
2. How was the erosion of the mountain simulated?
3. How was the total height of the mountain affected by the transfer of the sand?
4. Why does the ocean floor stay relatively stable in depth?

Explanation:
The thick block of wood and the sand on top of it represent land with a mountain on it. The rain water and the weathering erodes the mountain and transports the eroded material to the ocean floor: represented by the thin block of wood. The erosion process is simulated by the moving of the sand with the scoop. The total height of the mountain above the water surface stays about the same, even after taking some sand off, because the block gets lighter and floats higher above the water surface.

Similarly, the ocean floor (thin block) does not get much higher relative to the water surface, as it becomes heavier and thus floats lower.

EARTH SCIENCE

GEYSER ACTION &
VOLCANIC ACTION

15.6. HOW DOES A GEYSER WORK?

Materials:
1. A pyrex or aluminum pie plate.
2. A medium size Pyrex funnel, three bottle caps.
3. A hot plate or burner & stand, cardboard box.

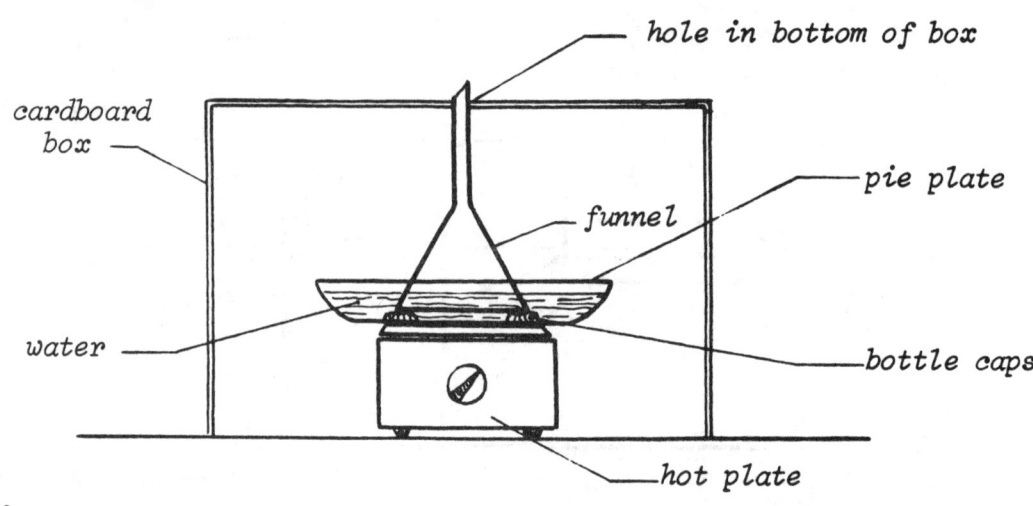

Procedure:
1. Fill the pie plate half full with water and place the funnel on it, resting on the three bottle caps.
2. Place the pie plate on the hot plate or over the burner (see sketch).
3. Heat the water in the pie plate until boiling. Make a hole in the bottom of the box, cover the whole set-up with it and let the funnel stem protrude through the hole. Observe geyser action!

Questions:
1. What did the water do when it was heated?
2. What made the water spurt out of the funnel?
3. What can we compare the cardboard surface to?
4. Why does the hot water inside the earth escape to the surface?
5. What will happen when the heat is turned off?
6. How can we let the geyser erupt again?
7. How similar is volcanic action compared to geyser action?

Explanation:
By heating the water, it expands. At the moment that it boils, water vapor is formed, pressure is built up under the funnel and this force pushes the water out of the funnel. When the heat is taken away, the water cools down and the eruptions stop. In order to start the geyser to erupt again, the water needs to be heated to boiling.

This demonstration illustrates clearly the similar working of the hot water under the earth's surface, the build-up of pressure, and the release of the water through small cracks in the earth's surface: called the **Geyser**. Volcanic action is quite similar except that instead of water, we have molten lava, which comes out of the mouths of volcanoes.

To make it inquiry oriented, conceal the set-up and let students find out the reason why the water spurts out.

EARTH SCIENCE   ROCKS & DENSITY

15.7. HOW CAN WE DETERMINE THE ROCK'S VOLUME?

Materials: 1. A medium size tin can.
            2. A measuring cylinder, a piece of thread.
            3. Different rocks, including: pumice, sandstone, granite.

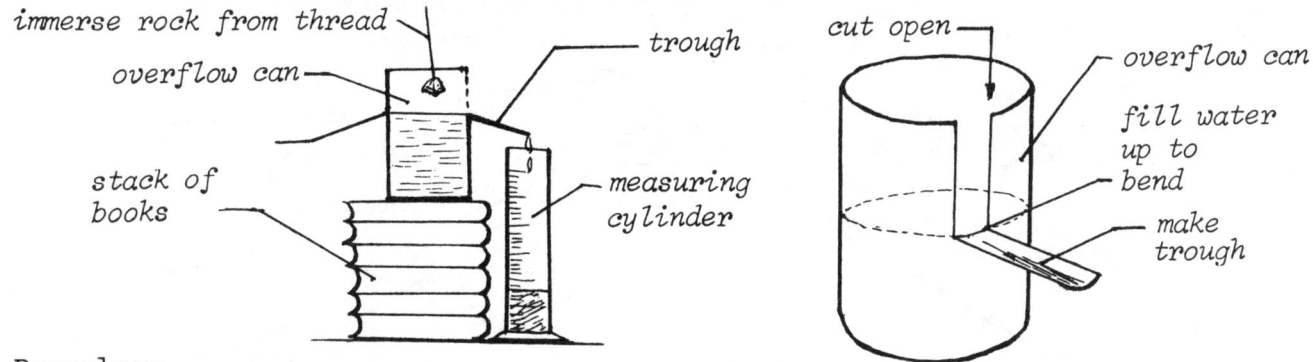

Procedure:
  1. Make an overflow can by cutting a 2 cm vertical strip of about 5 cm long
     out of the side of the can and bend this strip outward, making a small
     gutter out of the strip by bending the outer edges up (see Sketch).
  2. Place this overflow can on a stack of books and fill it with water until
     the water just overflows.
  3. Place the measuring cylinder underneath the end of the overflow trough to
     catch the overflowing water.
  4. Tie the rock to be measured to a string and lower it slowly into the over-
     flow can until it is completely submerged. Read off the volume of the
     water in the cylinder (for floating objects, like pumice, use a needle to
     immerse the whole object underneath the surface of the water).
  5. Weigh each of the rocks on a scale and let students determine the densities
     of each of the rocks (D = M / V).

Questions:
  1. What made the water overflow?
  2. What influences the amount of water overflowing?
  3. Why does the pumice stone have to be pushed under water?
  4. What result would we get if the pumice was not completely submerged?
  5. What are the variables to be controlled if we wanted to compare the water
     displaced by the different rocks?
  6. Does the shape of the rocks influence the amount of overflow?
  7. Would a round and a cubical rock of the same type and weight (mass) have
     different amounts of overflow?

Explanation:
     The different types of rocks have different densities, and as density is
defined as being mass of the object per unit of volume, it is necessary to
determine the volume and the mass of each of the rocks. The volume of an odd
shaped object (which rocks usually are) can easily be determined by submerging
it in water and measuring the displaced water. This can be done with the
overflow can.
     When rocks of the same weight but of different types are submerged in the
overflow can, different volumes of displaced water will be obtained. This is
caused by the different densities of the rocks. The amount of overflow for
the same type of rock will be the same, as long as the mass of the rock is
held constant (the same), no matter what shape it is in.

15.8. SIMULATE A VOLCANO ERUPTION

Materials:
1. Ordinary earth clay & a wooden board (about 1m x 1m).
2. Ammonium bichromate (50 g), magnesium powder (12 g).
3. Magnesium ribbon (about 1 m long).

CAUTION:
PERFORM OUTDOORS OR UNDER FUMEHOOD!

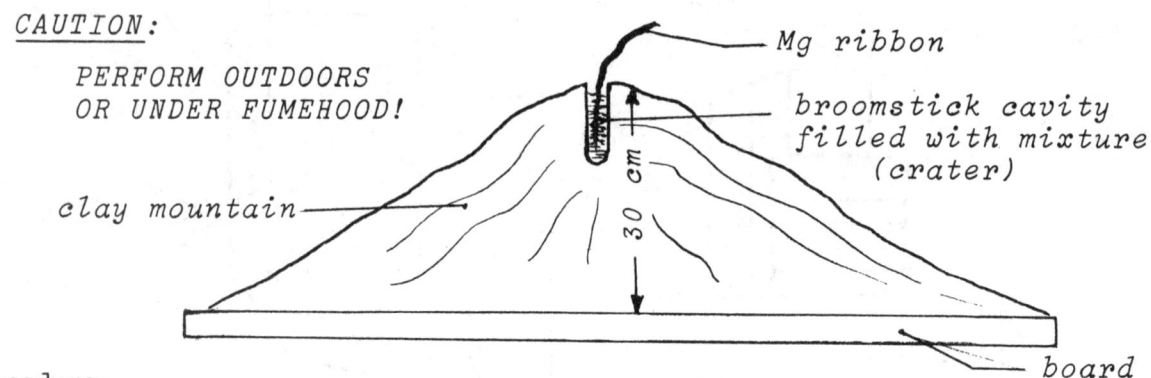

Procedure:
1. With the wood board as a base, have the students build a volcanic cone of about 30 cm high and a diameter of 60 cm at the base. Push the handle of a broom in the middle of the volcano down to about 6 cm deep to form the crater.
2. The 50 g ammonium bichromate is enough for about 3-4 eruptions. Mix about one third of the amount of bichromate with about 4 g of magnesium powder and pour the mixture in the crater.
3. Cut a 10 cm strip of magnesium ribbon and push one end into the mixture filling the crater, let the other end stick out of the crater for a fuse.
4. Light the magnesium ribbon fuse with a match and stand back! (If the eruption did not set off, wait a few moments, then insert a second fuse and try again).
5. After the eruption, a second eruption can be simulated by pouring more of the bichromate-magnesium powder mixture into the crater while it is still hot (use a long paper or cardboard chute to fill crater).

Questions:
1. What is the function of the magnesium ribbon?
2. What was the cause of the eruption (as a chemical reaction)?
3. What causes an actual volcano to erupt?
4. What comes out of the crater of an actual volcano?
5. How can actual volcanic eruptions be predicted?

Explanation:
Ammonium bichromate $(NH_4)_2Cr_2O_7$ is a very strong oxidant and unstable when heated. At the time that the magnesium ribbon flare touches the mixture of bichromate and magnesium powder, the bichromate decomposes into ammonia gas, chromium oxide, and oxygen. In the presence of the magnesium powder it also forms magnesium chromate. The sudden formation of the gases is the main cause for the upward spray and spewing out of the crater.

A real volcano erupts because of the pressure built up in the earth. The craters in the volcanoes are acting as valves through which this earth's pressure is released. When erupting, a volcano spews hot molten earth, called **magma** (underneath the earth's surface) or **lava** (above the earth's surface) out of the crater.

EARTH SCIENCE ASTRONOMY

15.9. WHAT CAUSES THE PHASES OF THE MOON?

Materials:
1. A large shoe box, carbon paper or other black paper.
2. A styrofoam ball (about 5 cm in diameter).
3. A small strong flashlight, masking tape, black thread.

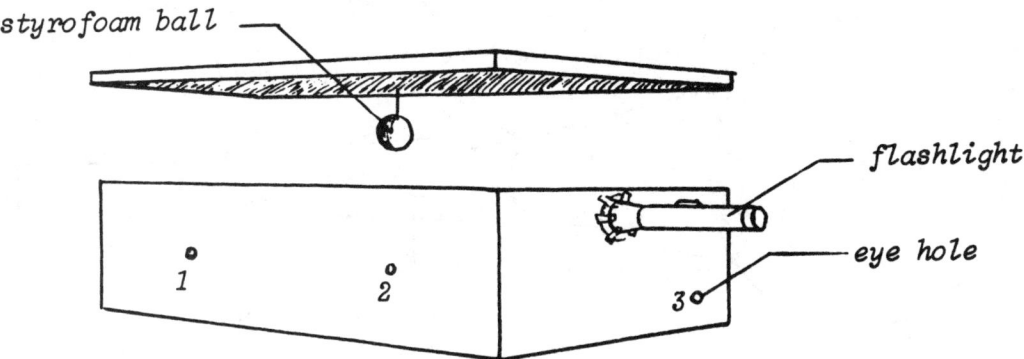

Procedure:
1. Cover the inside of the shoe box with black paper by gluing or placing small rolls of masking tape between the paper and the box.
2. Suspend the styrofoam ball from a black thread of about 2 cm long, and tape the thread against the center of the lid (see sketch).
3. Cut a hole the size of the flashlight at the end of the box and seal any space around it with masking tape.
4. Make five eye holes in the sides of the box: two on each long side and one at the end of the box obliquely under the flashlight.
5. Put the lid with the suspended ball on the box and seal the edges around it with masking tape.
6. Look through the eye holes and observe the ball (with the flashlight on) in the order numbered in the sketch.

Questions:
1. What do you see when you look through the eye holes?
2. Turn the flashlight off; what do you see through the eye holes?
3. Looking through which of the holes gives you a similar picture as looking at the **new moon**?
4. What is it that causes the phases of the moon?

Explanation:
Perhaps the best way to explain seeing the eight different phases of the moon is the illustration on the right. A quarter **new moon** would correspond with looking through hole 1, **half moon** and **three quarter moon** with looking through hole 2, **full moon** with looking through hole 3, etc.

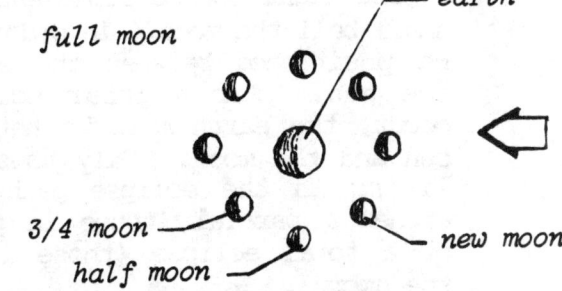

EARTH SCIENCE   ASTRONOMY

15.10. WHAT CAUSES AN ECLIPSE?

Materials: 1. A strong flashlight, slide projector, or other light source.
            2. A large (20 cm diameter) solid color sphere.
            3. A small (5 cm diameter) rubber or styrofoam ball.
            4. A knitting needle.

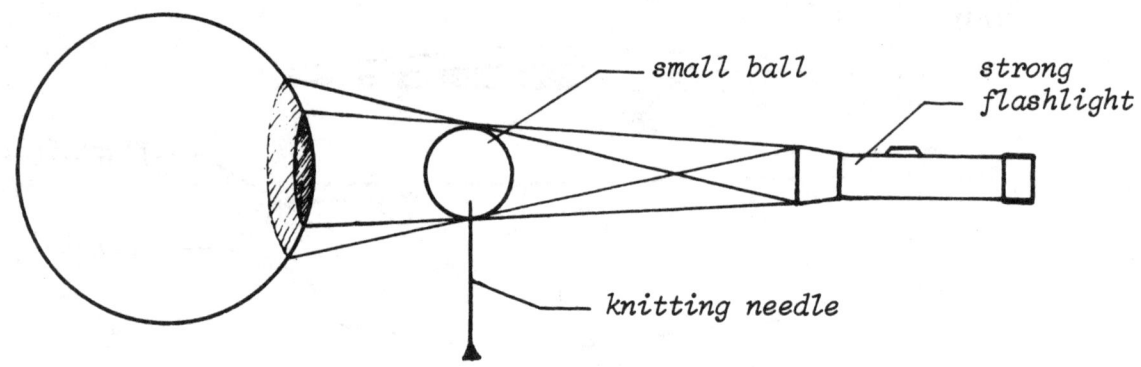

Procedure:
1. Stick the knitting needle into the small ball, so that you can hold it between the light source and the large ball without casting a shadow of your hand.
2. Hold the smaller ball between the light source and the large sphere and adjust the distance from the sphere, such that the small ball casts a dark shadow in the center and a greyish shadow on the edges.

Questions:
1. What does it mean if something is eclipsed?
2. For a solar eclipse to occur, how must the earth, sun, and moon be positioned in relation to each other?
3. Would all people on earth be able to see a solar eclipse when it occurs?
4. At what time (day or night) can a solar eclipse be observed?
5. How must the earth, sun, and moon be positioned in relation to each other, in order for a lunar eclipse to occur?
6. What are safe ways to observe a solar eclipse?

Explanation:
   The light source represents the sun, the large sphere the earth, and the small ball the moon. In order for a solar eclipse to occur, the moon must be positioned between the earth and the sun. For a lunar eclipse to occur, the earth must be between the sun and the moon. Only those people living in the eclipse path can see either a partial (those in **penumbra**) or a total eclipse (those living in the **umbra**).

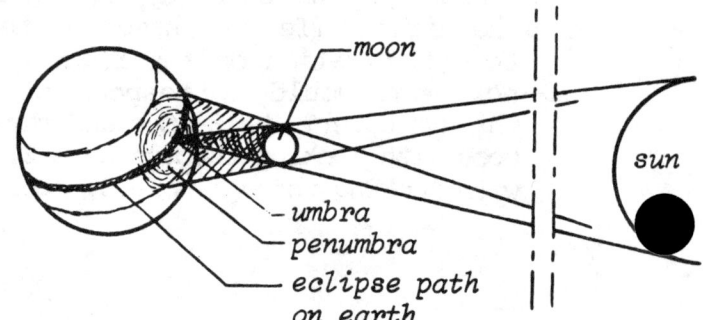

SECTION IV

LIVING THINGS

This section consists of two chapters, one dealing with plants and the different variables that are influencing their growth, and the last chapter contains demonstrations and activities about human biology.

Chapter 16 deals with the variables affecting the growth of a plant, like: light, gravity, the amount of water. Osmosis and capillary action, respiration and photosynthesis are also dealt with.

Chapter 17 contains demonstrations and activities to start off lessons about the senses, the nervous system, the circulatory, the respiratory, and digestive system in the human body.

Demonstrate .....

and
Enjoy Yourselves !

CHAPTER 16

WHAT VARIABLES ARE AFFECTING THE GROWTH OF PLANTS?

OBJECTIVES

After dealing with and studying the concepts and sub-concepts in this chapter, the students should be able to:

a. recognize the correct explanation of an observed event based on each of the sub-concepts;
b. explain in their own words which of the sub-concepts is determining the course of an event;
c. distinguish true from false statements concerning each one of the sub-concepts;
d. identify the correct explanation of an event in daily life applying one of the sub-concepts;

all in relation to the following sub-concepts:

-- Conditions for germination of seeds are: moisture, warm temperatures, and air supply.
-- Plants grow in the direction of the light source.
-- Plant roots grow down towards the center of the earth.
-- The parts of a plant that are above the ground grow vertically up.
-- Water moves into the plant roots by osmosis.
-- Water moves into higher parts of the plant by capillary action.
-- Leaves give off water vapor.
-- Green leaves exhale oxygen in sun light and carbon dioxide in the dark.
-- Photosynthesis is the production of sugar from water and carbon dioxide in chlorophyll of the plant leaf.
-- Chlorophyll production needs sunlight.

PLANTS   GERMINATION

16.1. HOW DO SEEDS GERMINATE?

Materials: 1. Small seeds (mustard, radish, green beans, etc.).
             2. Blotting paper or several layers of paper towel.
             3. A drinking glass or beaker.

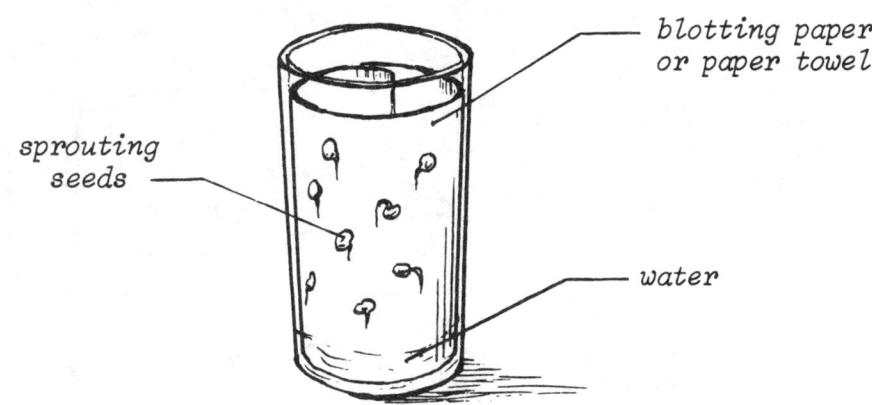

Procedure:
1. Cut a rectangular piece of blotting paper as wide as the glass is tall (if paper towel is used, triple or quadruple the layers).
2. Moisten the paper and place the seeds on it.
3. With the seeds sticking to the moist paper, roll the paper in a cylinder slightly smaller in diameter than the glass and insert the roll into the glass.
4. Let the paper stand against the glass wall and pour some water into the glass to keep the paper moist.
5. Put the glass in a warm place and cover loosely (make sure that some water stays in the glass at all times).

Questions:
1. Which seeds sprouted first? Which last?
2. In which direction did the rootlets grow?
3. How could the paper stay moist at all times?
4. Why did the glass have to be loosely covered?
5. What are the conditions for germination of seeds?

Explanation:
In three to six days, depending on what seeds, the seeds will sprout and send their rootlets in a downward direction: this is in the direction of the water. The blotting paper stayed moist, because the water worked itself up through **capillary action** in the paper fibres.

The glass has to be covered in order to prevent fast evaporation of the water, but it has to be done loosely so that the seeds are not cut off from the atmospheric air.

The ideal conditions for germination of seeds are: presence of moisture, warm temperatures, and supply of air.

16.2. THE BENDING PLANT

Materials:
1. Bean or radish seeds.
2. A small flower pot and soil.
3. A cardboard box.

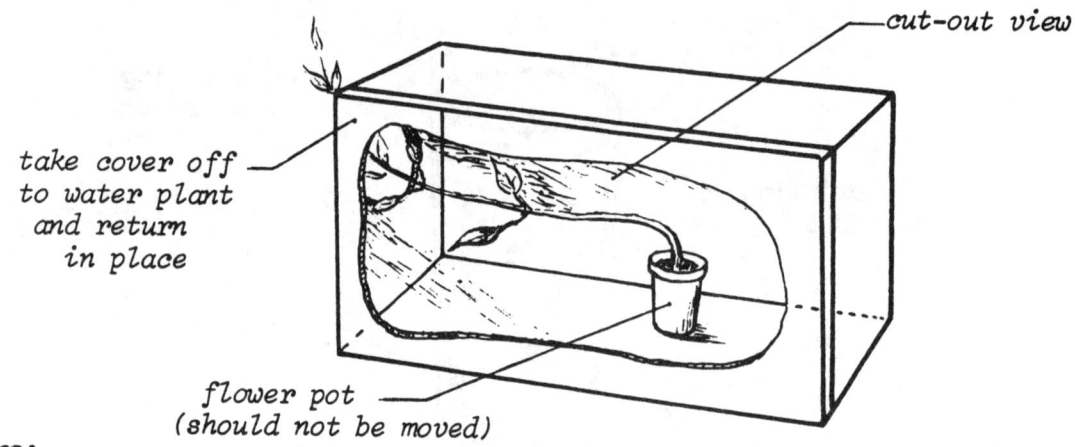

Procedure:
1. Plant one seed on one small flower pot and water daily.
2. After the seed has started to sprout, let it grow until it is about 5 cm high.
3. Now place the pot in the cardboard box in which a hole has been cut out in its end side (see sketch).
4. Keep the box close to a window to let light in through the hole.
5. Keep the pot in the same position, water daily for the next few days. (replace cover after watering), and observe plant growth!

Questions:
1. How did the plant grow? In which direction?
2. How would the plant grow, if the pot were turned around every day?
3. How would the plant grow, if the box were turned every day?
4. How would the plant grow, if the box cover were left open?
5. What do plants need in order to grow?

Explanation:
Plants grow in the direction from where the light comes, in this case towards the hole in the box, and even through the hole when left for a longer time. This shows that the plant needs light in order for it to grow. By turning toward the light (hole in the box) its leaves will catch more light and thus grow better.

If the plant were turned half way around every day, it would grow straight up, or keep bending back and forth. If the box were turned, the plant would still bend towards the hole. Without the box cover, the plant would just grow straight up (and slightly bend toward the window).

16.3. THE CROOKED ROOT

Materials:
1. A small young plant and potting soil.
2. A flat, square, transparent container or two 10 x 10 cm pieces of window glass & three 1 x 1 x 10 cm wooden blocks.
3. Cardboard or black paper.

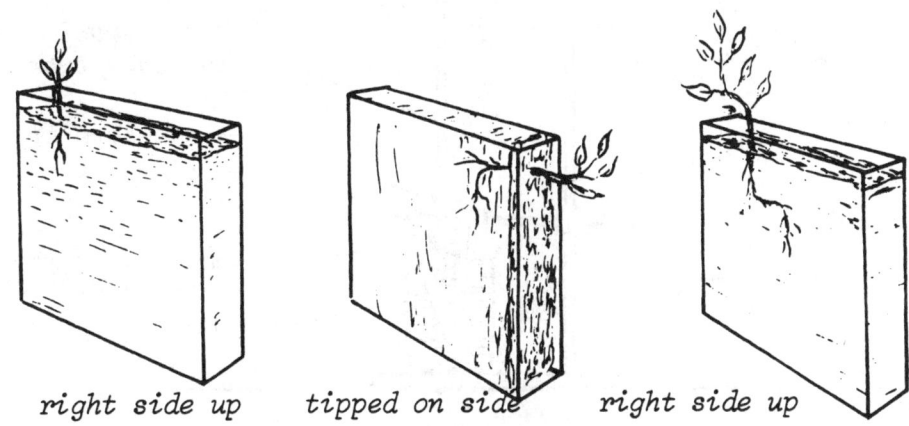

right side up   *tipped on side*   *right side up*

Procedure:
1. If a transparent thin square container (chemical cell) is not available, construct one by taping two pieces of window glass against each other with the three wooden blocks in between them.
2. Fill the container with potting soil and place the plant on one side in the soil of the container.
3. Cover the sides of the container with cardboard or black paper and let the plant grow for 3-4 days.
4. Tip the container on its side (cover the open top with cardboard if the soil tends to fall out), and let the plant further grow.
5. After another 3-4 days, turn the container right side up again.
6. Point 4 and 5 can be repeated once more, then remove the covering cardboard or black paper and observe the roots.

Questions:
1. How did the roots of the plant grow?
2. In what direction do roots grow?
3. Why did the sides of the container have to be covered?
4. On the day just before turning the container right side up, in which direction did the plant grow?

Explanation:
Plant roots grow downward in a vertical direction. They grow towards the center of the earth, opposite to the direction of the plant parts that are above the ground, which grow vertically upwards. Regardless of the position of the container, the roots will grow downward, this is why turning the container on its side makes the roots turn in their direction of growth. The covering of the sides of the container keeps the roots in the dark, simulating the earth or an opaque pot of soil.

PLANTS
GROWTH
OSMOSIS

16.4. THE WATER-SUCKING ROOTS

Materials:
1. A beaker (250 ml), a one-hole stopper, a glass tube.
2. A carrot or a cylinder shaped potato, syrup (sugar), candle wax.
3. A coring knife (apple corer), a stand and clamp.

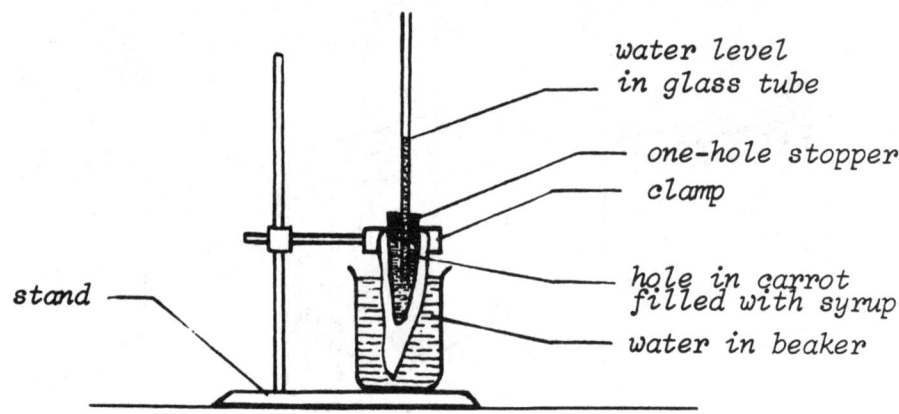

Procedure:
1. With the coring knife, cut a hole in the carrot or potato about three-quarters down its length, such that the one-hole stopper will fit in it and close it tightly (see sketch).
2. Insert a 20 cm long glass tube in the one-hole stopper.
3. Fill the hole in the carrot or potato with syrup or a concentrated solution of sugar in water brimful.
4. Push the stopper with the glass tube in the hole (liquid level should rise in the tube) and seal any openings between the stopper and the carrot or potato with candle wax (light a candle and let the melted wax drop on the places that you want sealed).
5. Mark the liquid level in the glass tube with a piece of masking tape, a grease pencil, or a rubber band.
6. Clamp the carrot or potato and immerse it in water. Observe the water level in the glass tube at the end of the period.

Questions:
1. What made the water level in the glass tube rise?
2. Would this water level also rise if the tube were filled with plain water? With salt water?
3. Why did the stopper have to be sealed with wax?
4. What would happen if the carrot and tube were filled with plain water and the beaker with sugar solution?

Explanation:
   The skin, tissue, and fibres of the carrot or potato act like a **semi-permeable membrane**, letting only the small water molecules through, but not the large sugar molecules. This makes the water move from the beaker into the carrot and up the tube. If the concentration of sugar is higher in the beaker compared to that inside the carrot, the water will move out of the carrot and thus the water level in the tube will go down.
   This action and migration of water molecules through a semi-permeable membrane is called: **osmosis**.

PLANTS
GROWTH
OSMOSIS

16.5. MAKE AN EGG OSMOMETER

Materials:
1. A fresh chicken egg, soda straw (transparent) glass tubing.
2. Dilute HCl or strong vinegar, household cement or sealing wax.
3. A saucer, a drinking glass, a thin wire.

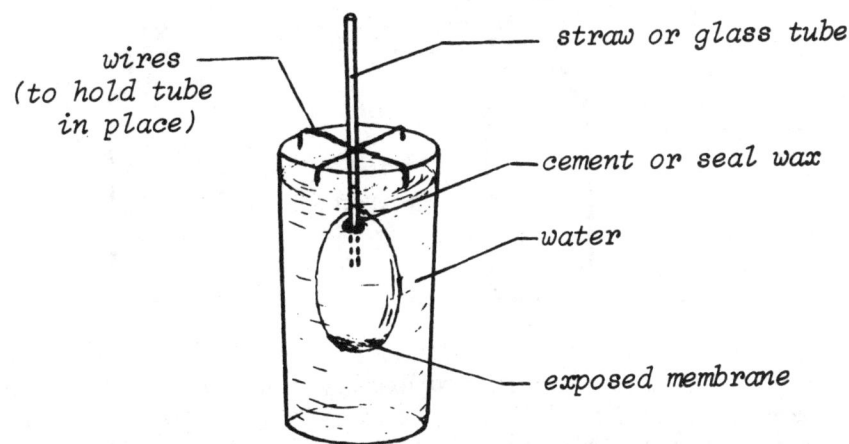

Procedure:
1. Put some dilute hydrochloric acid or strong vinegar on the saucer and hold the large end of the egg in the liquid until the shell has been eaten away, leaving the thin membrane exposed.
2. After an area of about 4 cm² of the thin membrane has been exposed, rinse the acid off with fresh water.
3. At the other end of the egg, make a hole the size of the straw or glass tube (take care that the egg does not crack).
4. Insert the straw or glass tube through the hole into the interior of the egg and seal any excess opening between the egg and the tube off with cement or wax (candle wax will do).
5. Place the osmometer in a glass of water and hold the tube in place (vertically up) by wrapping the thin wire around it and bending the wire over the rim of the glass (see sketch).
6. Let stand for a few hours and observe liquid level in the tube.

Questions:
1. What do you think the egg shell consists of?
2. Why does the seal between the egg and the tube have to be absolutely tight? What would happen if it were not?
3. Why did the liquid level in the tube rise?

Explanation:
The thin membrane of the egg acts like a **semi-permeable membrane**, which allows small size molecules of water to pass through it, but not the large protein molecules in the egg. The water thus **migrates** from the outside of the egg to the inside, and in so doing pushes the liquid level up the tube. This liquid level difference, between that in the tube and that of the water in the glass, is called **osmotic pressure**.

16.6. THE SWOLLEN EGG

Materials:
1. A fresh chicken egg.
2. An old fashioned milk bottle (or wine carafe or any bottle of which the mouth diameter is a bit smaller than the egg's).
3. Dilute HCl or strong vinegar, a cup or beaker.

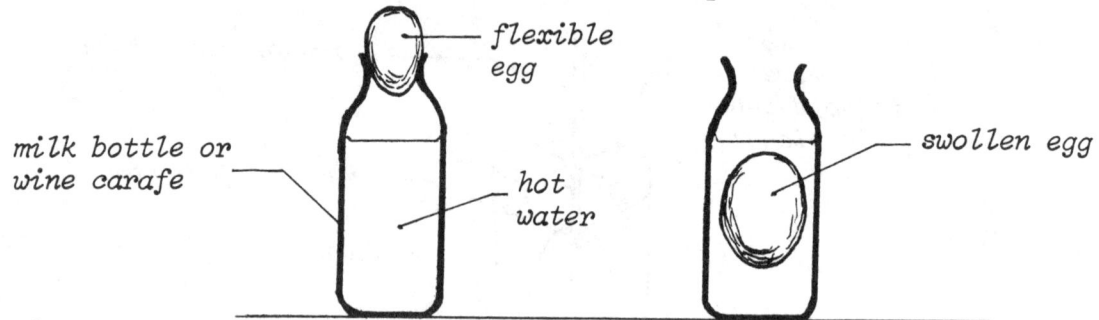

Procedure:
1. Place the egg in the cup and add dilute HCl (hydrochloric acid) or strong vinegar to it until it is completely immersed. Hold the egg under the surface of the liquid or keep rotating the egg. Do this for 10 minutes.
2. Check whether all the egg shell has dissolved by touching the egg and very carefully pressing on it. When only the membrane is left, it should feel soft and flexible.
3. Fill the milk bottle about 3/4 full with hot water (almost boiling) and place the egg immediately on the mouth of the bottle and let stand.
4. Observe! Leave for an hour or so and observe again!

Questions:
1. Why did the egg become soft and flexible after placing it in acid?
2. What did the hot water do to the air in the milk bottle?
3. What did the flexible egg do after it was placed on the bottle?
4. What did the egg do after an hour wait?
5. What made the egg grow larger in the bottle?
6. How can we get the egg out of the bottle without breaking it?
7. How can we reverse the process of osmosis?

Explanation:
The shell of an egg consists mainly of calcium carbonate. By placing the egg in dilute hydrochloric acid or strong vinegar, the calcium carbonate reacts with the acid to form a soluble calcium salt and carbondioxide gas. This leaves only the membrane around the egg, which makes the egg feel soft and flexible.

The steam of the hot water drives the air partly out of the bottle, thus by placing the flexible egg on the bottle, the egg is slowly sucked into the bottle while the water is cooling off.

After the egg is totally plunged in the water, **osmosis** is taking place. Water molecules are migrating through the **semi-permeable membrane** of the egg from the outside into the egg and thus making the egg swell. In order to get the egg back out, we have to shrink the egg, which can be done by pouring all the water out and letting the egg dry out and shrink. Then by heating the bottle while the egg sits in the neck, it will slowly be pushed out by the expanding air.

16.7. HOW DOES GRAVITY AFFECT GROWTH?

Materials: 1. Three small plants (identical in size and kind).
2. A ring stand, two pieces of wire, cardboard.

Procedure:
1. Turn one flower pot with the small plant in it on its side and face the plant away from the window.
2. Place another flower pot with the small plant upside down on the ring stand supported on the two wires (tape cardboard securely over soil to prevent it from falling out), and place this stand below the window.
3. Place the third flower pot with plant normally in front of the window.
4. Observe plant growth in all three pots and compare.

Questions:
1. How did the plant on its side grow?
2. How did the upside down plant grow?
3. In what direction did all three plants grow?
4. Why did the plant on its side have to be faced away from the window?
5. How did the light from the window affect the growth?
6. Can we be absolutely sure that the light was not the cause of the plants to bend upwards?
7. How would we set up an experiment to investigate the influence of **only** gravity on the growth of a plant?

Explanation:
Plants in general grow in the opposite direction of the growth of the roots, which is vertically upwards. They grow **in the opposite direction of the gravity force**. In nature, we can especially notice this when trees are growing on a steep hill or on the side of a mountain or cliff.

An experiment investigating the influence of gravity **only** would need to have all variables controlled, including the light variable, which has to come from all sides toward the plant. An open-ended question would be: How would a plant grow without gravity?

16.8. MAKE SOOT PRINTS OF LEAVES

Materials:
1. Two empty jars with tight lids.
2. A candle, grease or Vaseline.
3. Old newspaper, white blank paper, different leaves.

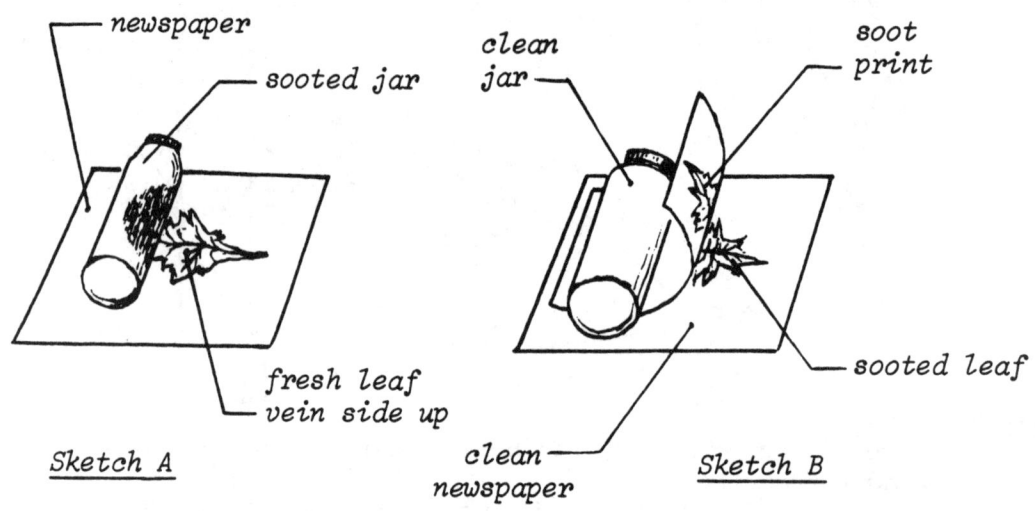

Sketch A — Sketch B

Procedure:
1. Cover the side of a smooth jar with a thin layer of grease or Vaseline. Fill the jar with cold water and screw the lid on tightly.
2. Cover the greased part of the jar with soot by holding it over a candle flame until it is evenly black.
3. Place a leaf with the vein side up on a newspaper and roll the soot-covered part of the jar over the leaf (see Sketch A).
4. Take the leaf and place it on a clean newspaper. Cover the leaf with a sheet of white paper and roll a clean jar over the paper and leaf. (see Sketch B).

Questions:
1. What is the purpose of the grease or Vaseline layer?
2. Why did we need cold water in the first jar?
3. Why was the leaf placed on the newspaper vein side up?
4. Which part of the leaf was covered with the black soot?
5. What type of leaves would give the best soot prints?

Explanation:
The jar with the grease and soot layer functioned as an ink roller to cover the veins of the leaf with a thin layer of soot. The leaf had to be placed vein side up, because the veins are bulging up (protruding) compared to the other parts of the leaf, and this makes it possible to color only the veins. When the white paper was placed on the sooted leaf, the veins only would give a black print on the paper. Leaves with strong bulging veins make the best soot prints.

PLANTS

LEAVES
RESPIRATION

16.9. DO LEAVES GIVE OFF WATER?

Materials: 1. A large wide leaf or small plant.
            2. Four large drinking glasses (of the same size).
            3. Two paper cards.

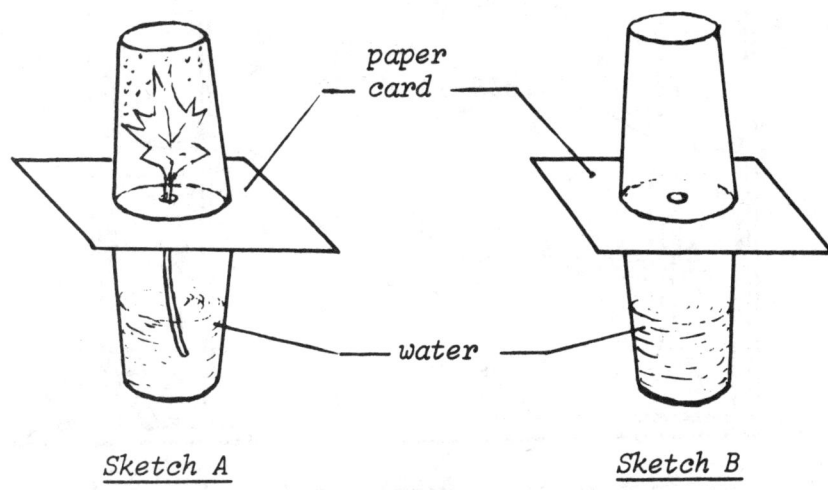

Sketch A   Sketch B

Procedure:
1. Fill two of the glasses halfway with water.
2. Cut a small hole in each of the two paper cards, such that the stem of the leaf or small plant would fit through it.
3. Cover the glasses with the paper cards, insert through one card the leaf stem, and cover with the inverted glasses (see sketch).
4. Observe and compare the inside of the upper glasses.

Questions:
1. In which glass did water droplets form?
2. Why did we need the set of glasses in Sketch B?
3. Where did the water droplets come from?
4. What kind of leaves would give off most water?
5. What variables would influence the amount of water transpired?
6. Would a whole plant covered in the same manner give off water too?

Explanation:
   The water **transpired** from the leaf is in the form of water vapor. This latter condenses on the cold surface of the upper glass in Sketch A and thus water droplets are formed. In order to show that these droplets are indeed coming from the leaf rather than the water in the lower glass, the set-up in A is being compared with one without a leaf in it.
   A small plant covered in the same way with a card around the stem in the pot and an inverted glass over the whole plant, would give the same results.
   Variables that would influence the amount of water transpired are: type of leaf, surface area of the leaf, age of leaf, etc.

PLANTS

LEAVES
RESPIRATION

16.10. WHAT DO GREEN LEAVES BREATHE OUT? (I)

Materials:
1. Green weed & wood split.
2. A large beaker, a funnel, a test tube.
3. A stand and clamp.

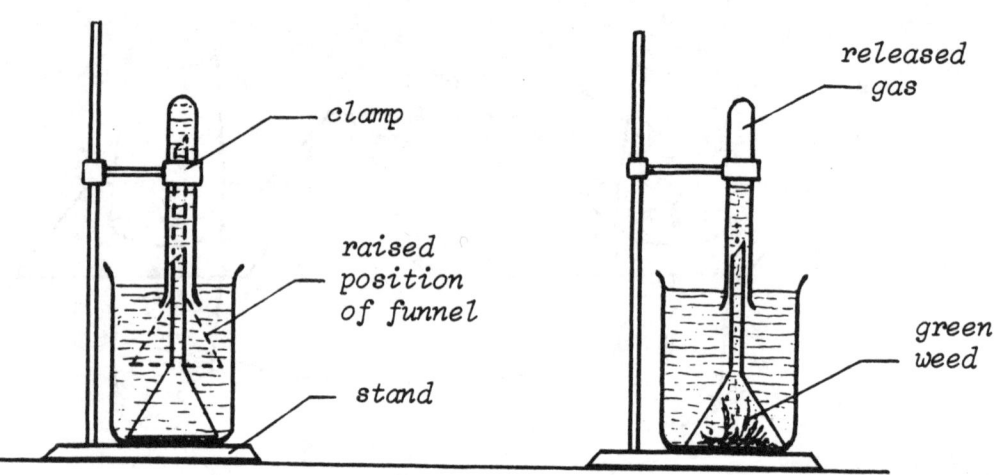

Procedure:
1. Fill the beaker with water, immerse the funnel and the test tube in the water, and set the apparatus up as in Sketch A.
2. Raise the funnel and place some green weed under it.
3. Leave the apparatus in strong sunlight or under a spotlight and observe the bubbles given off by the leaves.
4. After collecting almost a full test tube of gas, test it with a glowing wood splint.

Questions:
1. What gas is collected from the test tube?
2. What did the glowing wood splint do when lowered in the test tube?
3. What made the water in the test tube stand so much higher than the water level in the beaker? (Sketch A).

Explanation:
   The green in the leaves, which is chlorophyll, produces sugar and cellulose and starch in the plant. During this process of sugar production carbon dioxide and water, oxygen is released. This only occurs during daytime when the sunlight is shining on it. The purpose of the funnel is to bring all the bubbles released by the weed together under the test tube. As the glowing splint flares up into the bright flame in the gas, it indicates that the gas is oxygen.
   The fact that plants give off oxygen during the daytime makes having them in the living room a good thing. The air is enriched with oxygen and it is therefore healthy to have plants in the room.

PLANTS

LEAVES
RESPIRATION

16.11. WHAT DO GREEN LEAVES BREATHE OUT? (II)

Materials:
1. A wide-mouthed gallon jar & lid.
2. A small potted plant.
3. Limewater (calcium hydroxide).

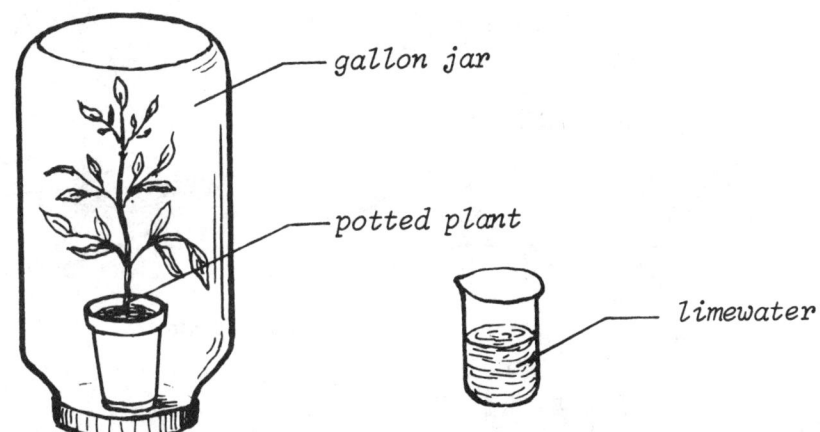

Procedure:
1. Place the potted plant on the inverted lid of the jar and invert the large jar over the whole plant, and turn the jar into the lid.
2. Leave the jar and plant in a completely dark place for several hours.
3. Turn the lid loose and remove the plant carefully.
4. Turn the jar right side up and pour about 100 ml of limewater in it, screw the lid on, and shake.
5. Observe the limewater (turning milky).

Questions:
1. What gas was released by the leaves in the dark?
2. What reaction did the gas give with the limewater?
3. How would this gas react with a glowing wooden splint?
4. With many plants in a room, how would the composition of the air change from daytime to nighttime?
5. Why is it better to take plants out of the bedroom at night?

Explanation:
Plants give off carbon dioxide when placed in the dark. This is indicated by the reaction with the limewater (calcium hydroxide): $CO_2 + Ca(OH)_2 = CaCO_3 + H_2O$, forming calcium carbonate which is insoluble in water. This is the cause for the limewater to turn milky.

When this gas produced in the dark is tested with the glowing splint, it would extinguish the glow completely, as carbon dioxide does not support the burning process. With many plants in a room, the air would be rich in oxygen content during the day and high in carbon dioxide content at night. This is why it is better to take plants out of the bedroom at nighttime.

PLANTS

LEAVES
PHOTOSYNTHESIS

16.12. HOW IS THE GREEN IN THE LEAVES PRODUCED?

Materials:
1. A plant with large wide leaves.
2. Carbon paper or black construction paper.
3. Paper clips or masking tape.

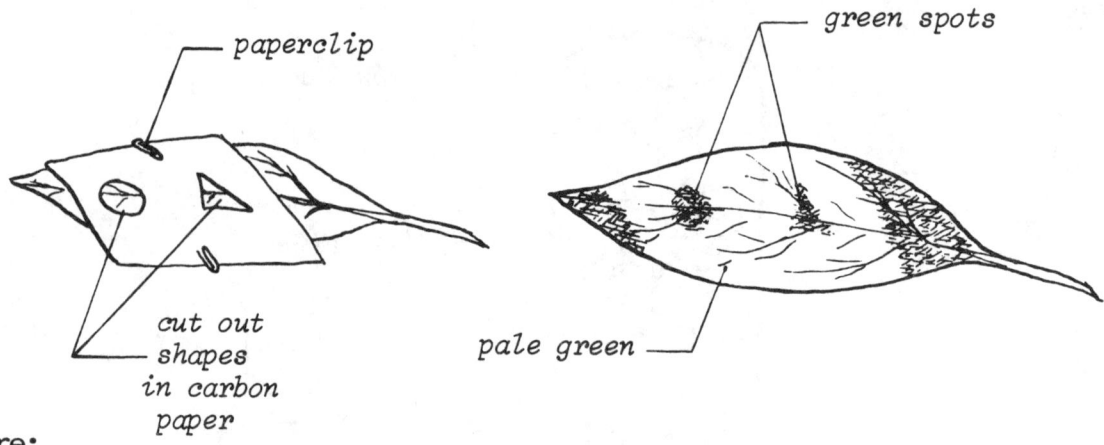

Procedure:
1. Cut out several patterns (circle, square, triangle) in several pieces of the carbon paper.
2. Cover three or more leaves as much as possible with the cut out carbon paper by attaching it to the leaves with the paper clip or masking tape.
3. Cover some leaves halfway with carbon paper close to the stem (or any other pattern of covering) and leave it attached for 2 or 3 days.
4. After leaving the black paper against the leaves for several days, remove the attached paper and observe the leaves.

Questions:
1. How did the covered areas of the leaves compare to the uncovered ones?
2. Do plants need sunshine to produce the green color?
3. What is the green color in the plant leaves called?
4. What is the process of production of the green color called?
5. What is the function of the chlorophyll in plant leaves?

Explanation:
   The covered areas of the leaves will become much paler. The longer it stays covered, the paler the color, because no sunshine is penetrating the green pigment that enables every plant that possesses it to combine water and carbon dioxide from the air to form sugar. This process in which sunshine is an essential ingredient is called **photosynthesis**. It is the sugar in the plants which gives animals and man the energy when it is consumed by them.
   The chlorophyll also produces cellulose, a much larger molecule than sugar, which is the basic building material in plants. Thus, without sunshine the leaves do not produce chlorophyll, no cellulose, and therefore plant do not grow.

16.13. CAN AIR ENTER THROUGH A LEAF?

Materials:
1. A leaf with a long stem.
2. A small Erlenmeyer flask.
3. A 2-hole stopper that fits in the flask.
4. A bent glass tube.
5. A candle & matches.

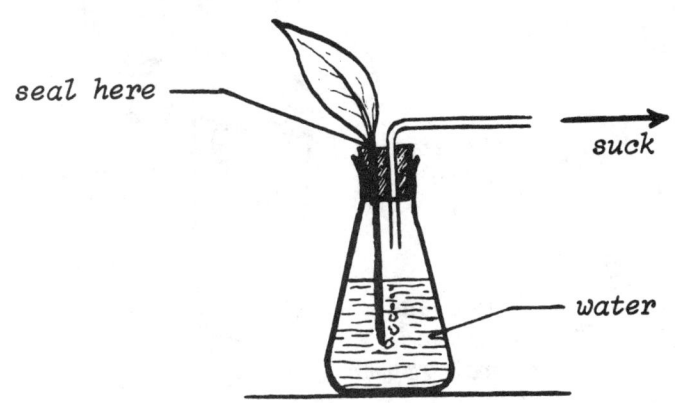

Procedure:
1. Stick the leaf stem through one of the holes in the 2-hole stopper and seal it with dripping wax from a lit candle.
2. Insert the bent glass tube in the other hole of the stopper.
3. Fill the Erlenmeyer flask with water to such a level that only the leaf stem immerses in it and not the glass tube.
4. Place the stopper tightly into the flask and stuck through the side tube.
5. Observe air bubbles issuing from the end of the stalk.

Questions:
1. Why does the stem have to be sealed in the stopper?
2. What would happen if the glass tube were also immersed in the water?
3. Are the leaf and stem actually that porous that air can go through them?
4. What is the actual structure of the leaves?

Explanation:
The sucking through the side tube lowered the pressure inside the flask, causing the atmospheric air to seep through the leaf and the stalk resulting in the bubbles issuing from the end of the stalk. When looking through a microscope and examining the underside of leaves, we can see breathing pores or stomata with two little guard cells on each side of the stomata (see sketch on right).

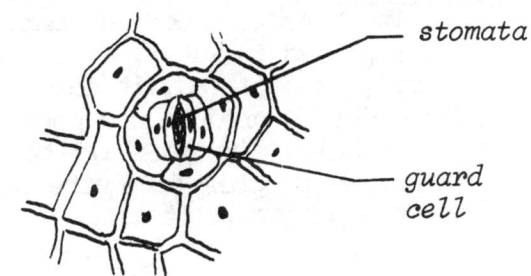

PLANTS — LEAVES

16.14. GROW SWEET POTATO AND CARROT LEAVES

Materials:
1. A sweet potato.
2. A drinking glass or beaker.
3. A shallow disk (petri dish).
4. Three toothpicks or thin nails.
5. Some pebbles or gravel.

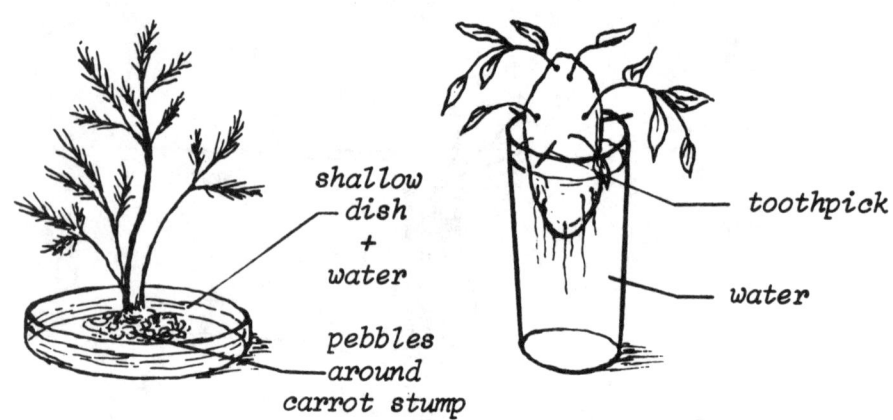

Procedure:
1. Hold the sweet potato with the eyes or buds on the top side.
2. Immerse the potato about one-third in the water contained in the drinking glass.
3. Support the potato by sticking three toothpicks or nails into its side and resting it on the glass rim.
4. Remove the old leaves from the top of the carrot and cut off all of the lower part except about 5 cm.
5. Place this stump in water in the shallow dish and support with pebbles or gravel around it.
6. Put sweet potato and carrot in a warm and sunny place and observe foliage growth.

Questions:
1. Where did the potato or carrot plant get its food from?
2. How does the potato leaf differ from that of the carrot?
3. Would these plants develop new potatoes or carrots?
4. What other plants could be grown this way?

Explanation:
The potato, like the carrot, beet or turnip, contains much stored food. When placed in water, it is enough to produce thick foliage. When this food supply is depleted, it needs to be placed in soil and receive other nutrients in order to produce new potatoes and carrots. When a pineapple is cut about 5 cm below the base of the leaves and placed in water, the leaves will continue to grow for quite some time drawing nutrients from its stored food in the pineapple itself.

PLANT KINGDOM

16.15. MAKE A RED-BLUE CARNATION

Materials:
1. A white carnation (with a long stem).
2. Red food coloring & blue ink.
3. Two small beakers or cups.

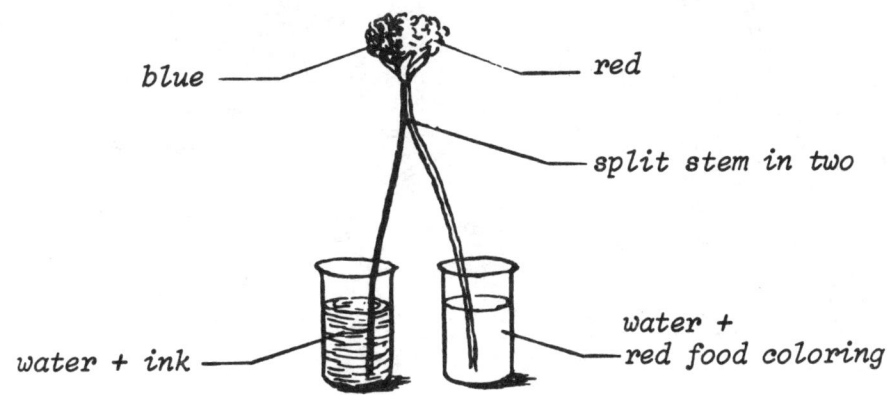

Procedure:
1. Take the carnation and cut the stem to leave about 25 cm on the flower.
2. Split the stem in half, starting a cut with a knife and further splitting it along the fibres (without breaking them!)
3. Fill the two small beakers with water, color one red with the food coloring and the other blue with the ink.
4. Place each half of the stem in each beaker and observe the flower.

Questions:
1. What was the purpose of splitting the stem in half?
2. Did the whole white carnation get colored?
3. Is it possible to color only part of the flower and leave another part white?
4. How would the stem have to be split to obtain three colors in the flower, say red, blue and white?
5. What force is pulling the colored solution up the stem?
6. Would leaves be able to be colored the same way?

Explanation:
The colored water is drawn up the stem of the carnation by osmosis and **capillary action**. The water molecules diffuse through the fibre membranes from a less to a larger concentration of plant sap (osmosis). The fibres are so tiny that the adhesive force of the water molecules to the fibre walls becomes very great. This capillary force in combination with the osmotic pressure sucks the water up the flower.

When the stem is split three ways, it is very likely that the flower will be three way colored. This shows that the fibres must somehow run all the way from the stem to the petals of the flowers.

PLANTS   CAPILLARY ACTION
                                                          ABSORPTION

16.16. THE TOOTHPICK STAR

Materials: 1. Five toothpicks (flat kind), 1 box for the whole class of 30.
             2. A cup of water.

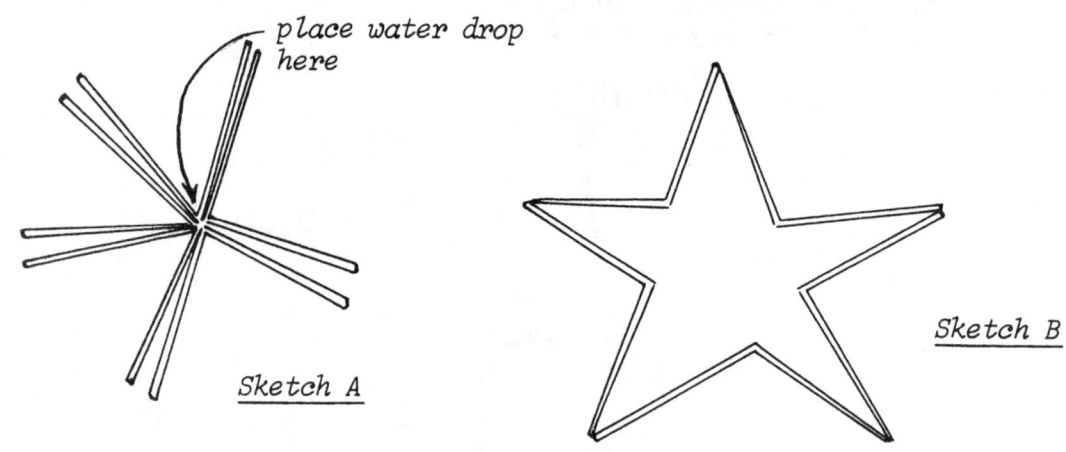

Sketch A

Sketch B

Procedure:
1. Break five toothpicks in half but leave the parts together, and leave them in a V-shape.
2. Place the five V-shaped toothpicks on a smooth surface with their points as close as possible together (see Sketch A).
3. Dip your finger tip in the cup of water and let one drop fall in the center of the toothpick configuration.
4. Observe carefully what is happening to the toothpicks!

Questions:
1. What happened to the toothpicks?
2. What did the water do with the toothpicks?
3. What material do the toothpicks consist of?
4. What will happen to picked flowers when they are not placed in a vase in water? Why do they need water?
5. How does water reach the top leaves in a tall tree?
6. Compare a dry broken toothpick with one that was wetted by holding one of the legs of the V-shape and pushing the other leg down: Which one stays closed/down?

Explanation:
   By breaking the toothpick in half, not all the vessels/fibers are cut off. The water is **absorbed** by the wood fibers by **capillary action**, which consist mostly of adhesive forces. Also **diffusion** through the fiber walls is taking place: water is thus filling the unbroken vessels or fibers. This makes the toothpick tend to straighten up and make it more springy than the dry one.
   This is the reason why picked flowers become limp when they are not placed in water. Capillary action and **osmotic pressure** are the forces that bring the water from the roots of a tree all the way up to the top leaves.
   When repeating the activity, dry toothpicks have to be used. The wet ones will not form and stay in a narrow V-shape. They become more springy than the dry toothpicks and form a much wider V-shape.

16.17. HOW DO MOLDS REPRODUCE?

Materials:
1. Eight slices of bread.
2. Saran Wrap or cellophane.
3. A magnifying glass.
4. A shallow dish (saucer or petri dish).

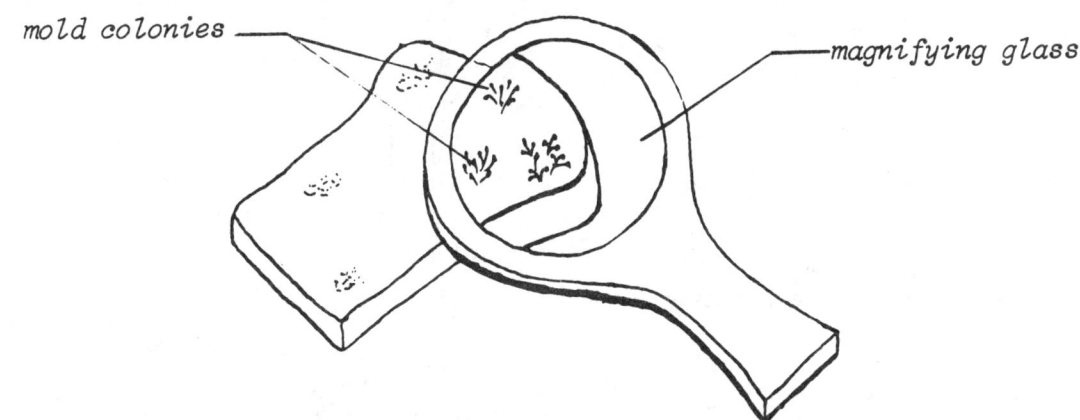

Procedure:
1. Take two slices of bread. Wrap one slice in the Saran Wrap and seal off from the air and leave the other one open to the air.
2. Take two other slices: place one in a dry spot and place the other over a shallow dish filled with water.
3. Expose two other slices: one to a strong light (flood light) and keep the other in a dark place.
4. Expose two more slices: keep one in a warm dark place and the other in a cold dark place.
5. Observe the surface of the bread slices daily through the magnifying glass and record the changes. Make a note of the mold colonies, the number of them and their growth rate.

Questions:
1. What variable was manipulated in step 1 of the procedure?
2. What variable did I choose to change in Steps 2, 3, and 4?
3. What was the responding variable in all four cases?
4. In which case did the mold develop best?
5. How should bread be stored to keep it mold free?
6. What other types of food will develop molds if kept over a longer period of time?

Explanation:
Mold grows best in moist, dark and warm places. Dry, light and cold conditions will discourage the growth of molds. Under step 1 of the procedure the presence or absence of air was the manipulated variable, under step 2, moisture, step 3, the presence of light, and step 4, the temperature. Other foods that will develop molds are: cheese, jam, leftover food or meats, etc.

16.18. PICK UP A CARAFE WITH A STRAW

Materials:
1. A wine carafe (or other long necked bottle).
2. A wheat straw.

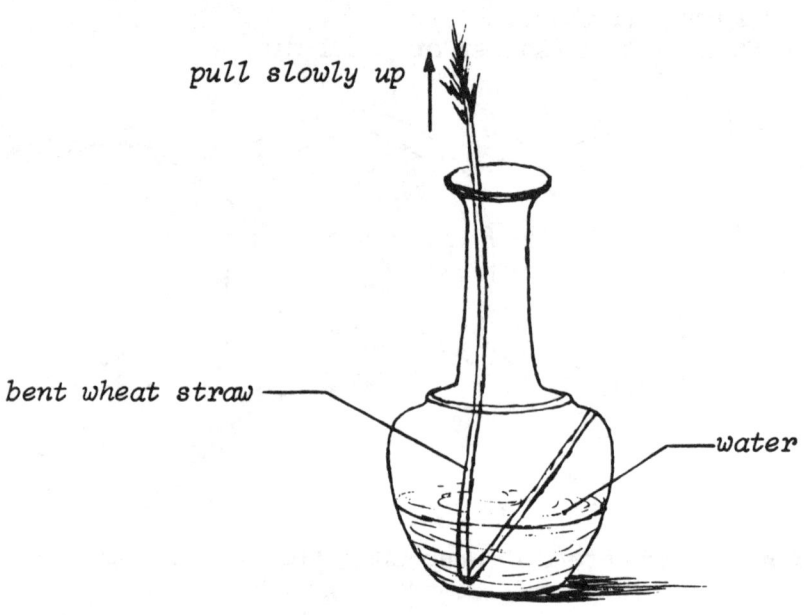

Procedure:
1. Fill the carafe half full with water.
2. Bend the straw such that the short bent part will fit in the wider part of the bottle (see sketch).
3. Lower the straw in the bottle and allow the bent part of the straw to open up.
4. Now pull the carafe slowly upwards by holding the protruding straw.

Questions:
1. Why did the straw not break?
2. What force was exerted on the straw?
3. What will happen to the straw if we wait a while before pulling the carafe up?
4. What will water do when plant material is soaked in it?
5. Where or in what direction lies the greatest strength of a wheat stalk?
6. What implications can we see in living plants?

Explanation:
It is especially critical where the fold in the straw is made. It should be folded at a spot such that the short part will just fit in the wider part of the carafe. By pulling the straw up, the bottom part of the straw should press on the side, and the top part should hook on the rather horizontal part of the bottle. The stress in the straw would be mostly lengthwise and as the tissue of the wheat stalk is made up of tubular capillaries and vessels, it can withstand a rather high stress. Over time, the water will penetrate into the straw tissue and weaken it. This will be more so the case when a dry straw is used rather than a fresh one.

CHAPTER 17

WHAT CAN WE LEARN ABOUT THE HUMAN BODY?

OBJECTIVES

After dealing with and studying the concepts and sub-concepts of this chapter, the students should be able to:

a. recognize the correct explanation of an observed event based on each of the sub-concepts;
b. explain in their own words which of the sub-concepts is determining the course of an event;
c. distinguish true from false statements concerning each one of the sub-concepts;
d. identify the correct explanation of an event in daily life applying one of the sub-concepts;

all in relation to the following sub-concepts:

-- All objects seen by the human eye cast an upside down image on the retina, which is interpreted by our brain as right side up.
-- People need two eyes to see objects in three dimensions.
-- The majority of people are right-sighted (possible correlation with right-handedness).
-- The cones in the human eye pair the following colors: red-green, yellow-blue, and white-black.
-- Illusion is an interpretation of the brain, which deviates from the true perception by the eyes.
-- The number of nerve endings is larger at the finger tips compared to other parts of the human body.
-- It takes time for an impulse to travel from the receiver (eyes, ears, or other hand) to the arm muscle.
-- The human brain is composed of two hemispheres: the left half controls the right-hand side of the body, while the right half controls the left side of the body.
-- The number of heartbeats increases with intensity of exercise.
-- Exhaled air contains more carbon dioxide than inhaled air.
-- Enzymes in saliva break down starch into sugar.
-- Food is pushed down the esophagus by peristaltic action.

17.1. THE REVERSED IMAGE

Materials:
1. A paper card (3x5"/8x12 cm).
2. A regular sewing straight pin.

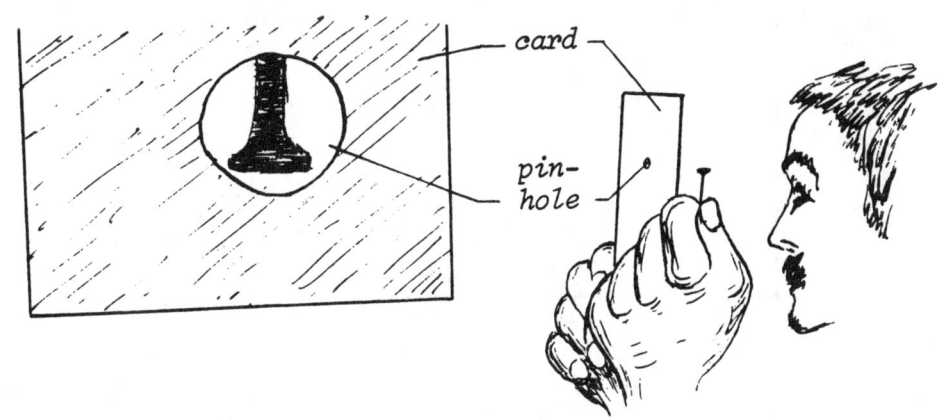

Procedure:
1. Pierce a hole with the pin in the center of the card.
2. Hold the card about 10 cm in front of one eye and hold the pin rightside up between the card and the eye (view the pin against a window or light): an upside down image of the pin will be observed.
3. Now hold the hold in the card about 3 cm from the eye and hold your eyelid almost closed: upside down images of the eyelashes will be seen!

Questions:
1. Why do we see an upside down image of the pin?
2. Why did the image of the pin disappear from in front of the card?
3. What did the eye actually focus on?
4. What was the purpose of the pinhole?
5. Why do we have to view the pin against a window or a light?

Explanation:
All objects will cast an upside down image onto the **retina** when the human eye is focussed on them, which the brain interprets as rightside up (see Sketch A).

The small hole in the card serves as a slit to admit the light, forming a small bundle of light that enters the eye (see Sketch B). This small light beam casts a shadow of the pin upside down onto the retina (actually rightside up, but interpreted by the brain as upside down).

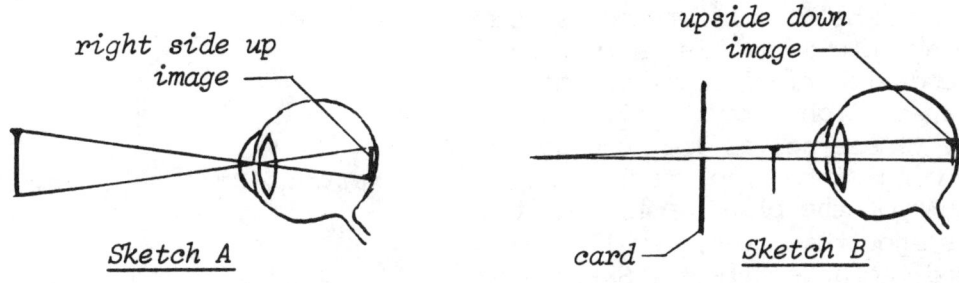

Sketch A   *Sketch B*

17.2. ARE WE PARTIALLY BLIND?

Materials: 1. A blank sheet of paper.
            2. A pencil or pen.

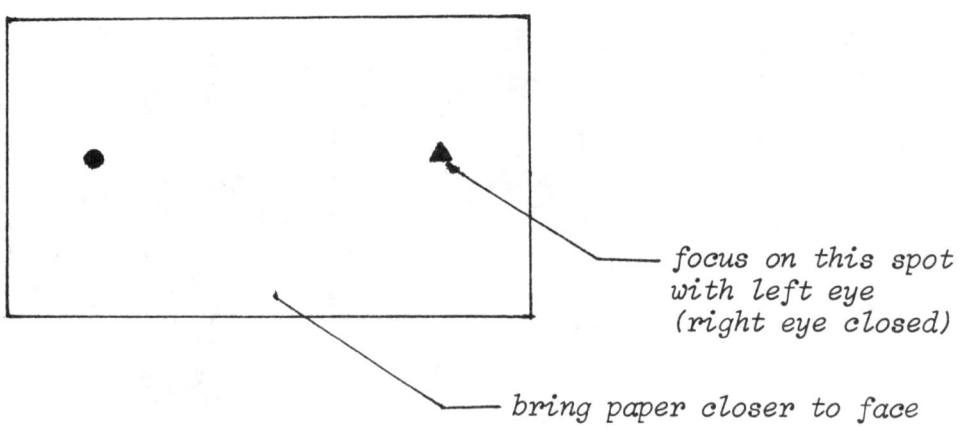

focus on this spot with left eye (right eye closed)

bring paper closer to face

Procedure:
1. Draw a small circle (about .5 cm diameter) and a small triangle (about the same size), about 15 cm apart on the white blank sheet, and completely darken them in.
2. Hold the paper about 30 cm in front of your eyes and shorten this distance very gradually (bring the paper closer to you slowly). While doing this look with your left eye (close right eye) and focus on the triangle or look with your right eye (left eye closed) and focus on the circle.
3. If you do not notice anything happening with the other dot, you may want to hold the paper somewhat higher or somewhat lower while approaching it.

Questions:
1. Why does the circle disappear at a certain distance from your left eye?
2. Is the angle or the distance between the circle and the left eye more important in getting it to disappear?
3. Does it make any difference if we had other shapes on the paper?
4. What makes us see the circle although we're focussing on the triangle?
5. How does the human eye function?
6. Is the retina of our eye totally complete and continuous?

Explanation:
   The retina of the human eye may be compared with the photographic film in a camera. The big difference is that in the human eye it is nerve cells that send electrical impulses to the brain when light rays fall on the retina. These nerves are bunched up and coming into the retina at a certain spot **the blind spot,** and it is at this spot that very small images are undetectable (see Sketch on right).

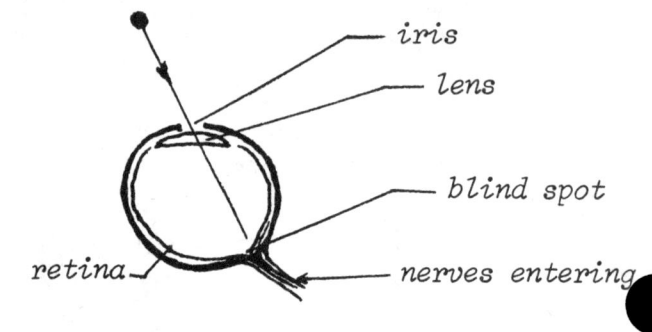

Top view of left eyeball

17.3. WHY DO WE NEED TWO EYES?

Materials: 1. A pencil & a piece of molding clay.

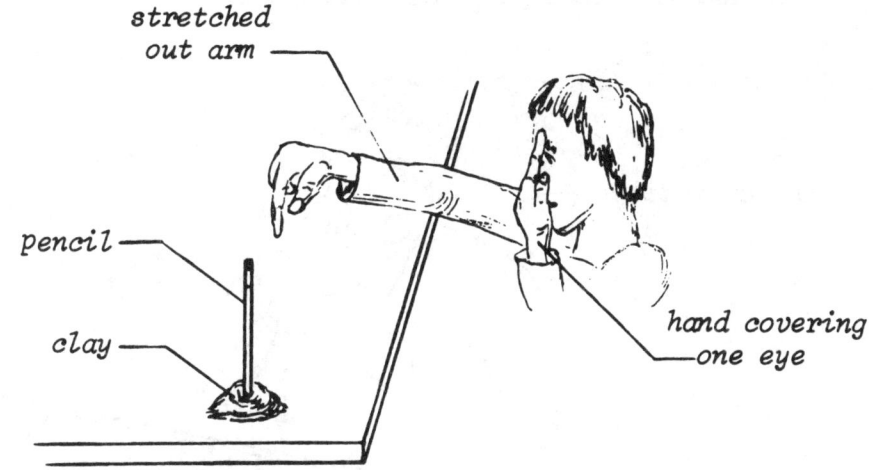

Procedure:
1. Place the pencil vertically in a piece of clay on the table top.
2. Let the students one by one try the following, while the rest of the class observes: Approach the pencil from the side about 3 - 4 meters away with one eye (cover the other eye with your hand). Hold the other hand stretched out and without hesitation, point down with the index finger and try to touch the pencil. Repeat again. Now repeat with both eyes open.

Questions:
1. Why did most students miss touching the pencil end?
2. After a student tried to do the trick several times, why did he/she get better at touching the pencil?
3. Do you think it would be easier to do by approaching the pencil slowly?
4. What happens when the experiment is done with both eyes open?
5. What do you lack when just one eye is used?
6. Why must the pencil be touched without hesitation?
7. If we see differently with each eye, why don't we see two images when both eyes are open?

Explanation:
Most people will not be able to touch the pencil on the first try, because they do not see with one eye, how far in front of them the pencil is located. **One cannot judge depth and distance as well with one eye** as with two. With one eye, one sees everything in the same plane (as in a picture). In other words, everything becomes two dimensional rather than three dimensional.

With a little practice one will get better at judging distances with only one eye.

HUMAN BIOLOGY   EYESIGHT

17.4. DO WE REALLY USE BOTH EYES?

Materials: 1. Two cardboard cylinders (about 30 cm long)--the core of wrapping paper or paper towel suits fine.
2. Two different pages of a newspaper.

Procedure:
1. Place the two sheets of newspaper on the desk in front of you.
2. Hold the cardboard tubes in front of each of your eyes, such that you can look at one sheet with one eye and at the other sheet with the other eye.
3. Keep both eyes open at all times--can you read both pages at the same time? Which do you read more easily?

Questions:
1. Which page do you read more readily?
2. Do you have to shut the other eye when reading one of the pages?
3. Why is it not possible to read both pages at the same time?
4. What does the preference of using one of your eyes indicate?
5. In what instances do we have to use just one eye?

Explanation:
We shall not be able to read both pages simultaneously, because we can only focus on one thing at a time. Although both eyes remain open, we can focus on one page or the other at will. It takes a simple act of will for one eye to see properly and for the other to stare blankly. It may thus be argued that we often make use of only one eye, even when we keep the other wide open. Evidently we do not have to cover one eye, either by holding a hand in front or closing the eye lid, when the use of only one eye is required, like when aiming a gun or looking through a microscope, or any other instrument requiring the use of only one eye.

People, however, do prefer to use either the left or the right eye when they have the option, indicating that **there are left-sighted or right-sighted people,** just as there are left or right-handed people. Which group do you belong to?

17.5. ARE YOU LEFT- OR RIGHT-SIGHTED? (I)

Materials:
1. Two blank sheets of white paper.
2. A pencil.

paper with hole in center

paper with dot in center

Procedure:
1. Pierce a hole in the center of one sheet of paper with a pencil.
2. Draw a black dot in the center of the other sheet of paper the size of a penny, and place this about 40 cm in front of you on the table.
3. With both eyes open, hold the sheet with the hole between your face and the sheet with the dot, and move the first sheet about until the black dot can be seen through the hole.
4. While holding this sheet steady (while seeing the dot), close first your left eye and then your right. When does the dot disappear?

Questions:
1. Does the dot disappear after closing your left or your right eye?
2. If the dot disappears when closing your left eye, are you left-sighted or right-sighted?
3. If the dot disappears when closing your right eye, are you left-sighted or right-sighted?
4. After determining that you are right-sighted, does it make any difference whether you are closing your left eye or not in looking at the dot?

Explanation:
For most people the dot will disappear when closing the right eye. This indicates that most people are **right-sighted**. It means that most people prefer to use their right eye over the left, if they are confronted with the option of using only one. In this case it means that those people can see the dot with both eyes open or with only the right eye open, but not with only the left eye. In other words, when both eyes are open, the left eye does not work. There is probably a connection between this phenomenon and the right-handedness of most people, although it would be hard to say which is the cause and which the effect.

17.6. ARE YOU LEFT- OR RIGHT-SIGHTED? (II)

Materials: 1. A pencil or straw or ruler (or any straight edge).
2. A vertical straight edge of the room (inside/outside corners).

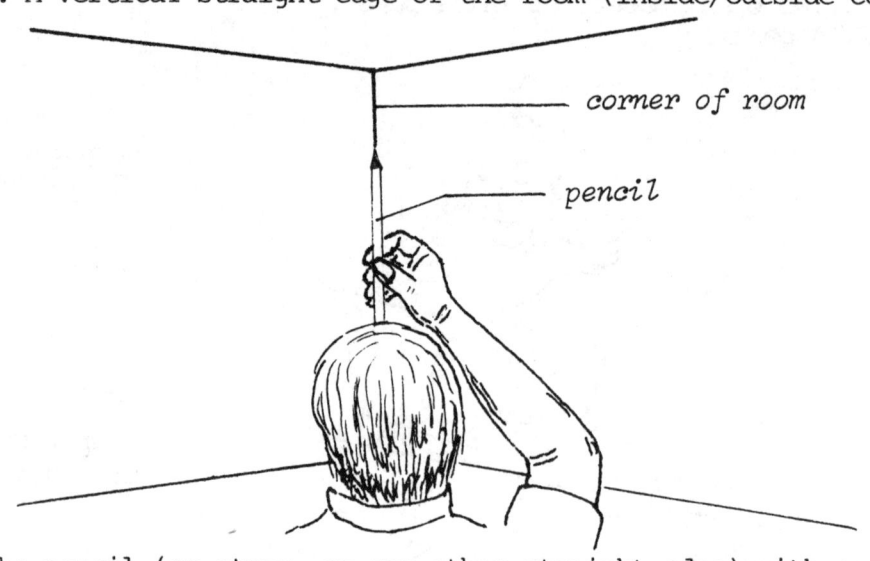

Procedure:
1. Hold the pencil (or straw, or any other straight edge) with an outstretched arm vertically in front of you.
2. Find a straight vertical edge (corner) of the room and align the pencil with it, holding both eyes open.
3. While keeping the pencil aligned with the straight edge of the room, close one eye at a time. Which eye do you close to make it look like the pencil jumps aside?

Questions:
1. Does the pencil jump aside when closing your left or right eye?
2. If the pencil jumps aside when closing your right eye, are you using more your left or your right eye?
3. If the pencil jumps aside when closing your left eye, are you left-sighted or right-sighted?
4. What makes the pencil jump aside when one of the eyes is closed?

Explanation:
For the majority of people the pencil or whatever straight edge is aligned with the vertical line of the room, will jump aside when the right eye is closed. This indicates that most people are right-sighted. The reason for the pencil to jump aside, is that the right-sighted person uses his/her right eye more than the left. Thus the pencil is lined up in a straight line between the straight edge of the room and the right eye. When this eye is closed, the left eye still sees the pencil, but it is not in line with the edge of the room.

This is one of the easiest ways to tell whether a person is left- or right-sighted. If a pencil is not available, a forefinger held vertically may be used just the same.

HUMAN BIOLOGY   COLOR PERCEPTION

17.7. HOW DO WE PERCEIVE COLOR?

Materials:
1. A drawing of a heart-shaped figure (or other shape) having a yellow border, a green interior, and a small black dot in the center (on a double page size paper).
2. A plain double page size paper with a small black dot in the center of it.

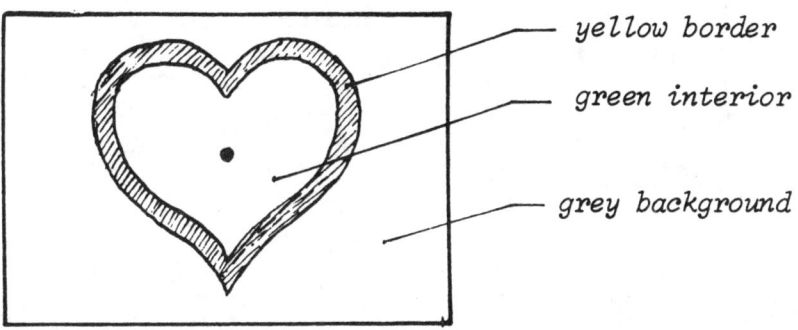

Procedure:
1. Let the students one by one stare at the colored heart for twenty seconds without blinking their eyes, focussing on the black dot.
2. Let them immediately after look at the plain piece of white paper, focussing on the black dot.
3. Ask the students to report what they see.

Questions:
1. What color is the interior of the heart? The border?
2. What did you see when looking at the plain white paper?
3. What were the colors as compared to the first heart you viewed?
4. Why do you think was the second heart not the same as the first one?
5. Was there any heart drawn on the blank sheet of paper?
6. What colors does the human brain pair together?

Explanation:
There appears to be an unusual relationship between the eye and the brain in the human being. There is a relationship among four primary colors: red, green, blue, and yellow. As the retina absorbs light, **coders** farther back in the eye discriminate between the colors. The black-white coders can send a combined grey image, but one coder relays signals for green or red, and another coder for blue or yellow. These colors oppose each other and will not mix. **Steady exposure to any color tends to weaken the brain's response:** a bleaching effect makes the color fade or makes it grey. After staring at the yellow and green heart, a clear after-image was seen of a red heart bordered in blue. There is a temporary switching of signals to the brain. Since red and green share a single coding mechanism, as do yellow and blue, withdrawal of the color stimulus shuts down one part of the mechanism and triggers the other part for a moment or two.

The cones in the human eye pair the following colors: red-green, yellow-blue, and white-black.

HUMAN BIOLOGY   PERCEPTION

17.8. WHAT GIVES US THE ILLUSION?

Materials: 1. Rulers (30 cm) & pictures or drawings of sketches below:

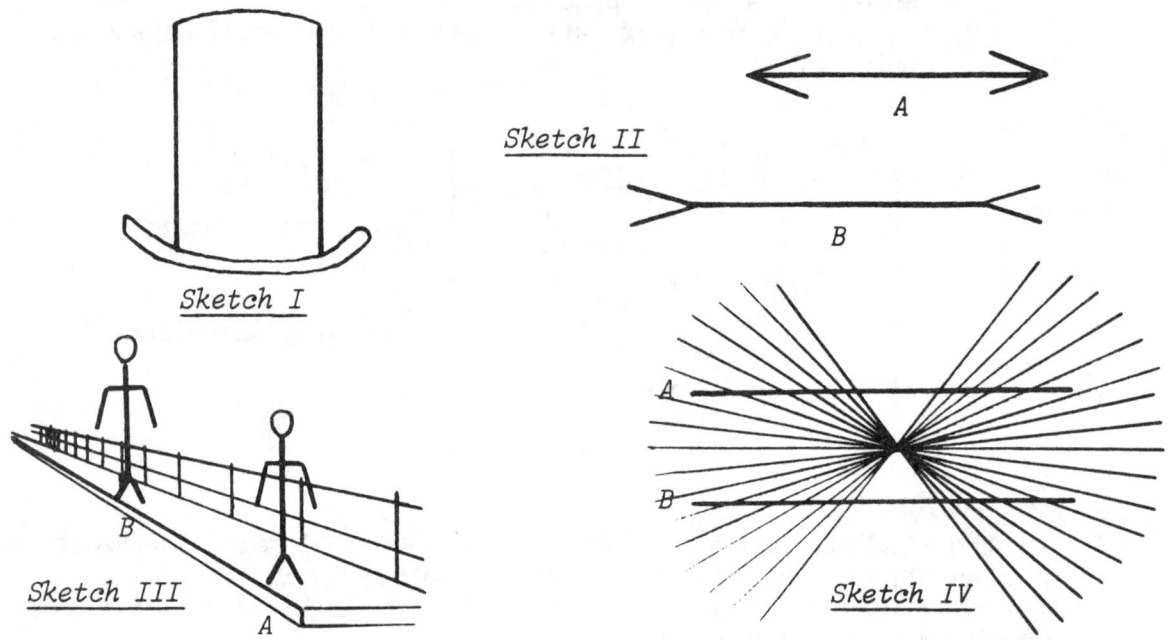

Procedure:
1. Let the students look at the pictures and ask the questions corresponding with each of the sketches.
2. Let the students measure or place the ruler along the lines.

Questions:
1. How tall is the top hat compared to its width? (Sketch I)
2. Which of the two lines A or B is longer? (Sketch II)
3. Which of the two puppets is taller? (Sketch III)
4. Are the two lines A and B straight or curved? (Sketch IV)
5. What gave us the illusion in each of the sketches?

Explanation:
   In Sketch I of the top hat, people do not take the brim of the hat into account, and this is why the hat looks taller than it is wide. In the second sketch, people tend to look at the picture as a whole: line plus arrows at the end, and then line B looks longer. In picture III, the two puppets are seen in a three dimensional environment, where puppet B looks farther away than puppet A, and thus puppet B seems much taller. In Sketch IV it is the environment again that gives us the illusion that the two heavy drawn lines are curved.
   By measuring each of the dimensions and by placing the ruler against the lines (for Sketch IV), we soon see that the dimensions of the hat, the two lines (in Sketch II), and the two puppets are exactly the same, and also that the two lines (in Sketch IV) are straight and running parallel to each other.

HUMAN BIOLOGY   PERCEPTION

17.9. THE SWAYING CARDBOARD

Materials: 1. A white 5 x 8" cardboard or heavy paper.

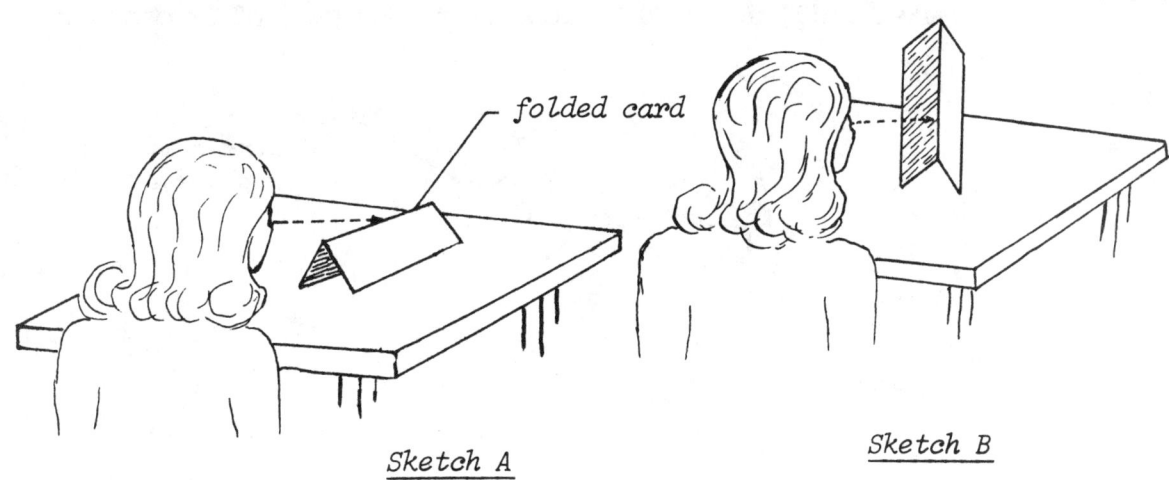

Sketch A   Sketch B

Procedure:
1. Fold the paper card along the long axis in half.
2. Place the folded card directly in front of you on the table with the ridge pointing in your direction (See Sketch A).
3. Select a spot in the center of the fold and stare at it steadily with one eye shut (if you're right-sighted cover your left with your hand).
4. First you will see the card like in Sketch A. Continue staring at it until suddenly you see the card in its second position (like in Sketch B).
5. When you see the card in its second position (standing up), move your head slowly from side to side (still keeping one eye shut). Do you observe the card swaying back and forth?

Questions:
1. How long did it take you to see the card in its standing position?
2. What did you see the card do when moving your head from side to side (with one eye shut)?
3. When seeing the card in its second position, are all the depth information and the way the light and shadow falls on paper and table, consistent?
4. What do you see when moving your head closer and farther from the card?

Explanation:
   This interesting double illusion demonstrates a phenomenon called **parallax**. When first seeing the folded card, it is lying down and you see it lying down as in Sketch A. All the information of the depth and how the light and shadows are falling, are consistent. When seeing it in the second position (like in Sketch B), the cues are all contradictory: the near point of the fold becomes the farthest point and the far corners of the paper become closest. Normally, when moving your head from side to side, near points will move more than far points of an object (try it with both eyes open). Seeing the card in its second position reverses these points and thus gives us an unusual distorted image.

HUMAN BIOLOGY   EYESIGHT PERCEPTION

17.10. SEE A HOLE IN YOUR HAND

Materials: 1. A cardboard cylinder (the core of paper towel or wrapping paper serves well) about 30 cm long (for each pair of students).

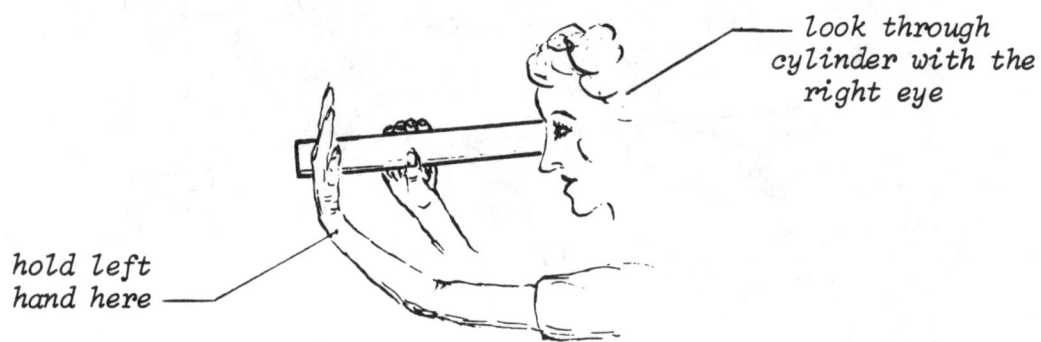

Procedure:
1. Pass the cardboard cylinders to the students (one for each pair or let two or three cylinders float around).
2. Demonstrate how to hold the cylinder and hand: Hold the cylinder with the right hand in front of the right eye, hold the left hand next to the cylinder close to the end of it, keep both eyes open and look (focus) at a distant point.
3. Now close each eye one at a time; open both eyes again. Do you see the hole in your left hand?

Questions:
1. What did your left eye see (when the right was closed)?
2. What did your right eye see (when the left was closed)?
3. How did you come to see a hole in your left hand?
4. What do you have to do to see a hole in your right hand?
5. What can you infer from this, that the brain is doing with the vision or perception of each individual eye?

Explanation:
By focussing at a distant point, the left eye in front of which the left hand is held, does register an image of the hand, but it is more the image of the right eye which is registered in the brain (because the eyes were focussed at a far point through the cylinder). When both eyes are open, it is the image of both eyes that are registered and combined by the brain. This is the reason why a hole is seen in the left hand. By switching the roles of the hands, it is possible to see a hole in the right hand.

HUMAN BIOLOGY

PERCEPTION
EYESIGHT

17.11. THE FLOATING PIECE OF FINGER

Materials: 1. Yourself and your two index fingers.

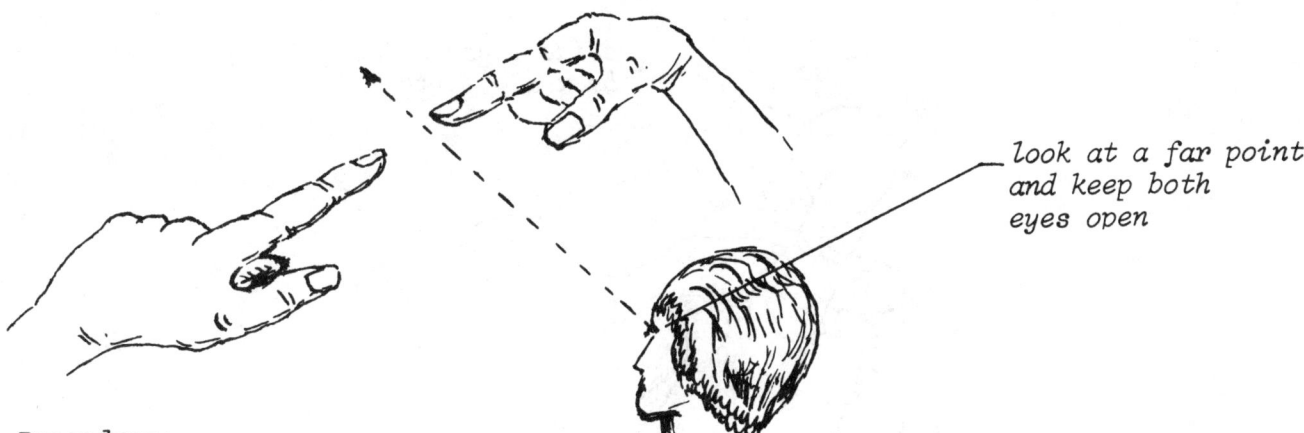

look at a far point
and keep both
eyes open

Procedure:
1. Hold your left and your right forefinger about 30 cm in front of you at the height of your eyes. Hold them horizontally about two-three cm apart (see Sketch above).
2. Do not focus your eyes on the fingers but look over them and focus at a far point.
3. Wiggle the fingers slightly up and down. Can you see a double-nailed piece of finger floating in the air?!

Questions:
1. What made you see only a piece of each finger?
2. What happens if you focus your eyes on the fingers?
3. What happens if you look under the fingers and focus at a far point?
4. What do you see if you put your thumbs in the same position?
5. How can you check what your left eye is actually seeing?
6. Try closing one eye at a time; what do you see?

Explanation:
Even though the eyes are focussed on a far point, we still see objects that are closer to the eyes. The image from the left eye and the image projected in the right eye are both combined in our brain.

This is the reason why we see only a piece of the finger. Whatever image overlaps is seen more clearly. You would never be able to see the same floating piece of finger with one eye closed. We are not able to see depth with only one eye. It is like looking at a two dimensional picture, if we use only one eye.

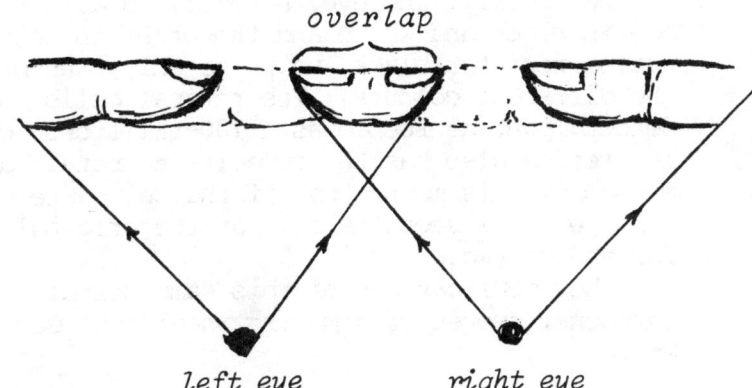

17.12. THE HAND IS QUICKER THAN THE EYE

Materials: 1. Two quarters or two other identical coins.

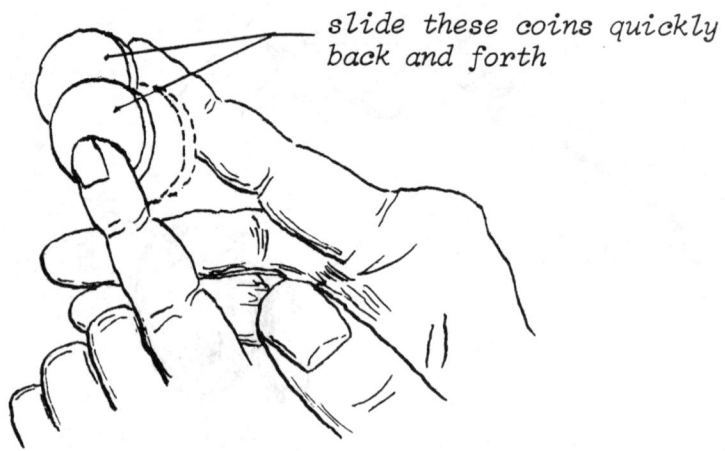

Procedure:
1. Place two coins between your two index fingers (conceal this from the students or audience, do not tell them how many coins you have).
2. Then rub the two coins back and forth as quickly as you can.
3. While rubbing, show them to the students (close up), and ask: "How many coins do you see?" Anticipated answer: "Three". If the answer is "Two", let them take a good look again at the moving coins.
4. After the majority of students (ideally all students) say that there are three coins between your fingers, go to the student that said "definitely three", and drop the two coins in his/her hand.

Questions:
1. How many coins were actually moving?
2. Why do we see three coins when the coins are rubbed against each other?
3. What does it mean: the hand is quicker than the eye?
4. What do you see when the rubbing is considerably slowed down?
5. What are daily applications of this same principle?

Explanation:
   The human eye is a marvelous instrument. It operates on the same principle as the camera. The eye, however, is able to take two simultaneous pictures, one in black and white and the other in colour. Cells called **rods** in the retina register black and white only; and other cells called **cones** register the different colours. The retinal cells are so sensitive that they can detect light as feeble as a 100-trillionth of a Watt. This is 1×10^{-11} Watt. The retina also has the capacity to retain an image a little shorter than half a second, this means that if the coins are moved faster than half a second, the eye still sees the coin at the original spot. Thus it sees three coins instead of two.
   Other applications of this same principle are: motion pictures, animated cartoons, spokes of a moving wheel that sometimes are seen moving in reverse, etc.

HUMAN BIOLOGY   ILLUSION
                                                                 EYESIGHT

17.13. PUT THE BIRD IN THE CAGE

Materials: 1. A half dollar or dollar coin (a quarter is OK too).
             2. Two large sewing needles.

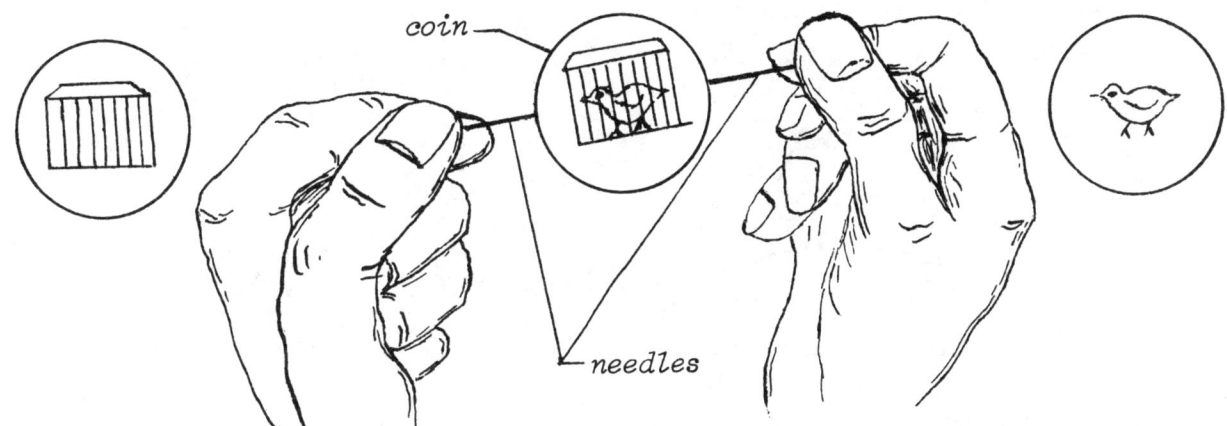

Procedure:
1. Cover both sides of the coin with tape that you can write on (masking or 3M magic tape) and trim the edges off.
2. First draw a cage on one side, then you draw a bird on the other side, making sure that you flip the coin upside down, and not sideways, before drawing the bird.
3. Show the drawings to the audience and ask them: "How can I put the bird in the cage?"
4. Find the diameter of the coin half way up the cage by cutting out a piece of paper the size of the coin and fold it in half.
5. Place the coin on top of the paper circle and pick the coin up with the two needles exactly along the diameter of the coin (see Sketch).
6. With the two needles bring the coin close to your mouth and blow against the upper half of the coin.
7. Repeat the blowing with short puffs.
   Observe the spinning coin: Bird in the cage!

Questions:
1. Why do we place the two needles along a diameter of the coin?
2. What would happen if they were not placed along a diameter?
3. Around what point do spinning objects rotate?
4. What made us see the bird in the cage?
5. Did the bird actually go in the cage?

Explanation:
   All spinning objects spin around their center of gravity. This is why it is necessary to find a true diameter of the coin and place the two needles along this diameter. If the needles were not placed along a diameter, the spinning would not go very smoothly.
   Seeing the bird in the cage is only an **illusion**, just like in moving pictures. The image of the cage or the bird is retained longer by our eyes than the time it takes for the coin to flip around. Thus both images are combined by our brain and we see the bird in the cage.

HUMAN BIOLOGY

ILLUSION
EYESIGHT

17.14. THE ELLIPTICAL PENDULUM SWING

Materials:
1. A thin string, a washer or nut (or other bob).
2. A polaroid filter, or one lens of a pair of sun glasses, or semi-dark film negative.

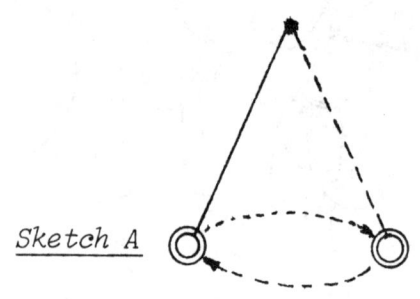
Sketch A
film over left eye

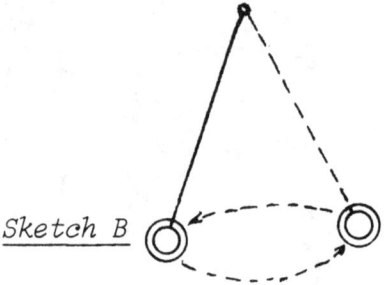
Sketch B
Film over right eye

Procedure:
1. Attach the washer or metal nut to the thin string to make a pendulum.
2. Let the pendulum swing from a fixed point (someone could hold the end of the pendulum with a steady hand).
3. Look at the swinging washer with both eyes open, but cover your left eye with the polaroid filter (or other screens). What do you observe?
4. Now look at the swinging pendulum with the film over your right eye. What do you observe? In which direction is the bob swinging?

Questions:
1. What is the reason for seeing the pendulum swinging in an eliptical path?
2. Would it be easier to see objects in a dark or in a bright room?
3. Would the eye perceive an object faster or slower if that object is brightly or less brightly illuminated?
4. If one eye is perceiving a moving object a fraction of a second slower, where would this eye actually see the object? In front or behind the actual position of the object?

Explanation:
The eye that is covered by the darkened filter perceives the pendulum a fraction of a second later than the open eye. This means that the left eye (say that this eye was covered with the filter) perceives the washer a little behind the actual position A (see Sketch below). Our brain interprets these two perceptions as coming from one object and combines the two images, thus we see it either a little farther or a little closer than the actual position.

The Sketch on the right illustrates the washer A moving to the right. The left eye (covered) perceives it at position B and the brain interprets it in position C. On the way back the brain interprets the position of the washer as being closer than the actual spot. Thus the motion is perceived as being eliptical.

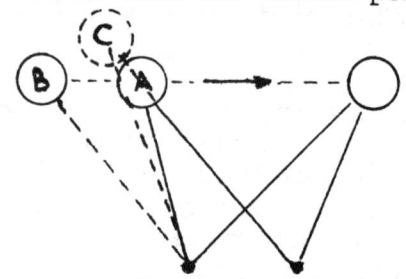
Left eye   Right eye

17.15. HOW MANY POINTS ARE TOUCHING?

Materials: 1. A bobby pin (hair pin) for each pair of students.

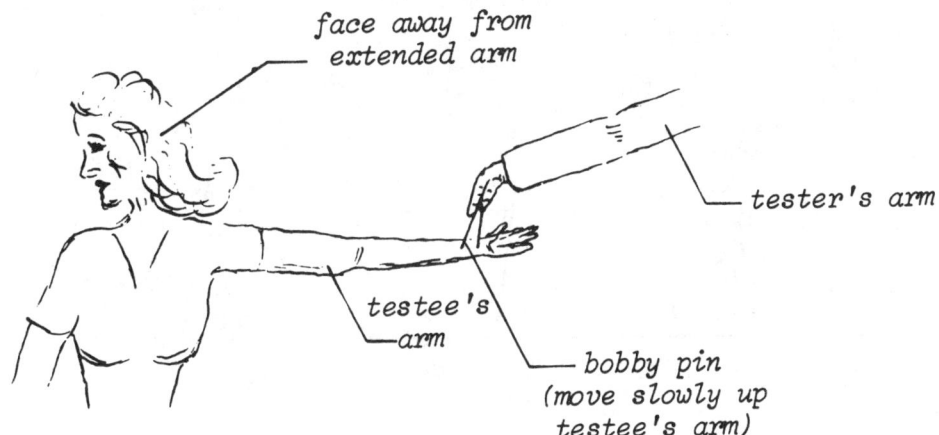

Procedure:
1. Let students work in pairs: one be the testee and the other the tester. The latter holding the pin.
2. Instruct the tester to spread the two tips of the bobby pin and place it on the testee's finger, hand, and arm, moving slowly up the arm starting from the middle finger, touching the skin randomly either with one or two tips.
3. The testee should look away from the touched arm (see sketch) and tell the tester how many tips he/she feels (without looking at the pin).
4. The tester should keep track of the number of mistakes made at the different parts of the arm. (Let tester and testee change roles).

Questions:
1. Where was it easier to tell how many tips were touching?
2. Why did we make more mistakes the higher up the arm the pin was touched to the skin?
3. What makes our touch sense more sensitive?
4. Where would a cut in the skin hurt most?

Explanation:
This activity demonstrates the distribution of the **nerve endings** in the human body, in our case from the fingertips to the shoulder. It is increasingly difficult to tell without looking how many tips are actually touching the skin, the higher up the arm the pin is placed. This is caused by the fact that there are many more nerve endings in the fingertips and hands compared to the upper arm. It would therefore be more painful to experience a cut in the skin closer to the fingertips than higher up the arm. For this same reason, our sense of touch is most sensitive at our fingertips compared to other parts of the body.

HUMAN BIOLOGY

NERVOUS SYSTEM
TOUCH

17.16. IS THE WATER WARM OR COLD?

Materials: 1. Three beakers (400 ml) or large cups.

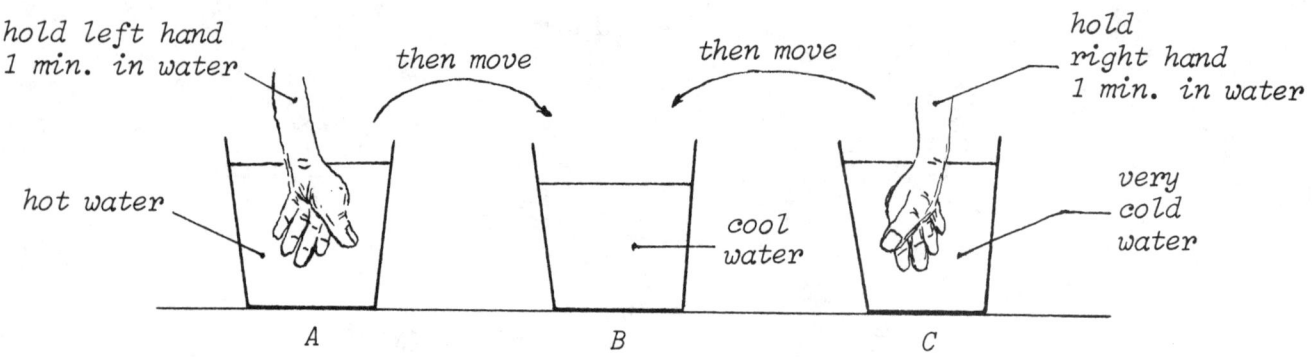

Procedure:
1. Fill a beaker (A) three quarters full with hot water (about 50°C) and another beaker (C) three quarters full with very cold water (about 5°C) (add a few ice cubes if water is not cold enough in the summer time).
2. Fill the third beaker (B) with regular water of room temperature (20°C) and place it in between the hot and the cold water beaker.
3. Immerse your left hand in the hot water and hold it under water for one minute, then move it in the center beaker. Does the water feel warm?
4. Immerse your right hand in the cold water and hold it under water for one minute, then move it in the center beaker. Does the water feel warm or cold?

Questions:
1. Did the water in the center beaker (B) change in temperature?
2. Why did the water in beaker B feel warm with one hand and cold with the other?
3. What gives us the sensation of warm or cold in our body?
4. What are other similar situations where this same principle applies?

Explanation:
   When the left hand is placed in the hot water, the **nerve endings** in our nervous system send a message to our brain, telling it that it feels warm. Over time, however, the nerve endings are dulled and the hand adapts itself to the warm sensation. When it is placed into the water of room temperature, there is a difference of about 30°C lower, and the water suddenly feels cold to this hand.
   The same interpretation can be put forth for the hand immersed in the cold water. This time it's only going the opposite way: from cold to warm.
   Similar situations are encountered when we take a shower with very cold hands and adjusting the water temperature with the touch of our cold hands; placing our body under the shower, the water still feels cold because to the hand the shower water already felt very warm.
   Someone used to living in a home with 18°C room temperature might find on visiting a friend's house with a 22°C room temperature quite warm, and of course visa versa.

17.17. CATCH THE DOLLAR BILL

Materials: 1. A crisp dollar bill (any denomination).
(ideally: one of each denomination up to $20).

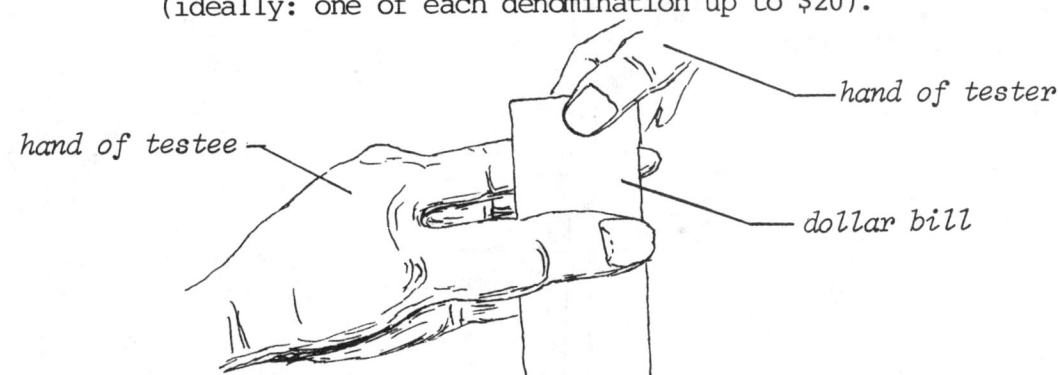

Procedure:
1. Hold your forefinger and thumb of your left hand about 5 cm (2 inches) apart, place the dollar bill about half way in between the two fingers.
2. Let the dollar bill fall and show the audience that you can easily catch it between your two fingers.
3. Ask the members of the audience to hold their forefinger and thumb in the same position to try to catch the dollar bill.
4. Hold the dollar bill between their fingers. The rule is that they may not go down with their hand, and the gap between the fingers should be not smaller than 5 cm.
5. Vary the time between placing the dollar bill between their fingers and the moment that you let go of the bill. Tell the audience that whoever catches the bill may keep it. Continue doing it with larger denominations, but remember to vary the dropping time, otherwise the bill might get caught by anticipation.

Questions:
1. Why can you catch the dollar bill so easily when you drop it yourself?
2. Why do you need to vary the time between placing the dollar bill and the releasing of it?
3. What is the reason that the bill can never be caught?
4. What makes muscles contract?
5. How did the finger muscles get the order to catch the bill?
6. Where did the initial stimulus come from?

Explanation:
The initial stimulus to catch the dollar bill was from seeing the falling of it. The eyes got the stimulus, they sent a signal to the brain and the brain in turn sent an electrical impulse to the finger muscles to contract and catch the bill. But all that took a longer time than the falling of the dollar bill, although both take only a fraction of a second.

The bill can only be caught by someone who anticipates the falling and actually start closing the fingers before the bill falls. This is the reason why we should vary the time between placing the bill between the fingers and the actual releasing of it.

HUMAN BIOLOGY

NERVOUS SYSTEM
REACTION TIME

17.18. HOW FAST CAN YOU REACT?

Materials: 1. A meter stick (15 meter sticks for a class of 30).

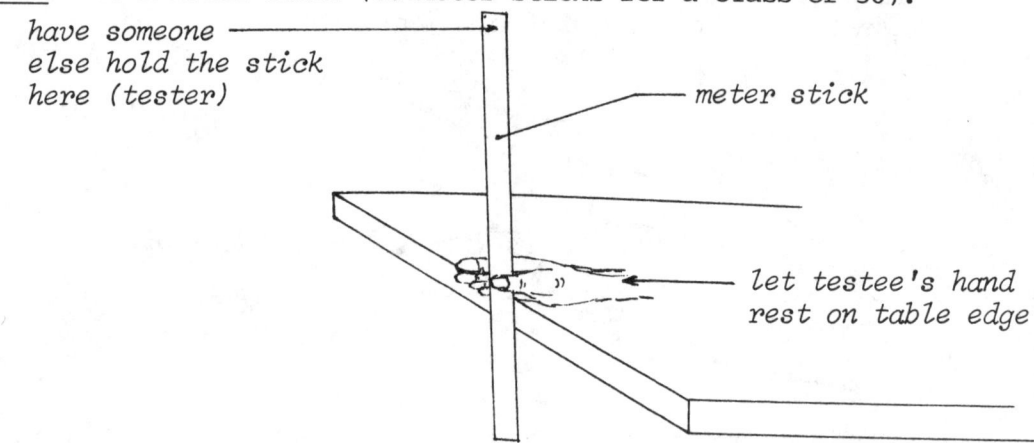

Procedure:
1. Place your hand over the edge of a table and leave about 3 cm opening between the thumb and forefingers.
2. Let your tester (someone else) hold the meter stick vertically from the top end of the stick.
3. Read off the spot where your thumb is on the meter stick before the tester drops the stick (start at an even number f.i. 10 or 20).
4. Let the tester drop the stick. The moment that you see the stick drop - catch the stick immediately and do another reading of where your thumb ended up. Record the difference (distance in drop).
5. To test the other stimuli: Hearing: close your eyes and let tester say: "Now" exactly at the moment that he/she drops the stick.
Testing for tactile (touch): close your eyes and let tester touch your other hand exactly at the moment that he/she let go of the stick.

Questions:
1. How do girls compare to boys in reaction time?
2. In doing the above comparison, what variables have to be controlled?
3. How do the different stimuli compare in your reaction time?
4. What is the actual falling time for the distance in the drop?
5. What are all the variables involved in this experiment?

Explanation:
The different variables in this experiment are: the **manipulated variable** is the stimulus: through the eyes, hearing, and touch (or sex: when comparing boys versus girls). The **responding variable** is the falling time or distance on the meter stick.
As the distance in free fall is: $d = 1/2\, gt^2$, the time derived from that is $t = $ square root of $2d$ over g, where g is the gravity acceleration.
All other variables, like: distance between fingers, the place where the tester holds the stick, catching with two fingers or with the whole hand, the accuracy of saying "now" and simultaneously letting go of the stick, etc. have to be held constant, in other words they have to be done the same way.

17.19. THE STIMULUS-RESPONSE ACTION

Materials: 1. A wooden rod or block (a ruler would do).
2. Paper and pencil for each student.

Procedure:
1. Stand in the rear of the class behind the students.
2. Instruct the students to make a tally mark every time you say "write."
3. Strike a desk with the wooden rod and simultaneously say "write" for a total of 20 times with 2 second intervals.
4. Continue striking the desk, but stop saying "write" until everyone has stopped writing.
5. Have the students count their tally marks.
6. Tell the students that you said "write" exactly 20 times.

Questions:
1. Why did most students make more than 20 tallies?
2. What was the stimulus and what was the response?
3. Why did I pair the strikes with the saying of "write"?
4. What was the purpose of the taps with the wooden rod?
5. Why did everyone stop writing eventually?

Explanation:
When a **response-eliciting stimulus** (the saying of "write") is paired with a **neutral stimulus** (taps with the wooden rod), the student becomes **conditioned to the neutral stimulus,** so that the neutral stimulus will elicit the response, even when it is presented alone. The conditioning will become extinguished after a certain time, if the presentation of the **conditioned stimulus** (tapping noise) is continued without pairing with the **unconditioned stimulus** (the saying of "write").

This conditioning of the human being to neutral stimuli occurs in daily life for example, when the preparation of food or a meal is constantly accompanied by the clatter of kitchen utensils; hearing only the clatter will stimulate the feeling of hunger or might even stimulate the production of saliva in the salivary glands (drooling).

HUMAN BIOLOGY   CIRCULATORY SYSTEM

17.20. HOW FAST DOES YOUR HEART BEAT?

Materials: 1. Low step stools.
            2. A stopwatch or a watch with a second hand.

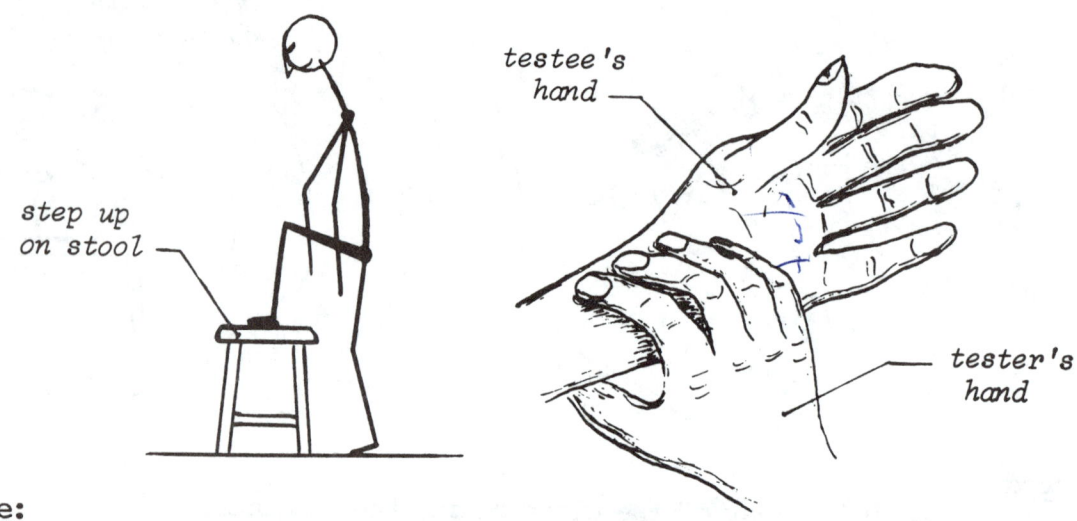

Procedure:
1. Let students work in pairs: one being the tester and the other the testee.
2. Instruct the tester to count the number of heartbeats of the testee by placing his fingers on the testee's wrist, and counting the pulses within 10 seconds (and multiply this by 6).
3. Now let the testee step up and down the stool 10 times, and immediately after that let the tester take the testee's pulse (repeat point 2).
4. The number of steps up and down the stool can now be increased to 20 and 30, immediately followed by a pulse count.
5. Let the tester and testee change roles.

Questions:
1. What was the pulse before the steps up and down the stool?
2. What was the pulse after 10, 20, and 30 steps up and down the stool?
3. What made the heart beat faster after the exercise?
4. What did the muscles need to do all that work?

Explanation:
   The pulse or heartbeat is caused by the **blood pressure impact on the arteries** as the heart muscles contract. It can be felt by placing a finger on the **radial artery** at the wrist. By doing vigorous exercise, like the steps up and down the stool, the leg muscles need more oxygen and thus more blood to carry this oxygen. The heartbeat thus automatically speeds up to pump more blood to the working muscles, from around 72 beats per minute to more than 120. Within three minutes this speeded pulse should return to the normal 72. With increasing number of steps up on the stool, an increase in the pulse should be observed in the testee. Some deviations from these values may be normal for certain individuals.

HUMAN BIOLOGY
RESPIRATION

17.21. WHICH CONTAINS MORE CARBON DIOXIDE?

Materials:
1. Two Erlenmeyer flasks.
2. Two sets of glass tubing in 2-hole stoppers (see sketch).
3. Saturated lime water (calcium hydroxide).

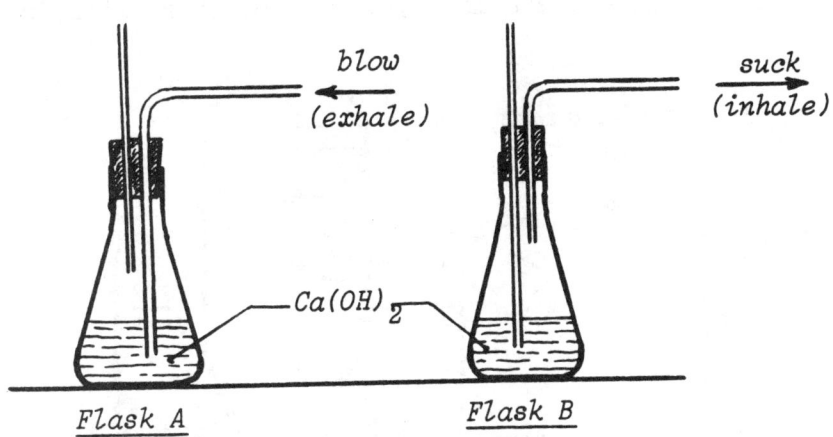

Procedure:
1. Make a saturated solution of calcium hydroxide by adding small amounts of the solid powder, a little at a time to warm water while stirring, until no more solid dissolves.
2. Fill each flask half full with the lime water and cover them with the two-hole rubber stoppers (and fitted glass tubing).
3. Suck through the tube which ends above the water surface in flask A and exhale through the tube that extends into the liquid in flask B (see sketch).

Questions:
1. Which of the two flasks is getting milky first?
2. What gas is actually led through the liquid in flask A?
3. What does it indicate if the liquid turns milky?
4. What chemical reaction is taking place?
5. Will flask A eventually also turn milky if the sucking is continued?
6. What will the liquid in flask B eventually do if the blowing through it is continued?
7. Which of the two, inhaled or exhaled air, contains more CO_2?

Explanation:
By sucking through the tube in flask A, atmospheric air is led through the saturated calcium hydroxide solution (this air is the same as the air we inhale). By blowing through the tube in flask B, exhaled air is led through the liquid, which turns milky sooner than that in flask A. This indicates that there is more CO_2 present in exhaled air. The reaction is as follows: $CO_2 + Ca(OH)_2 \rightarrow CaCO_3 + H_2O$, in which the calcium carbonate precipitates out and makes the solution appear milky. The CO_2 in the air will eventually turn the liquid in flask A also milky, and when the blowing is continued in flask B, the reaction: $CO_2 + CaCO_3 + H_2O \rightarrow Ca(HCO_3)_2$ will turn the liquid clear again (Ca-bicarbonate is soluble in water).

17.22. MEASURE THE CAPACITY OF YOUR LUNGS

Materials:
1. A large glass jar (pickle gallon jar).
2. An unused aquarium (or other transparent container).
3. A length of rubber tubing (about 40 cm long).
4. A 1 litre measuring cylinder (or other litre container).

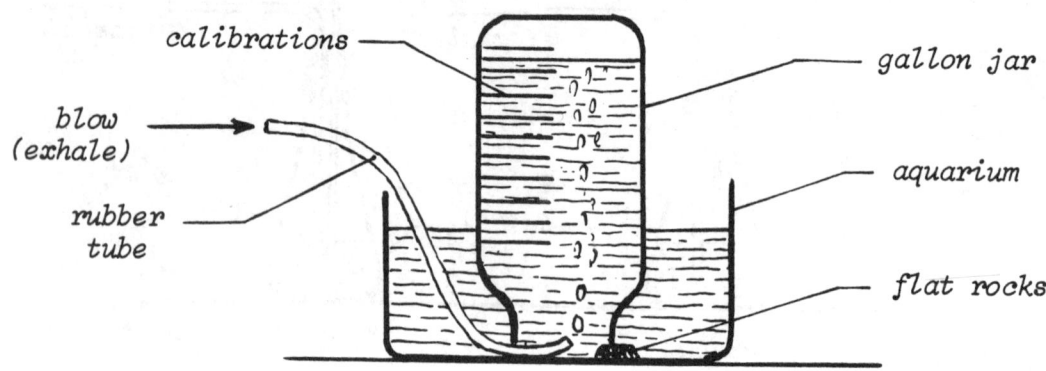

Procedure:
1. Calibrate the large jar by filling it with water, litre by litre, and marking the water level with a marker or masking tape.
2. Fill the gallon jar completely full and the aquarium 3/4 full with water.
3. Place three flat stones of the same thickness (or other heavy objects) on the bottom of the aquarium, and invert the gallon jar into the water-filled aquarium so that it rests on the three stones.
4. Insert one end of the rubber tubing under the mouth of the gallon jar and let the other end hang over the rim of the aquarium.
5. Let one student hold the inverted jar steady and another student, whose lung capacity is to be measured, exhale through the tube after inhaling as deeply as he/she can.
6. Measure volume of exhaled air in jar.

Questions:
1. What made the water stay up in the inverted jar?
2. Why did the student have to inhale as deeply as possible before inhaling?
3. In exhaling, what does the student have to do in order to obtain a true measure of his/her total lung capacity?

Explanation:
By inhaling as deeply as we can, we are actually filling our lungs full with air. When we blow all the air out through the tube and catch this exhaled air in the jar above the water, the volume or capacity of our lungs can thus be measured. This we do by reading off the volume of air in the jar. The larger this lung capacity, the more it is indicating that the individual involved is enjoying better health.

17.23. HOW DO WE BREATHE?

Materials:
1. A large bottle with a rather narrow neck.
2. A one-hole stopper (fitting in the bottle neck).
3. A Y-glass tube (or straight tube).
4. Two small balloons & rubber bands.
5. One large balloon (or beach ball balloon).

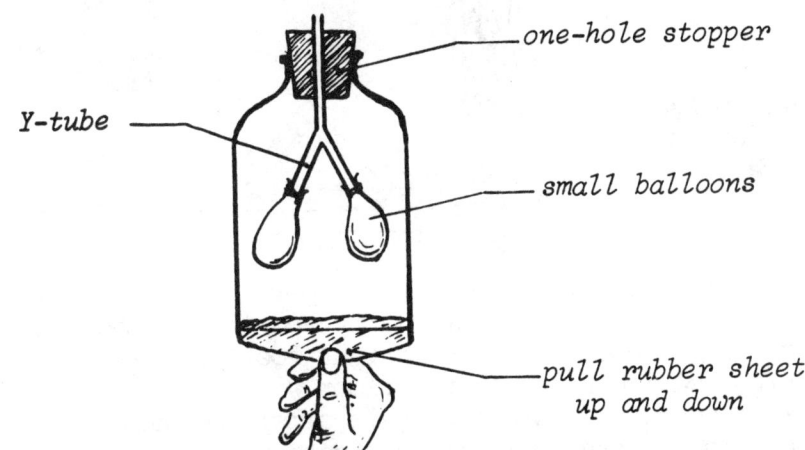

Procedure:
1. Cut the bottom of the large jar out, by covering it with a cm thick layer of hot oil, and touching the outside of the jar with an ice cube exactly at the surface of the oil (the jar will crack and the bottom will drop out). Smooth the edge by filing or firing.
2. Tie the two small balloons to the ends of the Y-tube with the rubber bands. Hold the long end of the Y-tube through the jar neck and insert the one-hole stopper over the Y-tube and in the jar neck.
3. Cut the large balloon in half, stretch it over the open bottom of the jar and tie or tape it around airtight (it represents the diaphragm).
4. Pull the center of the diaphragm up and down and observe the balloons expand and collapse.

Questions:
1. What does the Y-tube represent in this demonstration?
2. What do the small balloons and the stretched rubber sheet represent?
3. What made the small balloons expand?
4. What stage of the breathing could the expanded state of the balloons be compared with?
5. How different would it be to compare the glass jar with our chest cavity? How are they the same?

Explanation:
The Y-tube in our demonstration represents the bronchial tube, the small balloons, the lungs, the rubber sheet, the diaphragm, and the jar, the chest cavity. This latter one, however, can be expanded or contracted through the flexible rib joints, which is not the case with the glass jar. By pulling the rubber sheet down, the pressure inside the jar decreases, thus sucking the air from outside into the small balloons (inhaling).

17.24. THE UNCONTROLLABLE FOOT

Materials:
1. A blank sheet of paper
2. A pencil or pen.

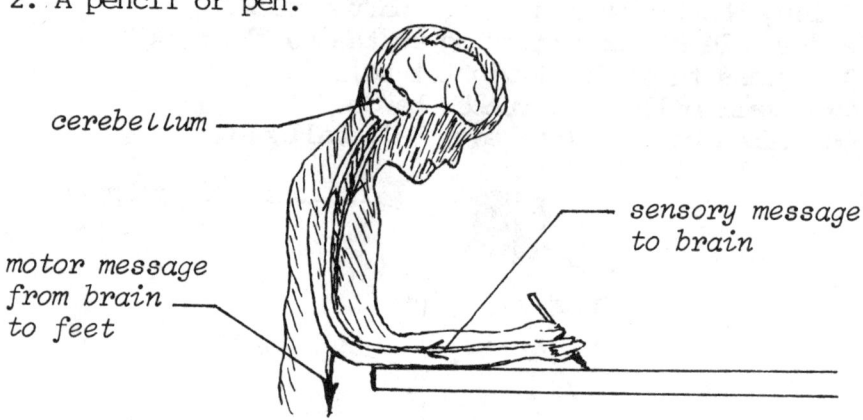

Procedure:
1. Stand up next to the table with the paper in front of you.
2. Hold the pen or pencil in your right hand (if you're right-handed).
3. If you're right-handed, let your right foot rotate in a clockwise direction (describing small circles on the floor with you right foot).
4. Try to keep your foot rotating while simultaneously writing a large number 6 on the paper. Observe what your foot is doing!
5. Now repeat point 3 and 4 but with your left foot making the clockwise motions. Try drawing the six now with your right foot making counter-clockwise motions. Then do the same with your left foot.

Questions:
1. Which of the four combinations was easiest to carry out?
2. When moving your foot in a clockwise direction, was it easier to do it with your left or right foot (while drawing the 6)?
3. Did you try to write the sixes with your left hand (if you're right-handed) while rotating your foot?
4. What makes it so difficult to rotate your right foot (when right-handed) or your left foot (when left-handed) in a clockwise direction?
5. What part of the human body controls the muscles?

Explanation:
The **cerebellum** located in the base of the **cerebrum** of the human brain controls the coordination of voluntary movements. The cerebellum is important for such activities as walking, dancing, playing ball, or even for such routine tasks as tying a shoelace or writing a figure 6.

Doctors have known for generations that nerve fibres from the right side of the body cross over in the brain stem to the left side of the brain. Similarly, nerve fibres from the left side of the body cross over to the right side of the brain. In other words, the whole left side of the body is controlled by the right half of the brain, and similarly, the whole right side of the body is controlled by the left half of the brain.

By writing a figure 6 with the right hand, the left half of the brain has instructed the right hand to make a counter-clockwise motion. The right foot could easily make the same counter-clockwise motion, but the opposite movement requires a special effort.

HUMAN BIOLOGY   MUSCLE COORDINATION

17.25. THE KICKING FROG LEGS

Materials: 1. A freshly cut frog leg.
            2. A copper (bronze) bolt, and two nuts.
            3. An iron stand or frame (galvanized iron strip will do).

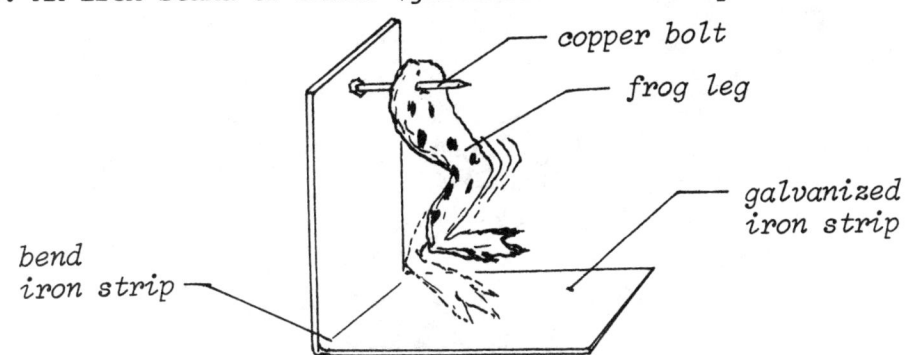

Procedure:
1. Drill a hole in one end of the iron strip and attach the copper bolt tightly to this strip with the two nuts.
2. Bend the iron strip at a spot in such a way that when the frog leg is hung from the bolt, the foot will touch the lower part of the iron strip.
3. Pierce the frog leg through the copper bolt in the thigh and let the leg droop down. Observe the spasms!

Questions:
1. What made the frog leg go into spasms?
2. What would be necessary for a muscle to contract?
3. Would an iron bolt do the same thing for the frog leg?
4. Would this demonstration work if you use an iron bolt in a copper stand?
5. What other metals would give this same contraction in the frog leg?
6. How can we find out which metals give the potential difference?
7. How can we compare this with human muscles and their contraction?
8. Why is it dangerous to grab a live wire on purpose?
9. How should we test a live wire with our hands?

Explanation:
   There is a potential difference between copper and iron. At the moment that the frog leg touches the iron frame, while it is suspended from the copper bolt, an electrical current (a flow of electrons) is running through the muscle and makes it to contract. After a while the leg muscles relax and the leg droops down, the foot touches the iron and the leg goes into spasms again.
   The further apart two metals are in the **electromotive series**, the larger the potential difference, in other words the larger the difference in tendencies to go into solution or give off electrons. Other combinations are: copper and zinc, magnesium and copper, iron and silver, etc.
   Human muscles are also controlled by electric currents. It is therefore very dangerous to deliberately grab a "live" wire, as you may not be able to let go of the wire. In order to test a live wire, use the back of your fingers or hand and touch it for a short moment. If the wire sends electricity through your muscle, your fingers will be moving away from the wire rather than clamping around it tighter.

HUMAN BIOLOGY   BODY BUILD OF
                                                           MEN VS WOMEN

17.26. ARE WOMEN MORE AGILE THEN MEN?

Materials: 1. A cigarette box (or any other object of the same dimensions).

Procedure:
1. Let one of the ladies in the audience sit on her knees and lower leg (like in a kneeling position) on the floor.
2. Place a cigarette box (or other small object) in front of her the length of her forearm away from her knees on the floor.
3. With arms behind her back, let her bend forward and knock the object over with her nose! Men cannot do this without falling forward!

Questions:
1. Why can women knock the object over without falling, but not men?
2. Are women actually more agile and lean compared to men?
3. What is different in the basic skeleton built of women compared to men?
4. Where is the center of gravity of women located?
5. Where is the center of gravity of men located as compared to women?
6. How would the exceptions of women that cannot do the trick be built?
7. How would the exceptions of men that can do the trick be built?

Explanation:
   In general, women's body structure is such that they have a lower **center of gravity**, because of their wider hips and heavier bone structure in the lower abdomen part of the skeleton as compared to men's structure. Similarly we can say for men in general, that they have wider shoulders as compared to women. This makes the center of gravity of men's bodies higher than women's.
   This lower center of gravity is the main cause for women to be able to bend forward in the kneeling position without falling over forward. When men try to bend over they will fall forward because of the location of their center of gravity. When this center of gravity passes beyond the knees, their bodies will topple forward. (See also Event 13.10 for a similar activity).

17.27. TURN A CRACKER INTO SUGAR

Materials: 1. Unsalted and unsweetened crackers.

chew cracker

Procedure:
1. Distribute one cracker to each student.
2. Let them chew the cracker without swallowing it for a minute or two.
3. Ask them the following questions.

Questions:
1. How did the cracker taste in the beginning of the chewing?
2. How did the cracker taste at the end of the chewing period?
3. What was the cracker mixed with in your mouth?
4. What does your saliva do to the cracker while chewing?
5. What agent in your saliva breaks down the starch molecules?
6. Are sugar molecules smaller or larger than starch molecules?
7. Why is it better to chew food a little longer before swallowing?
8. What would happen if we almost did not chew our food?

Explanation:
The cracker that was put in the mouth consists of carbohydrates (starch). This being unsalted and unsweetened, will taste quite bland in the beginning of the chewing period. While the cracker is being pulverized into the pulpy mass called a **bolus**, the **digestive juice** of the **salivary glands** begins a breakdown of the carbohydrates. An **enzyme (amylase)** in saliva splits the molecules of starch into smaller molecules of sugar. This enzyme cannot break down starch particles that are still enclosed in their natural cellulose envelopes, therefore starches should be cooked before eating.

Another purpose of the saliva is to provide moisture needed by the taste buds. Saliva also has a cleansing action on the teeth. It washes away food particles that otherwise might provide a home for bacteria.

By chewing food a little longer before swallowing, it gets mixed more thoroughly with our saliva, providing an opportunity for the enzymes to break down the large starch molecules, and by the time the food reaches the stomach and the intestines, it can be easily digested.

HUMAN BIOLOGY   DIGESTIVE SYSTEM

17.28. TRY DRINKING WHILE STANDING ON YOUR HEAD

Materials: 1. A glass of drinking water or juice.
             2. A bent drinking tube or flexible drinking straw.

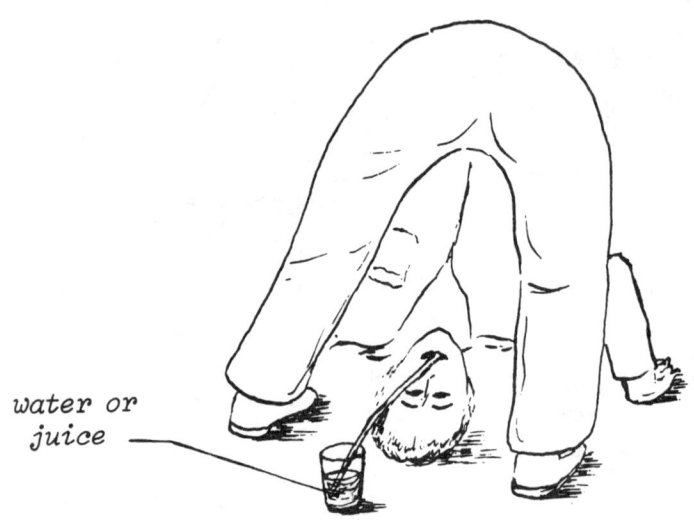

water or juice

Procedure:
1. Ask one of the students to assist you, to put the straw in your mouth as soon as you are ready to drink.
2. Try to stand on your head against the wall (if this is not possible, bend from your waist down until your head touches the floor).
3. Have your student assistant bring you the glass and straw so that you can drink from it.
4. Suck through the straw while your body (or upper body) is upside down, and empty the whole glass.

Questions:
1. Is gravity needed to make fluids come down the esophagus?
2. Can we drink water while we stand on our head?
3. How does food go down the esophagus into the stomach?
4. Is the esophagus like a glass or rubber tube going in the stomach?
5. Why did the fluid not flow out of the mouth when drinking upside down?
6. What is the muscle action called, which pushes food into the stomach?

Explanation:
After food is swallowed, it enters the **esophagus**. Here it is pushed along by **muscle action**, which is called **peristalsis**. This movement of the food is carried out by **involuntary muscles**. There are two layers of muscles: the inner layer forms a series of circles around the tube, and the outer layer is longitudinal. When the inner layer contracts, the tube becomes smaller at that point, and when they relax the longitudinal muscles contract. This **alternate contraction and relaxation** of the two sets of muscles push the food along the tube in **peristaltic waves**. This is the reason why food, whether it is in solid or liquid form, may be swallowed with the body positioned in any direction. Gravity has little or no influence on this.

APPENDIX

Table 1. HUMIDITY CHART

Difference between Dry- and Wet-bulb Thermometers

Degrees - C	0.5	1.0	1.5	2.0	2.5	3.0	3.5	4.0	4.5	5.0	5.5	6.0	6.5	7.0

Reading of Dry-bulb Thermometer Degrees - C — Percent Humidity

Dry-bulb °C	0.5	1.0	1.5	2.0	2.5	3.0	3.5	4.0	4.5	5.0	5.5	6.0	6.5	7.0
17.5	95	89	84	79	74	69	64	60	55	51	46	42	38	33
18	95	89	84	79	74	70	65	60	56	51	47	43	38	34
18.5	95	90	85	80	75	70	65	61	56	52	48	44	39	35
19	95	90	85	80	75	71	66	61	57	53	49	45	40	36
19.5	95	90	85	80	76	71	66	62	58	53	49	45	41	37
20	95	90	85	81	76	71	67	63	58	54	50	46	42	38
20.5	95	90	86	81	76	72	67	63	59	55	51	47	43	39
21	95	90	86	81	77	72	68	64	60	55	52	48	44	40
21.5	95	91	86	81	77	72	68	64	60	56	52	48	45	41
22	95	91	86	82	77	73	69	65	61	57	53	49	45	42
22.5	95	91	86	82	78	73	69	65	61	57	53	50	46	42
23	95	91	86	82	78	74	70	66	62	58	54	50	47	43
23.5	95	91	87	82	78	74	70	66	62	58	55	51	47	44

Table 2. DENSITIES OF SUBSTANCES

Substance	Density (g/cc)
Mercury	13.6
Lead	11.3
Steel	7.6 - 7.8
Aluminum	2.70
Glass	2.4 - 2.8
Glycerine	1.27
The Human Body	1.07 (Average)
Water	1
Ice	0.917
Alcohol	0.794
Cork	0.24
Room Air	1.2×10^{-3}
Hydrogen	8.9×10^{-5}

Table 3. INDEXES OF REFRACTION

Substance	Index of Refraction
Diamond	2.4173
Glass, dense flint	1.655
, crown	1.517
Carbondisulfide	1.6255 (20°)
Carbontetrachloride	1.460 (20°)
Ethyl alcohol	1.361 (20°)
Water	1.33299 (20°)
Ethyl ether	1.00152
Carbondioxide	1.000450
Air	1.000293

FURTHER READINGS

Alyea, Hubert N. and Dutton, Frederic B.
"Tested Demonstrations in Chemistry," Reprinted from the Journal of Chemical Education, 1966. Publ.: Division of Chemical Education of the American Chemical Society.

Carin, Arthur A. and Sund, Robert B.
"Teaching Science Through Discovery," Charles E. Merrill, 1970.

Friedl, Alfred E.
"Teaching Science to Children: the Inquiry Approach Applied," Random House, N.Y. 1972.

Lowery, Lawrence F.
"The Everyday Science Sourcebook. Ideas for Teaching in the Elementary and Middle School," Allyn and Bacon, Inc. 1976.

Nelson, L.W. and Lorbeer, G.C.
"Science Activities for Elementary Children," Wm. C. Brown, 1980.

Parker, Bertha Morris
"Science Experiences, Elementary School," Row, Peterson and Co., 1958.

Physics Fun and Demonstrations with Professor Julius Sumner Miller, Fred H. Gunzel, Jr., Central Scientific Co. (Ed), Cenco, 1968.

Schmidt, V.E. and Rockcastle, V.N.
"Teaching Science with Everyday Things," McGraw-Hill, 1968.

UNESCO, Unesco Source Book for Science Teaching, Published by Unesco, Paris, 1969.

Viorst, Judith
150 Science Experiments Step-by-Step, Bantam Science and Mathematics, 1963.

Walker, Jearl
"The Flying Circus of Physics with Answers," John Wiley and Sons, 1977.

INDEX

ACETONE, 5.19
ACETALDEHYDE, 5.19
ACIDS & BASES, 5.3, 5.4
ACTION, capillary, 4.19, 16.15, 16.16
    force in, 14.10, 14.15
ADHESION, 4.19-4.21
AIR, 1.1-1.43, 5.14, 5.30, 16.13
AIR, burning of, 5.30
    exhaled, 5.14, 17.21
    expansion of, 1.25-1.34
    flowing, 2.1-2.8
    inhaled, 17.21
    lift, 1.39
    occupying space, 1.1-1.6, 1.16
    pressure of, 1.7-1.27
ALCOHOL, 4.3, 4.8, 4.32, 7.20
    burner, 1.11, 1.13, 1.14,
        4.6, 5.3, 5.12, 5.19,
        5.24, 5.25, 7.3, 7.4,
        7.5, 7.9, 7.15, 7.17,
        7.18, 7.19,
ALKA SELTZER, 4.34
ALUMINUM, 5.18
ALUMINUM BAR, 12.14
    FOIL, 9.8, 10.6, 14.11
    PIPE, 12.15
AMMONIA, 1.42, 5.10, 5.15
AMMONIUM CHLORIDE, 1.42, 5.10, 5.15
AMPLIFIER, sound, 12.1
ANEMOMETER, 3.6, 3.7
APPLE, 14.8
ARCHIMEDES, 4.37
ARM, 13.41, 17.15
ASTRONOMY, 15.9, 15.10

BAG, paper, 1.28
    plastic, 1.1, 1.2, 1.39
BAKING SODA, 5.3
BALANCE, 1.28
BALL, 1.38, 2.3, 2.7, 3.17
    styrofoam, 13.6, 15.9
BALLOON, 1.13, 1.22, 1.23, 1.32,
    1.37, 4.2, 9.1, 14.10, 17.23
BALLOON, charged, 9.1, 9.10, 9.12
BARIUM NITRATE, 5.23
BAROMETER, 3.3, 3.4

BATTERY, 10.1-10.9, 10.14, 10.15, 10.17,
    10.18, 14.14
BATTERY, coin, 10.12
    liquid, 10.10
BATTERY POLES, 10.3
BELL, 10.18, 12.10
    soundless, 12.10
BELT HANGER, 13.19
BERNOULLI, principle of, 2.1-2.7, 2.9,
    12.16
BICEP, 13.41
BIOLOGY, human, 17.1-17.28
BIONIC FINGER, 1.16, 1.17
BLACKBOARD, 9.1
BOILING POINT, 4.15-4.17
BOLT, hot, 6.1
BOOK, 6.13, 7.16, 9.3, 10.10, 12.8, 13.2,
    13.42, 14.6
    swinging, 12.8
BOOMERANG, 14.27
BOTTLE, confused, 7.8
    glass, 1.29, 6.6, 12.2, 13.24
    medicine, 1.27
    singing, 12.9
    soda-pop, 3.1, 7.8, 14.5
    wide mouth, 1.2, 3.2, 3.3
BOX, 1.5, 15.6, 15.9, 16.2
BREEZES, 6.3
BUBBLES, popping, 12.11
BUCKET, angry, 5.26
BUGLE, 12.16
BULB, 10.1-10.8, 10.11, 10.17
    flash, 10.10
BULB POLES, 10.3
BUOYANT FORCE, 4.36, 4.38, 4.39
BURGLAR ALARM, 10.2
BURNER, Bunsen, 7.6
    propane, 12.5

CABBAGE, 5.3
CALCIUM BICARBONATE, 5.14, 17.21
    CARBONATE, 4.7, 5.8, 5.14, 15.2,
        16.11, 17.21
    HYDROXIDE, 5.14, 16.11, 17.21
CAMERA, 11.1
CAN, come-back, 13.30
    falling, 13.4

CAN, soda-pop, 2.5, 6.5, 12.12
    tin, 1.12, 1.18, 3.12, 14.12
CANDLE, 1.33-1.35, 4.32, 5.26
    7.16, 16.18
CANDLE FLAME, 2.9, 2.10, 7.7
    HOLDER, 9.12
    SNUFFER, 1.5, 7.1
CANNISTER, 14.9
CANNON, 6.5
CAPACITOR, static, 9.13-9.14
CAPILLARY ACTION, 4.19, 16.15
CARAFE, 16.18
CARBON, 5.1
CARBON DIOXIDE, 5.8, 5.14, 17.21
    DISULFIDE, 5.21, 5.22, 7.20
    TETRACHLORIDE, 5.13, 7.20
CARD, paper, 1.7, 2.2, 2.6, 4.25
    13.1, 13.3, 14.27, 17.1
CARNATION, 16.15
CARROT, 16.4, 16.14
CATALYST, 5.18-5.20
CELL, guard, 16.13
CHALK, 14.5
CHANGE, physical & chemical, 5.1, 5.2
CHARCOAL, 5.1
CHARGE, induced static, 9.11
CHEMISTRY, 5.1-5.39
CHLOROPHYLL, 16.10-16.12
CIGARETTE, 7.2
CIRCLES, wobbling, 13.16
CIRCUIT BREAKER, 10.18
CIRCUIT, parallel & series, 10.5
CIRCUITS, 10.2, 10.3, 10.5-10.9,
    10.18
CIRCULATORY SYSTEM, human, 17.20
CLAY, 4.36-4.7, 4.39, 17.3
CLIMATE, 3.17, 3.18
CLOTH, 1.21, 15.1
    damp, 12.14
CLOUD, 3.1, 3.2, 3.11
COAT HANGER, 6.7, 12.13, 14.16
CODERS, in human eye, 17.7
COHESION, 4.18, 4.20, 4.21
COIL, wire, 7.1, 10.13, 10.14, 10.17
COIN, 1.31, 7.2, 10.12
COLOR, 5.4, 5.11, 5.23, 7.11, 17.7
    warm & cold, 7.11
COMB, 9.6-9.9, 12.1
COMBUSTION, 5.30

COMBUSTION, spontaneous, 5.21-5.25
COMPASS, magnetic, 8.8-8.10, 10.15,
    10.16
CONDENSATION, 3.10, 3.11, 4.10
CONDITIONING, 17.19
CONDUCTION, heat, 7.1-7.6
CONDUCTIVITY, current, 10.1-10.4
CONDUCTOR, 10.1, 10.2, 10.4
    semi-, 10.4
CONVECTION, heat, 7.6-7.10
COPPER, 5.17, 5.19, 10.10, 10.11
COPPER NITRATE, 5.16, 5.23
    SULFATE, 5.17
    TUBING, 7.3
CORK, 2.13, 4.21, 7.4, 13.6
COTTON, charless, 7.2
CRACKER, 5.7
CROWN, 4.37
CRYSTALS, 4.10
CUP, 1.15, 1.22, 4.26, 9.14
    paper, 7.19
CURRENT, fluid, 7.9
CURRENT SOURCES, 10.11-10.13
CYLINDER, cardboard, 17.4, 17.10
    glass, 1.36, 4.3, 4.27, 5.20,
    5.22

DAYS, 3.17
DENSITY, 4.32-4.42
DEPTH, 17.3
DETERGENT, 4.24, 4.25, 4.27, 5.20
DEW POINT, 3.10
DIFFUSION, 7.8
DIGESTIVE SYSTEM, 17.27, 17.28
DIME, 4.26
DISH, Petrie, 9.4, 9.5
DISPERSION, 11.16-11.19
DISPLACED WATER, 4.36-4.42
DISTILLATION, 4.17
DIVER, 1.40
DOG, barking, 5.22
DOORBELL, 10.8
DOWEL, wooden, 7.3
DYNAMO, 10.13

EARTH, 15.9, 15.10
EARTH SCIENCE, 15.1-15.8
ECLIPSE, 15.10
EFFORT, 13.39-13.42

EGG, 1.26, 4.38, 16.5
ELECTRICITY, current, 10.1-10.8
    static, 9.1-9.14
ELECTRODE, 10.10, 10.11
ELECTROLYTE, 10.10, 10.12
ELECTROMAGNET, 10.14, 10.17, 10.18
ELECTROMAGNETISM, 10.14-10.18
ELECTROPHORUS, 9.12
ELECTROSCOPE, 9.8, 9.10
ELEVATOR, 14.32
ENERGY, 6.1-6.17
    chemical sources, 6.4, 6.5
    heat sources, 6.8
    magnetic sources, 6.9
    mechanical sources, 6.6, 6.7
    nuclear sources, 6.10
    potential vs. kinetic, 6.13-6.16
ENZYME, 17.27
EROSION, 15.1-15.5
ESOPHAGUS, 17.28
ETHER, 5.13
EVAPORATION, 3.13, 4.8
EXHAUST, rocket, 5.25
EXPANSION by heat, 7.13-7.18
EXPANSION of water, 15.4
EXPLOSION, 5.7, 5.26, 6.5
EYE, human, 17.1-17.14

FAN, electric, 3.5, 6.3
    paper, 2.14
FERROUS SULFATE, 5.17
FINGER(S), 1.15-1.18, 9.9, 9.11
    9.12, 13.13, 13.19, 13.21, 14.16
FINGER, bionic, 1.16, 1.17
FISHING LINE, 13.4, 14.10
FISSION, 6.10
FLAG, chemical, 5.21
FLAME, 2.9, 2.10, 4.7, 5.23, 7.7
    cool, 7.20
FLAME EXTINGUISHER, 4.7
FLASHLIGHT, 15.19
FLASK, distillation, 4.17, 5.30
    Erlenmeyer, 1.13, 4.9, 4.10
    5.33, 7.14
    round-bottom, 1.14, 7.13, 12.10, 12.11
FORCE, buoyant, 4.36, 4.38, 4.39

FORCE, centripetal, 14.16-14.20
    centrifugal, 14.16-14.20
    direction of, 14.5
FORCES, 13.1-13.43
    latent, 13.30
FORK, 13.7, 13.20
FOUNTAIN, 1.14, 1.19
FREEZING, 3.12, 4.10, 6.2
FREEZING POINT, 6.2
FROST, 3.12
FULCRUM, 13.6-13.9, 13.20, 13.39-13.41
FUNNEL, 1.6, 2.3, 5.26, 15.6
FUSE, 10.6
FUSION, 6.10

GALVANOMETER, 10.12, 10.13, 10.16
    simple, 10.13, 10.16
GAS, 4.9, 4.10
    burning of, 5.30
    properties of, 5.8
GAS PRODUCTION, 5.7-5.9
GENERATOR, 10.13
GERMINATION, 16.1
GEYSER ACTION, 15.6
GLASS, drinking, 1.7, 13.20, 16.1, 16.9
    magnifying, 11.12, 16.17
    singing, 12.6
    window, 9.3, 15.3, 16.3
    wine, 12.6, 13.31, 14.17
GLYCERINE, 4.28, 5.20, 5.25, 6.10
GRANITE, 15.2
GRAPE, 4.35
GRAVITATIONAL PULL, 13.3, 13.4, 14.19-20
GRAVITY, 13.1-13.5, 14.19, 14.20, 16.7
    center of, 13.6-13.18, 13.22, 13.31
GROWTH, plant, 16.2-16.7

HAIR, 3.15
HAMMER, 13.9, 13.16, 13.40
HAND, 17.10-17.12, 17.16
HARMONICS, 12.16
HEARTBEAT, 17.20
HEAT, 6.8, 7.1-7.20
    heavy, 7.17
HEAT CAPACITY, 6.1
    OF FUSION, 6.2
    OF VAPORIZATION, 6.3
    RACE, 7.4, 7.5

HEAT vs. TEMPERATURE, 6.1
HOLE, 17.10
HOOP, 14.5
HOTPLATE, 15.6
HUMIDITY, 3.13-3.15
HYDROCHLORIC ACID, 4.7, 5.4, 5.8,
    5.10, 5.17, 15.2
HYDROGEN, 5.8, 5.29
HYDROGEN PEROXIDE, 5.11, 5.20
HYGROMETER, 3.14, 3.15

ICE, 4.6, 4.13, 4.14, 6.2, 7.6
ICE-WATER, 7.6
ILLUSION, 17.8-17.14
IMAGE, 11.12, 17.1
INDICATOR, 5.3, 5.4
INFERENCE BOARDS, 10.8
INERTIA, 14.1-14.6
INSULATION, 7.1, 10.3
IODINE, 4.9, 5.11, 5.13, 5.18
IRON FILINGS, 5.17, 8.5, 8.8
ISOSTATICS, 15.5

JAR, gallon, 3.2, 3.3, 3.14, 9.13,
    17.22
    glass, 15.4
JUICES, 7.13, 7.14
JUMPING JACKS, 9.3

KETTLE, 3.11
KINDLING POINT, 7.19, 7.20
KNIFE, 13.7, 13.31, 14.8

LAVA, 15.6, 15.8
LEAVES, 16.8-16.14
LEMON, 10.11
LEFT HAND RULE, 10.15
LEVERS, first class, 13.29
    second class, 13.40
    third class, 13.41
LIGHT, 11.1-11.20, 16.2
    traavel of, 11.1
LIGHTER FUEL, 6.5
LIMESTONE, 15.2
LIMEWATER, 5.14, 16.11, 17.21
LIQUID, 4.8
LOUDNESS, sound, 12.1
LUNGS, human, 17.22, 17.23

MAGIC, 1.25, 5.24, 8.3, 13.23, 13.30
MAGNET, bar, 8.1, 8.2, 8.5-8.11
    disc, 8.3, 8.4
    make-up of, 8.8
    ring, 8.4
    temporary, 8.11
MAGNETIC CONFUSION, 8.9
    INDUCTION, 8.11
    LINES, 8.5, 8.6, 10.14
    MATERIALS, 8.1, 8.8
    POLE, 8.2
    RULE, 8.2, 8.9
MAGNETISM, 8.1-8.11
    attraction in, 8.3, 8.4
    repulsion in, 8.3, 8.4
    earth, 8.9, 8.10
MANGANESE DIOXIDE, 5.20, 5.25
MARBLE CHIPS, 4.7
    ORE, 15.2
MARBLES, 4.1, 14.15, 14.21
MASS, center of, 13.5-13.16, 13.21, 13.22,
    13.31, 14.6
MATCHES, 7.7, 14.11
MATTER, characteristics of, 4.1-4.42
MEDIUM OF TRAVEL, sound, 12.10, 12.13
MELTING, 4.6, 4.10-4.14, 6.2
MELTING POINT, 6.2
MEMBRANE, semi-permeable, 16.4, 16.5
MEN, 13.10
METAL, 4.23
METERSTICK, 6.13, 13.22, 13.31
MIDDLE, golden, 13.22
MILK, 11.19
MILK BOTTLE, 1.26
    CARTON, 4.18, 14.12
MIRROR, 11.4, 11.5
MIXTURE, 4.3
MOLD, 16.17
MOLECULAR SPACING, 4.1-4.4
MOMENTUM, 14.15, 14.21, 14.27
    angular, 14.26, 14.27
    conservation of, 14.15,
        14.21-14.25
MOON, 15.9, 15.10
    phases of, 15.9
MOTHBALLS, 4.10, 4.34
MOTION, accelerated, 14.9
MOUNTAIN, 15.5
MUSCLE, 13.41

MUSCLE, involuntary, 17.28

NAIL, 8.11, 9.13, 10.13, 10.14, 10.17
NAIL MAGNET, 10.14
NAPKIN, rising, 7.10
NEEDLE, 4.23, 8.10
NERVE ENDINGS, 17.15
NERVOUS SYSTEM, human, 17.15
NEWSPAPER, 1.24, 17.4
NEWTON'S FIRST LAW, 14.6-14.8
        SECOND LAW, 13.4, 14.9
        THIRD LAW, 14.10-14.15
NICKEL-CHROME WIRE, 10.9
NODES & ANTINODES, 12.15

OBOE, straw, 12.3
OIL, 4.11, 4.31, 11.20
OPAQUENESS, 11.20
OSMOMETER, 16.5
OSMOSIS, 16.4-16.6, 16.15
OXYDANT, strong, 5.23
OXYGEN, 5.8, 16.10

PAPER, 1.4, 1.30, 2.1, 7.3, 9.2, 11.20, 13.2, 13.31, 17.5
    blotting, 10.12, 16.1
    carbon, 7.11, 14.23, 16.12
    scorching, 7.3
PAPER AIRPLANE, 2.12
    CARD, 1.7, 2.2, 2.6, 4.25, 13.1, 13.3, 14.27, 17.1
PAPERCLIP, 8.11, 9.13, 10.1, 10.14, 14.11
    floating, 8.1
PAPER FAN, 2.14
    SNIPPERS, 9.3, 12.16
    WING, 2.11
PEN, ink, 6.14
PENCIL, 10.4, 12.13, 17.3
PENDULUM, 6.16, 12.18, 13.28, 13.29, 17.3
PENNY, 1.31, 5.19, 13.3, 14.16
PENUMBRA, 15.10
PEPPER, 4.24, 9.4, 9.5
PERCEPTION, 17.7-17.11
    color, 17.7
PERISTALSIS, 17.28
PHENOLPHTALEINE, 5.4
PHOSPHORUS, white, 5.21, 5.22

PHOTOSYNTHESIS, 16.12
PIN, bobby, 17.15
    knitting, 3.17, 13.6
PINHOLE, 11.1, 17.1
PIPE, water, 13.21, 13.39
    weighted, 13.21
PITCH, reversing, 12.2
PITHBALL, 9.9, 9.11
PIVOT POINT, 13.6-13.9, 13.20
PLANE, inclined, 13.42, 14.30
    paper, 2.12
PLANET, 14.26
PLANT, bending, 16.2
PLANTS, 15.3, 16.1-16.18
PLASTER OF PARIS, 15.3
POT, flower, 15.3, 16.2, 16.7
POTASSIUM CHLORATE, 5.7, 5.8, 5.23, 5.25, 5.28, 6.4
    IODIDE, 5.11
    PERMANGANATE, 5.23, 5.24, 7.9
POTATO, sweet, 16.14
POURING AIR, 1.3
POWDER, 4.27, 5.26
PRECIPITATE, 5.14
PRINT, soot, 16.8
PROJECTOR, overhead, 5.16, 5.29
PROPELLER, 14.14
PULLEYS, 13.43

RACE, balloon, 14.10
    cork, 2.13
    heat, 7.4, 7.5
    smoke ring, 1.43
    straw drinking, 1.8
RADIATION, heat, 7.11
RAZOR BLADE, 4.23, 7.17
REACTION, catalytic, 5.18-5.20
    exothermic, 5.24-5.26, 6.4, 6.5
    endothermic, 5.35
    force of, 14.10-14.15
    precipitous, 5.14
    rate of, 5.11
    replacement, 5.16, 5.17
REFLECTION, 11.2-11.5, 11.13-11.15, 11.17
REFRACTION, 11.6-11.18
REPRODUCTION, 9.5
REPRODUCTION IN PLANTS, 16.17
REPULSION, static, 9.8-9.10
RESISTANCE, electric, 10.4-10.6, 10.9
    in forces, 13.29, 13.41
    variable, 10.4

RESONANCE, 12.8, 12.9, 12.16
RESPIRATION OF PLANTS, 16.9-16.11
RESPIRATORY SYSTEM, human, 17.21-17.23
RETINA, 17.1
RING, 6.15
     paper, 13.31
ROCK, 15.3, 15.4
ROCKET, 5.25
ROD, brass, 7.17
    glass, 7.5
    metal, 7.16
    wooden, 14.8
ROOT, 16.3, 16.4
ROPE, 13.43, 14.7
ROTATION, 13.16, 13.31
       plane of, 14.26, 14.27
RUBBER BAND, 4.5, 13.30, 13.41, 14.15
RULER, 9.3, 9.4, 9.8, 10.13, 10.16,
     13.9, 13.42, 14.21, 14.23
     magnetic, 9.2

SAIL, 14.14
SALIVA, 17.27
SALT, 4.4, 4.13, 4.39, 5.12, 9.4,
    10.12
    epsom, 15.1
SALT FORMATION, 5.10, 15.1
SAND, 4.1, 6.6, 15.5
SAUSAGE, charcoal, 5.1
SCALE, bathroom, 14.32
     hardness, 15.2
     spring, 13.42
SEEDS, 15.3, 16.1
SHOEBOX, 1.5
SIGHT, human eye, 17.1-17.10
SILVER NITRATE, 5.16
SIPHON, broken, 1.19
SKATE, roller, 14.14, 14.15, 14.22
SMOKE SCREEN, 5.10
SOAP, 4.25, 4.27-4.29, 6.10
SOAP FILM, 4.29, 6.10
SODA-POP, 4.35
SODIUM HYDROXIDE, 5.4
      NITRATE, 5.23
       THIOSULFATE, 5.11, 11.18
SOIL, 15.3
SOLUBILITY, 5.12-5.15
SOLUTION, saturated, 5.12, 15.1
SOUND, 12.1-12.20

SOUND ECHO, 12.19
    PITCH, 12.1-12.6
    TRAVEL, 12.10-12.13
    VELOCITY, 12.18, 12.19
SOURCES, current, 10.9-10.13
SPACE SCIENCE, 14.1-14.32
SPHERE, 2.4
SPINNING, 6.15, 13.31
SPLINT, wood, 5.8
SPOON, 13.20
SPRING SCALE, 13.42
STALAGMITE & STALACTITE, 15.1
STARCH, 5.11, 5.26
STATES OF MATTER, 4.6-4.12
STEEL BALL, 13.4, 14.23
STEEL BAR, 8.6
STEELWOOL, 1.36
STICK, wooden, 1.24, 13.31, 13.29
STIMULUS-RESPONSE, 17.19
STOMATA, 16.13
STONE, 13.2, 14.15, 15.2
STOPPER, one-hole, 13.6, 14.18
STORAGE, electrostatic, 9.13, 9.14
STRAW, drinking, 1.8, 1.20, 1.37, 2.9, 3.5,
       3.15, 5.4, 7.16, 12.3, 12.5,
       14.10
    wheat, 16.18
STRENGTH, tensile, 16.18
STRESS, sudden, 13.31
STRING, 14.6, 14.7, 14.12
STRONTIUM NITRATE, 5.23
SUDS, 5.20
SUGAR, 5.1, 5.23, 6.4, 16.4
SULFUR, 5.7
SULFURIC ACID, 5.1, 5.23, 5.24, 6.4, 10.10,
     11.18
SUN, 3.18, 6.12, 15.9, 15.10
SUNSET, 11.18
SURFACE TENSION, 4.22-4.31
SWITCH, simple, 10.6, 10.9, 10.17, 10.18

TAPE, curving, 7.15
TELEGRAPH, 10.17
TELEPHONE, 12.12
TEMPERATURE, 3.9-3.12, 6.1
TEMPERATURE vs. HEAT, 6.1
TENDON, 13.41
TESTER, conductivity, 10.1
     convection, 7.7

TEST TUBE, 1.9, 1.11, 8.8
THERMOMETER, 3.9, 3.13, 4.6, 4.15-
    4.17, 6.1-6.3, 6.6
THREAD, 14.7, 14.9
TORQUE, 13.19-13.21
TRANSLUCENCY, 11.20
TRAY, plastic, 14.17
TREE, silver, 5.16
TROMBONE, straw, 12.5
TUBE, plastic rigid, 12.17
TUNING FORK, 12.9
TURBINE, 6.14
TURBULENCE, 2.10

UMBRA, 15.10

VACUUM, 13.3, 13.4
    partial, 1.10-1.14, 1.16,
        1.17, 12.10
VINEGAR, 5.3, 12.6, 15.2
VEINS, leaf, 16.8
VOLCANIC ACTION, 15.6

WAND, magic, 5.24
WASHER, metal, 12.7, 14.9, 14.18
WATER, 1.3, 1.4, 1.7-1.15, 1.18-1.21,
    1.29, 1.33, 1.34, 1.36, 1.40,
    3.11, 4.15, 4.27, 4.32, 4.39,
    5.18, 6.14, 7.13, 7.14, 9.14,
    10.10, 15.4, 16.4, 16.9
WATER, boiling, 1.10-1.14, 4.16,
        7.6, 7.19
    muddy, 4.17
    rising, 1.10, 1.11, 1.14, 1.19,
        1.29, 1.33, 1.34, 1.36, 2.8
WATER CYCLE, 3.11
    STREAM, 4.18, 4.20, 9.6
WATERTIGHT, 1.21
WAVES, closed & open end, 12.17
    longitudinal, 12.12, 12.14
    standing, 12.8, 12.15, 12.17
    transverse, 12.14, 12.15
WAX, candle, 7.5, 16.5
    seal, 16.5
WEATHER, 3.1-3.18
    hi-lo pressure, 3.1-3.4
WEATHERING, 15.1-15.4
WEIGHT, 1.38, 7.6, 13.22, 13.30,
    13.42, 14.7, 14.22

WEIGHTLESSNESS, 14.32
WHEEL, 6.8
WIND, in weather, 3.5-3.8
WIND CURRENT, 3.8
WIND DIRECTION, 3.5, 3.7
WIND ENERGY, 2.13, 2.14
WIND SPEED, 3.6, 3.7
WINDVANE, 3.5
WINTER, 3.16, 3.17
WIRE, aluminum, 7.5
    copper, 5.16, 7.4, 7.5, 7.18, 9.14,
        10.1-10.8
    iron, 4.14, 6.7, 7.5, 13.19, 14.23
WIRE FRAME, 4.14, 6.10
WOOD, 13.19, 13.41, 15.1, 15.5
WOMEN, 13.10

ZINC, 5.8, 10.10, 10.11

PRESENTING

HOW TO TURN STUDENTS ON TO SCIENCE

BY USING DISCREPANT EVENTS

A ONE DAY HANDS-ON WORKSHOP

FOR

ELEMENTARY, JUNIOR, OR SENIOR HIGH SCHOOL TEACHERS

Are you looking for ways to capture the interest of your students? What is the secret to create in your students that **eager want** to know more about science?

Find out how **DISCREPANT EVENTS** can work for you and help you put more spark and spunk in your teaching. Dr. Liem will demonstrate a selected number of **DISCREPANT EVENTS**, involve you in doing the same with very simple materials, and leave you totally motivated in the teaching of science!

For more information on how to get Dr. Liem to come over to conduct an Inservice Workshop, please contact:

SCIENCE INQUIRY ENT.
14358 VILLAGE VIEW LANE
CHINO HILLS, CA 91709
TEL: 714-590-4618

If you could choose anyone in North America to make a dynamic and motivating presentation or lead a workshop on the teaching of science at any level, you would probably find Dr. Liem to be heading the list.

Dr. Liem is Chinese-Indonesian of origin and received most of his education in the Dutch language in his native country Indonesia (formerly the Dutch East Indies). He obtained his Bachelor and Master of Chemical Engineering from the University of Indonesia, and his Ph.D. degree in Science Education from Cornell University. He has had 28 years of teaching experience: 6 years at a Teacher's College in Indonesia, 21 years at St. Francis Xavier University in Canada, and one year as an adjunct professor at San Diego State University. He is presently an independent Science Education Consultant with Science Inquiry Enterprises.

In 1981 he authored a 500 page teacher's guide entitled: **Invitations to Science Inquiry**, now in its ninth printing and second edition, which is widely acclaimed and used by science teachers all over the world.

Within the past fifteen years Dr. Liem has been doing over 300 presentations and workshops to professional Science Teachers Associations and inservice teachers at all levels, in Canada, the USA, the Netherlands, and even Australia. His discrepant events are arousing so much interest that they not only appeal to science students, teachers and educators, but also to general educators and the general public as well.

FROM: (Please print clearly)

_____

_____

_____

Phone:(   ) _____

Place stamp here.

SCIENCE INQUIRY ENT.
14358 VILLAGE VIEW LANE
CHINO HILLS, CA 91709

✓ I'd like more information about the items checked on the following section.

SCIENCE EDUCATION CONSULTING SERVICES

- ☐ LECTURE-DEMONSTRATION — Presentation to a large group (unlimited number) of inservice teachers. (1½ - 2 hrs)
- ☐ HANDS-ON WORKSHOP — Inservice workshop for teachers K - 12.
  - ☐ Half-day Workshop — Lecture-Demo plus one 2-hr session.
  - ☐ Full-day Workshop — Lecture-Demo plus two 2-hr sessions.
    (and/or Multiple-day)
  - ☐ Long term staff development, dealing with science concepts and processes.
- ☐ STUDENT-DEMONSTRATION — Presentation to large assembly of students. K - 12, College Pre-service Teachers.

SCIENCE INQUIRY ENTERPRISES
DR. TIK L. LIEM (DIRECTOR/OWNER)
14358 VILLAGE VIEW LANE
CHINO HILLS, CA 91709
TEL: 714-590-4618

AVAILABLE SCIENCE EDUCATIONAL MATERIALS

<u>Teachers' Guide</u>: "INVITATIONS TO SCIENCE INQUIRY"- 2nd Ed $40.00
500 pages, with over 400 discrepant events
to motivate students in learning science.
(Buy in multiples of 16 and get 10% discount!)

<u>Video Tapes</u>:
1. "DISCREPANT EVENTS - THE SCIENCE MOTIVATOR" (60 min)..$49.00
   The cognitive dissonance rationale; an assortment
   of discrepant events presented to teachers.
2. "TURNING KIDS ON TO SCIENCE" (60 min).................. 49.00
   A different assortment of discrepant events
   demonstrated to science students of grade 7 level.
   Both video tapes for 80.00
   Package of above text plus either one video 80.00
   Package of above text plus both videos 110.00

<u>Materials</u>:

EVENT NO.	CHAPTER	TITLE/DESCRIPTION	PRICE
ASSORTMENT A			
8.10	MAGNETISM	MAKE A NEEDLE COMPASS	$ 2.00
12.16	SOUND	THE TWIRLING BUGLE	5.00
13.19	FORCES	THE FLOATING BELT HANGER	4.00
13.38	FORCES	THE YIP-YIP STICK	4.00
14.3	SPACE SCIENCE	THE IMMOVABLE PENNY	2.00
14.5	SPACE SCIENCE	GET THE CHALK IN THE BOTTLE	3.00
Package of all six items above $15.00			
ASSORTMENT B			
1.28	AIR	THE INVERTED PAPER BAG BALANCE	$ 2.50
4.7	CHAR OF MATTER	THE INVISIBLE FLAME EXTINGUISHER	2.00
6.5	ENERGY	THE TIN CAN BAZOOKA (THE LIGHTER-FUEL CANNON)	8.00
8.4	MAGNETISM	THE FLOATING DISCS	4.00
13.6	FORCES	THE BALANCING PINS	5.50
14.21	SPACE SCIENCE	THE FUNNY MARBLES	3.00
Package of all six items above $20.00			
ASSORTMENT C			
1.37	AIR	THE BALANCING BALLOONS	$ 2.00
2.3	FLOWING AIR	THE FUNNEL AND THE BALL	3.00
4.20	CHAR OF MATTER	POUR WATER ALONG A STRING	2.00
7.11	HEAT	WHICH IS THE WARMER COLOR?	6.00
4.29	CHAR OF MATTER	THE THREAD CIRCLE IN THE SOAP FILM)	
4.30	CHAR OF MATTER	THE STRONG SOAP FILM	)15.00
6.10	ENERGY	FISSION AND FUSION)	
Package of all five sets above $25.00			

MINIMUM ORDER OF MATERIALS: $10.00
ADD SHIPPING & HANDLING: FOR MATERIALS: $3.50 PER PACKAGE
FOR TEXT: $4.75 PER BOOK
FOR VIDEO TAPES: $4.00 PER VIDEO

O R D E R   F O R M

Please send the following items:

```
_____copy/copies of INVITATIONS TO SCIENCE INQUIRY..$40.ea......$_____
_____set(s) of BOOK + DISCREPANT EVENTS (VIDEO).....$80.ea......  _____
_____set(s) of BOOK + TURNING KIDS ON (VIDEO).......$80.ea......  _____
_____set(s) of BOOK + BOTH VIDEOS ..................$110.ea.....  _____
_____copy/copies of DISCREPANT EVENTS - VIDEO ONLY..$49.ea......  _____
_____copy/copies of TURNING KIDS ON - VIDEO ONLY....$49.ea......  _____
_____set(s) of both VIDEO TAPES ....................$80.ea......  _____
_____Individual set(s) of MATERIALS FOR EVENT NO._____.   _____
                         _____    _____
                         _____    _____
                         _____    _____

_____set(s) of PACKAGE OF MATERIALS,(ASSORTMENT A)..$15.ea......  _____
_____set(s) of     "       "       ,(ASSORTMENT B)..$20.ea......  _____
_____set(s) of     "       "       ,(ASSORTMENT C)..$25.ea......  _____

     Add Shipping & Handling accordingly......................   _____
     Add 7% Tax for California residents......................   _____

                                    Total Remitted.....
                                                         ==========
```

To: **Name:**_____

    Address:_____
                (No. & Street)

            _____
                (City)

            _____
                (State)   (Zip)

Please make checks payable to:

SCIENCE INQUIRY ENT.
14358 Village View Lane
Chino Hills, CA 91709

HUMAN BIOLOGY

NERVOUS SYSTEM
REACTION TIME

17.17. CATCH THE DOLLAR BILL

Materials: 1. A crisp dollar bill (any denomination).
            (ideally: one of each denomination up to $20).

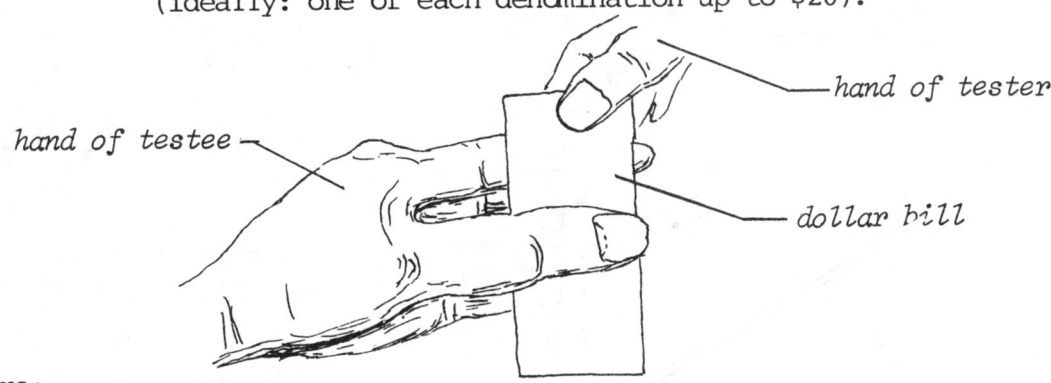

Procedure:
1. Hold your forefinger and thumb of your left hand about 5 cm (2 inches) apart, place the dollar bill about half way in between the two fingers.
2. Let the dollar bill fall and show the audience that you can easily catch it between your two fingers.
3. Ask the members of the audience to hold their forefinger and thumb in the same position to try to catch the dollar bill.
4. Hold the dollar bill between their fingers. The rule is that they may not go down with their hand, and the gap between the fingers should be not smaller than 5 cm.
5. Vary the time between placing the dollar bill between their fingers and the moment that you let go of the bill. Tell the audience that whoever catches the bill may keep it. Continue doing it with larger denominations, but remember to vary the dropping time, otherwise the bill might get caught by anticipation.

Questions:
1. Why can you catch the dollar bill so easily when you drop it yourself?
2. Why do you need to vary the time between placing the dollar bill and the releasing of it?
3. What is the reason that the bill can never be caught?
4. What makes muscles contract?
5. How did the finger muscles get the order to catch the bill?
6. Where did the initial stimulus come from?

Explanation:
   The initial stimulus to catch the dollar bill was from seeing the falling of it. The eyes got the stimulus, they sent a signal to the brain and the brain in turn sent an electrical impulse to the finger muscles to contract and catch the bill. But all that took a longer time than the falling of the dollar bill, although both take only a fraction of a second.
   The bill can only be caught by someone who anticipates the falling and actually start closing the fingers before the bill falls. This is the reason why we should vary the time between placing the bill between the fingers and the actual releasing of it.

HUMAN BIOGY

NERVOUS SYSTEM
REACTION TIME

17.18. HOW FAST CAN YOU REACT?

Materials: 1. A meter stick (15 meter sticks for a class of 30).

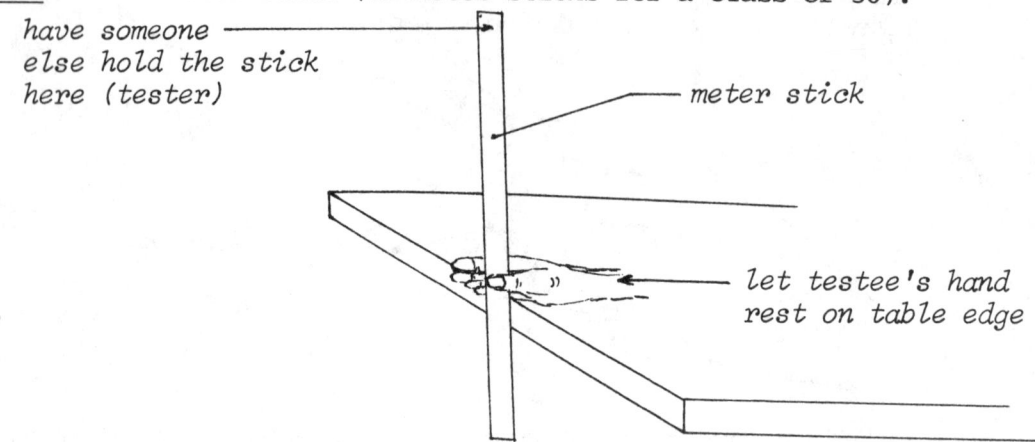

Procedure:
1. Place your hand over the edge of a table and leave about 3 cm opening between the thumb and forefingers.
2. Let your tester (someone else) hold the meter stick vertically from the top end of the stick.
3. Read off the spot where your thumb is on the meter stick before the tester drops the stick (start at an even number f.i. 10 or 20).
4. Let the tester drop the stick. The moment that you see the stick drop - catch the stick immediately and do another reading of where your thumb ended up. Record the difference (distance in drop).
5. To test the other stimuli: Hearing: close your eyes and let tester say: "Now" exactly at the moment that he/she drops the stick.
Testing for tactile (touch): close your eyes and let tester touch your other hand exactly at the moment that he/she let go of the stick.

Questions:
1. How do girls compare to boys in reaction time?
2. In doing the above comparison, what variables have to be controlled?
3. How do the different stimuli compare in your reaction time?
4. What is the actual falling time for the distance in the drop?
5. What are all the variables involved in this experiment?

Explanation:
   The different variables in this experiment are: the **manipulated variable** is the stimulus: through the eyes, hearing, and touch (or sex: when comparing boys versus girls). The **responding variable** is the falling time or distance on the meter stick.
   As the distance in free fall is: $d = 1/2\, gt^2$, the time derived from that is $t =$ square root of $2d$ over g, where g is the gravity acceleration.
   All other variables, like: distance between fingers, the place where the tester holds the stick, catching with two fingers or with the whole hand, the accuracy of saying "now" and simultaneously letting go of the stick, etc. have to be held constant, in other words they have to be done the same way.